Biomarkers of Oxidative and Radical Stress

Biomarkers of Oxidative and Radical Stress

Editor

Chryssostomos Chatgilialoglu

MDPI

Basel • Beijing • Wuhan • Barcelona • Belgrade • Novi Sad • Cluj • Manchester

Editor
Chryssostomos Chatgilialoglu
Institute for Organic
Synthesis and Photoreactivity
National Research Council of
Italy
Bologna
Italy

Editorial Office
MDPI
St. Alban-Anlage 66
4052 Basel, Switzerland

This is a reprint of articles from the Special Issue published online in the open access journal *Biomolecules* (ISSN 2218-273X) (available at: www.mdpi.com/journal/biomolecules/special_issues/biomarkers_oxidative_radical_stress).

For citation purposes, cite each article independently as indicated on the article page online and as indicated below:

Lastname, A.A.; Lastname, B.B. Article Title. *Journal Name* **Year**, *Volume Number*, Page Range.

ISBN 978-3-7258-0328-6 (Hbk)
ISBN 978-3-7258-0327-9 (PDF)
doi.org/10.3390/books978-3-7258-0327-9

Contents

About the Editor

Chryssostomos Chatgilialoglu

Dr. Chryssostomos Chatgilialoglu is an Emeritus Researcher at the Italian National Research Council (CNR) in Bologna and Visiting Professor at the Center for Advanced Technologies, Adam Mickiewicz University, Poznan (Poland). He is also the Co-Founder and past President (until 30 May 2023) of the company Lipinutragen. He achieved his doctorate degree in Industrial Chemistry at Bologna University in 1976 and completed his postdoctoral studies at York University (UK) and the National Research Council of Canada, Ottawa. From March 2014 to May 2016, he was appointed as the Director of the Institute of Nanoscience and Nanotechnology at the NCSR "Demokritos" in Athens (Greece). He was Chairman of the COST Actions Free Radicals in Chemical Biology (2007–2011) and Biomimetic Radical Chemistry (2013–2016).

He is the author or co-author of more than 320 publications in peer-reviewed journals and 36 book chapters, and the author and editor of several books, including *Organosilanes in Radical Chemistry*, Wiley 2004; *Encyclopedia of Radicals in Chemistry, Biology, and Materials*, Wiley 2012; and *Membrane Lipidomics for Personalized Health*, Wiley 2015. He was invited to be the Guest Editor for Special Issues in several scientific journals and to speak over 260 times at congresses and institutions.

His research group is active in the field of free radical chemistry, addressing applications in the life sciences. In recent years, he has developed the biomimetic chemistry of radical stress and associated biomarkers. The discovery of endogenous formation of trans-lipids and research on 5′,8-cyclopurine DNA lesions and fatty-acid-based lipidomics have attracted worldwide attention. He is responsible for introducing tris(trimethylsilyl)silane as a radical-based reducing agent, and for this he was the winner of the Fluka Prize "Reagent of the Year 1990".

biomolecules

MDPI

Editorial

Biomarkers of Oxidative and Radical Stress

Chryssostomos Chatgilialoglu [1,2]

[1] Institute for Organic Synthesis and Photoreactivity, National Research Council (CNR), 40129 Bologna, Italy; chrys@isof.cnr.it

[2] Center for Advanced Technologies, Adam Mickiewicz University, 61–614 Poznań, Poland

1. Reactive Species, Oxidative Damage, and Biomarkers

Reactive oxygen and nitrogen species (ROS/RNS) are generated as a result of normal intracellular metabolism. Their physiological roles in biology related to redox signaling mean that they participate in the modulation of apoptosis, stress responses, and proliferation [1–3]. ROS/RNS species include small molecules such as hydrogen peroxide (H_2O_2), peroxynitrite ($ONOO^-$), and hypohalous acids (HOCl or HOBr), as well as radicals such as superoxide radical anion ($O_2^{\bullet-}$), nitric oxide ($NO^\bullet$), hydroxyl radical ($HO^\bullet$), nitrogen dioxide radical ($NO_2^\bullet$), and carbonate radical anion ($CO_3^{\bullet-}$) [4,5]. The two progenitors of the ROS network are $O_2^{\bullet-}$ and $NO^\bullet$ (Figure 1). Their concentrations are ~0.1 nM $O_2^{\bullet-}$ and ~10 nM $NO^\bullet$ under physiological conditions, but these concentrations can increase by up to 100-fold during inflammatory response, having a negative effect by causing damage to biomolecules [6]. The enzymes superoxide dismutase (SOD) and catalase (CAT) control the production of $O_2^{\bullet-}$, first by dismutation to H_2O_2 and O_2 and, secondly, by transforming H_2O_2 to water and oxygen (see Figure 1—red color) [7]. Figure 1 also shows the main pathways via which other biologically important free radicals can be produced, either via H_2O_2 or as a consequence of peroxynitrite $ONOO^-$ formation [8]. H_2O_2 is at the crossroads of several pathways for the formation of $HO^\bullet$, with the main ones being Fenton reaction (Fe^{2+} and H_2O_2) [9], Haber–Weiss reaction ($O_2^{\bullet-}$ and H_2O_2), and the reduction of previously formed HOCl or HOBr by superoxide radicals. The spontaneous decomposition of ONOOH also produces $HO^\bullet$. $ONOO^-$ reacts with CO_2, and the resulting adduct rapidly decomposes to $NO_2^\bullet$ and $CO_3^{\bullet-}$. The regulation of ROS and related species, as well as oxidative repair, becomes less efficient with aging, resulting in the accumulation of ROS-derived damage. The *free radical theory of aging* claims that ROS-derived damage contributes to the functional decline of organ systems and predisposes those affected to pathologies such as cancer, as well as cardiovascular and neurodegenerative diseases [10].

It is also worth mentioning that small reactive sulfur species (RSS) like H_2S and its congeners or H_2S-derived radicals are emerging players in biological processes, particularly those related to mitochondria chemistry [11]. In chemical biology, it is believed that molecule modification/damage is due mainly to the activation of ROS through RSS generation. Regarding the reactivity of free radicals and, in particular, thiyl radicals ($RS^\bullet$) towards biomolecules, there is the well-known cis–trans isomerization of double bonds in unsaturated fatty acids moieties, which leads to the alteration of these biomolecules without resulting in oxidative stress products [12]. Figure 2 shows the reaction mechanism for this transformation, which consists of the reversible addition of $RS^\bullet$ to the cis double bond, forming the trans geometry as the thermodynamically favorable structure. It is worth noting that the radical $RS^\bullet$ acts as a catalyst for cis–trans isomerization, and the isomerization mechanism occurs on polyunsaturated substrates as a step-by-step process (i.e., each isolated double bond behaves independently).

Biomarkers and their products are substances, structures, or processes that can be measured in the body. They can be used to influence or predict the incidence of an outcome or a disease. Some of the reactive species mentioned above can damage organs, tissues, and

Citation: Chatgilialoglu, C. Biomarkers of Oxidative and Radical Stress. *Biomolecules* **2024**, *14*, 194. https://doi.org/10.3390/biom14020194

Received: 24 January 2024
Accepted: 31 January 2024
Published: 5 February 2024

cells by oxidizing DNA, proteins, and lipids, thereby resulting in diseases. The enormous importance of oxidative and free radical chemistry for a variety of biological processes, including aging and inflammation, has motivated researchers to try to understand the related mechanistic steps at the molecular level with the development of related biomarkers. The identification of modified biomolecules has a diagnostic value for the evaluation of in vivo damage. Therefore, the development of biomarkers through biomolecule modification and characterization by analytical protocols, followed by biomarker validation and extension to clinical research, have important applications in medicine and therapeutical approaches (Figure 3).

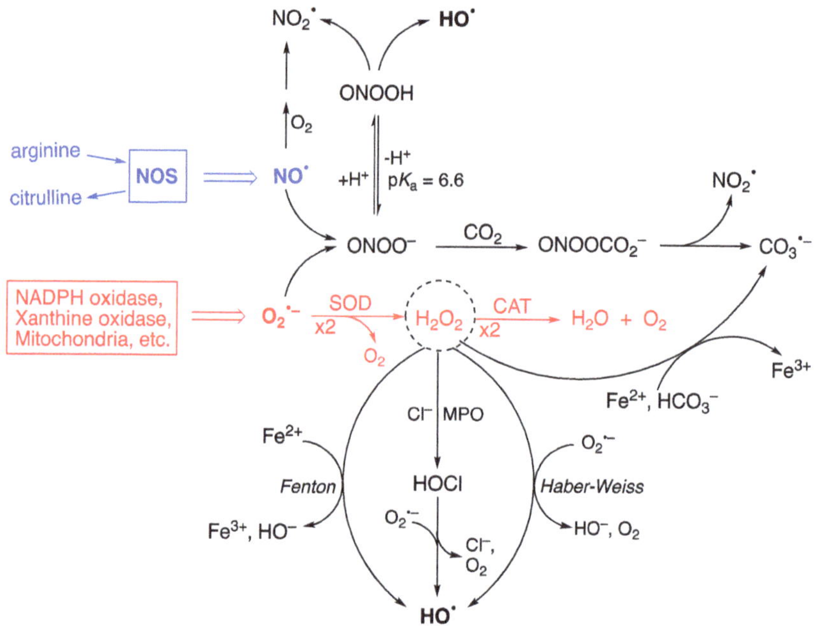

Figure 1. The enzymes SOD and CAT control the production of superoxide radical anion ($O_2^{\bullet-}$) (see red color). The processes that generate $HO^\bullet$ are the Fenton and Haber–Weiss reactions of H_2O_2, the reduction of HOCl by $O_2^{\bullet-}$, and the spontaneous decomposition of ONOOH, whereas the processes that generate $CO_3^{\bullet-}$ are the decomposition of ONOOH and H_2O_2 reacting with Fe^{2+} and bicarbonate.

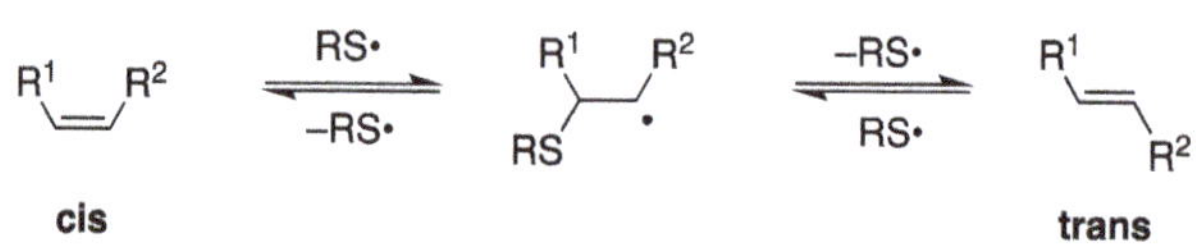

Figure 2. Thiyl radical-catalyzed cis–trans isomerization of a monounsaturated fatty acid moiety.

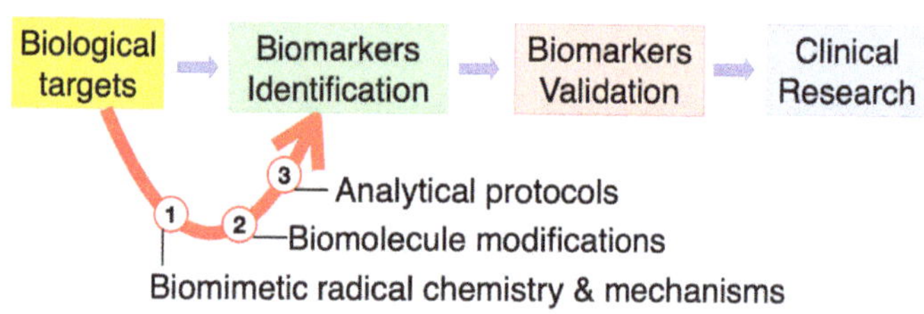

Figure 3. Omics technologies and the role of biomimetic radical chemistry in biomarker discovery.

This Special Issue covers various aspects of biomarker research, from biomarker identification, including chemical reactivity and analytical procedures, to biomarker validation and pre-clinical applications. Examples include DNA oxidation products, peptide and protein modifications, lipid peroxidation and isomerization, and defense and repair strategies.

2. Brief Overview of the Special Issue

2.1. Biomarkers of DNA Damage

Purine 5′,8-cyclo-2′-deoxynucleosides (cPu) are solely generated by the attack of $HO^{\bullet}$ radicals on purine moieties via C5′-radical chemistry, resulting in the formation of an additional C5′−C8 covalent bond; 5′,8-cyclo-2′-deoxyadenosine (cdA) and 5′,8-cyclo-2′-deoxyguanosine (cdG) exist in the 5′*R* and 5′*S* configurations (Figure 4). cPu can be removed only by the nucleotide excision repair (NER) pathway, and different repair efficiencies of *R* and *S* diastereoisomers have been detected [8]. On the contrary, the well-known 8-oxo-purine (8-oxo-Pu) lesions, namely, 8-oxo-7,8-dihydro-2′-deoxyadenosine (8-oxo-dA) and 8-oxo-7,8-dihydro-2′-deoxyguanosine (8-oxo-dG), derive from the oxidation at the C8 position of adenine and guanine caused by a variety of reactive oxygen species (ROS) and can be repaired by base excision repair (BER) [8].

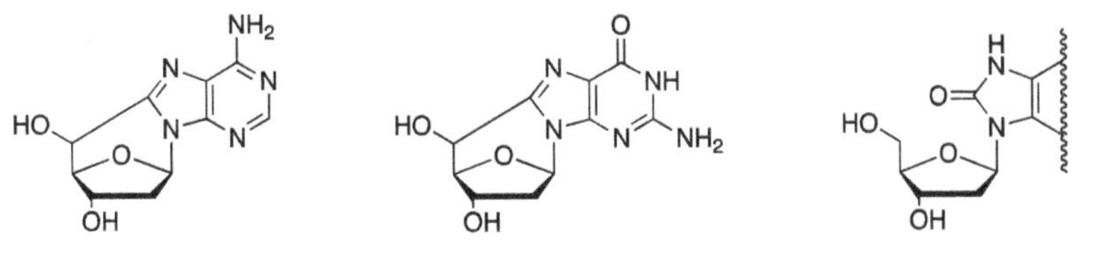

Figure 4. The structures of 5′,8-cyclo-2′-deoxyadenosine (cdA), 5′,8-cyclo-2′-deoxyguanosine (cdG), and 8-oxo-7,8-dihydro-2′-deoxypurine (8-oxo-dG or 8-oxo-dG).

Two articles in this Special Issue deal with the simultaneous measurement of six purine lesions in DNA samples using LC-MS/MS protocol. The first of these articles reports on Cockayne syndrome (CS) cell lines [13]. In particular, the six purine lesions were ascertained in the mtDNA of wild-type CSA and CSB cells and defective counterparts in comparison with the corresponding total nDNA. The 8–oxo–Pu levels were found to be in the range of 25–50 lesions/10^7 nucleotides in both mtDNA and nDNA. The four cPu were undetectable in mtDNA both in defective cells and in the wt counterparts (CSA and CSB), contrary to their detection in nDNA, indicating the absence of $HO^{\bullet}$ reactivity within mtDNA. Additional tailored in vitro experiments demonstrated a higher resistance to $HO^{\bullet}$ attack for mtDNA in comparison with nDNA associated with their different DNA helical topologies. The second of these articles reports on the brain tissue of mice [14]. A murine model of immunodeficient (SCID) xenografted young (4 weeks old) and old (17 weeks old) mice was compared with corresponding controls without tumor implantation. Both cPu and 8-oxo-cPu formation was evaluated in this study to compare the effect of tumor development and gather information on the aging process. Progressive DNA damage due to age and tumoral conditions was confirmed by raised levels of 5′*S*-cdG and 5′*S*-cdA.

2.2. Biomarkers of Lipid Damage

The above-mentioned brain tissue of young (4 weeks old) and old (17 weeks old) mice, in parallel to DNA damage, was also evaluated to ascertain changes in lipid content [14]. The remodeling of fatty acids involved a diminution of palmitic acid accompanied by an increase in arachidonic acid, along both age and tumor progressions, causing increases in the unsaturation index, the peroxidation index, and total TFA, indicators of increased oxidative and free radical reactivity. The authors concluded that under aging and disease progressions, membranes are not spectators, and targeted strategies are needed for preserving their molecular integrity [14].

Lipid peroxidation (LP), an important type of oxidative/radical damage in biological systems, is associated with a large number of pathological conditions, from atherosclerosis and cardiovascular diseases to neurological disorders and cancer [15]. Moreover, ferroptosis is a caspase-independent type of cell death triggered by iron-dependent LP [16]. The chemistry of LP, with radical chain-propagation reactions, is reviewed in [17] in this Special Issue, with a focus on the kinetics of various processes which help us to understand the mechanisms and efficacy of antioxidant strategies. The author of this article also describes the LP products that are commonly used as biomarkers to monitor the damage of biological systems, as well as their ability as electrophiles to alter other biomolecules such as proteins. An overview of the biological consequences of LP is presented, with an emphasis on membrane integrity and function, cell signaling, cancer, neurological disorders, and ferroptosis [17].

4-hydroxynonenal (4-HNE) is the quintessential biomarker for the oxidative degradation of polyunsaturated fatty acids (PUFA). The relationship between 4-HNE and the metabolomic profiling of patients with prostate cancer is reported on in [18]. In this study, measurements of 4-HNE–protein adducts in prostate cancer tissues and plasma samples from prostate cancer patients were obtained and compared with samples collected from healthy controls. The results indicated an absence of 4-HNE–protein adducts in prostate carcinoma tissues but increased 4-HNE–protein levels in the plasma of these patients, associated with different long-chain and medium-chain fatty acids with the presence of prostate cancer. The metabolic pathway of unsaturated fatty acid biosynthesis was found to be significantly affected by 4-HNE 18].

Plasmalogens are an important class of membrane phospholipids; in red blood cells (RBCs), their content reaches 15–20% and increases in tissues like the heart (32–50%), brain (20–50%), and spermatozoa (55%). Plasmalogens are characterized by having the following two fatty acids: one unsaturated fatty acid linked to the glycerol moiety through a cis-vinyl ether function and the other PUFA residue linked through an acyl function. An article in this Special Issue reports on the radical reactivity and identification of TFA [19]. In addition to the reported optimal transesterification procedure, it was demonstrated that the vinyl ether reactivity is similar to the reactivity of the arachidonic moiety that contains four C–C double bonds in cis–trans isomerization by thiyl radicals. Furthermore, a biomimetic Fenton-like model involving plasmalogen-containing liposomes or RBC ghosts, was used to compare peroxidation and isomerization processes, allowing for a full picture of plasmalogen reactivity under free radical conditions [19]. Trans fatty acids can also be produced endogenously from free radicals and thiols, which makes them valuable biomarkers for free radical activity in the human lipidome [12]. It was recently reported that resistance to ferroptosis induced by the LP microenvironment is based on the membrane enrichment of combined SFA and TFA during remodeling [20].

2.3. *Antioxidant Protection*

The main types of antioxidants for LP are discussed in terms of structure–activity rationalization, with a focus on mechanisms, kinetics, and their potential role in modulating ferroptosis, in [17]. In this particular study, antioxidants were classified as preventive, chain-breaking, or termination-enhancing and varied from small molecules to complex enzyme systems.

A review paper in this Special Issue focuses on red blood cells (RBCs) as a biological target of oxidative reactivity due to their content of both proteins (in particular, hemoglobin) and lipids (in particular, polyunsaturated components of RBC membranes) [21]. After presenting a summary of the most relevant mechanisms and effects of oxidants in RBCs, along with a summary of the production of modified biomolecules that can act as biomarkers, the review describes an antioxidant armoire with low-molecular-weight compounds (such as ascorbate, alpha-tocopherol, and glutathione) and a large network of antioxidant proteins that can react and deactivate oxidant species (such as superoxide dismutase, glutathione peroxidase, glutathione reductase, peroxiredoxin 2, thioredoxin, thioredoxin reductase, glutaredoxins, and catalase). An excursus of the main pathological consequences of oxidative stress in RBCs is also provided. Enzymatic defense deficiencies, which cause this otherwise very resistant and flexible cell type to become fragile and degrade, are also explored [21].

Another review paper in this Special Issue focuses on diabetes mellitus, a pathological condition characterized by oxidative stress and inflammation; thus, knowledge on the molecular mechanisms of polyphenols is required to ensure their utility as nutritional antioxidants that can be used to improve the efficiency of antidiabetic treatments [22]. Diabetes mellitus includes the occurrence of this condition during pregnancy; therefore, the use of nutrition as a tool for pharmacological improvements and to optimize fetus resistance is even more justified. After a thorough description of the molecular mechanisms of reactive oxygen species generation in cells, DNA, and lipid- and protein-damaging processes, along with the individuation of biomarkers of this reactivity (such as advanced glycation-end—AGE—products), this overview of diabetes mellitus not only includes results from clinical trials but also concludes by describing pancreatic cell function, glucose homeostasis (also controlled by antioxidants), and the enzymatic antioxidant defense system. This review offers a complete evaluation of polyphenols as antioxidants, focusing on nutritional, metabolic, and clinical aspects and supporting their use in diabetes mellitus treatment [22].

Another article in this Special Issue focuses on the toxic effects of cadmium (Cd) as a potentially toxic element (PTE) deriving from pollution or industrial processes that humans can be involved with, thus causing a human health concern and the need for individuating a good antioxidant treatment [23]. In particular, liver cells are targets of Cd-induced oxidative stress, especially when this element is combined with the exposure of free fatty acids, which are known as the main compounds processed in liver cells. Moreover, melatonin (MLT), a hormonal regulator of the circadian rhythm that, as an antioxidant and cytoprotective agent, is endowed with pharmacological properties, was considered to efficiently counteract the oxidative damage. This article shows a proof of concept of the synergic activity between Cd and fatty acids for reactive oxygen species (ROS) generation in an in vitro model of liver cells (HepaRG) and intestinal epithelial cancer cells (CACO-2 cells), thus enhancing the potential for fatty acids to produce lipotoxicity and cell damage. Melatonin demonstrated its utility as a cytoprotective agent, having specific molecular mechanisms involving ERK1/2 agonism, halting cell signaling via the MAPK-ERK1/2 pathway and SAPK activation in the hepatic cell line. Cd and fatty acids induced the accumulation of cellular lipids, whereas MLT lessened the cell death induction effects both in HepaRG and CACO-2 cell lines [23]. These results should encourage further pre-clinical and clinical studies on using MLT as an antioxidant intervention option in the context of Cd exposure.

2.4. Oxidative and Radical Stress: Measurements and Protein Studies

Another review in this Special Issue deals with important issues related to the use of fluorescent or luminescent probes and other chemical reagents to measure oxidative and radical stress [24]. This review emphasizes the need to understand reaction pathways and, in particular, to quantify the kinetic parameters of key reactions and measure the intracellular levels and localization of probes if such reagents have to be used. Very useful information is given, including (i) a discussion on ROS in general terms, (ii) the most widely used probes for ROS, and (iii) key points to consider when using chemical probes in free radical biology.

The structural properties of some proteins (keratin, collagen, bovine and fish gelatins) exposed to gamma radiation (10 kGy) have been studied by researchers via a variety of analytical and spectroscopical techniques [25]. Protein denaturation and changes in the secondary structures have been observed. Proteins with higher degrees of ordered structures are more stable against gamma radiation and temperature. The presence of riboflavin has different effects on their secondary structures, with a stabilizing effect for keratin and fish gelatin and a destabilizing effect for bovine gelatin, observed in both irradiated and non-irradiated samples. Riboflavin facilitates the release of free radicals that further interact with the protein chains.

Despite the therapeutic and commercial success of protein therapeutics, the development of stable protein formulations can present challenges. Indeed, proteins are subjected to physical and chemical degradation. Another review in this Special Issue focuses on the primary processes of free radical formation that are relevant to pharmaceutical formulations [26]. In this review, it is reported that oxidation and the formation of free radicals represent the major pathways for the chemical degradation of pharmaceutical formulations. In particular, emphasis is placed on autoxidation, metal-catalyzed oxidation (Fenton and Fenton-like reactions), photo-degradation, and radical generation from cavitation as result of mechanical stress and various stabilizing additives like antimicrobial preservatives.

3. Conclusions

The articles/reviews that comprise this Special Issue and are described above focus on modified endogenous biological molecules as markers of oxidative changes or other changes initiated by free radicals, as well the structural properties of proteins formulated for storage. A broader understanding of the mechanisms and effects of oxidative stress generation in a biological context can be obtained by reading the articles of this Special Issue. The strong multidisciplinarity of this research area is well represented, allowing for readers to obtain a broad perspective of chemical, mechanistic, analytical, molecular, and structural aspects, together with an understanding of applications in biological, pharmaceutical, preclinical, and clinical fields. It is hoped that through exploring the articles/reviews within this collection, the readers of this Special Issue can find interesting sources of inspiration for conducting further research on free radicals and oxidative stress.

Funding: This research received no external funding.

Institutional Review Board Statement: Not applicable.

Informed Consent Statement: Not applicable.

Conflicts of Interest: The author declares no conflicts of interest.

References

1. Sies, H.; Belousov, V.V.; Chandel, N.S.; Davies, M.J.; Jones, D.P.; Mann, G.E.; Murphy, M.P.; Yamamoto, M.; Winterbourn, C. Defining roles of specific reactive oxygen species (ROS) in cell biology and physiology. *Nat. Rev. Mol. Cell Biol.* **2022**, *23*, 499–515. [CrossRef]
2. Lennicke, C.; Cochemé, H.M. Redox metabolism: ROS as specific molecular regulators of cell signaling and function. *Mol. Cell* **2021**, *81*, 3691–3707. [CrossRef]
3. D'Autreáux, B.; Toledano, M.B. ROS as signalling molecules: Mechanisms that generate specificity in ROS homeostasis. *Nat. Rev. Mol. Cell Biol.* **2007**, *8*, 813–824. [CrossRef] [PubMed]
4. Murphy, M.P.; Bayir, H.; Belousov, V.; Chang, C.J.; Davies, K.J.A.; Davies, M.J.; Dick, T.P.; Finkel, T.; Forman, H.J.; Janssen-Heininger, Y.; et al. Guidelines for measuring reactive oxygen species and oxidative damage in cells and in vivo. *Nat. Metab.* **2022**, *4*, 651–662. [CrossRef] [PubMed]
5. Weidinger, A.; Kozlov, A.V. Biological Activities of Reactive Oxygen and Nitrogen Species: Oxidative Stress versus Signal Transduction. *Biomolecules* **2015**, *5*, 472–484. [CrossRef] [PubMed]
6. Geacintov, N.E.; Shafirovich, V. Reactions of small reactive species with DNA. In *Encyclopedia of Radicals in Chemistry, Biology and Materials*; Chatgilialoglu, C., Studer, A., Eds.; Wiley: Chichester, UK, 2012; Volume 3, pp. 1284–1317.
7. Fridovich, I. Superoxide radical and superoxide dismutases. *Annu. Rev. Biochem.* **1995**, *64*, 97–112. [CrossRef] [PubMed]
8. Chatgilialoglu, C.; Ferreri, C.; Krokidis, M.G.; Masi, A.; Terzidis, M.A. On the relevance of hydroxyl radical to purine DNA damage. *Free Radic. Res.* **2021**, *55*, 384–404. [CrossRef] [PubMed]
9. Zhao, Z. Hydroxyl radical generation from the physiologically relevant Fenton-like reactions. *Free Radic. Biol. Med.* **2023**, *208*, 510–515. [CrossRef] [PubMed]
10. Finkel, T.; Holbrook, N.J. Oxidants, oxidative stress and the biology of ageing. *Nature* **2000**, *408*, 239–247. [CrossRef] [PubMed]

11. Filipovic, M.R.; Zivanovic, J.; Alvarez, B.; Banerjee, R. Chemical biology of H_2S signaling through persulfidation. *Chem. Rev.* **2018**, *118*, 1253–1337. [CrossRef]
12. Chatgilialoglu, C.; Ferreri, C.; Melchiorre, M.; Sansone, A.; Torreggiani, A. Lipid geometrical isomerism: From chemistry to biology and diagnostics. *Chem. Rev.* **2014**, *114*, 255–284. [CrossRef] [PubMed]
13. Chatgilialoglu, C.; Krokidis, M.G.; Masi, A.; Barata-Vallejo, S.; Ferreri, C.; Pascucci, B.; D'Errico, M. Assessing the Formation of Purine Lesions in Mitochondrial DNA of Cockayne Syndrome Cells. *Biomolecules* **2022**, *12*, 1630. [CrossRef] [PubMed]
14. Krokidis, M.G.; Prasinou, P.; Efthimiadou, E.K.; Boari, A.; Ferreri, C.; Chatgilialoglu, C. Effects of Aging and Disease Conditions in Brain of Tumor-Bearing Mice: Evaluation of Purine DNA Damages and Fatty Acid Pool Changes. *Biomolecules* **2022**, *12*, 1075. [CrossRef] [PubMed]
15. Foret, M.K.; Lincoln, R.; Do Cormo, S.; Cuello, A.C.; Cosa, G. Connecting the "Dots": From Free Radical Lipid Autoxidation to Cell Pathology and Disease. *Chem. Rev.* **2020**, *120*, 12757–12787. [CrossRef] [PubMed]
16. Stockwell, B.R. Ferroptosis turns 10: Emerging mechanisms, physiological functions, and therapeutic applications. *Cell* **2022**, *185*, 2401–2421. [CrossRef] [PubMed]
17. Valgimigli, L. Lipid Peroxidation and Antioxidant Protection. *Biomolecules* **2023**, *13*, 1291. [CrossRef]
18. Perkovic, M.N.; Jaganjac, M.; Milkovic, L.; Horvat, T.; Rojo, D.; Zarkovic, K.; Ćorić, M.; Hudolin, T.; Waeg, G.; Orehovec, B.; et al. Relationship between 4-Hydroxynonenal (4-HNE) as Systemic Biomarker of Lipid Peroxidation and Metabolomic Profiling of Patients with Prostate Cancer. *Biomolecules* **2023**, *13*, 145. [CrossRef]
19. Ferreri, C.; Ferocino, A.; Batani, G.; Chatgilialoglu, C.; Randi, V.; Riontino, M.V.; Vetica, F.; Sansone, A. Plasmalogens: Free Radical Reactivity and Identification of Trans Isomers Relevant to Biological Membranes. *Biomolecules* **2023**, *13*, 730. [CrossRef]
20. Hirata, Y.; Ferreri, C.; Yamada, Y.; Inoue, A.; Sansone, A.; Vetica, F.; Suzuki, W.; Saya Takano, S.; Noguchi, T.; Matsuzawa, A.; et al. Geometrical isomerization of arachidonic acid during lipid peroxidation Interferes with ferroptosis. *Free Radic. Biol. Med.* **2023**, *204*, 374–384. [CrossRef]
21. Orrico, F.; Laurance, S.; Lopez, A.C.; Lefevre, S.D.; Thomson, L.; Möller, M.N.; Ostuni, M.A. Oxidative Stress in Healthy and Pathological Red Blood Cells. *Biomolecules* **2023**, *13*, 1262. [CrossRef] [PubMed]
22. Krawczyk, M.; Burzynska-Pedziwiatr, I.; Wozniak, L.A.; Bukowiecka-Matusiak, M. Impact of Polyphenols on Inflammatory and Oxidative Stress Factors in Diabetes Mellitus: Nutritional Antioxidants and Their Application in Improving Antidiabetic Therapy. *Biomolecules* **2023**, *13*, 1402. [CrossRef] [PubMed]
23. Migni, A.; Mancuso, F.; Baroni, T.; Di Sante, G.; Rende, M.; Galli, F.; Bartolini, D. Melatonin as a Repairing Agent in Cadmium- and Free Fatty Acid-Induced Lipotoxicity. *Biomolecules* **2023**, *13*, 1758. [CrossRef] [PubMed]
24. Wardman, P. Factors Important in the Use of Fluorescent or Luminescent Probes and Other Chemical Reagents to Measure Oxidative and Radical Stress. *Biomolecules* **2023**, *13*, 1041. [CrossRef] [PubMed]
25. Stanca, M.; Gaidau, C.; Zaharescu, T.; Balan, G.-A.; Matei, I.; Precupas, A.; Leonties, A.R.; Ionita, G. Physico-Chemical Changes Induced by Gamma Irradiation on Some Structural Protein Extracts. *Biomolecules* **2023**, *13*, 774. [CrossRef]
26. Schöneich, C. Primary Processes of Free Radical Formation in Pharmaceutical Formulations of Therapeutic Proteins. *Biomolecules* **2023**, *13*, 1142. [CrossRef]

biomolecules

MDPI

Article

Assessing the Formation of Purine Lesions in Mitochondrial DNA of Cockayne Syndrome Cells

Chryssostomos Chatgilialoglu [1,2,*], Marios G. Krokidis [3], Annalisa Masi [4], Sebastian Barata-Vallejo [1,5], Carla Ferreri [1], Barbara Pascucci [4,6] and Mariarosaria D'Errico [6]

1 Istituto per la Sintesi Organica e la Fotoreattività, Consiglio Nazionale delle Ricerche, Via P. Gobetti 101, 40129 Bologna, Italy
2 Center for Advanced Technologies, Adam Mickiewicz University, 61-614 Poznań, Poland
3 Institute of Nanoscience and Nanotechnology, N.C.S.R. "Demokritos", Agia Paraskevi Attikis, 15310 Athens, Greece
4 Institute of Crystallography, Consiglio Nazionale delle Ricerche, Monterotondo Stazione, 00015 Rome, Italy
5 Departamento de Ciencias Químicas, Facultad de Farmacia y Bioquimíca, Universidad de Buenos Aires, Junin 954, Buenos Aires CP 1113, Argentina
6 Department of Environment and Health, Istituto Superiore di Sanità, Viale Regina Elena 299, 00161 Rome, Italy
* Correspondence: chrys@isof.cnr.it

Abstract: Mitochondrial (mt) DNA and nuclear (n) DNA have known structures and roles in cells; however, they are rarely compared under specific conditions such as oxidative or degenerative environments that can create damage to the DNA base moieties. Six purine lesions were ascertained in the mtDNA of wild type (wt) CSA (CS3BE–wtCSA) and wtCSB (CS1AN–wtCSB) cells and defective counterparts CS3BE and CS1AN in comparison with the corresponding total (t) DNA (t = n + mt). In particular, the four 5′,8–cyclopurine (cPu) and the two 8–oxo–purine (8–oxo–Pu) lesions were accurately quantified by LC–MS/MS analysis using isotopomeric internal standards after an enzymatic digestion procedure. The 8–oxo–Pu levels were found to be in the range of 25–50 lesions/10^7 nucleotides in both the mtDNA and tDNA. The four cPu were undetectable in the mtDNA both in defective cells and in the wt counterparts (CSA and CSB), contrary to their detection in tDNA, indicating a nonappearance of hydroxyl radical (HO•) reactivity within the mtDNA. In order to assess the HO• reactivity towards purine nucleobases in the two genetic materials, we performed γ–radiolysis experiments coupled with the 8–oxo–Pu and cPu quantifications on isolated mtDNA and tDNA from wtCSB cells. In the latter experiments, all six purine lesions were detected in both of the DNA, showing a higher resistance to HO• attack in the case of mtDNA compared with tDNA, likely due to their different DNA helical topology influencing the relative abundance of the lesions.

Keywords: mitochondrial and nuclear DNA damage; 5′,8–cyclopurines; 8–oxo–dG; gamma radiolysis; hydroxyl radical; isotope dilution LC–MS/MS; cockayne syndrome

Citation: Chatgilialoglu, C.; Krokidis, M.G.; Masi, A.; Barata-Vallejo, S.; Ferreri, C.; Pascucci, B.; D'Errico, M. Assessing the Formation of Purine Lesions in Mitochondrial DNA of Cockayne Syndrome Cells. *Biomolecules* **2022**, *12*, 1630. https://doi.org/10.3390/biom12111630

Academic Editors: Anthony J. Berdis and Vladimir N. Uversky

Received: 27 September 2022
Accepted: 30 October 2022
Published: 3 November 2022

1. Introduction

Molecular oxygen (O_2) is used for the production of reactive oxygen species (ROS) that are involved in the signaling pathways of various basal and adaptive physiological responses controlling organism homeostasis [1–4]. However, ROS are also responsible for a variety of pathological processes, as their overproduction contributes to biomolecule damage, which has been linked with the etiology of various diseases [4–6]. Under physiological conditions, most human resting cells experience ca. 5% oxygen tension; however, the [O_2] gradient occurring between the extracellular environment and mitochondria, where oxygen is consumed by cytochrome c oxidase, results in a significantly lower [O_2] exposition of mitochondria [7–9]. It is estimated that up to 1% to 5% of the consumed mitochondrial oxygen is converted to ROS [10]. In 1972, Harman proposed that mitochondria were the primary source of cellular free radicals, and were thus responsible for the free–radical–based ageing process [11].

Human mitochondrial DNA (mtDNA) is a circular molecule of ~16.5 kb and must be compacted in order to fit within a mitochondrion [12,13]. mtDNA exists in a compacted DNA–protein complex known as the mitochondrial nucleoid that may be protective towards the source of mitochondrial free radicals [14]. As cellular ROS are produced by the mitochondrial respiratory chain, in the form of superoxide, much attention has been focused on the putative role of ROS in mitochondrial mutagenesis [15]. mtDNA such as nuclear DNA (nDNA) is highly susceptible to ROS, and is easily oxidized to accumulate DNA modifications [16–18]. Increased oxidative damage in mtDNA has been associated with neurological degeneration, inflammasomes, tumorigenesis, and malignant progression [19,20]. Among the mtDNA repair pathways, the base excision repair (BER) pathway has been extensively characterized to remove some of oxidative DNA damage in the mitochondria as efficiently as in the nuclei [17]. Implications of other repair pathways remain unclear, although the absence of nucleotide excision repair (NER) in the mitochondria is well documented [17]. Indeed, despite the permanent exposure to ROS and the less protective pathways in the mitochondria, it is not clarified how the integrity of genetic information is maintained in this compartment [16–18].

ROS include radicals such as the superoxide radical anion ($O_2^{\bullet -}$), nitric oxide ($NO^{\bullet}$), hydroxyl radical ($HO^{\bullet}$), nitrogen dioxide ($NO_2^{\bullet}$), and the carbonate radical anion ($CO_3^{\bullet -}$), as well as molecules such as hydrogen peroxide (H_2O_2), hypochlorous acid (HOCl), and peroxynitrite ($ONOO^-$) [6,21,22]. In quantitative terms, $O_2^{\bullet -}$ is the most abundant radical formed in aerobic organisms and the main entry to the ROS network. Aerobic life would not be possible without the enzymes known as superoxide dismutases (SODs) and catalase (CAT), which transform $O_2^{\bullet -}$ to water [23]. As shown in Figure 1, H_2O_2 is at the crossroad of several pathways: H_2O_2 transformation to highly reactive $HO^{\bullet}$ occurs by the Fenton reaction (Fe^{2+} + H_2O_2) and Haber–Weiss reaction ($O_2^{\bullet -}$ + H_2O_2) [22,24]; myeloperoxidase (MPO) uses H_2O_2 and Cl^- to generate HOCl, which further reacts with $O_2^{\bullet -}$ to produce $HO^{\bullet}$ [25]. Therefore, $O_2^{\bullet -}$ being quite unreactive in typical free radical reactions, such as hydrogen atom abstraction or addition, is converted to H_2O_2, which is able to diffuse and generate the most reactive $HO^{\bullet}$ radical. The diffusion distance of $HO^{\bullet}$ is very small because of their high reactivity with all types of biomolecules (DNA is not an exception) and, consequently, there is a low probability to be intercepted by antioxidants [26]. $HO^{\bullet}$ is able to react with DNA, causing single strand breaks, abasic sites, DNA–DNA intrastrand adducts, DNA–protein crosslinks, and base damage [27]. Evidence has been provided that in human fibroblasts, mtDNA may be more vulnerable to H_2O_2 compared with nDNA, showing a higher frequency of H_2O_2–driven lesions in cell culture models, despite it being a well–known H_2O_2 scavenging system (cf. Figure 1) [28]. It has been reported that H_2O_2 treatment results in strand breaks or abasic sites that are converted to strand breaks [29]. It has been suggested that the relationship between free radicals and mtDNA mutations is not as straightforward as it is often portrayed [17]. Indeed, it has been reported that the absence of oxidative stress induced mutations in the mitochondrial genome may be due to the rapid degradation of oxidized DNA molecules [30]. Consistently, the mechanism of damaged mitochondrial DNA degradation has been recently characterized [31,32].

In the present work, we considered the simultaneous measurement of the six purine lesions in mtDNA shown in Figure 2. 5′,8–cyclopurines (cPu) represent a very interesting and peculiar family of DNA lesions because they are exclusively generated by the reaction of $HO^{\bullet}$ radicals with genetic material via C5′ radical chemistry of the purine moieties [33]. They consist of 5′,8–cyclo–2′–deoxyadenosine (cdA) and 5′,8–cyclo–2′–deoxyguanosine (cdG), existing as 5′*R* and 5′*S* diastereoisomeric forms (Figure 2) [34–36]. On the other hand, the 8–oxo–Pu family, which consists of 8–oxo–dG and 8–oxo–dA, is generated by oxidation at the C8 position by a variety of ROS, such as $HO^{\bullet}$ and $ROO^{\bullet}$ radicals, H_2O_2, singlet oxygen or $ONOO^-$ [33,37]. cPu lesions are substrates of NER, whereas 8–oxo–Pu lesions are substrates of BER [38–41].

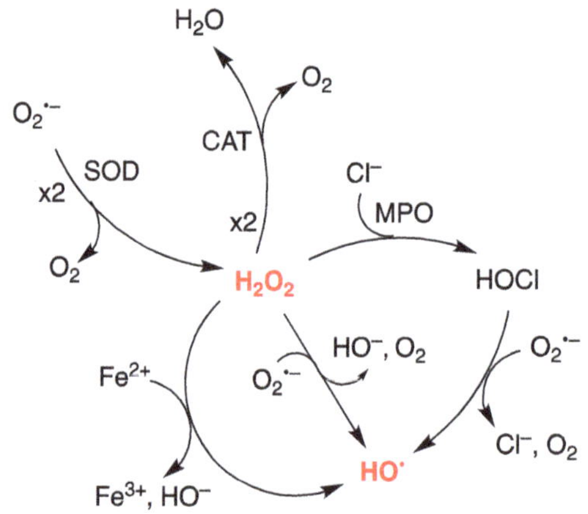

Figure 1. Relevant pathways of the reactive oxygen species (ROS) network: hydroxyl radical ($HO^{\bullet}$) formation from superoxide radical anion ($O_2^{\bullet -}$) in the mitochondria. SOD: superoxide dismutase; CAT: catalase; MPO: myeloperoxidase.

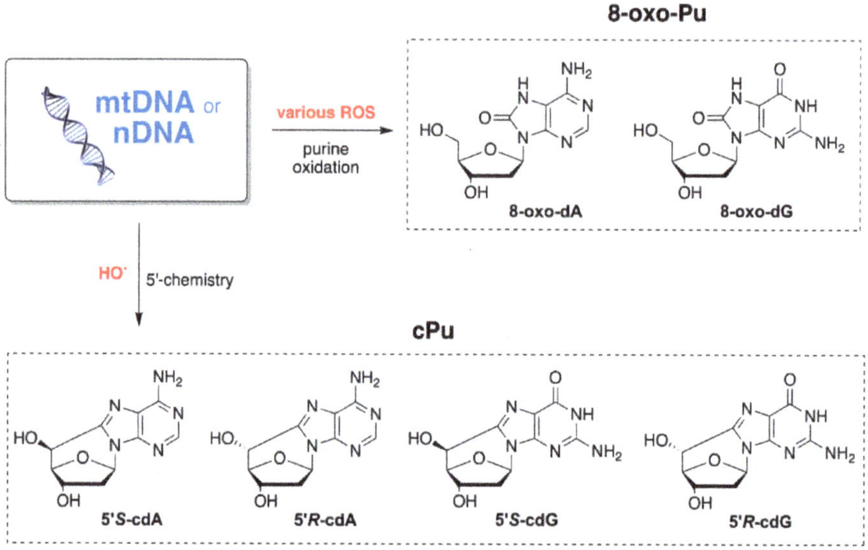

Figure 2. Structures of cPu lesions generated by H–atom abstraction from H5′ position by $HO^{\bullet}$ radical and 8–oxo–Pu lesions generated by purine oxidation from $HO^{\bullet}$ radical and other ROS species.

cPu, as transcriptional blocking lesions, have been identified as molecular defects in neurodegenerative processes [42]. In particular, Cockayne syndrome (CS) is an autosomal recessive neurodegenerative premature aging disorder associated with defects in NER. Over 90% of CS cases are due to mutations in either the CSA or CSB genes, responsible for the defect in the transcription coupled nucleotide excision repair (TC–NER) observed in CS cells. The lack of this repair mechanism makes CS cells hypersensitive to UV light. Moreover, cells from CS patients present elevated levels of ROS and are also defective in the repair of a variety of oxidatively generated DNA lesions [43–45]. Additionally, elevated levels of mitochondrial DNA damage, hypersensitivity to bioenergetic inhibitors, redox unbalance due to an increase of mitochondrial ROS, and mitochondrial dysfunction have been reported in CS cells [20]. We recently reported two studies on the oxygen–

dependent accumulation of purine lesions in total (t) DNA (t = n + mt) [46] and membrane lipidome remodeling [47] in wild type and defective CSA and CSB cell lines. Based on our interest in clarifying the DNA damage scenario, in the present work, we evaluated in this cell system mtDNA damage of purine with a very sensitive protocol (LC–ESI–MS/MS system with isotopomeric internal standards) [33,36,48]. We addressed the following chemical/biological points: (i) the simultaneous measurement of the six purine lesions shown in Figure 2, carried out in the mtDNA of wild type and defective CSA and CSB cell lines, grown under atmospheric oxygen tension; (ii) comparison of the six purine lesions between mtDNA and tDNA, and role of $HO^{\bullet}$ radicals in the oxidatively–induced damage; and (iii) the model reactivity of mtDNA and tDNA in an "isolated" context, using the reaction of genetic material with $HO^{\bullet}$ radicals under biomimetic conditions [49] and measuring the levels of six lesions. The results contribute to a better understanding of the genome integrity features under $HO^{\bullet}$ radical reactivity estimating the contribution of different helical topology in distinct genetic pools such as mtDNA and nDNA.

2. Materials and Methods

2.1. *Cell Lines and DNA Isolation*

CSA and CSB SV40–transformed cell lines were established and cultured as previously described [50]. More precisely, an isogenic cell line that expresses the wtCSA protein tagged with the Flag and HA epitopes (CS3BE–wtCSA) was used. The defective counterpart is CS3BE [51]. For CSB cell lines, we used CS1AN–wtCSB and CS1AN (defective CSB cells) [52]. Defective cell lines carry the empty vector. Cell culture studies are grown under standard atmospheric oxygen tension. The mitochondria were isolated by a non–mechanical, reagent–based method according to the procedure of The Mitochondria Isolation Kit (Thermo Fisher Scientific, Waltham, MA, USA). Then, mtDNA was extracted using a high–salt extraction procedure [46,50]. A similar procedure was used for the isolation of tDNA.

2.2. *γ–Radiolysis Experiments*

Each sample of mtDNA and tDNA from CSA1N–wtCSB was dissolved in double distilled water (ddH_2O) with a concentration of 0.5 mg/mL; in particular, 33.5 µg of tDNA was suspended in 67 µL and 45.9 µg of mt–DNA was suspended in 91.8 µL. The solution was placed in a glass vial of 2 mL containing a 300 µL glass insert, flushed with N_2O for 10 min and exposed to gamma rays at room temperature (22 ± 2 °C) using a ^{60}Co–Gammacell apparatus at different doses (dose rates: 1.44 Gy/min). The exact absorbed radiation dose was determined with the Fricke chemical dosimeter, by taking G(Fe^{3+}) 1.61 µmol J^{-1} [53]. The irradiation doses used were 0, 20, and 40 Gy for tDNA and 0, 15, 30, and 45 Gy for mtDNA. The experiments were performed in triplicate. The samples were lyophilized after the irradiation experiments.

2.3. *Enzymatic Digestion and Quantification of Modified Nucleosides by Stable Isotope LC–MS/MS*

The purine DNA lesions levels were quantified as described previously [33,36,46,54] and are summarized in Figure 3. Briefly, 10 µg of DNA were enzymatically digested in a reaction mixture including 0.2 mM pentostatin, 5 µM BHT, 3 mM deferoxamine, and the internal standards ([$^{15}N_5$]–5′*S*–cdA, [$^{15}N_5$]–5′*R*–cdA, [$^{15}N_5$]–5′*S*–cdG, [$^{15}N_5$]–5′*R*–cdG, [$^{15}N_5$]–8–oxo–dG and [$^{15}N_5$]–8–oxo–dA), the samples were filtered off by centrifugation through a 3 kDa microspin filter, and were cleaned up and enriched by an HPLC–UV system coupled with a sample collector and injected into the LC–MS/MS system. The quantification of the modified nucleosides was carried out by a triple–stage quadrupole mass spectrometer (Thermo, Waltham, MA, USA) using positive electrospray ionization (ESI) following a gradient program (2 mM ammonium formate, acetonitrile, and methanol) and the detection was executed in multiple reaction monitoring mode (MRM) using the two most intense and characteristic precursor/product ion transitions for each lesion [55,56].

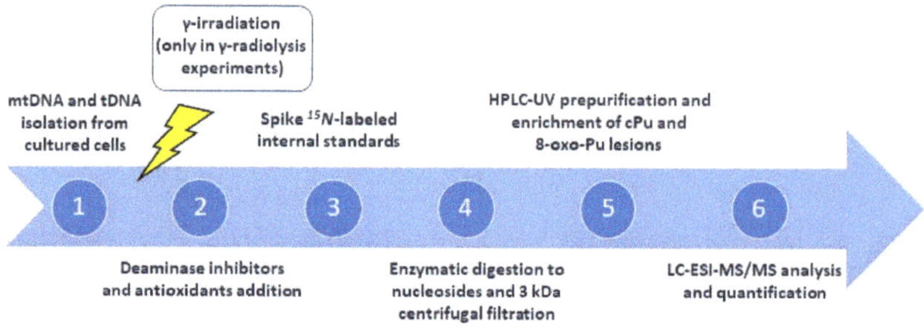

Figure 3. Flow diagram showing protocol steps for the quantification of cPu and 8–oxo–Pu lesions via isotope–dilution LC–ESI–MS/MS.

2.4. Statistical Analysis

All of the measurements were performed in triplicate and the data were expressed as mean ± standard deviation (SD). The unpaired *t*-test was used for the statistical analysis and a two–tailed *p*-value < 0.05 and *p*-value < 0.01 were considered to indicate a statistically significant difference.

3. Results and Discussion

3.1. Purine mtDNA Lesions Levels in Wild Type and Defective CSA and CSB Cells

The mtDNA from wtCSA (CS3BE–wtCSA), wtCSB (CS1AN–wtCSB), and defective counterparts CS3BE and CS1AN cell lines cultivated under standard atmospheric oxygen tension have been isolated. After hydrolysis of the genetic material to single nucleosides by an enzymatic cocktail containing nucleases, analysis by liquid chromatography with tandem mass spectrometry (LC–MS/MS) was performed for the determination of the modified nucleosides (four cPu and two 8–oxo–Pu), in accordance with a recently optimized protocol [33,36,46,55,56]. The levels of 8–oxo–dG and 8–oxo–dA are reported in Table 1. Unexpectedly, none of the four cPu lesions were detected. 8–oxo–dG was found to be significantly raised in defective CSB cells compared with the wild type cell line (p = 0.011).

Table 1. The levels (lesions/10^7 nucleosides) of 8–oxo–dG and 8–oxo–dA in mtDNA isolated from CSA and CSB (wt and defective) cells.

mtDNA	8–oxo–dG [1]	8–oxo–dA [1]
CS3BE–wtCSA	28.30 ± 0.13	7.32 ± 0.13
CS3BE	29.31 ± 0.22	7.58 ± 0.07
CS1AN–wtCSB	39.00 ± 0.18 *	8.08 ± 0.15
CS1AN	40.90 ± 0.22 *	8.71 ± 0.17

[1] The numbers represent the mean value (±standard deviation) of the DNA lesions levels from the measurement of triplicate; statistical significance: * (p < 0.05) was observed between CS1AN–wtCSB and CS1AN cell samples (see Table S1).

Comparing the 8–oxo–dG and 8–oxo–dA levels of mtDNA (Table 1) with the corresponding values of tDNA of the same cellular lines reported recently by us (see Table S2) [46], we observed an increase in lesions in all four cellular lines going from tDNA to mtDNA, e.g., the increase in 8–oxo–dG was ~40% in wt cells and ~10% in defective cells. Similar trends of 8–oxo–dG was previously reported in CS cells by the less sensitivity HPLC–ED detection method [52]. On the other hand, the calculated ratio 8–oxo–dG/8–oxo–dA of 3.9 and 4.8 for CSA and CSB cells, respectively (from the data of Table 1), was very similar to the analogous ratio calculated in the tDNA of the same cellular lines (cf. Table S2), indicating comparable reactivities of the two purine bases (dG and dA) towards the ROS independently of the two genetic pools.

Table 2 shows the levels of total 8–oxo–Pu and cPu in mtDNA in comparison with analogous data calculated for tDNA of the same cellular lines, and Figure 4 also illustrates the overall behavior. The 8–oxo–Pu levels were ~ 40% higher in the mtDNA than the tDNA of the wt cells, whereas the increase was limited to ~10% in defective cells. Table 2 and Figure 4 indicate the absence of cPu lesions in the mtDNA, whereas the cPu levels in tDNA were in the same order of 8–oxo–Pu. The four cPu in the mtDNA were undetectable in defective cells and in the wt counterparts (CSA and CSB). Based on the limit of detection of our analytical methodology, we can infer that the total number of cPu lesions was at least two orders of magnitude lower in the mtDNA compared with the tDNA.

Table 2. Total amount of 8–oxo–Pu and cPu (lesions/10^7 nucleosides) in mtDNA in comparison with values of tDNA isolated from CSA and CSB (wt and defective) cells.

	8–oxo–Pu [1]		cPu [1]	
	mtDNA [2]	**tDNA** [3]	**mtDNA** [2]	**tDNA** [3]
CS3BE–wtCSA	35.62 ± 0.26	26.32 ± 1.22	N/D	20.56 ± 0.30 *
CS3BE	36.89 ± 0.28	33.98 ± 2.40	N/D	22.04 ± 0.55 *
CS1AN–wtCSB	47.08 ± 0.21 ***	32.49 ± 3.04	N/D	22.84 ± 1.39
CS1AN	49.61 ± 0.13 ***	43.73 ± 1.86	N/D	23.67 ± 0.62

[1] The numbers represent the mean value (± standard deviation) of DNA lesions levels from the measurement of triplicate. [2] Present work; N/D = not detected. [3] From [46]. Statistically significant samples * (CS1AN–wtCSB vs. CS1AN; $p = 0.035$); *** (CS1AN–wtCSB vs. CS1AN; $p = 0.0005$).

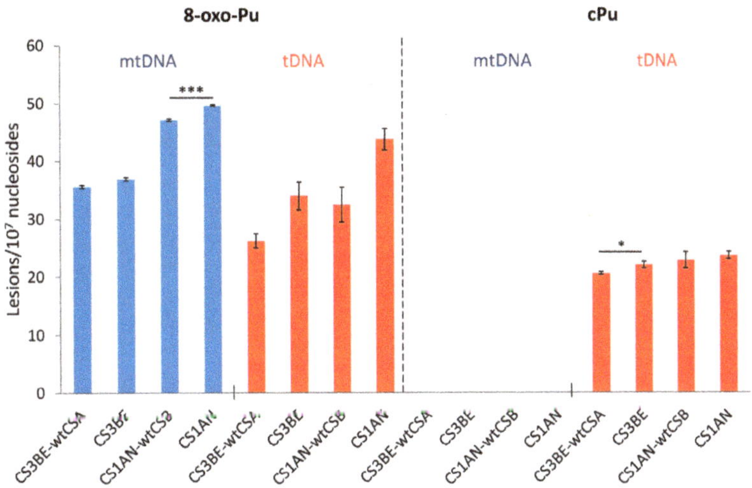

Figure 4. Comparison between 8–oxo–Pu and cPu lesions in the mtDNA and tDNA samples. The levels (lesions/10^7 Nu) of 8–oxo–Pu and cPu lesions in mtDNA (blue) and tDNA (red) extracted from CS1AN–wtCSB, CS1AN, CS3BE–wtCSA, and CS3BE cells. The error bars represent the standard deviation of the mean, calculated from three independent samples, * denotes a statistically significant difference ($p < 0.05$) between the groups, *** denotes a statistically significant difference ($p < 0.001$) between the groups (see Table S3).

The absence of cPu lesions in the mtDNA in both defective cells and in the wt counterparts (CSA and CSB) may indicate the absence of reactivity with $HO^{\bullet}$ radicals towards mtDNA. The presence of 8–oxo–Pu in the absence of cPu should be informative of the occurrence of the molecular rather than the radical reactivity. Indeed, similar to $HO^{\bullet}$ radicals, other oxidizing species such as H_2O_2 or $ONOO^-$ are also able to generate 8–oxo–Pu as DNA lesions [29,37]. It is interesting to note that these oxidizing species are increased in CSA defective cells. Treatment with catalase, a H_2O_2 scavenger, has shown that high levels of H_2O_2 are present in CSA defective cells [57]. Moreover, CSA defective cells are characterized by increased levels of reactive nitrogen species and peroxynitrite, and decreased levels of NO [57,58].

In humans, mitochondrial DNA represents about 1–10% of total cellular DNA (about 1000 to 10,000 copies per cell). In contrast with the invariable copy number of the nuclear genome (diploid), a single cell can contain many copies of mtDNA dependent on different processes involved in the mitochondrial homeostasis, such as mitochondrial replication, mitochondrial dynamics, and mitophagy. It has been estimated that each human cell contains from hundreds to thousand mitochondria [59]. Cells with a higher energy expenditure have a higher number of mitochondria and, consequently, more copies of mitochondrial DNA. A recent study estimated that cardiac and skeletal muscle contained between 4000 and 6000 copies of mtDNA per cell, while the liver, kidney, and lung tissues averaged between 500 and 2000 copies [60].

Previous work in human HeLa cell extracts indicated that cdA and cdG lesions are excised with a similar efficiency by NER and that the *R*–diastereoisomers of both cdA and cdG cause greater distortion of the DNA backbone and are better substrates of NER than the corresponding *S* ones [41,61]. However, the absence of NER in the mitochondria is well documented; therefore, the nonappearance of cPu lesions in the mtDNA cannot be due to their repair [17]. In order to understand the $HO^•$ reactivity towards DNA helical topology better, we selectively generated $HO^•$ radicals by ionizing irradiations in the presence of isolated mtDNA or tDNA samples from wtCSB cells and carried out the quantification of the six purine lesions as described in the next section.

3.2. Hydroxyl Radical–Induced Formation of Purine Lesions: tDNA vs. mtDNA

The findings in the cell cultures motivated our interest to investigate the reactivity of mtDNA and tDNA taken out of their biological contexts. $HO^•$ radicals are known for their reactivity and ability to cause chemical modifications to DNA, the site of attack being both the base moieties (85–90%) and the 2–deoxyribose units [62]. Therefore, we exposed tDNA and mtDNA, isolated from CS1AN–wtCSB, to $HO^•$ radicals generated by irradiation.

Radiolysis of neutral water leads to the reactive species e_{aq}^-, $HO^•$, and $H^•$ as shown in Reaction 1, together with H^+ and H_2O_2. The values in parentheses represent the radiation chemical yields (*G*) in units of μmol J^{-1}. In an N_2O–saturated solution (~0.02 M of N_2O), e_{aq}^- are converted into the $HO^•$ radical via Reaction 2 ($k_2 = 9.1 \times 10^9\ M^{-1}\ s^{-1}$), with $G(HO^•) = 0.55$ μmol J^{-1}, i.e., $HO^•$ radicals and $H^•$ atoms account for 90 and 10%, respectively, of the reactive species [63,64]. The rate constants for the reactions of $HO^•$ radicals and $H^•$ atoms with DNA (Reactions 3 and 4) have been reported to be ca. $2.5 \times 10^8\ M^{-1}\ s^{-1}$ and $6 \times 10^7\ M^{-1}\ s^{-1}$, respectively [62,63].

$$H_2O + \gamma\text{-}irr \rightarrow e_{aq}^-(0.27), HO^•(0.28), H^•(0.062), H^+(0.27), H_2O_2(0.07) \quad (1)$$

$$e_{aq}^- + N_2O + H_2O \rightarrow HO^• + N_2 + HO^- \quad (2)$$

$$HO^• + DNA \rightarrow \text{radical product} \quad (3)$$

$$H^• + DNA \rightarrow \text{radical product} \quad (4)$$

Figure 5 summarizes our findings and the resulting formation of 8–oxo–dG, 8–oxo–dA, 5′*R*–cdG, 5′*R*–cdA, 5′*S*–cdG, and 5′*S*–cdA in the function of irradiation doses in both tDNA and mtDNA (for specific values, see Table S4). As expected, the number of the lesions studied increased with the increment of the dose [55,56]. 8–oxo–dG is the main detected lesion, whereas 8–oxo–dA is formed in lower yields and similarly for each of the four cPu (5′*S*–cdG, 5′*R*–cdG, 5′*S*–cdA and 5′*R*–cdA). Furthermore, the slope of the obtained lines represents the number of lesions formed per Gy, as summarized in Table 3 (cf. Table S5).

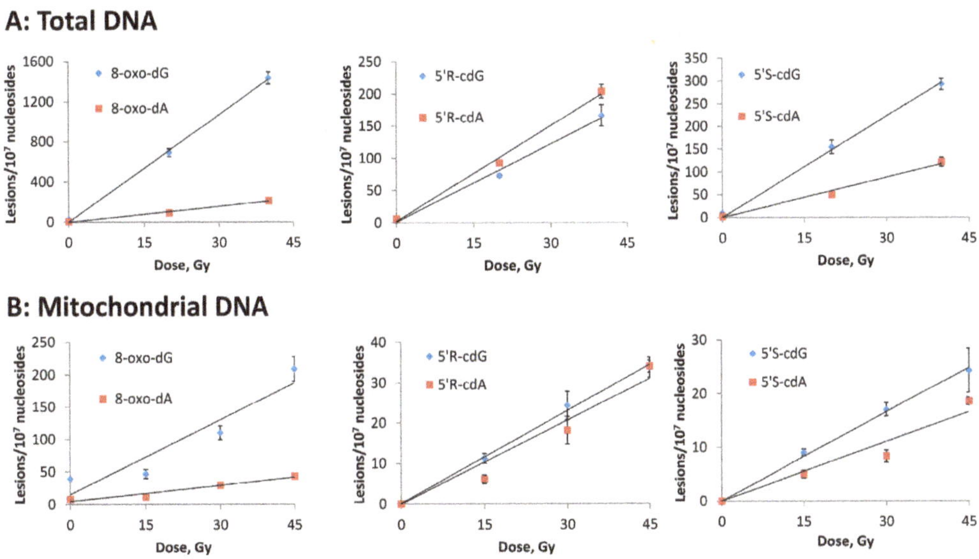

Figure 5. γ–Irradiation induced formation of 8–oxo–dG, 8–oxo–dA, 5′*R*–cdG, 5′*R*–cdA, 5′*S*–cdG, and 5′*S*–cdA in (**A**) tDNA and (**B**) mtDNA samples; the values represent the mean ± SD of $n = 3$ independent experiments.

Table 3. The levels of 8–oxo–dG, 8–oxo–dA, 5′*R*–cdG, 5′*R*–cdA, 5′*S*–cdG, and 5′*S*–cdA (Lesions/10^7 nu/Gy) from the irradiation of N_2O saturated tDNA or mtDNA (0.5 mg/mL) of CS1AN–wtCSB in aqueous solutions.

Lesion	tDNA	mtDNA
8–oxo–dG	35.2	3.8
8–oxo–dA	5.1	0.8
5′*R*–cdG	4.0	0.8
5′*R*–cdA	4.9	0.8
5′*S*–cdG	7.1	0.5
5′*S*–cdA	3.0	0.4

From the analysis of the data reported in Table 3 in terms of lesions/10^7 nu/Gy, the ratios 8–oxo–dG/8–oxo–dA were 6.9 and 4.8 in tDNA and mtDNA, respectively. It is worth mentioning that the same ratio in calf–thymus DNA was found to be 7.7 under similar experimental conditions, although the number of lesions/Gy was four to five times higher with respect to tDNA [55]. The mechanism of the formation of 8–oxo–dG through the reaction of $HO^{\bullet}$ radical with ds–oligonucleotide [55,56] and calf–thymus DNA [65,66] has been investigated in detail. It was demonstrated that the addition of $HO^{\bullet}$ to the C8 position of the guanine moiety accounts for a minor percentage (~10%), whereas the main yield of 8–oxo–dG is produced by a one–electron oxidation reaction involving DNA radicals [56, 65,66]. The lower yield of formation of 8–oxo–dA was attributed, to a minor extent, to the latter path in the case of adenine moieties. However, the $HO^{\bullet}$–adducts of guanine and adenine moieties afforded a variety of products, including 8–oxo–dG and 8–oxo–dA [33,67]. It is worth underlining that the lesions/Gy decreased substantially when going from tDNA to mtDNA (9.3 times for 8–oxo–dG and 6.4 times for 8–oxo–dA, see Table 3).

Intramolecular cyclization products 5′*R*–cdG, 5′*R*–cdA, 5′*S*–cdG, and 5′*S*–cdA were also formed in the same fashion, but in lower yields (Figure 4). The slope of the lines obtained represents the number of lesions formed per Gy, and these data are also reported in Table 3. The attack at the H5′ of DNA by $HO^{\bullet}$ was estimated to be 55% for all possible sugar positions and the resulting C5′ radical in the purine nucleotide moieties likely evolved with an internal cyclization onto the C8 position of the base with the formation of cPu as the final product (Scheme 1) [33–36]. The most important finding was the significant reduction in lesions/Gy going from tDNA to mtDNA (5.0, 6.1, 14.2, and 7.5 times for 5′*R*–cdG, 5′*R*–cdA, 5′*S*–cdG, and 5′*S*–cdA, respectively), similar to the case of 8–oxo–dG and 8–oxo–dA (9.3 and 6.4 times, respectively). The observed differences between tDNA and mtDNA in the reaction with HO• were likely due to the different topology of the nuclear and mitochondrial genome [68].

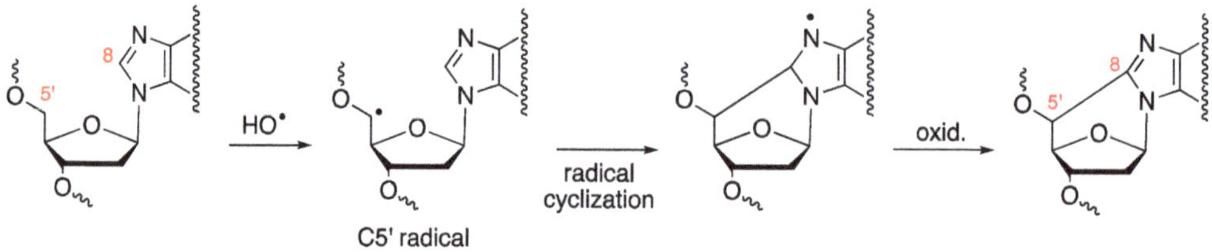

Scheme 1. Purine 2′–deoxynucleoside reacts with HO• yielding cPu via the cyclization of the C5′ radical followed by oxidation.

We also considered the total amount of 8–oxo–Pu and cPu lesions (lesions/10^7 nu) in both the tDNA and mtDNA of CS1AN–wtCSB cells irradiated at different doses. The scale referring to tDNA (Figure 6A) was six times higher than the scale referring to mtDNA (Figure 6B), clearly indicating the different amounts of accumulated lesions in the two genetic materials (Table S6). The 8–oxo–Pu was nearly twice that of the cPu values in both genomic materials (Figure 6): in units of lesions/10^7 nu/Gy, we calculated 40.3 for 8–oxo–Pu and 19.0 for cPu in tDNA, and 4.6 for 8–oxo–Pu and 2.5 for cPu in mtDNA. These values also indicate that 8–oxo–Pu and cPu were 8.8 and 7.6 times, respectively, higher in the tDNA than in the mtDNA.

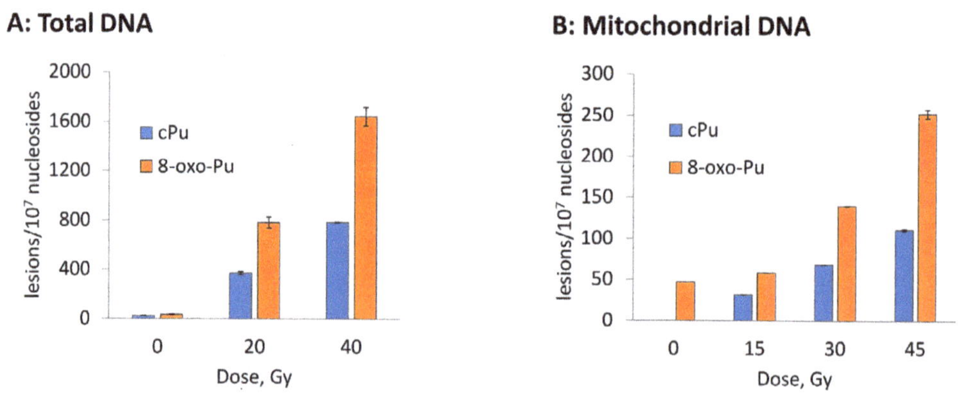

Figure 6. Total amount of cPu and 8–oxo–Pu lesions (lesions/10^7 nu) in irradiated samples of tDNA (**A**) and mtDNA (**B**). The numbers represent the mean value (± standard deviation) in triplicate.

From the reaction of the HO• radicals with tDNA and mtDNA, it is clear that (i) the formation 8–oxo–Pu was twice that of cPu, indicating a true competition of the two pathways in Figure 2, although we cannot exclude a small percentage of 8–oxo–Pu formation by the reaction of H_2O_2 generated from the irradiation of water (see reaction (1)), and (ii) the values of both 8–oxo–Pu and cPu in tDNA were ~8 times higher than in mtDNA, revealing a different accessibility of HO• to mtDNA and nDNA structures. The structure of mtDNA is arranged in a loop, which is loosely supercoiled [12]. The comparison of 8–oxo–Pu lesions arising from γ–radiation sourced HO• on tertiary DNA helical forms of supercoiled

(SC), open circular (OC), and linear (L) conformation was known [68]. Purine oxidation in dsDNA follows L > OC >>SC, indicating increased damage towards the extended B–DNA topology, where 8–oxo–dG and 8–oxo–dA levels increase ≥10–fold in both circular and linear conformers.

The cPu lesions can be present in two diasteroisomeric forms, $5'R$ and $5'S$, for each 2′–deoxyadenosine and 2′–deoxyguanosine moieties. The diastereomeric ratio (R/S) attracts interest as it can inform on mechanistic issues that are related to structural conformations of both isomers in association with their abundance. Figure 7 shows the R/S ratios for cdG and cdA lesions of irradiated tDNA and mtDNA samples. The reported data indicate that the ratios are similar within tDNA or mtDNA and independent of the irradiation dose, i.e., the R/S average ratios are 0.52 for cdG and 1.75 for cdA in tDNA, whereas in mtDNA they are 1.35 for cdG and 1.74 for cdA (Table S7).

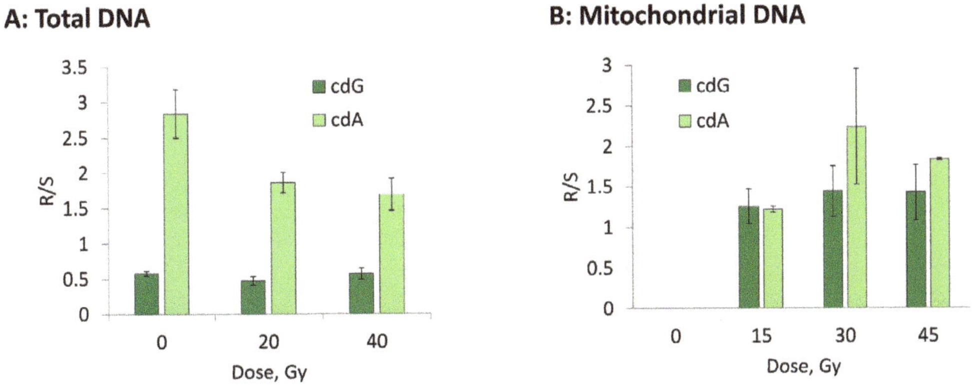

Figure 7. Diastereoisomeric ratios ($5'R/5'S$) for cdG and cdA lesions in irradiated samples of tDNA (**A**) and mtDNA (**B**). The numbers represent the mean value (± standard deviation) in triplicate.

It is worth underlining that in previous studies using extracted tDNA from various animal tissues, the *S* form was found to always be more abundant than the *R* form in cdG, whereas in cdA, the *R* form was always more abundant than the *S* form [36], this is in accordance with our present results. For example, the *R/S* levels of the cdG and cdA lesions in the tDNA of the liver and kidney of normal Swiss mice associated with age–related processes [69] are similar to the values of Figure 7A. In the brain of a rat model of Wilson disease, the *S*-cdG was always higher than the *S*-cdA (1.5 times) [70] similar to the ratios in the brains of normal, SCID, and tumor–bearing mice [71]; in the latter, the *R/S* cdA was always higher that the cdG ratio (approximately 1.3-1.6 times). Moreover, in various tissues (such as the brain, spleen, and liver) of prdx1$^{-/-}$ mice, the *R/S* cdG was found to be 0.23, 0.14, and 0.40 while the *R/S* cdA was 0.11, 0.23, and 0.08 [72]. As Figure 6B shows, in the mtDNA in both cdG and cdA, the *R* form is always more abundant in both cdG and cdA. In this respect, it is worth recalling that the *R/S* ratios of 8.3 for cdG and 6 for cdA were obtained in water upon the irradiation of free nucleosides (*R* form is always more abundant) [73,74], indicating that the diastereomer ratio is dependent on the molecular complexity. The *R/S* ratios can be used to support further biological implications in the formation and/or repair of these lesions, although a clear scenario for the *R/S* formation is still missing [36]. The structure of mtDNA is arranged in a loop, with one strand called H (heavy; purine rich) and the other strand called L (light; pyrimidine rich) [12]. We are suggesting that this arrangement strongly influences the local conformations at the reactive sites prior to C5′ radical cyclization, which makes one type of diastereoisomer more prevalent.

4. Conclusions

In this work, we measured the purine lesions in mtDNA of four cellular lines, i.e., wtCSA, wtCSB, and their defective counterparts, and compared with their analogous data of tDNA. The 8–oxo–Pu lesions were comparable in mtDNA and tDNA, although they were found to be constantly higher in mtDNA for all four cellular lines. The cPu lesions were undetectable in mtDNA, suggesting at least a 100 times lower level than in tDNA. We evidenced, for the first time, the DNA damage scenario as a contribution of two distinctive pathways, i.e., the molecular ROS and the radical ROS species, which can be clearly distinguished by comparing our results in the cell cultures with the cPu levels of the irradiation experiments. Indeed, we evaluated that the absence of cPu lesions in the mtDNA of the cellular experiment may indicate a nonappearance of $HO^{\bullet}$ radical reactivity and that the amount of 8–oxo–Pu in mtDNA was consistent with the contribution of DNA reactivity with oxidizing species, such as H_2O_2 or $ONOO^-$. Moreover, a suggestive hypothesis to explain the absent accumulation of cPu adducts in the mtDNA of CS cells could play a crucial role in specific mechanisms devoted to the maintenance of the mitochondrial genome integrity. In particular, in mammalian cells, a degradation mechanism of damaged mtDNA molecules has been recently discovered [32]. Moreover, mitophagy and mitochondrial fission can further contribute to the removal of dysfunctional/damaged organelles reducing the cPu levels up to undetectability. In the reaction of γ–irradiation generated $HO^{\bullet}$ radicals with isolated tDNA and mtDNA, the values of all six purine lesions in the tDNA are ~8 times higher than in mtDNA. We evidenced a different accessibility of $HO^{\bullet}$ into mtDNA and nDNA structures which were associated with different helical topologies.

Supplementary Materials: The following supporting information can be downloaded at: https://www.mdpi.com/article/10.3390/biom12111630/s1, Table S1: The levels of 8–oxo–dG, 8–oxo–dA, 5′*R*–cdG, 5′*S*–cdG, 5′*R*–cdA, and 5′*S*–cdA lesions in mtDNA isolated from CSA and CSB (wt and defective) cells; Table S2 and Table 3: Statistical analysis by comparing the means of lesions in mtDNA isolated from CSA and CSB cells (wt vs. defective); Table S4: Total amount of 8–oxo–dG, 8–oxo–dA, 5′*R*–cdG, 5′*R*–cdA, 5′*S*–cdG, and 5′*S*–cdA (lesions/10^7 nu) from the of N_2O saturated tDNA or mtDNA of CS1AN–wtCSB in aqueous solutions; Table S5: Equation and R–squared value of Figure 5 plots; Table S6: The levels of cPu and 8–oxo–Pu lesions from the irradiation experiments; Table S7: 5′*R*/5′*S* ratio for both cdG and cdA from the irradiation experiments.

Author Contributions: C.C. conceptualization and supervision; M.D. and B.P. performed cell cultures; B.P. and M.D. collected the biological samples; A.M and S.B.-V. irradiation experiments; M.G.K., A.M. and S.B.-V. worked up the samples; M.G.K. performed the LC–MS/MS analyses; C.C. and M.G.K. data analysis; C.C. and C.F. writing—original draft preparation; C.C. and M.G.K. contributed to the figures; C.C. and C.F. funding acquisition. All of the authors reviewed the manuscript. All authors have read and agreed to the published version of the manuscript.

Funding: C.C. acknowledge funding from Marie Skłodowska–Curie European Training Network (ETN) ClickGene: Click Chemistry for Future Gene Therapies to Benefit Citizens, Researchers, and Industry [H2020–MSCAETN–2014–642023]; This study was supported by "Bando Ricerca Indipendente ISS 2020–2022, project code: ISS20–4aeae96c8911" to D.M.

Institutional Review Board Statement: Not applicable.

Informed Consent Statement: Not applicable.

Data Availability Statement: The data presented in this study are available in this article and the Supplementary Material.

Acknowledgments: We thank Paola Fortini and Eleonora Parlanti for the critical revision of the manuscript.

Conflicts of Interest: The authors declare no conflict of interest.

References

1. Dröge, W. Free radicals in the physiological control of cell function. *Physiol. Rev.* **2002**, *82*, 47–95. [CrossRef] [PubMed]
2. Shadel, G.S.; Horvath, T.L. Mitochondrial ROS signaling in organismal homeostasis. *Cell* **2015**, *163*, 560–569. [CrossRef] [PubMed]
3. Sies, H.; Berndt, C.; Jones, D.P. Oxidative stress. *Annu. Rev. Biochem.* **2017**, *86*, 715–748. [CrossRef]
4. Sies, H.; Jones, D.P. Reactive oxygen species (ROS) as pleiotropic physiological signalling agents. *Nat. Rev. Mol. Cell. Biol.* **2020**, *21*, 363–383. [CrossRef]
5. Gammella, E.; Recalcati, S.; Cairo, G. Dual role of ROS as signal and stress agents: Iron tips the balance in favor of toxic effects. *Oxid. Med. Cell. Longev.* **2016**, *2016*, 1–9. [CrossRef]
6. Murphy, M.P. How mitochondria produce reactive oxygen species. *Biochem. J.* **2009**, *417*, 1–13. [CrossRef]
7. Solaini, G.; Baracca, A.; Lenaz, G.; Sgarbi, G. Hypoxia and mitochondrial oxidative metabolism. *Biochim. Biophys. Acta* **2010**, *1797*, 1171–1177. [CrossRef] [PubMed]
8. Mailloux, R.J. Teaching the fundamentals of electron transfer reactions in mitochondria and the production and detection of reactive oxygen species. *Redox Biol.* **2015**, *4*, 381–398. [CrossRef] [PubMed]
9. Mori, M.P.; Penjweini, R.; Knutson, J.R.; Wang, P.; Hwang, P.M. Mitochondria and oxygen homeostasis. *FEBS J.* 2021; *online ahead of print*. [CrossRef]
10. Richter, C. Oxidative damage to mitochondrial DNA and its relationship to ageing. *Int. J. Biochem. Cell Biol.* **1995**, *27*, 647–653. [CrossRef]
11. Harman, D. The biologic clock: The mitochondria? *J. Am. Geriatr. Soc.* **1972**, *20*, 145–147. [CrossRef]
12. Anderson, S.; Bankier, A.T.; Barrell, B.G.; de Bruijn, M.H.; Coulson, A.R.; Drouin, J.; Eperon, I.C.; Nierlich, D.P.; Roe, B.A.; Sanger, F.; et al. Sequence and organization of the human mitochondrial genome. *Nature* **1981**, *290*, 457–465. [CrossRef] [PubMed]
13. Nunnari, J.; Suomalainen, A. Mitochondria: In sickness and in health. *Cell* **2012**, *148*, 1145–1159. [CrossRef] [PubMed]
14. Farge, G.; Falkenberg, M. Organization of DNA in mammalian mitochondria. *Int. J. Mol. Sci.* **2019**, *20*, 2770. [CrossRef] [PubMed]
15. Mandavilli, B.S.; Santos, J.H.; Van Houten, B. Mitochondrial DNA repair and ageing. *Mutat. Res.* **2002**, *509*, 127–151. [CrossRef]
16. Huang, Z.; Chen, Y.; Zhang, Y. Mitochondrial reactive oxygen species cause major oxidative mitochondrial DNA damages and repair pathways. *J. Biosci.* **2020**, *45*, 84. [CrossRef]
17. Muftuoglu, M.; Mori, M.P.; de Souza–Pinto, N.C. Formation and repair of oxidative damage in the mitochondrial DNA. *Mitochondrion* **2014**, *17*, 164–181. [CrossRef]
18. Kauppila, J.H.K.; Stewart, J.B. Mitochondrial DNA: Radically free of free–radical driven mutations. *Biochim. Biophys. Acta* **2015**, *1847*, 1354–1361. [CrossRef]
19. Kopinski, P.K.; Singh, L.N.; Zhang, S.; Lott, M.T.; Wallace, D.C. Mitochondrial DNA variation and cancer. *Nat. Rev.* **2021**, *21*, 431–445. [CrossRef]
20. D'Errico, M.; Parlanti, E.; Pascucci, B.; Filomeni, G.; Mastroberardino, P.G.; Dogliotti, E. The interplay between mitochondrial functionality and genome integrity in the prevention of human neurologic diseases. *Arch. Biochem. Biophys.* **2021**, *710*, 108977. [CrossRef]
21. Chatgilialoglu, C.; Studer, A. (Eds.) *Encyclopedia of Radicals in Chemistry, Biology and Materials*; Wiley: Chichester, UK, 2012.
22. Halliwell, B.; Gutteridge, J.M.C. *Free Radicals in Biology and Medicine*, 5th ed.; Oxford University Press: Oxford, UK, 2015.
23. Fridovich, I. Superoxide radical and superoxide dismutases. *Annu. Rev. Biochem.* **1995**, *64*, 97–112. [CrossRef]
24. Stohs, S.J.; Bagchi, D. Oxidative mechanisms in the toxicity of metal ions. *Free Radic. Biol. Med.* **1995**, *18*, 321–336. [CrossRef]
25. Davies, M.J.; Hawkins, C.L. The role of myeloperoxidase in biomolecule modification, chronic inflammation, and disease. *Antioxid. Redox Signal.* **2020**, *32*, 957–981. [CrossRef] [PubMed]
26. Winterbourn, C.C. Reconciling the chemistry and biology of reactive oxygen species. *Nat. Chem. Biol.* **2008**, *4*, 278–286. [CrossRef]
27. Dizdaroglu, M.; Lloyd, R.S. *DNA Damage, DNA Repair and Disease*; Royal Society of Chemistry: Croydon, UK, 2021.
28. Alexeyev, M.; Shokolenko, I.; Wilson, G.; Ledoux, S. The maintenance of mitochondrial DNA integrity—Critical analysis and update. *Cold Spring Harb. Perspect. Biol.* **2013**, *5*, a012641. [CrossRef] [PubMed]
29. Shokolenko, I.; Venediktova, N.; Bochkareva, A.; Wilson, G.L.; Alexeyev, M.F. Oxidative stress induces degradation of mitochondrial DNA. *Nucleic Acids Res.* **2009**, *37*, 2539–2548. [CrossRef] [PubMed]
30. Furda, A.M.; Marrangoni, A.M.; Lokshin, A.; Van Houten, B. Oxidants and not alkylating agents induce rapid mtDNA loss and mitochondrial dysfunction. *DNA Repair.* **2012**, *11*, 684–692. [CrossRef]
31. Moretton, A.; Morel, F.; Macao, B.; Lachaume, P.; Ishak, L.; Lefebvre, M.; Garreau-Balandier, I.; Vernet, P.; Falkenberg, M.; Farge, G. Selective mitochondrial DNA degradation following double-strand breaks. *PLoS ONE* **2017**, *12*, e0176795. [CrossRef]
32. Peeva, V.; Blei, D.; Trombly, G.; Corsi, S.; Szukszto, M.J.; Rebelo-Guiomar, P.; Gammage, P.A.; Kudin, A.P.; Becker, C.; Altmüller, J.; et al. Linear mitochondrial DNA is rapidly degraded by components of the replication machinery. *Nat. Commun.* **2018**, *9*, 1727. [CrossRef]
33. Chatgilialoglu, C.; Ferreri, C.; Krokidis, M.G.; Masi, A.; Terzidis, M.A. On the relevance of hydroxyl radical to purine DNA damage. *Free Radic. Res.* **2021**, *55*, 384–404. [CrossRef]
34. Chatgilialoglu, C.; Ferreri, C.; Terzidis, M.A. Purine 5′,8–cyclonucleoside lesions: Chemistry and biology. *Chem. Soc. Rev.* **2011**, *40*, 1368–1382. [CrossRef]
35. Dizdaroglu, M.; Jaruga, P. Mechanisms of free radical–induced damage to DNA. *Free Radic. Res.* **2012**, *46*, 382–419. [CrossRef] [PubMed]

36. Chatgilialoglu, C.; Ferreri, C.; Geacintov, N.E.; Krokidis, M.G.; Liu, Y.; Masi, A.; Shafirovich, N.; Terzidis, M.A.; Tsegay, P.S. 5′,8–Cyclopurine lesions in DNA damage: Chemical, analytical, biological and diagnostic significance. *Cells* **2019**, *8*, 513. [CrossRef] [PubMed]
37. Cui, L.; Ye, W.; Prestwich, E.G.; Wishnok, J.S.; Taghizadeh, K.; Dedon, P.C.; Tannenbaum, S.R. Comparative analysis of four oxidized guanine lesions from reactions of DNA with peroxynitrite, single oxygen, and γ–radiation. *Chem. Res. Toxicol.* **2013**, *26*, 195–202. [CrossRef]
38. Brooks, P.J.; Wise, D.S.; Berry, D.A.; Kosmoski, J.V.; Smerdon, M.J.; Somers, R.L.; Mackie, H.; Spoonde, A.Y.; Ackerman, E.J.; Coleman, K.; et al. The oxidative DNA lesion 8,5′–(S)–cyclo–2′–deoxyadenosine is repaired by the nucleotide excision repair pathway and blocks gene expression in mammalian cells. *J. Biol. Chem.* **2000**, *275*, 22355–22362. [CrossRef]
39. Kuraoka, I.; Bender, C.; Romieu, A.; Cadet, J.; Wood, R.D.; Lindahl, T. Removal of oxygen free–radical–induced 5′,8–purine cyclodeoxynucleosides from DNA by the nucleotide excision– repair pathway in human cells. *Proc. Natl. Acad. Sci. USA* **2000**, *97*, 3832–3837. [CrossRef] [PubMed]
40. Pande, P.; Das, R.S.; Sheppard, C.; Kow, Y.W.; Basu, A.K. Repair efficiency of (5′S)–8,5–cyclo–2′–deoxyguanosine and (5′ S)–8,5′–cyclo–2′–deoxyadenosine depends on the complementary base. *DNA Repair.* **2012**, *11*, 926–931. [CrossRef]
41. Kropachev, K.; Ding, S.; Terzidis, M.A.; Masi, A.; Liu, Z.; Cai, Y.; Kolbanovskiy, M.; Chatgilialoglu, C.; Broyde, S.; Geacintov, N.E.; et al. Structural basis for the recognition of diastereomeric 5′,8–cyclo–2′–deoxypurine lesions by the human nucleotide excision repair system. *Nucleic Acids Res.* **2014**, *42*, 5020–5032. [CrossRef]
42. Krasikova, Y.; Rechkunova, N.; Lavrik, O. Nucleotide excision repair: From molecular defects to neurological abnormalities. *Int. J. Mol. Sci.* **2021**, *22*, 6220. [CrossRef]
43. Hanawalt, P.C.; Spivak, G. Transcription–coupled DNA repair: Two decades of progress and surprises. *Nat. Rev. Mol. Cell. Biol.* **2008**, *9*, 958–970. [CrossRef]
44. D'Errico, M.; Pascucci, B.; Iorio, E.; Van Houten, B.; Dogliotti, E. The role of CSA and CSB protein in the oxidative stress response. *Mech. Ageing Dev.* **2013**, *134*, 261–269. [CrossRef]
45. Pascucci, B.; Fragale, A.; Marabitti, V.; Leuzzi, G.; Calcagnile, A.S.; Parlanti, E.; Franchitto, A.; Dogliotti, E.; D'Errico, M. CSA and CSB play a role in the response to DNA breaks. *Oncotarget* **2018**, *9*, 11581–11591. [CrossRef] [PubMed]
46. Krokidis, M.G.; D'Errico, M.; Pascucci, B.; Parlanti, E.; Masi, A.; Ferreri, C.; Chatgilialoglu, C. Oxygen–Dependent Accumulation of Purine DNA Lesions in Cockayne Syndrome Cells. *Cells* **2021**, *9*, 1671, Correction in *Cells* **2021**, *10*, 41. [CrossRef] [PubMed]
47. Ferreri, C.; Sansone, A.; Krokidis, M.G.; Masi, A.; Pascucci, B.; D'Errico, M.; Chatgilialoglu, C. Effects of Oxygen Tension for Membrane Lipidome Remodeling of Cockayne Syndrome Cell Models. *Cells* **2022**, *11*, 1286. [CrossRef] [PubMed]
48. Chatgilialoglu, C. Cyclopurine (cPu) lesions: What, how and why? *Free Radic. Res.* **2019**, *53*, 941–943. [CrossRef] [PubMed]
49. Chatgilialoglu, C. Biomimetic Radical Chemistry and Applications. *Molecules* **2022**, *27*, 2042. [CrossRef]
50. D'Errico, M.; Teson, M.; Calcagnile, A.; Nardo, T.; De Luca, N.; Lazzari, C.; Soddu, S.; Zambruno, G.; Stefanini, M.; Dogliotti, E. Differential role of transcription–coupled repair in UVB–induced response of human fibroblasts and keratinocytes. *Cancer Res.* **2005**, *65*, 232–238. [CrossRef]
51. Pascucci, B.; D'Errico, M.; Romagnoli, A.; De Nuccio, C.; Savino, M.; Pietraforte, D.; Lanzafame, M.; Calcagnile, A.S.; Fortini, P.; Baccarini, S.; et al. Overexpression of parkin rescues the defective mitochondrial phenotype and the increased apoptosis of Cockayne syndrome A cells. *Oncotarget* **2016**, *8*, 102852–102867. [CrossRef]
52. Pascucci, B.; Lemma, T.; Iorio, E.; Giovannini, S.; Vaz, B.; Iavarone, I.; Calcagnile, A.; Narciso, L.; Degan, P.; Podo, F.; et al. An altered redox balance mediates the hypersensitivity of Cockayne syndrome primary fibroblasts to oxidative stress. *Aging Cell* **2012**, *11*, 520–529. [CrossRef]
53. Spinks, J.W.T.; Woods, R.J. *An Introduction to Radiation Chemistry*, 3rd ed.; John–Wiley and Sons, Inc.: New York, NY, USA, 1990; p. 100.
54. Terzidis, M.A.; Chatgilialoglu, C. An ameliorative protocol for the quantification of purine 5′,8–cyclo–2′–deoxynucleosides in oxidized DNA. *Front. Chem.* **2015**, *3*, 47. [CrossRef]
55. Chatgilialoglu, C.; Krokidis, M.G.; Masi, A.; Barata–Vallejo, S.; Ferreri, C.; Terzidis, M.A.; Szreder, T.; Bobrowski, K. New insights into the reaction paths of hydroxyl radicals with purine moieties in DNA and double-stranded oligonucleotides. *Molecules* **2019**, *24*, 3860. [CrossRef]
56. Chatgilialoglu, C.; Eriksson, L.A.; Krokidis, M.G.; Masi, A.; Wang, S.; Zhang, R. Oxygen dependent purine lesions in double-stranded oligodeoxynucleotides: Kinetic and computational studies highlight the mechanism for 5′,8–cyclopurine formation. *J. Am. Chem. Soc.* **2020**, *142*, 5825–5833. [CrossRef] [PubMed]
57. Pascucci, B.; Spadaro, F.; Pietraforte, D.; De Nuccio, C.; Visentin, S.; Giglio, P.; Dogliotti, E.; D'Errico, M. DRP1 Inhibition Rescues Mitochondrial Integrity and Excessive Apoptosis in CS–A Disease Cell Models. *Int. J. Mol. Sci.* **2021**, *22*, 7123. [CrossRef] [PubMed]
58. Chatre, L.; Biard, D.S.F.; Sarasin, A.; Ricchetti, M. Reversal of mitochondrial defects with CSB–dependent serine protease inhibitors in patient cells of the progeroid Cockayne syndrome. *Proc. Natl. Acad. Sci. USA* **2015**, *112*, E2910–E2919. [CrossRef] [PubMed]
59. Wiesner, R.J.; Rüegg, C.; Morano, I. Counting target molecules by exponential polymerase chain reaction: Copy number of mitochondrial DNA in rat tissues. *Biochem. Biophys. Res. Commun.* **1992**, *183*, 553–559. [CrossRef]
60. D'Erchia, A.M.; Atlante, A.; Gadaleta, G.; Pavesi, G.; Chiara, M.; De Virgilio, C.; Manzari, C.; Mastropasqua, F.; Prazzoli, G.M.; Picardi, E.; et al. Tissue–specific mtDNA abundance from exome data and its correlation with mitochondrial transcription, mass and respiratory activity. *Mitochondrion* **2015**, *20*, 13–21. [CrossRef]

61. Shafirovich, V.; Kolbanovskiy, M.; Kropachev, K.; Liu, Z.; Cai, Y.; Terzidis, M.A.; Masi, A.; Chatgilialoglu, C.; Amin, S.; Dadali, A.; et al. Nucleotide excision repair and impact of site–specific 5′,8–cyclopurine and bulky DNA lesions on the physical properties of nucleosomes. *Biochemistry* **2019**, *58*, 561–574. [CrossRef] [PubMed]
62. von Sonntag, C. *Free–Radical–Induced DNA Damage and Its Repair. A Chemical Perspective*; Springer Science: Berlin/Heidelberg, Germany, 2006.
63. Buxton, G.V.; Greenstock, C.L.; Helman, W.P.; Ross, A.B. Critical review of rate constants for hydrated electrons, hydrogen atoms and hydroxyl radicals (OH/O−) in aqueous solution. *J. Phys. Chem. Ref. Data* **1988**, *17*, 513–886. [CrossRef]
64. Ross, A.B.; Mallard, W.G.; Helman, W.P.; Buxton, G.V.; Huie, R.E.; Neta, P. *NDRLNIST Solution Kinetic Database–Ver. 3*; Notre Dame Radiation Laboratory, Notre Dame, IN and NIST Standard Reference Data: Gaithersburg, MD, USA, 1998.
65. Bergeron, F.; Auvré, F.; Radicella, J.P.; Ravanat, J.-L. HO• radicals induce an unexpected high proportion of tandem base lesions refractory to repair by DNA glycosylases. *Proc. Natl. Acad. Sci. USA* **2010**, *107*, 5528–5533. [CrossRef]
66. Ravanat, J.-L. Endogenous natural and radiation–induced DNA lesions: Differences and similarities and possible implications for human health and radiological protection. *Radioprotection* **2018**, *53*, 241–248. [CrossRef]
67. Chatgilialoglu, C. The Two Faces of the Guanyl Radical: Molecular Context and Behavior. *Molecules* **2021**, *26*, 3511. [CrossRef]
68. Terzidis, M.A.; Prisecaru, A.; Molphy, Z.; Barron, N.; Randazzo, A.; Dumont, E.; Krokidis, M.G.; Kellett, A.; Chatgilialoglu, C. Radical–induced purine lesion formation is dependent on DNA helical topology. *Free Radic. Res.* **2016**, *50*, S91–S101. [CrossRef] [PubMed]
69. Krokidis, M.G.; Louka, M.; Efthimiadou, E.K.; Zervou, S.K.; Papadopoulos, K.; Hiskia, A.; Ferreri, C.; Chatgilialoglu, C. Membrane lipidome reorganization and accumulation of tissue DNA lesions in tumor–bearing mice: An exploratory study. *Cancers* **2019**, *11*, 480. [CrossRef] [PubMed]
70. Yu, Y.; Guerrero, C.R.; Liu, S.; Amato, N.J.; Sharma, Y.; Gupta, S.; Wang, Y. Comprehensive assessment of oxidatively induced modifications of DNA in a rat model of human Wilson's disease. *Mol. Cell. Proteom.* **2016**, *15*, 810–817. [CrossRef]
71. Krokidis, M.G.; Prasinou, P.; Efthimiadou, E.K.; Boari, A.; Ferreri, C.; Chatgilialoglu, C. Effects of Aging and Disease Conditions in Brain of Tumor–Bearing Mice: Evaluation of Purine DNA Damages and Fatty Acid Pool Changes. *Biomolecules* **2022**, *12*, 1075. [CrossRef] [PubMed]
72. Egler, R.A.; Fernandes, E.; Rothermund, K.; Sereika, S.; de Souza–Pinto, N.; Jaruga, P.; Dizdaroglu, M.; Prochownik, E.V. Regulation of reactive oxygen species, DNA damage, and c–Myc function by peroxiredoxin 1. *Oncogene* **2005**, *24*, 8038–8050. [CrossRef] [PubMed]
73. Chatgilialoglu, C.; Bazzanini, R.; Jimenez, L.B.; Miranda, M.A. (5′*S*)– and (5′*R*)–5′,8–cyclo–2′–deoxyguanosine: Mechanistic insights on the 2′–deoxyguanosin–5′–yl radical cyclization. *Chem. Res. Toxicol.* **2007**, *20*, 1820–1824. [CrossRef]
74. Boussicault, F.; Kaloudis, P.; Caminal, C.; Mulazzani, Q.G.; Chatgilialoglu, C. The fate of C5′ radicals of purine nucleosides under oxidative conditions. *J. Am. Chem. Soc.* **2008**, *130*, 8377–8385. [CrossRef]

biomolecules

MDPI

Article

Effects of Aging and Disease Conditions in Brain of Tumor-Bearing Mice: Evaluation of Purine DNA Damages and Fatty Acid Pool Changes

Marios G. Krokidis [1,2,*], Paraskevi Prasinou [3], Eleni K. Efthimiadou [2,4], Andrea Boari [3], Carla Ferreri [1] and Chryssostomos Chatgilialoglu [1,5,*]

1 Istituto per la Sintesi Organica e la Fotoreattività, Consiglio Nazionale delle Ricerche, Via Piero Gobetti 101, 40129 Bologna, Italy
2 Institute of Nanoscience and Nanotechnology, National Center for Scientific Research "Demokritos", 15310 Athens, Greece
3 Faculty of Veterinary Medicine, University of Teramo, 64100 Teramo, Italy
4 Department of Chemistry, National and Kapodistrian University of Athens, 15784 Athens, Greece
5 Center for Advanced Technologies, Adam Mickiewicz University, 61-614 Poznan, Poland
* Correspondence: m.krokidis@inn.demokritos.gr (M.G.K.); chrys@isof.cnr.it (C.C.)

Abstract: The consequences of aging and disease conditions in tissues involve reactive oxygen species (ROS) and related molecular alterations of different cellular compartments. We compared a murine model of immunodeficient (SCID) xenografted young (4 weeks old) and old (17 weeks old) mice with corresponding controls without tumor implantation and carried out a compositional evaluation of brain tissue for changes in parallel DNA and lipids compartments. DNA damage was measured by four purine 5′,8-cyclo-2′-deoxynucleosides, 8-oxo-7,8-dihydro-2′-deoxyguanosine (8-oxo-dG), and 8-oxo-7,8-dihydro-2′-deoxyadenosine (8-oxo-dA). In brain lipids, the twelve most representative fatty acid levels, which were mostly obtained from the transformation of glycerophospholipids, were followed up during the aging and disease progressions. The progressive DNA damage due to age and tumoral conditions was confirmed by raised levels of 5′*S*-cdG and 5′*S*-cdA. In the brain, the remodeling involved a diminution of palmitic acid accompanied by an increase in arachidonic acid, along both age and tumor progressions, causing increases in the unsaturation index, the peroxidation index, and total TFA as indicators of increased oxidative and free radical reactivity. Our results contribute to the ongoing debate on the central role of DNA and genome instability in the aging process, and on the need for a holistic vision, which implies choosing the best biomarkers for such monitoring. Furthermore, our data highlight brain tissue for its lipid remodeling response and inflammatory signaling, which seem to prevail over the effects of DNA damage.

Keywords: tumor-bearing mice; aging; hydroxyl radical; oxidatively-induced DNA lesions; brain fatty acids; age-induced remodeling

Citation: Krokidis, M.G.; Prasinou, P.; Efthimiadou, E.K.; Boari, A.; Ferreri, C.; Chatgilialoglu, C. Effects of Aging and Disease Conditions in Brain of Tumor-Bearing Mice: Evaluation of Purine DNA Damages and Fatty Acid Pool Changes. *Biomolecules* **2022**, *12*, 1075. https://doi.org/10.3390/biom12081075

Academic Editor: Undurti N. Das

Received: 8 June 2022
Accepted: 1 August 2022
Published: 4 August 2022

1. Introduction

The aging process or disease conditions cause free radical stress and can impair the molecular and enzymatic network that controls the redox balance in organisms [1,2]. Single DNA adducts or multiple lesions generated under stress, such as tandem or clustered lesions, are substrates of the different cellular repair systems that protect genome instability. However, due to the progressive loss of the protective machineries, depending either on the aging degenerative mechanisms or on enzymatic deficiencies, these adducts are poorly repaired and, therefore, may accumulate in the genome, causing damage to cellular components [3]. Purine 5′,8-cyclo-2′-deoxynucleosides (cPu) are solely generated by the attack of $HO^•$ radicals on purine moiety via C5′-radical chemistry, resulting in the formation of an additional C5′−C8 covalent bond; 5′,8-cyclo-2′-deoxyadenosine (cdA) and 5′,8-cyclo-2′-deoxyguanosine (cdG) exist in the 5′*R* and 5′*S* configurations (Figure 1) [4–6]. cPu can be removed only by the nucleotide excision repair (NER) pathway, and different repair efficiencies of the *R* and *S* diastereoisomers has been detected [7,8]. On the contrary, the well-known 8-oxo-purines (8-oxo-Pu) lesions, i.e., 8-oxo-7,8-dihydro-2′-deoxyadenosine (8-oxo-dA) and 8-oxo-7,8-dihydro-2′-deoxyguanosine (8-oxo-dG), are derived from the oxidation at the C8 position of adenine and guanine by a variety of reactive oxygen species (ROS) and can be repaired by base excision repair (BER) [9]. The role of oxygen is crucial in the formation of the four cPu. In recent model studies by our group [10,11], cPu levels decreased substantially by increasing oxygen tension, favoring products derived from the peroxyl radical. It should be emphasized that 8-oxo-Pu gradually increased by increasing oxygen concentration. Interestingly, a similar trend was not reported in a cellular environment, where reducing the oxygen incubation conditions raises the accumulation of both cPu and 8-oxo-Pu [12,13]. The brain presents an increased oxidative stress due to a high level of tissue oxygen consumption and a diminished ratio of antioxidant to pro-oxidant enzymes [14]. An age-dependent accumulation, with much higher levels of cdA, was observed in the brain as well as in liver and kidney of (XPA)-deficient mice, compared with wild-type animals, showing the strong involvement of NER in the effective repair of oxidative DNA damage in different tissue compartments [15].

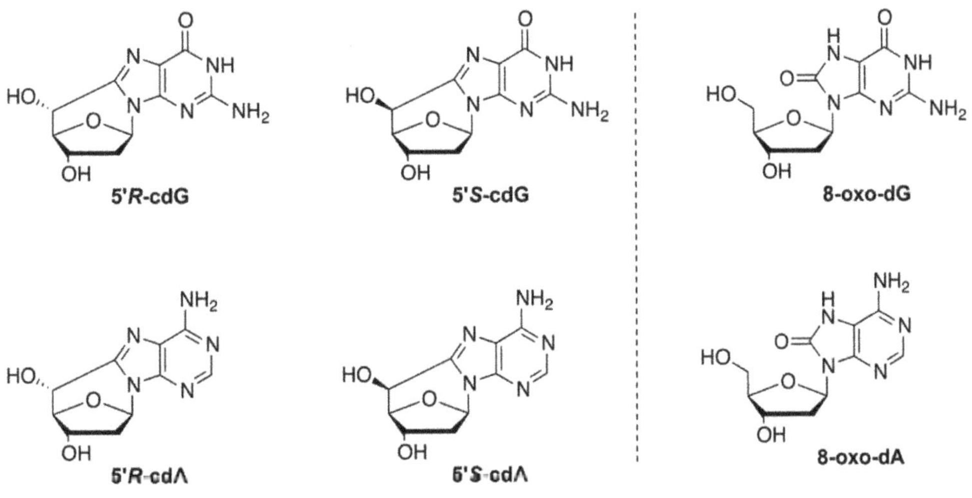

Figure 1. Structures of 5′,8-cyclo-2′-deoxyguanosine (cdG) and 5′,8-cyclo-2′-deoxyadenosine (cdA) in their 5′*R* and 5′*S* diastereomeric forms (**left**) and 8-oxo-7,8-dihydro-2′-deoxyguanosine (8-oxo-dG) and 8-oxo-7,8-dihydro-2′-deoxyadenosine (8-oxo-dA) (**right**).

Another important molecular contribution in the brain, under physiological and pathological conditions, comes from lipid homeostasis, which is strongly connected with chemical events due to the redox properties of several reactive species that can cause oxidative damage [16]. The brain is the tissue that is second-most rich in lipids, after adipose tissue, with necessary components that support structural, biochemical, and cell signaling functions. Its homeostasis corresponds to the correct exploitation of the various processes, including immunity, inflammation, and resolution. The phospholipid composition of neuronal and glial membranes in mammals, with about 75% of lipids exclusive to neural tissue, is a clear example of the role of the molecular characteristics that are needed to accomplish the wide variety of organization and shaping of brain tissue [17]. Changes in brain lipidome are reported in the modulation of neuronal functions, as well as in neurodegenerative diseases and aging, with glycerophospholipids, sphingolipids, and sterols representing the majority of the brain's lipids [18,19]. A fatty acid-based analysis of neural tissue showed evidence of the presence of saturated and monounsaturated fatty acids

(SFA and MUFA) that are formed by de novo lipogenesis, together with polyunsaturated fatty acids (PUFA), which are supplied to neuronal cells by an exchange with blood lipids and biosynthesized in the brain cells after the intakes of essential PUFAs, i.e., linoleic acid (18:2 omega-6, LNA) and alpha-linolenic acid (18:3, omega-3), which cannot be formed in mammals (Figure 2) [20]. However, in rodents, the PUFA biosynthesis is not very active in the brain [21], which should receive a constant supply. Especially important for the correct brain functioning is the balance between the omega-6 PUFA arachidonic acid (ARA) and the omega-3 PUFA docosahexaenoic acid (DHA) in neuronal membrane phospholipids, to ensure the appropriate release of these fatty acids as precursors of lipid mediators with pro- and anti-inflammatory properties [20–22]. The occurrence of inadequate levels due to diet or consumption, resulting from oxidative damage to PUFA, is the basis for important neuronal impairments, as described in seminal papers on neuronal membrane expansion [23], cognitive development [24], Alzheimer's disease, and Parkinson disease [25–28].

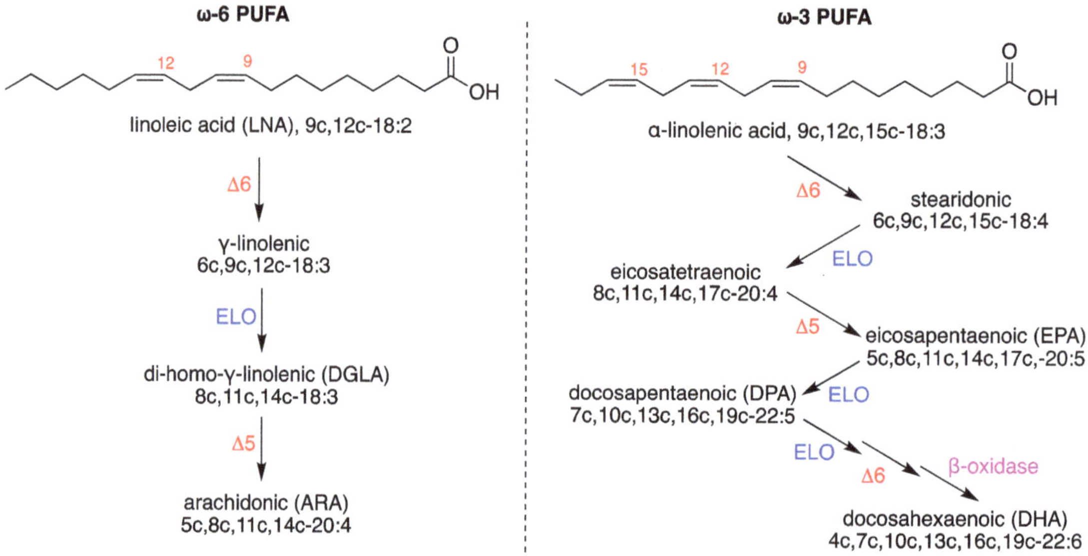

Figure 2. Biosynthesis of PUFA: (**left**) the omega-6 series starting linoleic acid; (**right**) the omega-3 series starting alpha-linolenic acid. Enzymes: ELO elongase; Δ5-, Δ6-, and Δ9-desaturase; β-oxidase. Numerical abbreviations describing the position and cis geometry of the double bonds (e.g., 9c), as well as the notation of the carbon chain length and total number of unsaturations (e.g., 18:2); in parentheses are the acronyms used in this work (e.g., ARA for arachidonic acid).

In this scenario of molecular contributions to tissue health, oxidation cannot be seen as having a damaging effect, because oxidative metabolism in the brain is necessary, with 20% of the total basal oxygen budget used to support not only the ATP cellular production but also the production of reactive oxygen species and oxygenated metabolites, allowing for adaptation and signaling in neuronal tissues, as described in a recent review [29]. Therefore, molecular damage must be carefully evaluated in order to distinguish between damage and signals. The role of thiols is crucial, either as sulfur compounds—including the amino acid cysteine—or as a functional group in enzymes, as the thiol–disulfide redox balance allows the quenching of ROS reactivity, as shown for glutathione in Alzheimer's patients and in a Parkinson animal model [30,31].

In our holistic vision of oxidative stress, we explored specific biomarkers of free radical damage. In particular, for biomarkers for DNA reactivity, we envisaged a specific radical-based reaction of purine nucleosides that are transformed into cPu and 8-oxo-Pu lesions [4–6], and for biomarkers of membrane reactivity, we envisaged both the formation of trans fatty acids (TFA) and the remodeling of membrane lipidome as interesting markers of tissue transformation [32–34]. In TFA formation, thiyl radicals generated from the antioxidant reactivity of thiols are involved in the addition-elimination reaction on the unsaturated double bonds of MUFA and PUFA. In membrane remodeling, MUFA and PUFA residues respond to different conditions, such as aging or disease progression, and their balance can be helpful in distinguishing between free radical or metabolic events in the specific pathological condition in models and clinical studies [35,36].

We recently used an experimental diseased animal model—severe combined immunodeficient (SCID) female mice inoculated with U87MG human glioblastoma cells—to assess differences in the cPu levels of the liver and the kidney at the early and final stages of tumorigenesis (4-week-old and 17-week-old tumor-bearing mice, respectively). A profound enrichment of oxidatively induced DNA damage lesions in both tissues of 17-week-old xenografts, compared with the early stage of tumorigenesis, was observed [37]. In the present paper, we report the examination of brain tissue coupling for the content of both cPu and 8-oxo-Pu with those of fatty acid components in the brains of young and older tumor-bearing mice, with a parallel evaluation of the same lesions in control SCID mice without tumor implantation (4 weeks old and 17 weeks old, respectively). The aim of our work is to unravel potential differences in tumor-bearing animals and to obtain important insights that are related to immunodeficiency and aging.

2. Materials and Methods

2.1. Materials

All reagents were obtained from Sigma–Aldrich (Steinheim, Germany) and solvents (chloroform, methanol, n-hexane) were purchased from Fisher Scientific (HPLC grade). Nuclease P1 from Penicillium citrinum, phosphodieasterase I and II, alkaline phosphatase from bovine intestinal mucosa, DNase I and DNase II, benzonase 99%, BHT, deferoxamine mesylate, and pentostatin were purchased from Sigma-Aldrich (Steinheim, Germany). RNase T1 was obtained from Thermo Fisher Scientific (Waltham, MA, USA) and RNase A was obtained from Roche Diagnostic GmbH (Mannheim, Germany). 2′-Deoxyadenosine monohydrate were purchased from Berry & Associates Inc. (Dexter, NY, USA). All fatty acid methyl esters (FAME) used as references were commercially available from Supelco (Bellefonte, PA, USA) or Sigma–Aldrich. U87MG brain glioblastoma was obtained from the American Type Culture Collection (ATCC). High glucose Dulbecco's modified Eagle Medium (DMEM) was purchased from Sigma. Trypsin-EDTA, L-glutamine, penicillin–streptomycin solution, and heat inactivated fetal bovine serum (FBS) were obtained from Biochrom KG. Ultrapure water (18.3 MΩ·cm) and deionized water (Milli-Q water) were purified by a Milli-Q system (Merck–Millipore, Bedford, PA, USA).

2.2. Animal Studies and Xenografts Construction

Female SCID and normal Swiss mice were housed at the SOL-GEL laboratory at the NCSR "Demokritos" and SCID mice were xenografted at two weeks of age, subcutaneously just above the right flank, with U87MG cells that were previously grown in DMEM, as previously described [37]. The tumor volume, mice weights, and survival rates were calculated in different time intervals. Mice were housed in groups of three per cage under positive pressure in polysulfone type IIL individual ventilated cages (Sealsafe, Tecniplast, Buguggiate, Italy) and had ad libitum access to water and food. Room temperature and relative humidity were 24 ± 2 °C and 55 ± 10% respectively. All animals in the facility were screened regularly according to the Federation of European Laboratory Animal Science Associations' recommendations and were found to be free of the respective pathogens. Animals were sacrificed under deep ether anesthesia and the brain tissues were rapidly extracted, placed in a polypropylene tube, immediately snap-frozen in liquid nitrogen, and stored at −80 °C.

2.3. DNA Isolation and Quantification of Modified Nucleosides by Stable Isotope Dilution LC-MS/MS

Genomic DNA from frozen tissues was isolated using a high-salt extraction procedure [37,38], enzymatically digested in the presence of 100 μM deferoxamine, 100 μM butylated hydroxytoluene. And the internal standards ([$^{15}N_5$]-5′S-cdA, [$^{15}N_5$]-5′R-cdA, [$^{15}N_5$]-5′S-cdG, [$^{15}N_5$]-5′R-cdG, [$^{15}N_5$]-8-oxo-dG and [$^{15}N_5$]-8-oxo-dA), and lesions were quantified as described previously [4–6,10–13,39]. The samples were filtered by centrifugation through a 3 kDa microspin filter (Millipore; Bedford, OH, USA), cleaned and enriched by an HPLC-UV system coupled with a sample collector, and injected into the LC-MS/MS system. The quantification of the modified nucleosides was carried out by a triple-stage quadrupole mass spectrometer using positive electrospray ionization (ESI), following a gradient program (2 mM ammonium formate, acetonitrile, and methanol), and the detection was executed in multiple reaction monitoring mode (MRM) using the two most intense and characteristic precursor/product ion transitions for each lesion [11,38].

2.4. Fatty Acid Analysis of Brain Tissues

Tissues were disrupted in 2:1 chloroform:methanol using a mechanical homogenizer submerged in a cooling bath of acetone and dry ice to maintain a temperature of approximately −80 °C. The lipid extract, after solvent evaporation to dryness, was then treated with 0.5 M KOH/MeOH for 10 min at room temperature under stirring for the derivatization of fatty acid residues of the glycerol esters-containing lipids into their corresponding fatty acid methyl esters (FAME) [40]. After this transesterification step, FAME were extracted with n-hexane; n-hexane phase was dehydrated with anhydrous Na_2SO_4, evaporated and analyzed via an Agilent 7890B CG system equipped with a 60 m × 0.25 mm × 0.25 μm (50%-cyanopropyl)-methylpolysiloxane column (DB23, Agilent, USA) and a flame ionization detector (FID), with an injector temperature at 230 °C and split injection of 50:1. Oven temperature started at 165 °C, was held for 3 min, then increased by 1 °C/min up to 195 °C, held again for 40 min, then increased by 10 °C/min up to 240 °C and held for 10 min. A constant pressure mode (29 psi), with helium as the carrier gas, was used. Methyl esters were identified by comparison with the retention times of commercially available standards or trans fatty acid references, which were obtained as described elsewhere [41,42].

2.5. Statistical Analysis

The mice groups were divided into groups of six and the results were expressed as mean ± standard deviation (SD). The statistical significance (p values) of the results was calculated by unpaired two-tailed Student's t-test using GraphPad Prism™ software version 6.01 for Windows (GraphPad Software Inc., La Jolla, CA, USA). A multiple comparison test was applied to compare the differences among the distinct pairs of groups.

3. Results

3.1. Protocol Outline

Human tumor xenografts were obtained by inoculating U87MG human brain glioblastoma cells subcutaneously in two-week-old SCID mice. Approximately 2 weeks post-injection, the first set of animals was sacrificed (at 4 weeks of age). The second set of tumor-bearing mice was sacrificed at 17 weeks of age. In this article, the first group is referred as being in the early stage of tumorigenesis, while the second group is referred to as being in the final stage of tumor presence and characterized by very poor condition.

We also evaluated DNA lesions and the fatty acids pool in the brain tissue of the control SCID mice without tumor implantation, at 4 weeks of age and 17 weeks of age, to identify their potential differences with tumor-bearing animals. Because human xenograft were selected and the tumors originated by way of exogenous inoculation with human malignant cells, the tumor tissues were not analyzed, following an approach described for a genetically engineered mouse [43]. The accumulation of DNA damage in tissues that are distant from a tumor site can be induced by tumors of different origin, sometimes as a consequence of a cancer-related chronic inflammatory response in vivo [44]. Parallel DNA and lipid analyses were also performed in normal Swiss mice to evaluate healthy conditions as well as to recognize the potent differences between a healthy state and an immune-deficient condition.

Genomic DNA was isolated from the brains, hydrolyzed to single nucleosides by an enzymatic cocktail containing nucleases, and analyzed by LC-MS/MS for the determination of the modified nucleosides (the four cPu and two 8-oxo-Pu). The evaluation of the fatty acid pool was performed in the lipid extract, under conditions in which the fatty acid esterified to glycerol moieties (mainly brain glycerophospholipids) were transformed to fatty acid methyl esters (FAMEs) and could be analyzed by GC, as described elsewhere (for details, see the Materials and Methods section).

3.2. *Evaluation of Purine DNA Lesions*

An LC-MS/MS analysis of the brain DNA nucleosides provided the four cPu and the two 8-oxo-Pu lesions of our animal cohorts. In general, as shown in Table 1 and Figure 3, the levels of cPu and 8-oxo-dA were found to be similar, while 8-oxo-dG was one order of magnitude higher (for specific values for normal mice, see Table S1). Examining the SCID groups, we found the following: (i) a significant enhancement of 5′*S*-cdG in 17-week-old xenografted mice, compared with control SCID animals of the same age ($p = 0.0370$, cf. Table S2 for p-values); (ii) increased levels of 5′*S*-cdG in 17-week-old control SCID mice, compared with younger xenografts (($p = 0.0354$); (iii) 5′*S*-cdA was significantly raised in old tumor-bearing SCID mice, compared with the 4-week-old ones ($p = 0.0342$); and (iv) 5′*S*-cdA was also at higher levels in 17-week-old control animals, compared to 4-week-old ones ($p = 0.0102$).

Table 1. The levels (lesions/10^6 nucleosides) of 5′*R*-cdG, 5′*S*-cdG, 5′*R*-cdA, 5′*S*-cdA, 8-oxo-dG, and 8-oxo-dA in the brain tissues of control SCID and tumor-bearing SCID, 4 weeks of age- and 17 weeks of age, respectively (mean ± standard deviation of six sample measurements).

	5′*R*-cdG	5′*R*-cdA	5′*S*-cdG	5′*S*-cdA	8-oxo-dG	8-oxo-dA
control 4w	0.176 ± 0.011	0.140 ± 0.016	0.205 ± 0.028	0.116 ± 0.014	1.828 ± 0.183	0.202 ± 0.041
control 17w	0.179 ± 0.036	0.146 ± 0.019	0.211 ± 0.024 [1,*]	0.137 ± 0.012 [2,*]	1.783 ± 0.208	0.201 ± 0.046
tumor-bearing 4w	0.185 ± 0.018	0.153 ± 0.014	0.213 ± 0.034	0.112 ± 0.014	1.730 ± 0.276	0.201 ± 0.012
tumor-bearing 17w	0.178 ± 0.031	0.140 ± 0.020	0.245 ± 0.025 [3,*]	0.136 ± 0.012 [3,*]	1.903 ± 0.234	0.216 ± 0.046

[1] Comparison between control 17-week-old mice vs. tumor-bearing 17-week-old mice, [2] Comparison between control 4-week-old mice vs. control 17-week-old mice, [3] Comparison between tumor-bearing 4-week-old mice vs. tumor-bearing 17-week-old mice. Statistical significance: * ($p < 0.05$).

Examining the effects of age and tumor progression in our SCID cohorts, alterations that were not statistically significant were exhibited at the levels of total cPu and 8-oxo-Pu, as highlighted in Figure 4A (for specific values, see Tables S3 and S4). To estimate the healthy conditions, Swiss mice were examined for their brains' accumulation of DNA lesions at the ages of 4 weeks and 17 weeks, to show whether any significant change was present (Figure 4B and Table S4; see also Figure S1 and Tables S1 and S2 for the single lesions). Notably, a comparison between the Swiss mice and the control SCID mice showed evidence of a significant enhancement of 5′*S*-cdA at 17 weeks of age in immunodeficient animals ($p = 0.0049$, Table S2). In general, the Swiss mice presented with similar levels of cPu and with lower 8-oxo-Pu levels, although the difference was not significant, compared with control SCID mice (Tables S3 and S4).

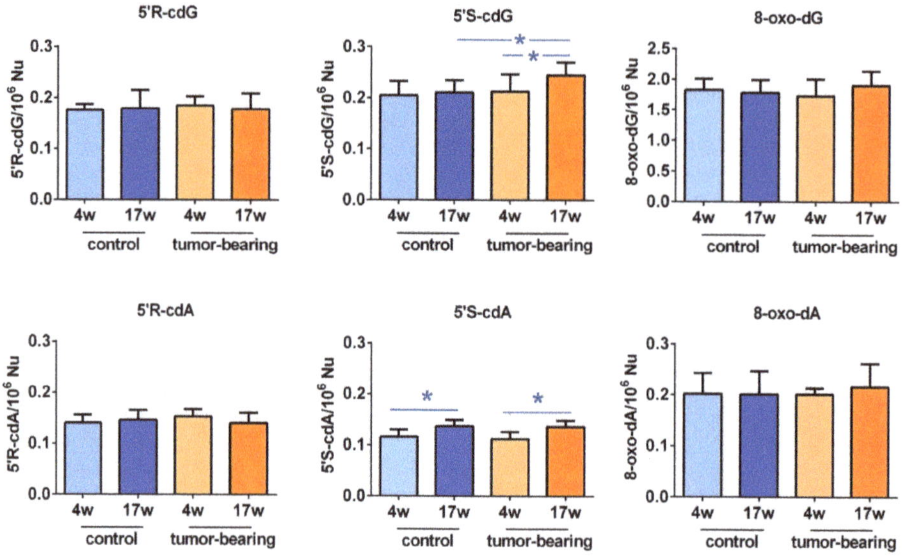

Figure 3. Purine DNA lesions in the brain of SCID mice: the levels (lesions/10^6 nucleosides) of 5′*R*-cdG, 5′*S*-cdG, 5′*R*-cdA, 5′*S*-cdA, 8-oxo-dG, and 8-oxo-dA in the brain tissue of control SCID and tumor-bearing SCID mice. For specific values, see Tables S1 and S2. The values are given as mean ± SD ($n = 6$); * ($p < 0.05$).

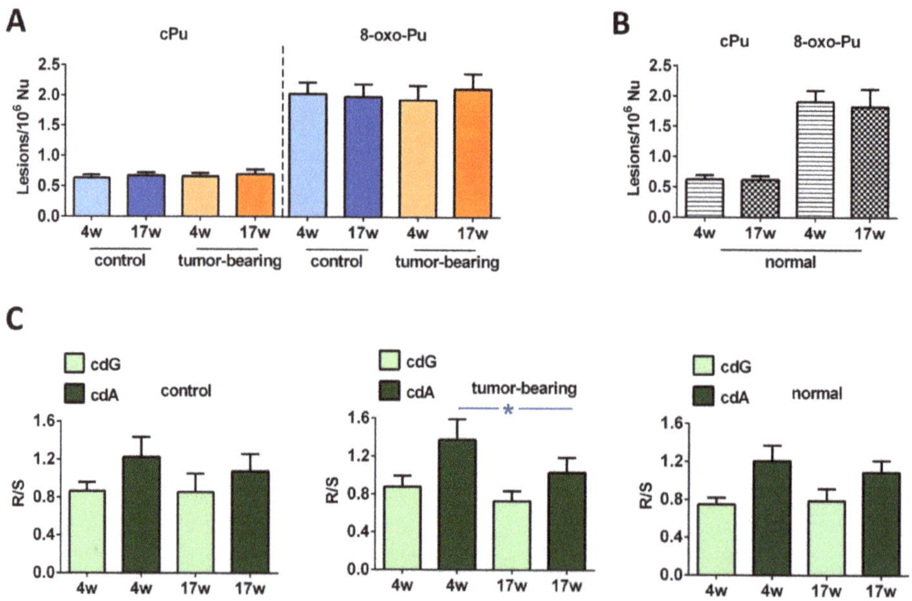

Figure 4. LC-MS/MS analysis of purine DNA lesions. (**A**) The levels (lesions/10^6 nucleosides) of cPu and 8-oxo-Pu in the brain tissue of control SCID and tumor-bearing SCID mice; for specific values see Tables S3 and S4. (**B**) The levels (lesions/10^6 nucleosides) of cPu and 8-oxo-Pu in the brain tissue of normal Swiss mice; for specific values see Tables S3 and S4. (**C**) The 5′*R*/5′*S* ratio of cdG and cdA in control SCID, tumor-bearing SCID, and normal Swiss mice; for specific values see Tables S5 and S6. The values are given as mean ± SD ($n = 6$); * ($p < 0.05$).

Regarding the ratio of diastereoisomeric lesions of cPu, it was reported that this indicator provided important information on the structural conformation of both isomers associated with the repair process [7,8]. The $5'R/5'S$ ratio of cdA was found to be higher than the ratio of corresponding cdG in all groups of samples, as shown in Figure 4C. A decrease in the R/S ratio of cdG and a significant diminution of the R/S ratio of cdA ($p = 0.0211$) were noted in 17-week-old tumor-bearing mice, compared with younger xenograted animals (Tables S5 and S6). Differences that were not statistically significant were observed in the control SCID as well as in the normal Swiss animals (Tables S5 and S6).

3.3. Evaluation of Brain Fatty Acid Pool

After isolation of the brain lipid content, we proceeded to the transformation of the fatty acid-containing lipids to their fatty acid methyl ester (FAME) derivatives, as described in the Materials and Methods section of this article, following a previously reported procedure [41,42]. The FAME separation and identification was carried out by the gold standard of gas chromatography (GC) analysis under known conditions, and we focused our attention to 12 fatty acid methyl esters (corresponding to >97% of the peaks that were present in the analysis), which were calibrated using commercially available materials for quantitative purposes. The values of this fatty acid cluster are reported in Table 2 as relative quantitative percentages (% rel. quant.), obtained by the quantitative data and reported as percentages of each FAME over the total FAME quantities resulting from the GC areas, following a known procedure (see Table S7 for the p-values of Table 2) [42]. The relevant results in our cohorts are reported in Figure 5, with statistically significant FAME levels comparing the control SCID and the tumor-bearing SCID mice at 4 weeks of age and 17 weeks of age (in Figure S2, all of the data are graphically represented). Palmitic acid (16:0) diminished in the control SCID and the tumor-bearing mice during aging (from 4 weeks of age to 17 weeks of age), along with increases in ARA. MUFA oleic acid (9c-18:1) decreased during aging (from 4 weeks to 17 weeks) in both the control and the tumor-bearing mice. ARA in tumor-bearing mice increased significantly, compared with control SCID, and oleic acid diminished, comparing the 4-week-old and 17-week-old mice of both series.

Table 2. Relative quantitative percentages (% rel. quant.) of fatty acid methyl esters (FAME) obtained from brain tissues of control SCID mice and tumor-bearing SCID mice at different ages [1].

FAME	FA Family	Control (4w)	Control (17w)	Tumor-Bearing (4w)	Tumor-Bearing (17w)
Palmitic acid (16:0)	SFA	23.67 ± 1.27 *	20.53 ± 0.54 **	25.87 ± 0.60	22.67 ± 0.51 **
Stearic acid (18:0)		26.20 ± 1.77	26.24 ± 0.40	23.58 ± 0.04	24.22 ± 0.88 *
Palmitoleic acid (9c-16:1)	MUFA	0.70 ± 0.25	0.79 ± 0.22	0.86 ± 0.01	0.67 ± 0.14
Oleic acid (9c-18:1)		24.02 ± 0.41	22.96 ± 0.75	20.19 ± 1.60 *	19.32 ± 0.91 **
Vaccenic acid (11c-18:1)		8.01 ± 1.31	6.76 ± 0.83	6.12 ± 0.83	6.09 ± 0.09
LNA (18:2-ω6)	PUFA ω6	1.02 ± 0.16	0.95 ± 0.12	0.92 ± 0.02	1.14 ± 0.15
DGLA (20:3-ω6)		1.18 ± 0.07	0.87 ± 0.21	0.82 ± 0.07	1.04 ± 0.12
ARA (20:4-ω6)		11.06 ± 1.26 **	15.63 ± 0.71 *	14.96 ± 0.07 *	19.62 ± 1.18 **
EPA (20:5-ω3)	PUFA ω3	0.57 ± 0.50	0.53 ± 0.18	1.13 ± 0.29	1.08 ± 0.61
DHA (22:6-ω3)		3.40 ± 0.16 **	4.60 ± 0.20	4.04 ± 0.18 *	3.67 ± 0.65
9t-18:1	TFA	0.03 ± 0.02	0.06 ± 0.04	0.06 ± 0.03	0.10 ± 0.05
mt-ARA 20:4		0.11 ± 0.05	0.09 ± 0.03	0.47 ± 0.05 **	0.37 ± 0.19 *

[1] The values are given as mean ± SD ($n = 3$). Statistical significances: * ($p < 0.05$), ** ($p < 0.01$).

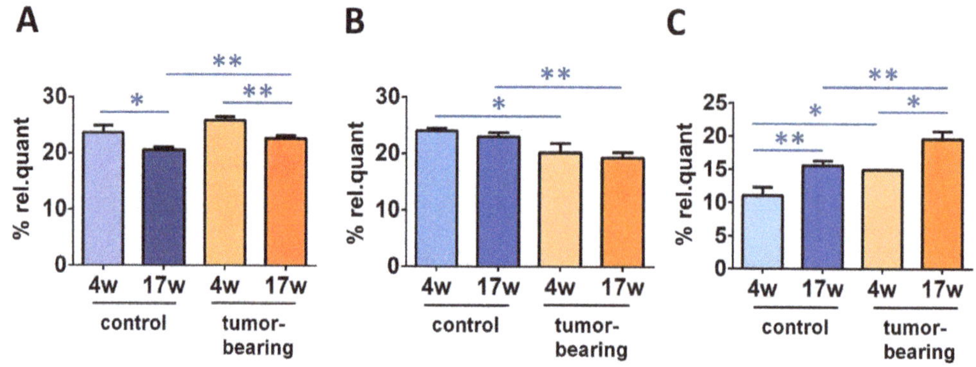

Figure 5. Some relevant fatty acid changes in SCID mice cohorts. Comparison of % rel. quant. of (**A**) C16:0; (**B**) 9c-18:1; (**C**) ARA in the brain tissue among control SCID mice and tumor-bearing SCID mice at 4 weeks of age and 17 weeks of age (for specific values, see Table 2). Significance: * ($p < 0.05$), ** ($p < 0.01$).

From the FAME quantities, relevant fatty acid calculations, such as sums, ratios, and equations could be applied, such as SFA, MUFA, PUFA, SFA/MUFA, SFA/PUFA, ω6/ω3, the unsaturation index (UI), and the peroxidation index (PI). Table 3 (see Table S8 for *p*-values) shows these data in triplicate for each mice group in this work: control SCID mice and tumor-bearing SCID mice at 4 weeks of age and 17 weeks of age. The above reported changes in FAME influenced the total SFA and PUFA ω6 levels, accordingly (SFA control; $p = 0.0051$ and tumor-bearing; $p = 0.0112$. PUFA ω6 control; $p = 0.0053$; tumor-bearing; $p = 0.0152$).

Table 3. Relative quantitative percentages (% rel. quant.) of fatty acid methyl families and ratios/indices obtained from brain tissues of control SCID mice and tumor-bearing SCID mice at different ages [1].

FA Family	Index	Control (4w)	Control (17w)	Tumor-Bearing (4w)	Tumor-Bearing (17w)
SFA		49.87 ± 0.50 **	46.77 ± 0.83 *	49.45 ± 0.56	46.89 ± 0.47
MUFA		32.74 ± 1.68	30.51 ± 0.48	27.16 ± 2.44	26.09 ± 0.83 **
PUFA ω6		13.26 ± 1.16 **	17.45 ± 0.61 *	16.70 ± 0.01 *	21.79 ± 1.36 **
PUFA ω3		3.98 ± 0.66	5.13 ± 0.33	5.17 ± 0.46	4.76 ± 0.42
PUFA [1]		17.24 ± 1.36 **	22.58 ± 0.51 **	21.87 ± 0.45 *	26.55 ± 0.94 **
TFA		0.15 ± 0.04	0.14 ± 0.04	0.52 ± 0.01 **	0.47 ± 0.23 **
	SFA/MUFA	1.53 ± 0.09	1.53 ± 0.05	1.83 ± 0.19	1.80 ± 0.05 **
	SFA/PUFA	2.90 ± 0.22 **	2.07 ± 0.08 **	2.26 ± 0.02 *	1.77 ± 0.08 **
	ω6/ω3	3.39 ± 0.58	3.41 ± 0.29	3.25 ± 0.29	4.62 ± 0.72
	Unsaturation Index (UI)	105.87 ± 4.48 **	127.78 ± 2.38 **	121.17 ± 0.11 *	137.40 ± 2.14 **
	Peroxidation Index (PI)	79.13 ± 6.76 **	105.94 ± 2.22 **	102.15 ± 3.22 *	118.22 ± 1.88 **

[1] PUFA = % PUFA ω-3+ % PUFA ω-6. Statistical significance: * ($p < 0.05$), ** ($p < 0.01$).

The calculations for the unsaturation index (UI) and the peroxidation index (PI) were carried out by Equations (1) and (2), respectively:

$$UI = (\%MUFA \times 1) + (\%LNA \times 2) + (\%DGLA \times 3) + (\%ARA \times 4) + (\%EPA \times 5) + (\%DHA \times 6) \quad (1)$$

$$PI = (\%MUFA \times 0.025) + (\%LNA \times 1) + (\%DGLA \times 2) + (\%ARA \times 4) + (\%EPA \times 6) + (\%DHA \times 8) \quad (2)$$

UI and PI indicate the content of unsaturated lipids that impact the membrane properties, as MUFA and PUFA double bonds, and the chemical oxidative reactivity, mainly as PUFA double bonds, respectively [45,46]. The PI and UI values for control and tumor-bearing mice are reported in Table 3 (see Table S8 for the *p*-values) and are graphically reported in Figure 6A, B, respectively.

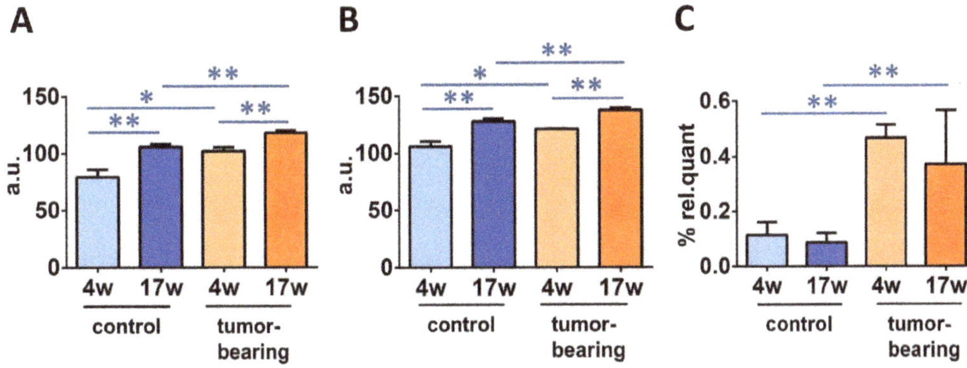

Figure 6. Significantly different fatty acid indices and families in SCID mice cohorts at different ages: (**A**) unsaturation index (UI); (**B**) peroxidation index (PI); (**C**) total TFA (for specific values, see Table 3). Significance: * ($p < 0.05$), ** ($p < 0.01$).

By following the formation of TFA, it is possible to identify a peculiar transformation of the natural cis unsaturated fatty acids into their geometrical isomers catalyzed by free radicals [33,34,47]. Figure 6C shows a statistical enhancement of TFA, in particular mono-trans ARA (see Figure S2), observed in young and older xenografted animals, compared with control SCID (4 weeks of age; $p = 0.004$, 17 weeks of age; $p = 0.0124$). It is worth noting that the free radical stress, expressed by the formation of TFA, reached the highest level in older mice, regardless of whether or not they were tumor-bearing.

When normal Swiss mice at 4 weeks of age and 17 weeks of age were used to isolate brain lipids and to identify the fatty acid pool differences, we noticed that a few significantly different values were found by comparing the two ages, i.e.,: the increase of palmitic acid ($p = 0.0011$) and the ω6/ω3 ratio ($p = 0.048$), the decrease in the PUFA ω3 EPA ($p = 0.0040$), and the decrease in the total TFA ($p = 0.0417$) (see Table S9). A comparison between the normal Swiss mice and the control SCID mice was also carried out in order to obtain valuable information related to immunodeficiency and radical-based reactivity in the brain tissue which was, especially useful in evaluating the age effect (Table S10). A significant reduction in SFA and an increase in PUFA ω3 and total PUFA was observed in 17-week-old SCID, compared with normal mice (SFA; $p = 0.0256$, ω-3; $p = 0.0190$, PUFA; $p = 0.0087$). Furthermore, the SFA/MUFA and SFA/PUFA ratios were statistically decreased (SFA/MUFA; $p = 0.0442$, SFA/PUFA; $p = 0.0132$) and the indices of unsaturation and peroxidation were enhanced (UI; $p = 0.0112$, PI; $p = 0.0045$). Significantly higher levels of DHA and lower levels of palmitic acid were also found for 17-week-old SCID animals (DHA; $p = 0.0072$, 16:0; $p = 0.0009$). Although further differences of the membrane fatty acid content between the two groups were present, such as total TFA and EPA decreases in younger mice and increases in older ones, as well as MUFA increases at both ages, there were no significances among these distinct conditions. In the younger animals, only 9t-18:1 was significantly reduced in 4-week-old diseased mice, compared to healthy mice of the same age ($p = 0.0221$, Table S10).

4. Discussion

Brain is a tissue with one of the highest levels of oxidative metabolism that is consistently associated with the production of oxidative damage. Some studies indicate that specific brain regions affect the responses to DNA damage, with single strand breaks (SSB) considered to be a serious threat to the aging brain [48]. Moreover, there is a strong connection between brain mitochondrial ROS production, glutamate excitotoxicity, and neuronal cell damage [49]. High ROS production, severe DNA damage, inflammatory stress, and heterochromatinization were found in postmitotic neurons from old C57Bl/6 mice, revealing that mature neurons develop a senescence-like phenotype with aging [50]. DNA damage increases with age and in multiple neurological and neurodegenerative disorders, such as Alzheimer's disease, which is a disorder that is characterized by the accumulation of double strand breaks (DSB) in both neuronal and glial cells [51], as well as a diminution in the expression of UDG1, polb, and bOGG1 glycosylases in AD brains, compared with age-matched controls [52]. Apart from these enzymes and the BER pathway, NER also participates in the progression of age-related cognitive decline, as accumulating DNA damage and reduced synaptic plasticity were observed in the hippocampus of *Ercc1*$^{\Delta/-}$ mice [53].

cPu lesions are unique lesions that are generated by H-atom abstraction from the C5′–H position of purine moieties by HO$^{\bullet}$ radicals, eventually resulting in the formation of an additional C5′−C8 covalent bond [4–6,54]. On the other hand, 8-oxo-Pu lesions are derived from oxidation at the C8 position of purine moieties by a variety of ROS, such as H_2O_2, singlet oxygen, and $ONOO^-$, in addition to HO$^{\bullet}$ radical [55,56]. The observation of lesions is connected with the efficiency of the repair. Damaged DNA activates different responses in a cellular environment, depending on the kind of damage, and cells may undergo apoptosis to avoid the propagation of defective cells. However, robust DNA repair and damage-bypass mechanisms protect the stability of the human genome, either by removing the damage or permitting the damage to ensure genome integrity. Together with determining the levels the oxidatively-induced 8-oxo-Pu, we determined the levels of cPu lesions. Both cPu and 8-oxo-cPu formation were evaluated in this study in young and old immunodeficient mice, with and without tumor implantation, to compare the effect of tumor development and to gather information on the aging process. This work is a continuation of an exploratory study of two groups of mice that were examined for the accumulations of liver and kidney DNA lesions [37]. In the present study, we deepened the oxidative DNA damage of the brain, as this tissue maintains a particularly high basal metabolic rate to fulfill the high-energy demand and produces increased levels of ROS [52]. Our results showed that the levels of the four cPu and 8-oxo-dA lesions are similar, in the range of 0.1–0.25 lesions/10^6 Nu, whereas 8-oxo-dG is one order of magnitude higher, independently of being SCID control, tumor-bearing SCID or normal Swiss mice in their lifetimes (4 weeks old or 17 weeks old) (Figure 3 and Table S1). In the cPu series, the following order was observed in all cases: 5′*S*-cdG > 5′*R*-cdG > 5′*R*-cdA > 5′*S*-cdA.

The detection of cPu in brain tissue was also reported by Wang et al., who analyzed the levels of these adducts in the brain of wild type and *Ercc1*$^{-/\Delta}$ mice of different ages using capillary high-performance liquid chromatography-tandem mass spectrometry [57]. No significant differences were found between wild type animals of 10 weeks, 21 weeks, and 3 years of age, with the levels of the four cPu lesions being in the range of 0.1–0.5 lesions/10^6 Nu. On the other hand, a diminution of all four cPu lesions was found in 21-week-old *Ercc1*$^{-/\Delta}$ mice, compared with 10-week-old mice. Furthermore, it should be noted that the significantly raised amounts of 5′*S*-cdG in 10-week-old deficient-progeroid animals compared to wild-type animals of the same age demonstrated the role of ERRC-1 protein in NER involvement [57]. cdA was also present in the brains of the XP group A gene-knockout (Xpa$^{-/-}$) and significant accretion were detected in mice at 6, 24, and 29 months of age, compared to wild-type animals, through an improved enzyme-linked immunosorbent assay (ELISA) [15] using a novel monoclonal antibody (cdA-1) specific for cdA in single-stranded DNA [58]. However, the cdA levels in the brain tissue of 4-week-old (one month), 12-week-old (three months), and 24-week-old (6 months) mice were fully raised through ELISA quantification, with detectable amounts of ~3.5–4/10^6 [15]. Immunoassays methods are characterized by adequate simplicity and reproducibility and they can be applied to DNA adduct detection, such as screening the lesions of 8-oxo-2′-deoxyadenosine [59], benzo[a]pyrene-dioleperoxide [60], and 4-hydroxyequilenin [61].

Competitive approaches with the presence of monoclonal antibodies offer higher specificity and effectiveness [58]. On the other hand, only chromatography-based methodologies coupled with isotope dilution mass spectrometry can provide the structural specificity of the studied adduct and the detailed molecular composition for accurate quantification [62,63].

The diastereomeric ratio ($5'R/5'S$) can be an important indicator for mechanistic issues that are related to structural conformation of both isomeric forms in association with their abundance and repair. cdA and cdG lesions are excised with similar efficiency by NER machinery in human HeLa cell extracts; however, the $5'R$-diastereoisomers of both cdA and cdG cause greater distortion of the DNA backbone, thereby being better substrates of NER than the corresponding $5'S$ ones [7]. As shown in Figure 4C, *R* is always more abundant than *S* in the cdA in each group of mice, whereas in the cdG, *S* always exceeds the *R* diastereoisomer. Another important aspect is derived by comparing these diastereomeric ratios between young and older organisms. Lower level of R/S cdG and cdA ratios were depicted in 17-week-old brain tissues of both control and tumor-bearing animals compared to 4-week-old tissues (Table S5). Significantly raised levels of cdA and cdG in both the *R* and *S* isoforms were also found in the brain tissues of the Long-Evans Cinnamon (LEA) rat, an animal model for human Wilson's disease, and the Long-Evans Agouti rat (LEC), healthy rats, using NanoLC-NSI-MS/MS analysis [64]. Although the R/S ratios constitute an index of cPu lesions' repair efficiency, differences in analytical procedures across distinct research groups did not provide a clear scenario for R/S formation and biological significance. Previous studies through LC/MS or GC/MS showed a high accumulation of cPu in knockout mice, such as both cdG and cdA in the brains of prdx1$^{-/-}$ mice [65]. Increased levels of $5'S$-cdA were measured in different organs of wild-type and csb knockout mice, with a significant enrichment of this unrepaired adduct in the brains of csb(-/-) animals, as well as in the livers and the kidneys, suggesting the important role of CSB protein in the DNA repair process [66].

At this point, it is worth mentioning that we reported an earlier detection of the four cPu levels of liver and kidney tissues at initial and final stages of tumorigenesis in the same animal cohorts (4-week-old and 17-week-old tumor-bearing mice, respectively) [37]. In both tissues of 17-week-old xenografts, we found increased cPu levels. In Figure 7A, a comparison of the total cPu lesions in brain, liver, and kidney tissues is presented. Unlike in brain tissue, statistically significant alterations were exhibited for total cPu in liver and kidney tissues during cancer progression, as highlighted (see also Table S11). A significant enhancement of cPu was found in the liver tissue of 17-week-old control mice, compared with 17-week-old tumor-bearing SCID mice ($p = 0.0014$, Table S12). Furthermore, a comparison between 4-week-old and 17-week-old xenografted animals showed a significant increase of cPu in kidney tissues ($p = 0.00395$). Higher levels of cPu were also found in the kidney tissues of 17-week-old control mice, compared with tumor-bearing SCID mice of the same age ($p = 0.0365$,). 8-oxo-dA levels were previously reported in liver and kidney tissues, and here we measured this lesion in brain tissue. Figure 7 shows the much lower extent of this lesion in the brain, compared with the other tissues. The data presented in our previous [37] and present exploratory studies indicate a differential involvement of genomic instability in mice cohorts, and may contribute to the ongoing debate about the central role of DNA and genome instability in the aging process, which also implies choosing the best biomarkers for such monitoring [67].

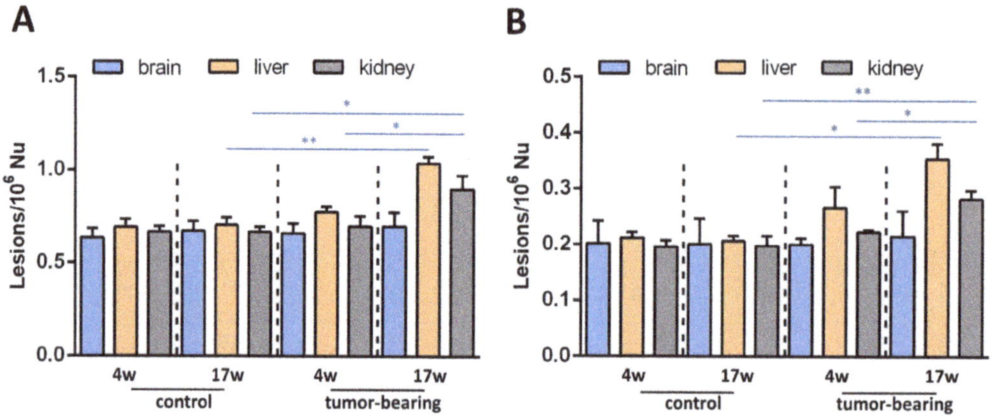

Figure 7. (**A**) cPu levels and (**B**) 8-oxo-dA levels in the brain tissue of control SCID and tumor-bearing SCID mice compared with previous tissue-specific patterns from the liver and the kidney [37] by isotope dilution liquid chromatography-tandem mass spectrometry * ($p < 0.05$), ** ($p < 0.01$).

A parallel examination of brain lipids can widen the perspectives in the assessment of aging and health conditions, such as tumoral progression, offering opportunities to obtain the holistic approach that is needed to fully understand complex organisms. Our results can be coupled with our previously reported data on the same mice cohorts, which were examined for their erythrocyte membrane phospholipid contents [37]. Evaluating mice brain fatty acid-containing lipids, our data are concerned mostly with glycerophospholipids, which reach almost 70% of the total lipid composition in this tissue [68]. This is the lipid class where the fatty acid composition is influenced by the so-called remodeling mechanism and reports the systemic effects of fatty acids introduced by different diets or changed in response to life conditions [34,36,69,70]. We are aware that the powerful shotgun lipidomic analysis is used to go into the details of all lipid classes [71]; however, it is worth emphasizing that by using the gas chromatographic analysis, we can obtain a precise and reliable separation of geometrical and positional fatty acid isomers, and individuate trans fatty acids that are crucial indicators of free radical stress (thiyl radicals, in particular) [32,33,72]. The role of oxygen in the formation of cPu was recently investigated in cellular models of defective CSA- and CSB-transformed fibroblasts and their normal counterparts, cultured under various oxygen tensions and increased levels of cPu, were detected under hypoxic conditions in both CSA- and CSB-defective cell lines, compared with normal cells [13]. In the same cellular models, the analysis of fatty residues in membrane phospholipids was studied, due to oxygen tensions [42]. A parallel examination of purine adduct formation and membrane molecular transformations in human embryonic epithelial cells, silenced for XPA under hyperoxic conditions, was performed, showing that oxygen promotes enzymatic transformations of the fatty acid pool [12].

Two results can be highlighted in the brain remodeling process: (a) the decrease in palmitic acid accompanied with the increase in arachidonic acid in both groups of immunodeficient (SCID) mice and tumor-bearing SCID mice, comparing 4-week-old to 17-week-old mice (cf. Figure 5); and (b) the increases in UI, PI, and total TFA going from 4-week-old control SCID mice to the tumoral groups (cf. Figure 6). The first data point attention to brain-signaling that starts from membrane phospholipids, and to the detachment of fatty acids for the production of bioactive lipids.The ARA membrane enrichment can lead to the disruption of an equilibrium with the omega-3 counterparts, especially DHA, which regulates neuroinflammation [20]. In control SCID mice, DHA is actually increased, whereas it diminishes in tumor-bearing animals, involving complications when disease conditions add to the aging process. It is worth highlighting that the ARA increase was also reported in the erythrocyte membrane analysis of the same animal cohorts [37], confirming the reporting role of this blood cell for a systemic condition. The combination of lipid metabolism and inflammation is also important in the development of therapeutical strategies for aging and diseases [73]. The second set of data, about UI and PI increases, is connected with an oxidatively-prone lipid environment that is, indeed, in connection with aging and disease progression. As discussed in the Introduction, neurodegeneration is strictly connected as a cause or as a consequence of oxidative reactivity [19,21,29], which is favored by the enrichment of PUFA, as seen in our animal cohorts. It is worth stating that no dietary changes were applied to the animals and that the old tumor-bearing animals were characterized by poor health conditions, so that the PUFA increase in the brain surely did not derive from higher intakes. These observations, together with information about increased TFA in tumor-bearing mice, are relevant in determining that in this particular health condition, the formation of radical reactive species, such as thiyl radicals, occurs, because the formation of TFA is a marker of this specific reactivity [47]. We can recall that thiol-disulfide homeostasis is a very important element of brain metabolism, starting from the hydrogen sulfide production in this tissue [74]. The overall scenario obtained by this study of DNA and membrane data indicates that in the brain, unlike in the liver and the kidney tissues, molecular changes of aging and disease progressions caused by free radicals occur mostly at the level of lipidome, rather than involving all molecular inventory (DNA in particular). Our results highlight that more research is needed to gain combined information on free radical reactivity in different cellular compartments, using both in vitro and in vivo models. Moreover, due to the analytical conditions that were used in this study, it was not possible to define whether specific areas of the brain were most involved in the effects of the aging or disease, as is well known in differential evaluations [75].

5. Conclusions

In the present work, the simultaneous measurement of two important molecular contributions, genetic (DNA) and lipid components, was performed under specific conditions of aging and disease in a murine model, completing the picture with information from brain tissue. In the brain tissue of young and older tumor-bearing animals, the analysis of four cPu and two 8-oxo-Pu lesions showed a slight but significant progressive age-dependent accumulation only for the 5′*S*-cdA and 5′*S*-cdG lesions. cPu were detected in various tissue types and clinical specimens and did not suffer from stability issues and artifacts, unlike the 8-oxo-Pu that are formed by oxidizing species. At the same time, fatty acid pool remodeling due to aging and tumoral conditions occurred, specifically involving SFA and PUFA, with TFA as biomarker of free radical stress. The overview of the molecular contributions of DNA and lipids adds new insights into the consequences of aging and disease, highlighting the brain's prevalent lipid remodeling response and inflammatory signaling, which seem to prevail over the effects of DNA damage. These results can inspire further and deeper in vitro and in vivo model investigations on protective and therapeutic strategies of neurodegenerative disorders, taking into account the extensive involvement of the membrane lipids seen in our model. Under aging and disease progressions, membranes are not spectators [76,77]. Membrane-targeted strategies tailored to the specific lipidome profile are needed to preserve the molecular integrity of the membrane and to demonstrate the effects of delaying degenerative processes as a whole.

Supplementary Materials: The following supporting information can be downloaded at: https://www.mdpi.com/article/10.3390/biom12081075/s1, Figure S1: Graphical presentation of of cPu and 8-oxo-Pu lesions in the brain tissues of 4-week-old and 17-week-old normal mice; Figure S2: Graphical presentation of FAME in control SCID mice and tumor-bearing SCID mice at 4 weeks and 17 weeks of age; Tables S1 and S2: The levels of 5′*R*-cdG, 5′*S*-cdG, 5′*R*-cdA, 5′*S*-cdA, 8-oxo-dG, and 8-oxo-dA lesions and statistical analysis in the brain tissues of control SCID, tumor-bearing SCID, and normal Swiss mice; Tables S3 and S4: The levels of cPu and 8-oxo-Pu lesions and statistical analysis in the brain tissues of control SCID, tumor-bearing SCID, and normal Swiss mice; Tables S5 and S6: 5′*R*/5′*S* ratio in DNA-isolated and statistical analysis from the brain tissues of control SCID, tumor-bearing SCID, and normal Swiss mice; Tables S7 and S8: Statistical analysis of fatty acids, families, and lipid indices in the brain tissues of control SCID and tumor-bearing SCID mice at 4 weeks and 17 weeks of age; Tables S9 and S10: Fatty acid methyl esters, families, and lipid indices and statistical analysis in the brain tissues of normal mice at 4 weeks and 17 weeks of age; Tables S11 and S12: cPu and 8-oxo-dA levels and statistical analysis in the brain, liver, and kidney tissue of control SCID, tumor-bearing SCID, and normal Swiss mice at 4 weeks and 17 weeks of age.

Author Contributions: Conceptualization, M.G.K. and C.C.; methodology, M.G.K., E.K.E. and P.P.; xenograft construction, E.K.E.; formal analysis, M.G.K. and P.P.; investigation, M.G.K. and P.P.; resources, C.C, C.F., E.K.E. and A.B.; data curation, M.G.K., C.F. and C.C.; writing—original draft preparation, M.G.K., C.F. and C.C.; writing—review and editing, M.G.K., C.F. and C.C.; supervision, C.C.; funding acquisition, C.C. and A.B. All authors have read and agreed to the published version of the manuscript.

Funding: This research was funded by the Marie Skłodowska-Curie European Training Network (ETN) ClickGene: Click Chemistry for Future Gene Therapies to Benefit Citizens, Researchers and Industry [H2020-MSCAETN-2014-642023]. PP's secondment was supported within the frame of the Rep-Eat Doctorate Programme and funded by the European Union's Horizon 2020 under the Marie Skłodowska-Curie Grant Agreement No. 713714.

Institutional Review Board Statement: This study was conducted according to the guidelines of the Declaration of Helsinki. All protocols and all animal procedures were approved by the General Directorate of Veterinary Services, Athens, Greece, and by the Bioethical Committee of that institution (Permit number EL 25 BIO 042, Protocol No. 1427, 5 May 2015), in accordance with Greek legislation (Presidential Decree 160/1991) and in compliance with Directive 2010/63/EU of the European Parliament, Directive 86/609/EEC on the protection of vertebrate animals used for experimental or other scientific purpose, and Regulation (EU) 2015/1992.

Informed Consent Statement: Not applicable.

Data Availability Statement: The data presented in this study are provided in this article and in the Supplementary Materials.

Conflicts of Interest: The authors declare no conflict of interest.

Abbreviations

ARA: arachidonic acid; BER, base excision repair; cdA, 5′,8-cyclo-2′-deoxyadenosine; cdG, 5′,8-cyclo-2′-deoxyguanosine; cPu, purine 5′,8-cyclo-2′-deoxynucleoside; DHA, docosahexaenoic acid; DGLA, dihomo-gamma-linolenic acid; DPA, docosapentaenoic acid; ELO, elongase; EPA, eicosapentaenoic acid; FA, fatty acid; FAME, fatty acid methyl esters; GC, gas chromatography; HPLC, high-pressure liquid chromatography; LC-MS/MS, liquid chromatography tandem mass spectrometry; LNA, linoleic acid; MUFA, monounsaturated fatty acids; NER, nucleotide excision repair; 8-oxo-dA, 8-oxo-7,8-dihydro-2′-deoxyadenosine; 8-oxo-dG, 8-oxo-7,8-dihydro-2′-deoxyguanosine; 8-oxo-Pu, purine 8-oxo-7,8-dihydro-2′-deoxynucleoside; PI, peroxidation index; PUFA, polyunsaturated fatty acids; ROS, reactive oxygen species; SCID, severe combined immunodeficient; SFA, saturated fatty acids; TFA, trans fatty acids; UI, unsaturation index.

References

1. Sies, H.; Berndt, C.; Jones, D.P. Oxidative stress. *Annu. Rev. Biochem.* **2017**, *86*, 715–748. [CrossRef]
2. Sies, H.; Jones, D.P. Reactive oxygen species (ROS) as pleiotropic physiological signalling agents. *Nat. Rev. Mol. Cell Biol.* **2020**, *21*, 363–383. [CrossRef] [PubMed]
3. Maynard, S.; Fang, E.F.; Scheibye-Knudsen, M.; Croteau, D.L.; Bohr, V.A. DNA Damage, DNA Repair, Aging, and Neurodegeneration. *Cold Spring Harb. Perspect. Med.* **2015**, *5*, a025130. [CrossRef] [PubMed]
4. Chatgilialoglu, C.; Ferreri, C.; Geacintov, N.E.; Krokidis, M.G.; Liu, Y.; Masi, A.; Shafirovich, N.; Terzidis, M.A.; Tsegay, P.S. 5′,8-Cyclopurine lesions in DNA damage: Chemical, analytical, biological and diagnostic significance. *Cells* **2019**, *8*, 513. [CrossRef] [PubMed]
5. Chatgilialoglu, C.; Ferreri, C.; Krokidis, M.G.; Masi, A.; Terzidis, M. On the relevance of hydroxyl radical to purine DNA damage. *Free Radic. Res.* **2021**, *55*, 384–404. [CrossRef] [PubMed]
6. Jaruga, P.; Dizdaroglu, M. 8,5′-Cyclopurine-2′-deoxynucleosides in DNA: Mechanisms of formation, measurement, repair and biological effects. *DNA Repair* **2008**, *7*, 1413–1425. [CrossRef] [PubMed]

7. Kropachev, K.; Ding, S.; Terzidis, M.A.; Masi, A.; Liu, Z.; Cai, Y.; Kolbanovskiy, M.; Chatgilialoglu, C.; Broyde, S.; Geancitov, N.E.; et al. Structural basis for the recognition of diastereomeric 5′,8-cyclo-2′-deoxypurine lesions by the human nucleotide excision repair system. *Nucl. Acids Res.* **2014**, *42*, 5020–5032. [CrossRef] [PubMed]
8. Kuraoka, I.; Bender, C.; Romieu, A.; Cadet, J.; Wood, R.D.; Lindahl, T. Removal of oxygen free-radical-induced 5′,8-purine cyclodeoxynucleosides from DNA by the nucleotide excision-repair pathway in human cells. *Proc. Natl. Acad. Sci. USA* **2000**, *97*, 3832–3837. [CrossRef]
9. Yu, Y.; Wang, P.; Cui, Y.; Wang, Y. Chemical analysis of DNA damage. *Anal. Chem.* **2018**, *90*, 556–576. [CrossRef] [PubMed]
10. Chatgilialoglu, C.; Eriksson, L.A.; Krokidis, M.G.; Masi, A.; Wang, S.-D.; Zhang, R. Oxygen dependent purine lesions in double-stranded oligodeoxynucleotides: Kinetic and computational studies highlight the mechanism for 5′,8-cyplopurine formation. *J. Am. Chem. Soc.* **2020**, *142*, 5825–5833. [CrossRef]
11. Chatgilialoglu, C.; Krokidis, M.G.; Masi, A.; Barata-Vallejo, S.; Ferreri, C.; Terzidis, M.A.; Szreder, T.; Bobrowski, K. New insights into the reaction paths of hydroxyl radicals with purine moieties in DNA and double-stranded oligonucleotides. *Molecules* **2019**, *24*, 3860. [CrossRef] [PubMed]
12. Krokidis, M.G.; Parlanti, E.; D'Errico, M.; Pascucci, B.; Pino, A.; Alimonti, A.; Pietraforte, D.; Masi, A.; Ferreri, C.; Chatgilialoglu, C. Purine DNA lesions at different oxygen concentration in DNA repair-impaired human cells (EUE-siXPA). *Cells* **2019**, *8*, 1377. [CrossRef] [PubMed]
13. Krokidis, M.G.; D'Errico, M.; Pascucci, B.; Parlanti, E.; Masi, A.; Ferreri, C.; Chatgilialoglu, C. Oxygen-Dependent Accumulation of Purine DNA Lesions in Cockayne Syndrome Cells. *Cells* **2020**, *9*, 1671. [CrossRef]
14. A Floyd, R.; Hensley, K. Oxidative stress in brain aging: Implications for therapeutics of neurodegenerative diseases. *Neurobiol. Aging* **2002**, *23*, 795–807. [CrossRef]
15. Mori, T.; Nakane, H.; Iwamoto, T.; Krokidis, M.G.; Chatgilialoglu, C.; Tanaka, K.; Kaidoh, T.; Hasegawa, M.; Sugiura, S. High levels of oxidatively generated DNA damage 8,5′-cyclo-2′-deoxyadenosine accumulate in the brain tissues of xeroderma pigmentosum group A gene-knockout mice. *DNA Repair* **2019**, *80*, 52–58. [CrossRef] [PubMed]
16. Malard, E.; Valable, S.; Bernaudin, M.; Pérès, E.; Chatre, L. The reactive species interactome in the brain. *Antioxid. Redox Signal.* **2021**, *35*, 1176–1206. [CrossRef]
17. Bozek, K.; Wei, Y.; Yan, Z.; Liu, X.; Xiong, J.; Sugimoto, M.; Tomita, M.; Pääbo, S.; Sherwood, C.C.; Hof, P.R.; et al. Organization and evolution of brain lipidome revealed by large-scale analysis of human, chimpanzee, macaque, and mouse tissues. *Neuron* **2015**, *85*, 695–702. [CrossRef] [PubMed]
18. Jiang, W.; Chen, J.; Gong, L.; Liu, F.; Zhao, H.; Mu, J. Alteration of Glycerophospholipid Metabolism in Hippocampus of Post-stroke Depression Rats. *J. Neurochem. Res.* **2022**, *47*, 2052–2063. [CrossRef]
19. Castellanos, D.B.; Martín-Jiménez, C.A.; Rojas-Rodríguez, F.; Barreto, G.E.; González, J. Brain lipidomics as a rising field in neurodegenerative contexts: Perspectives with Machine Learning approaches. *Front. Neuroendocrin.* **2021**, *61*, 100899. [CrossRef]
20. Bazinet, R.P.; Layé, S. Polyunsaturated fatty acids and their metabolites in brain function and disease. *Nat. Rev. Neurosci.* **2014**, *15*, 771–785. [CrossRef]
21. Contreras, M.A.; Rapoport, S.I. Recent studies on interactions between n-3 and n-6 polyunsaturated fatty acids in brain and other tissues. *Curr. Opin. Lipidol.* **2002**, *13*, 267–272. [CrossRef]
22. Farooqui, A.A.; Horrocks, L.A.; Farooqui, T. Modulation of inflammation in brain: A matter of fat. *J. Neurochem.* **2007**, *101*, 577–599. [CrossRef] [PubMed]
23. Darios, F.; Davletov, B. ω-3 and ω-6 fatty acids stimulate cell membrane expansion by acting on syntaxin 3. *Nature* **2006**, *440*, 813–817. [CrossRef] [PubMed]
24. Innis, S.M. Dietary ω 3 fatty acids and the developing brain. *Brain Res.* **2008**, *1237*, 35–43. [CrossRef] [PubMed]
25. Calon, F.; Lim, G.P.; Yang, F.; Morihara, T.; Teter, B.; Ubeda, O.; Rostaing, P.; Triller, A.; Salem, N., Jr.; Ashe, K.H.; et al. Docosahexaenoic acid protects from dendritic pathology in an Alzheimer's disease mouse model. *Neuron* **2004**, *43*, 633–645. [CrossRef] [PubMed]
26. Robson, L.G.; Dyall, S.; Sidloff, D.; Michael-Titus, A.T. Omega-3 polyunsaturated fatty acids increase the neurite outgrowth of rat sensory neurones throughout development and in aged animals. *Neurobiol. Aging* **2010**, *31*, 678–687. [CrossRef]
27. Bousquet, M.; Calon, F.; Cicchetti, F. Impact of ω-3 fatty acids in Parkinson's disease. *Ageing Res. Rev.* **2011**, *10*, 453–463. [CrossRef]
28. Hernando, S.; Requejo, C.; Herran, E.; Ruiz-Ortega, J.A.; Morera-Herreras, T.; Lafuente, J.V.; Ugedo, L.; Gainza, E.; Pedraz, J.L.; Igartua, M.; et al. Beneficial effects of n-3 polyunsaturated fatty acids administration in a partial lesion model of Parkinson's disease: The role of glia and NRf2 regulation. *Neurobiol. Dis.* **2019**, *121*, 252–262. [CrossRef]
29. Cobley, J.N.; Fiorello, M.L.; Bailey, D.M. 13 reasons why the brain is susceptible to oxidative stress. *Redox Biol.* **2018**, *15*, 490–503. [CrossRef]
30. Gumusyayla, S.; Vural, G.; Bektas, H.; Deniz, O.; Neselioglu, S.; Erel, O. A novel oxidative stress marker in patients with Alzheimer's disease: Dynamic thiol–disulphide homeostasis. *Acta Neuropsychiatr.* **2016**, *28*, 315–320. [CrossRef] [PubMed]
31. Zanganehnejad, Z.; Setorki, M. Effect of Biarum carduchrum extract on brain tissue thiol level in rat model of 6-hydroxydopamine-induced Parkinson's disease. *J. Herbmed. Pharmacol.* **2018**, *7*, 136–140. [CrossRef]
32. Chatgilialoglu, C.; Ferreri, C. Trans lipids: The free radical path. *Acc. Chem. Res.* **2005**, *38*, 441–448. [CrossRef]
33. Ferreri, C.; Chatgilialoglu, C. Geometrical trans lipid isomers: A new target for lipidomics. *ChemBioChem* **2005**, *6*, 1722–1734. [CrossRef] [PubMed]

34. Ferreri, C.; Chatgilialoglu, C. Role of fatty acid-based functional lipidomics in the development of molecular diagnostic tools. *Expert Rev. Mol. Diagn.* **2012**, *12*, 767–780. [CrossRef] [PubMed]
35. Pamplona, R. Membrane phospholipids, lipoxidative damage and molecular integrity: A causal role in aging and longevity. *Biochim. Biophys. Acta-Bioenergy* **2008**, *1777*, 1249–1262. [CrossRef]
36. Ferreri, C.; Masi, A.; Sansone, A.; Giacometti, G.; Larocca, A.; Menounou, G.; Scanferlato, R.; Tortorella, S.; Rota, D.; Chatgilialoglu, C. Fatty acids in Membranes as Homeostatic, Metabolic and Nutritional Biomarkers: Recent Advancements in Analytics and Diagnostics. *Diagnostics* **2017**, *7*, 1. [CrossRef]
37. Krokidis, M.G.; Louka, M.; Efthimiadou, E.K.; Zervou, S.-K.; Papadopoulos, K.; Hiskia, A.; Ferreri, C.; Chatgilialoglu, C. Membrane lipidome reorganization and accumulation of tissue DNA lesions in tumor-bearing mice: An exploratory study. *Cancers* **2019**, *11*, 480. [CrossRef]
38. Terzidis, M.A.; Chatgilialoglu, C. An ameliorative protocol for the quantification of purine 5′,8-cyclo-2′-deoxynucleosides in oxidized DNA. *Front. Chem.* **2015**, *3*, 47. [CrossRef]
39. Chatgilialoglu, C. Cyclopurine (cPu) lesions: What, how and why? *Free Radic. Res.* **2019**, *53*, 941–943. [CrossRef] [PubMed]
40. Folch, J.; Lees, M.; Sloane Stanley, G.H. A simple method for the isolation and purification of total lipids from animal tissues. *J. Biol. Chem.* **1957**, *226*, 497–509. [CrossRef]
41. Prasinou, P.; Dafnis, I.; Giacometti, G.; Ferreri, C.; Chroni, A.; Chatgilialoglu, C. Fatty acid-based lipidomics and membrane remodeling induced by apoE3 and apoE4 in human neuroblastoma cells. *Biochim. Biophys. Acta-Biomembr.* **2017**, *1859*, 1967–1973. [CrossRef] [PubMed]
42. Ferreri, C.; Sansone, A.; Krokidis, M.G.; Masi, A.; Pascucci, B.; D'Errico, M.; Chatgilialoglu, C. Effects of Oxygen Tension for Membrane Lipidome Remodeling of Cockayne Syndrome Cell Models. *Cells* **2022**, *11*, 1286. [CrossRef] [PubMed]
43. Morton, C.L.; Houghton, P.J. Establishment of human tumor xenografts in immunodeficient mice. *Nat. Protoc.* **2007**, *2*, 247–250. [CrossRef] [PubMed]
44. Redon, C.E.; Dickey, J.S.; Nakamura, A.J.; Kareva, I.G.; Naf, D.; Nowsheen, S.; Kryston, T.B.; Bonner, W.M.; Georgakilas, A.G.; Sedelnikova, O.A. Tumors induce complex DNA damage in distant proliferative tissues in vivo. *Proc. Natl. Acad. Sci. USA* **2010**, *107*, 17992–17997. [CrossRef] [PubMed]
45. Hulbert, A.J. Explaining longevity of different animals: Is membrane fatty acid composition the missing link? *Age* **2008**, *30*, 89–97. [CrossRef]
46. Puca, A.A.; Andrew, P.; Novelli, V.; Viviani, C.; Somalvico, F.; Cirillo, N.A.; Chatgilialoglu, C.; Ferreri, C. Fatty acids profile of erythrocyte membranes as possible biomarker of longevity. *Rejuvenation Res.* **2008**, *11*, 63–72. [CrossRef]
47. Hung, W.-L.; Whang, L.S.; Shahidi, F.; Pan, M.-H. Endogenous formation of trans fatty acids: Health implications and potential dietary intervention. *J. Funct. Foods* **2014**, *25*, 14–24. [CrossRef]
48. Narciso, L.; Parlanti, E.; Racaniello, M.; Simonelli, V.; Cardinale, A.; Merlo, D.; Dogliotti, E. The Response to Oxidative DNA Damage in Neurons: Mechanisms and Disease. *Neural Plast.* **2016**, *2016*, 3619274. [CrossRef]
49. Nguyen, D.; Alavi, M.V.; Kim, K.Y.; Kang, T.; Scott, R.T.; Noh, Y.H.; Lindsey, J.D.; Wissinger, B.; Ellisman, M.H.; Weinreb, R.N.; et al. A new vicious cycle involving glutamate excitotoxicity, oxidative stress and mitochondrial dynamics. *Cell Death Dis.* **2011**, *2*, e240. [CrossRef]
50. Jurk, D.; Wang, C.; Miwa, S.; Maddick, M.; Korolchuk, V.; Tsolou, A.; Gonos, E.S.; Thrasivoulou, C.; Saffrey, M.J.; Cameron, K.; et al. Postmitotic neurons develop a p21-dependent senescence-like phenotype driven by a DNA damage response. *Aging Cell.* **2012**, *11*, 996–1004. [CrossRef]
51. Shanbhag, N.M.; Evans, M.D.; Mao, W.; Nana, A.L.; Seeley, W.W.; Adame, A.; Rissman, R.A.; Masliah, E.; Mucke, L. Early neuronal accumulation of DNA double strand breaks in Alzheimer's disease. *Acta Neuropathol. Commun.* **2019**, *7*, 77. [CrossRef] [PubMed]
52. Canugovi, C.; Misiak, M.; Ferrarelli, L.K.; Croteau, D.L.; Bohr, V.A. The role of DNA repair in brain related disease pathology. *DNA Repair* **2013**, *12*, 578–587. [CrossRef] [PubMed]
53. Borgesius, N.Z.; de Waard, M.C.; van der Pluijm, I.; Omrani, A.; Zondag, G.C.; van der Horst, G.T.; Melton, D.W.; Hoeijmakers, J.H.; Jaarsma, D.; Elgersma, Y. Accelerated age-related cognitive decline and neurodegeneration, caused by deficient DNA repair. *J. Neurosci.* **2011**, *31*, 12543–12553. [CrossRef] [PubMed]
54. Flyunt, R.; Bazzanini, R.; Chatgilialoglu, C.; Mulazzani, Q.G. Fate of the 2′-Deoxyadenosin-5′-yl Radical under Anaerobic Conditions. *J. Am. Chem. Soc.* **2000**, *122*, 4225–4226. [CrossRef]
55. Cui, L.; Ye, W.; Prestwich, E.G.; Wishnok, J.S.; Taghizadeh, K.; Dedon, P.C.; Tannenbaum, S.R. Comparative analysis of four oxidized guanine lesions from reactions of DNA with peroxynitrite, single oxygen, and γ-radiation. *Chem. Res. Toxicol.* **2013**, *26*, 195–202. [CrossRef]
56. Chatgilialoglu, C. Biomimetic radical chemistry and applications. *Molecules* **2022**, *27*, 2042. [CrossRef]
57. Wang, J.; Clauson, C.L.; Robbins, P.D.; Niedernhofer, L.J.; Wang, Y. The oxidative DNA lesions 8,5′-cyclopurines accumulate with aging in a tissue-specific manner. *Aging Cell* **2012**, *11*, 714–716. [CrossRef]
58. Iwamoto, T.; Brooks, P.J.; Nishiwaki, T.; Nishimura, K.; Kobayashi, N.; Sugiura, S.; Mori, T. Quantitative and in situ detection of oxidatively generated DNA damage 8,5′-cyclo-2′-deoxyadenosine using an immunoassay with a novel monoclonal antibody. *Photochem. Photobiol.* **2014**, *90*, 829–836.

59. Cooke, M.S.; Olinski, R.; Loft, S. European standards committee on urinary (DNA) lesion analysis. *Cancer Epidemiol. Biomark. Prev.* **2008**, *17*, 3–14. [CrossRef]
60. Chung, M.K.; Riby, J.; Li, H.; Iavarone, A.T.; Williams, E.R.; Zheng, Y.; Rappaport, S.M. A sandwich enzyme-linked immunosorbent assay for adducts of polycyclic aromatic hydrocarbons with human serum albumin. *Anal. Biochem.* **2010**, *400*, 123–129. [CrossRef]
61. Okahashi, Y.T.; Iwamoto, N.; Suzuki, S.; Shibutani, S.; Sugiura, S.; Itoh, T.; Nishiwaki, S.; Ueno, T.; Mori, T. Quantitative detection of 4-hydroxyequilenin-DNA adducts in mammalian cells using an immunoassay with a novel monoclonal antibody. *Nucl. Acids Res.* **2010**, *38*, e133. [CrossRef] [PubMed]
62. Liu, S.; Wang, Y. Mass spectrometry for the assessment of the occurrence and biological consequences of DNA adducts. *Chem. Soc. Rev.* **2015**, *44*, 7829–7854. [CrossRef] [PubMed]
63. Dizdaroglu, M.; Coskun, E.; Jaruga, P. Measurement of oxidatively induced DNA damage and its repair, by mass spectrometric techniques. *Free. Radic. Res.* **2015**, *49*, 525–548. [CrossRef]
64. Wang, J.; Yuan, B.; Guerrero, C.; Bahde, R.; Gupta, S.; Wang, Y. Quantification of oxidative DNA lesions in tissues of Long-Evans Cinnamon rats by capillary high-performance liquid chromatography-tandem mass spectrometry coupled with stable isotope dilution method. *Anal. Chem.* **2011**, *83*, 2201–2209. [CrossRef] [PubMed]
65. Egler, R.A.; Fernandes, E.; Rothermund, K.; Sereika, S.; de Souza-Pinto, N.; Jaruga, P.; Dizdaroglu, M.; Prochownik, E.V. Regulation of reactive oxygen species, DNA damage, and c-Myc function by peroxiredoxin1. *Oncogene* **2005**, *24*, 8038–8050. [CrossRef] [PubMed]
66. Kirkali, G.; de Souza-Pinto, N.C.; Jaruga, P.; Bohr, V.A.; Dizdaroglu, M. Accumulation of (5′S)-8,5′-cyclo-2′-deoxyadenosine in organs of Cockayne syndrome complementation group B gene knockout mice. *DNA Repair* **2009**, *8*, 274–278. [CrossRef] [PubMed]
67. Schumacher, B.; Pothof, J.; Vijg, J.; Hoeijmakers, J.H.J. The central role of DNA damage in the ageing process. *Nature* **2021**, *592*, 695–703. [CrossRef]
68. Kaluzny, M.A.; Duncan, L.A.; Merritt, M.V.; Epps, D.E. Rapid separation of lipid classes in high yield and purity using bonded phase columns. *J. Lipid Res.* **1985**, *26*, 135–140. [CrossRef]
69. Fragopoulou, A.F.; Polyzos, A.; Papadopoulou, M.D.; Sansone, A.; Manta, A.K.; Balafas, E.; Kostomitsopoulos, N.; Skouroliakou, A.; Chatgilialoglu, C.; Georgakilas, A.; et al. Hippocampal lipidome and transcriptome profile alterations triggered by acute exposure of mice to GSM 1800 MHz mobile phone radiation: An exploratory study. *Brain Behav.* **2018**, *8*, e01001. [CrossRef]
70. Pakiet, A.; Jakubiak, A.; Czumaj, A.; Sledzinski, T.; Mika, A. The effect of western diet on mice brain lipid composition. *Nutr. Metab.* **2019**, *16*, 81. [CrossRef]
71. Chan, R.B.; Oliveira, T.G.; Cortes, E.P.; Honig, L.S.; Duff, K.E.; Small, S.A.; Wenk, M.R.; Shui, G.; Di Paolo, G. Comparative Lipidomic Analysis of Mouse and Human Brain with Alzheimer Disease. *J. Biol. Chem.* **2012**, *287*, 2678–2688. [CrossRef]
72. Ferreri, C.; Samadi, A.; Sassatelli, F.; Landi, L.; Chatgilialoglu, C. Regioselective Cis-Trans Isomerization of Arachidonic Double Bonds by Thiyl Radicals: The Influence of Phospholipid Supramolecular Organization. *J. Am. Chem. Soc.* **2004**, *126*, 1063–1072. [CrossRef] [PubMed]
73. van Diepen, J.A.; Berbée, J.F.; Havekes, L.M.; Rensen, P.C. Interactions between inflammation and lipid metabolism: Relevance for efficacy of anti-inflammatory drugs in the treatment of atherosclerosis. *Atherosclerosis* **2013**, *228*, 306–315. [CrossRef] [PubMed]
74. Zhang, J.; Zhang, S.; Shan, H.; Zhang, M. Biologic Effect of Hydrogen Sulfide and Its Role in Traumatic Brain Injury. *Oxid. Med. Cell. Longev.* **2020**, *2020*, 7301615. [CrossRef] [PubMed]
75. Wang, X.; Michaelis, E.K. Selective neuronal vulnerability to oxidative stress in the brain. *Front. Aging Neurosci.* **2010**, *2*, 12. [CrossRef] [PubMed]
76. Else, P.L.; Hulbert, A.J. Membranes as metabolic pacemakers. *Clin. Exp. Pharmacol. Physiol.* **2003**, *30*, 559–564. [CrossRef]
77. Xu, J.; Huang, X. Lipid metabolism at membrane contacts: Dynamics and functions beyond lipid homeostasis. *Front. Cell Dev. Biol.* **2020**, *8*, 615856. [CrossRef]

biomolecules

MDPI

Review

Lipid Peroxidation and Antioxidant Protection

Luca Valgimigli

Department of Chemistry "G. Ciamician", University of Bologna, Via Piero Gobetti 85, 40129 Bologna, Italy; luca.valgimigli@unibo.it; Tel.: +39-0512095683

Abstract: Lipid peroxidation (LP) is the most important type of oxidative-radical damage in biological systems, owing to its interplay with ferroptosis and to its role in secondary damage to other biomolecules, such as proteins. The chemistry of LP and its biological consequences are reviewed with focus on the kinetics of the various processes, which helps understand the mechanisms and efficacy of antioxidant strategies. The main types of antioxidants are discussed in terms of structure–activity rationalization, with focus on mechanism and kinetics, as well as on their potential role in modulating ferroptosis. Phenols, pyri(mi)dinols, antioxidants based on heavy chalcogens (Se and Te), diarylamines, ascorbate and others are addressed, along with the latest unconventional antioxidant strategies based on the double-sided role of the superoxide/hydroperoxyl radical system.

Keywords: autoxidation; peroxyl radicals; hydroperoxyl radicals; kinetics; antioxidants; phenols; catechols; nitroxides; thiols; pyridinols

Citation: Valgimigli, L. Lipid Peroxidation and Antioxidant Protection. *Biomolecules* **2023**, *13*, 1291. https://doi.org/10.3390/biom13091291

Academic Editor: Chryssostomos Chatgilialoglu

Received: 13 July 2023
Revised: 17 August 2023
Accepted: 21 August 2023
Published: 24 August 2023

1. Introduction

Lipid peroxidation (LP) is a complex phenomenon, first investigated in the early 20th century, consisting in the uptake of molecular oxygen by lipids exposed to air, which was soon recognized as bearing remarkable similarity to hydrocarbon autoxidation [1,2], the formal insertion of one molecule of oxygen in the C-H bond of a hydrocarbon to afford a hydroperoxide: R-H + O_2 → R-OO-H. Indeed, lipid peroxidation is one embodiment of hydrocarbon autoxidation.

While the direct reaction with ground state (triplet) oxygen is spin restricted and too slow to occur, the transformation of hydrocarbons (or lipids) by oxygen to hydroperoxides and further oxidized products occurs rapidly and efficiently via the intermediation of peroxyl radicals (ROO•), in a chain reaction that can be triggered by a multitude of events in any chemical system, such as in food or in living organisms, and it can be blocked or prevented by *antioxidants* [1–3].

Over the last 70 years, extensive research efforts have shown the association of lipid peroxidation with an impressive number of pathological conditions, from arteriosclerosis and cardiovascular diseases to neurological disorders and cancer [1–7], stemming from the "Free Radical Theory of Aging", which attributes the progressive decline of functionality associated with aging to the progressive accumulation of damage to biomolecules and essential biological structures, caused by radicals' overproduction, also referred to as "oxidative stress" [6,7]. Yet the interest in lipid peroxidation has recently seen a boost after the formal recognition in 2012 of ferroptosis as a new form of programmed cell death [8], driven by LP and specifically linked to the pathophysiology of several degenerative diseases [9–11]. Even more interesting are pioneering observations that antioxidants can effectively inhibit ferroptosis [12,13] and that either antioxidants or pro-oxidant molecules can be used to modulate it, promising a handle on the connected pathologies [10,13].

Although several valuable reviews have been dedicated to LP and ferroptosis [14,15], the importance of lipid peroxidation is not limited to them. Damage to membrane lipids by radicals and reactive oxygen species (ROS) is perhaps the most important type of oxidative/radical damage in biological systems; not only because of the susceptibility of lipids to radical attack and because it produces the majority of biomarkers commonly used to monitor such damage but because such biomarkers—the LP secondary products—are also major effectors of the damage itself [16], owing to their toxicity and to their ability to alter other biomolecules such as proteins [17], thereby extending the radical/oxidative damage [18].

The aim of this review is to offer a brief up-to-date overview on lipid peroxidation, including the main processes leading to secondary products, with focus on the kinetic aspects regulating the different reactions, so to guide the understanding of the efficacy of the different antioxidant protection strategies. The main type of antioxidants will also be briefly reviewed addressing their efficacy, on kinetic bases, and their mechanism of action.

2. The Chemistry of Lipid Peroxidation

2.1. The Three Stages of Lipid Peroxidation

LP is a radical chain reaction composed of the canonical three stages of initiation, propagation and termination, summarized in Figure 1 using PUFA as the prototypical oxidizable substrate.

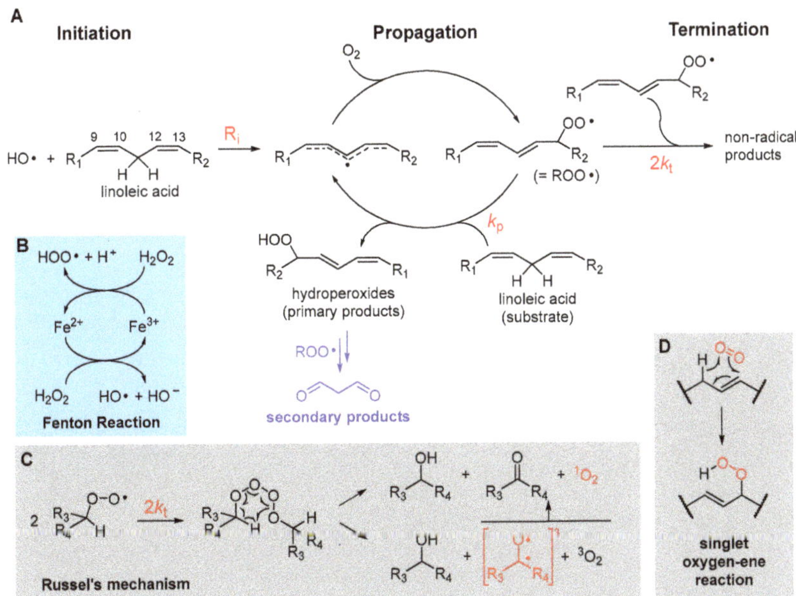

Figure 1. The main steps of lipid peroxidation (**A**), with indication of one main mechanism of initiation (**B**), details of Russel's mechanism for termination (**C**) and reaction of singlet oxygen with unsaturated lipids (**D**).

Initiation, i.e., the formation of the first lipid-derived radical, can consist in attack by a variety of radical species. In biological systems, these would most commonly be an alkoxyl radical (RO•) or a hydroxyl radical (HO•), e.g., generated by Fenton-type decomposition of hydroperoxides or hydrogen peroxide, catalyzed by transition metals such as iron or copper. Other initiating species could be hydroperoxyl radicals (HOO•), the neutral form of metabolically produced superoxide radical anion ($O_2^{\bullet-}$), which, however, is not prevailing at pH 7 or higher—the pK_a is 4.8 in water at 20 °C [19]. Excited triplet states produced by UV irradiation of carbonyl compounds, e.g., in the skin, or radicals produced from water by ionizing radiations would also serve the purpose. Since the attack to lipids occurs typically by formal hydrogen-atom transfer (HAT) from the >CH_2 in allylic/bis-allylic positions, having a bond dissociation enthalpy (BDE) of about 77/87 kcal/mole, respectively, any radical species X• forming a product X-H with BDE higher than this value would be suitable for the task.

Propagation of the oxidative chain occurs by two alternating steps: the reaction of the lipid-derived C-centered radical (>C(•)H; =R•) with oxygen to form the alkylperoxyl radical (ROO•) is extremely fast (rate constant in the range 2–5 × 10^9 $M^{-1}s^{-1}$ [20]), hence the kinetics of propagation is governed by the second step, the reaction of ROO• with a new lipid molecule to afford a new lipid-derived R• radical. Its rate constant k_p is the most important parameter in evaluating antioxidant strategies (vide infra) and it depends dramatically on the structure of the lipid molecule. As summarized in Table 1, k_p is negligible for saturated fatty acids (or hydrocarbons) at close to physiologic temperature [21], but it grows by about two orders of magnitude in monounsaturated fatty acids (MUFAs, e.g., methyl oleate) and by about four orders of magnitude with two unsaturations, like in methyl linoleate, which reaches k_p = 62 $M^{-1}s^{-1}$ at 30 °C [22]. This is due to resonance stabilization of the allyl and, particularly, of the bis-allyl radical, resulting by HAT from MUFA and PUFA, respectively. On increasing the degree of unsaturation in fatty acids, the number of >CH_2 in the bis-allylic position also increases, which explains the almost linear increase in k_p with the number of double bonds [6], first reported by Porter's group [23] (Table 1), representing a significant challenge for antioxidant protection (vide infra).

Table 1. Rate constants for the peroxidation of lipids in bulk or in chlorobenzene at 303 K.

Lipid	$k_p/(2k_t)^{1/2}$ 10^{-5} $M^{-1/2}s^{-1/2}$	k_p $M^{-1}s^{-1}$	$2k_t$ 10^5 $M^{-1}s^{-1}$	Ref.
Methyl stearate (18:0) [1]	~0.8	~0.01	15	[21]
Methyl oleate (18:1)	89.0	0.89	10	[22]
Methyl linoleate (18:2)	2100	62.0	88	[22]
Methyl linolenate (18:3)	3900	236.0	360	[22]
Linoleic acid (18:2)	-	62	-	[23]
Arachidonic acid (20:4)	-	197	-	[23]
Eicosapentaenoic ac. (20:5)	-	249	-	[23]
Docosahexaenoic ac. (22:6)	-	334	-	[23]
Cholesterol	-	11	-	[23]
7-Dehydrocholesterol	-	2260	-	[23]
Squalene	2500	68.0	74.0	[24]
Sunflower oil (60% of 18:2)	3600	66.9	34.5	[24]
PLPC [2]	-	16.6	1.27	[25]
DLPC [3]	-	13.6 [4]	1.02	[25]

[1] Data for hexadecane. [2] 1-Palmitoyl-2-linoleoyl-*sn*-glycero-3-phosphocholine; data at 310 K. [3] 1,2-Dilinoleoyl-*sn*-glycero-3-phosphocholine; data at 310 K. [4] For intermolecular propagation, k_p (intramolecular) = 5.1 s^{-1}.

The main products of the overall propagation stage are hydroperoxides (ROOH), the *primary products* of LP [1], and new alkylperoxyl radicals, which are the chain-carrying species in LP. Depending on the lipid structure, formation of other side products like epoxides and endoperoxides might also gain importance (vide infra).

Termination of the chain reaction occurs when two radicals quench each other to afford non-radical (diamagnetic) species, which are unable to propagate the chain. Uninhibited LP is carried on by secondary peroxyl radicals, and the best-established termination process is Russel's mechanism [26], which consists in the formation of a tetroxide that decomposes to afford carbonyl compounds and molecular oxygen (Figure 1). It has lately been clarified that O_2 released by decomposition of the tetroxide is predominantly in the excited singlet state $^1O_2$ [27], meaning that oxygen molecules bear all electrons in pairs with antiparallel spin. While the singlet state is the lowest in energy for most molecules, oxygen is the exception and its ground state is a (paramagnetic) triplet biradical, i.e., it has two electrons with parallel spin in (degenerate) antibonding π^* orbitals. This, perhaps counterintuitively, makes it much less reactive toward "normal" diamagnetic molecules due to spin restriction, i.e., $^3O_2$ is much less oxidizing than $^1O_2$. Other processes have been shown to form singlet oxygen in biological systems. Among them, the reaction of hydrogen peroxide (H_2O_2) and lipid hydroperoxides (ROOH) by myeloperoxidase (MPO) in the presence of Cl^- ions, which is possibly a defensive mechanism occurring during phagocytosis. It was demonstrated that the process occurs by reaction of H_2O_2 or ROOH with HOCl produced by MPO [27]. Earlier work indicated that $^1O_2$ is quantitatively produced by HOCl reaction on H_2O_2 but it is not observed with ROOH [28]. However, subsequent work, addressing the ability of different hydroperoxides to produce $^1O_2$ upon HOCl reaction, showed that tertiary alkylhydroperoxides are ineffective, while typical lipid-derived secondary ROOH form $^1O_2$ with an approximately 10% yield via Russel's mechanism [29]. The decomposition of lipid hydroperoxides in the presence of metal ions or peroxynitrite was also shown to produce $^1O_2$ with an approximately 10% yield [27]. Decomposition of 7-α-OOH, 6-β-OOH and 5-α-OOH cholesterol hydroperoxides, formed upon cholesterol autoxidation in biomembranes, is also a relevant source of $^1O_2$ via a Russel's-type mechanism [27]. Cytochrome c causes the formation of $^1O_2$ in mitochondria by promoting the autoxidation of cardiolipin [30]. Additionally, singlet oxygen can be produced by photoexcitation of ground state triplet oxygen in the presence of a photosensitizer [6], e.g., in the skin, as summarized in Equation (1). The photosensitizer (PS), such as a carbonyl compound in its ground singlet state, absorbs visible or UV light and is promoted to the excited singlet state, which undergoes intersystem crossing to convert into the excited triplet state. This finally transfers energy to ground state triplet oxygen via a triplet annihilation process, to go back to the ground state converting oxygen into the excited singlet state.

$$^1PS + {}^3O_2 + \text{light} \rightarrow {}^1PS^* + {}^3O_2 \rightarrow {}^3PS^* + {}^3O_2 \rightarrow {}^1PS + {}^1O_2 \quad (1)$$

Singlet oxygen is unable to restart the autoxidation chain; however, it directly reacts with unsaturated lipids via an oxygen-ene reaction to afford the lipid hydroperoxides (Figure 1) [6].

2.2. The Rate of Lipid Peroxidation

Chain termination typically quenches two peroxyl radicals in a fast radical–radical reaction with rate constant $2k_t$. Although $2k_t$ is much higher than the rate constant k_p for propagation (see Table 1), the actual rate of termination also depends on the square of the steady-state concentration of the chain-carrying peroxyl radical, which is very low, typically in the micromolar-to-nanomolar range. Therefore, it can be estimated that, following an initiation event, >100 propagation cycles occur before termination takes place in LP of unsaturated lipids, implying that one single initiating radical can damage hundreds of lipid molecules, which explains why antioxidants are necessary.

In the absence of antioxidants, the rate of peroxidation of a lipid substrate RH, at a constant rate of chain initiation R_i, follows Equation (2) [21,22,24,31], i.e., it is dictated by the ratio of the rate constant for propagation over the square root of the rate constant for termination. The kinetic quantity $k_p/(2k_t)^{1/2}$ is named the "oxidizability" of the specific lipid substrate; it increases with unsaturation (Table 1) and it determines how important antioxidant protection is for the specific substrate.

$$-\frac{d[O_2]}{dt} = -\frac{d[RH]}{dt} = \frac{d[ROOH]}{dt} = \frac{k_p}{\sqrt{2k_t}} \times \sqrt{R_i} \times [RH] + R_i \quad (2)$$

2.3. Light as a Product of Lipid Peroxidation

The termination stage of LP produces high-energy (transient) species that can deactivate to ground state or lose excess energy by emitting light, which can therefore be regarded as a product of LP. One important example is singlet oxygen, $^1O_2$, which can return to a ground state triplet by giving weak phosphorescence in the near infrared region at 1270 nm [27]. This chemoluminescence (CL) phenomenon has been extensively exploited to probe the formation of $^1O_2$ during LP [27–29] and strategies have been developed to enhance CL in biological systems to improve detection sensitivity [32].

Emission of chemoluminescensce in the visible region is also observed during LP and it is attributed to excited carbonyls in the triplet state, such as those formed in Russel's termination mechanism (Figure 1C), although singlet oxygen, e.g., formed in the same process, also contributes [33]. It has also been proposed that part of the excited triplet carbonyls affording CL are formed by an alternative termination mechanism consisting in cyclization of the alkylperoxyl radical to a dioxyethane followed by its thermolysis [34]. Emission can be enhanced by additives like 9,10-dibromoanthracene and it is proportional to the rate of LP [33]. Interestingly, the time-based monitoring of CL has been developed as a method to monitor the kinetics of LP both in animal and vegetable lipids [33,35], which has also been fruitfully applied to study the activity of antioxidants [36,37]

2.4. Further Insights into Chain-Propagation Reactions

2.4.1. β-Fragmentation of the Peroxyl Radical

Addition of oxygen to lipid radical R• to afford the peroxyl radical ROO• is a reversible process and its back reaction, named β-fragmentation, has a rate constant k_β, which depends on the stability of the (reformed) carbon-centered radical (Figure 2). While it is generally too slow compared to the forward reaction to affect the rate of chain propagation, under some conditions it can compete with the subsequent propagation step, the reaction of ROO• with the lipid molecule RH to afford the hydroperoxide primary product ROOH [38]. Therefore, it determines the distribution of isomeric hydroperoxide products, since *cis,trans* hydroperoxides are formed when propagation outcompetes β-fragmentation, hence they are the kinetic products of LP, while *trans,trans* hydroperoxides are the thermodynamic products, forming when β-fragmentation allows isomerization to take place [23,38].

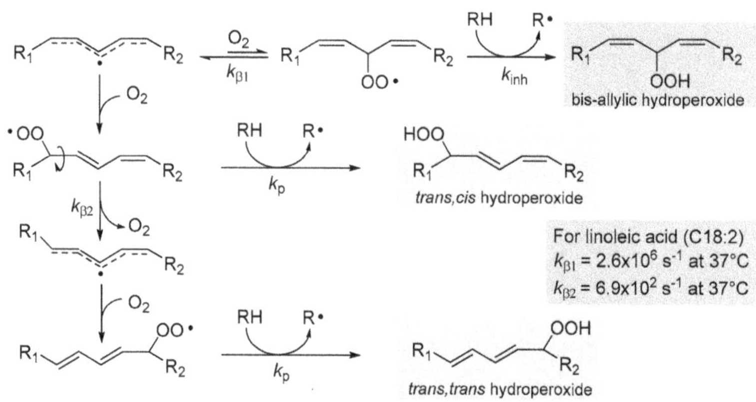

Figure 2. Schematic representation of β-fragmentation reactions in peroxyl radicals competing with H-atom abstraction from a donor RH, which is a lipid molecule (k_p) or an antioxidant (k_{inh}).

This reaction was thoroughly investigated in Porter's group who found that, upon calibration of the rate of β-fragmentations, these unimolecular reactions could be used as a "radical clock" to measure the kinetics of competing H-abstractions, by the analysis of the hydroperoxide product ratio from autoxidation [39]. For instance, they found that the rate constant for fragmentation of *cis,trans* hydroperoxides from linoleic acid (Figure 2) was $k_{\beta 2} = 6.9 \times 10^2$ s^{-1} (at 37 °C) and would be useful to measure rate constants for propagation (k_p). The fragmentation of bis-allyl hydroperoxide is instead much faster ($k_{\beta 1} = 2.6 \times 10^6$ s^{-1} at 37 °C) and the corresponding hydroperoxide would never be formed in the absence of a much faster H-atom donor such as α-tocopherol (α-TOH) or other antioxidants; therefore, they used this radical clock to measure the rate constant of inhibition k_{inh} for antioxidants (vide infra) [39].

2.4.2. Hydrogen-Atom Abstraction vs. Radical Addition: Formation of Primary Epoxides

Although chain propagation in PUFAs such as linoleic acid is carried out mainly by H-atom abstraction from the bis-allylic position to afford the stabilized pentadienyl-type radical, addition of the ROO• radical to one of the C=C double bonds can also compete, to some extent, being favored by the contribution of polar effects in the transition state [40]. The resulting alkylperoxyalkyl radical would rapidly undergo a homolytic substitution ($S_H{}^i$) reaction, forming an epoxide and releasing an alkoxyl radical RO• (Figure 3), which would rapidly attack another lipid molecule, propagating the chain. Therefore, from a kinetic perspective, H-abstraction and addition are difficult to distinguish, besides the fact they afford different primary oxidation products, hydroperoxides or epoxides, respectively.

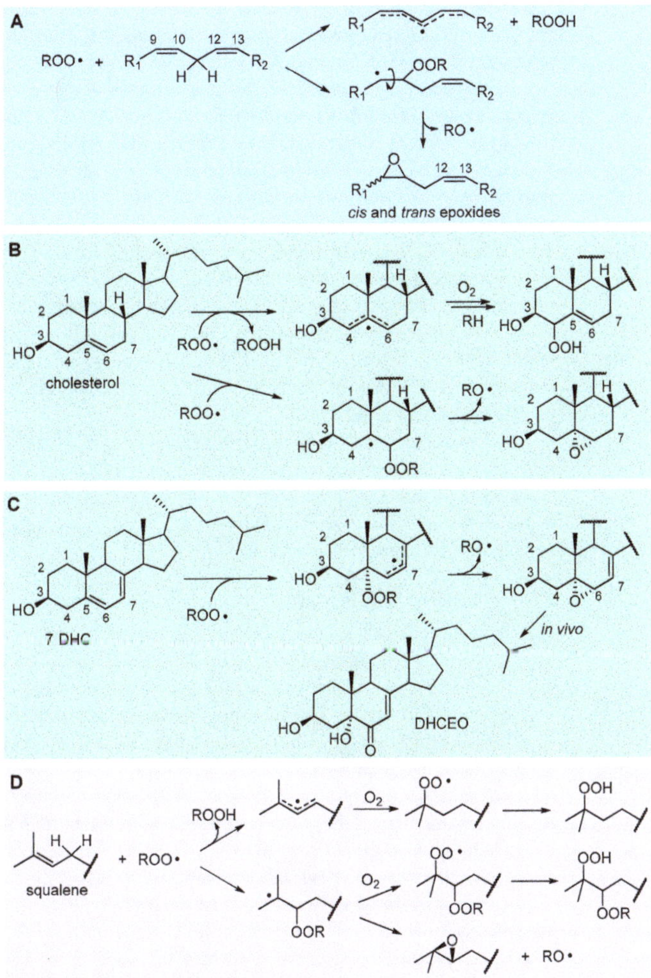

Figure 3. Formation of epoxides during the peroxidation of PUFA (**A**), of cholesterol (**B**), of 7-dehydrocholesterol (**C**) and of squalene (**D**).

The relative importance of addition might depend on the experimental conditions and would certainly depend on the lipid structure. Confirmation of its relevance comes from the identification of both 9,10- and 12,13-epoxides among the oxidation products of methyl linoleate [41], while epoxyeicosatrienoic acids formed via peroxyl radical addition were found in the peroxidation of arachidonic acid (C20:4) [42].

In the autoxidation of cholesterol, epoxides at C5,6 (α+β) were found to account for about 12% of the total oxidation products [43], demonstrating the importance of addition in chain propagation. A detailed study by Pratt's group showed that addition at C6 had similar activation energy ($\Delta G^{\ddagger}$ = 17.6 kcal/mol) to H-abstraction at C7-H (BDE = 83.2 kcal/mol, $\Delta G^{\ddagger}$ = 17.5 kcal/mol), being faster than H-abstraction at C4-H (BDE = 89.0 kcal/mol, $\Delta G^{\ddagger}$ = 19.4 kcal/mol) which would afford, respectively, the 7-OOH and 4-OOH hydroperoxides. Instead, the 6-OOR resulting from addition would undergo very rapid ($\Delta G^{\ddagger}$ = 7.3 kcal/mol) transformation to the 5,6-epoxide [43], as shown in Figure 3.

7-Dehydrocholesterol (7-HDC) is formed as the last step before cholesterol in the biosynthesis path that starts from squalene, and its metabolism is implicated in Smith−Lemli−Opitz syndrome (SLOS), a devastating autosomal disorder with a range of phenotypical expressions, including malformations and neurological deficits [38]. SLOS is characterized by much-elevated levels of 7-HDC (and low levels of cholesterol) due to defects in the gene encoding for 7-dehydrocholesterol reductase. 7-HDC is one the most oxidizable lipids in biological systems with a k_p 200-fold higher than cholesterol and 10- to 40-fold higher than PUFA (Table 1), which reacts with peroxyl radicals mainly H-abstraction, at C9-H or by addition at C5 (at variance with cholesterol) to afford a stabilized allylic radical [38]. This yields the 5α,6α-epoxide (Figure 3) that, in vivo, is converted to 3β,5α-dihydroxycholest-7-en-6-one (DHCEO), which has been used as a biomarker in SLOS [38,43].

The autoxidation of squalene, the polyunsaturated triterpenic precursor of cholesterol and phytosterols, has recently been kinetically characterized in our group, finding a k_p of 68 $M^{-1}s^{-1}$, just slightly higher than that of methyl linoleate and 6-fold that of cholesterol (Table 1), which is due mainly to H-abstraction at repeating allylic positions, with significant contribution of ROO• addition, to afford a tertiary alkyl radical developing into an epoxide, as illustrated in Figure 3 [24].

A peroxyl radical clock designed for measuring both H-atom abstraction and radical addition has recently been described and its application to a variety of unsaturated hydrocarbons indicates that addition grows in importance in conjugated dienes and polyenes; these include retinol (vitamin A) and carotenoids, conjugated linoleic acid (CLA 18:2) and conjugated linolenic acid (CLA 18:3) [44].

2.4.3. Release of HOO• and Chain-Transfer Processes

Alkoxyl radicals released via $S_H{}^i$ reaction upon epoxide formation have very high reactivity toward unsaturated lipids and, besides H-atom abstraction, they undergo very fast addition to C=C double bonds [45,46], forming a β-alkoxyalkyl radical that rapidly reacts with oxygen (Figure 4). The resulting β-alkoxyalkylperoxyl can undergo intramolecular H-atom abstraction (1,4-HAT) either in Cα to the RO- group, to form a radical stabilized by resonance with the oxygen lone pair (Path A), or from the allylic >CH_2 to form a radical stabilized by resonance with the double bond (Path B). In both cases, the resulting hydroperoxide can undergo fragmentation to release the hydroperoxyl radical HOO• and reform the C=C double bond in the lipid structure. The release of HOO• has important consequences in the antioxidant protection of lipids (vide infra).

Although this process has never been demonstrated to occur in lipids like PUFA, similar reactions are known for simpler unsaturated hydrocarbons. The best established is 1,4-cyclohexadiene, which upon attack by alkylperoxyl radicals releases HOO• and converts to benzene [47], and the monoterpene γ-terpinene, which undergoes identical chemistry affording HOO• and *p*-cymene [48]. In those hydrocarbons, the two steps, 1,4-HAT and elimination, are concerted and occur in a single transition state with a low barrier [49]. Pratt and coworkers were able to demonstrate that the reaction also occurs for simple monounsaturated hydrocarbons (like cyclooctene), albeit it does so stepwise, as depicted in Path A of Figure 4, and its kinetics is favored by quantum tunneling in the 1,4-HAT [49].

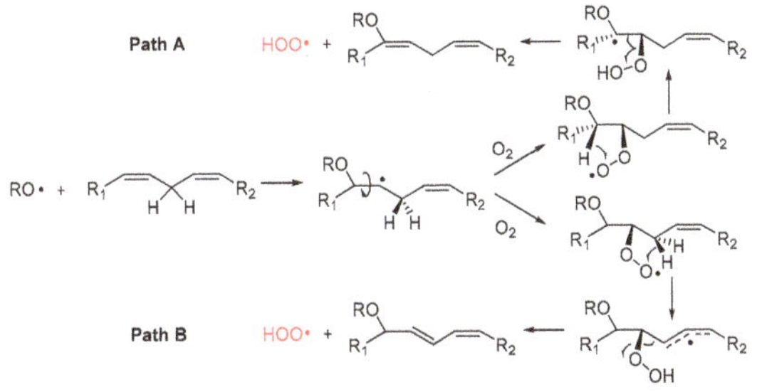

Figure 4. Mechanistic proposal for the release of HOO• during the autoxidation of methyl linoleate.

We have recently suggested the same mechanism to account for some non-classical behavior in the inhibited autoxidation of squalene [24], and ongoing studies in our group indicate a similar chemistry would take place during the autoxidation of methyl linoleate in micelles. In the case of PUFA, however, the possibility to form conjugated double bonds in the product suggests that Path B could also be a viable mechanism.

The release of HOO• has kinetic consequences, as these radicals can propagate the oxidative chain in bulk lipids—i.e., it represents a chain-transfer process—but, owing to their hydrosolubility, they could export the unpaired electron in the aqueous phase in heterogenous systems. As will be discussed in Section 5, it also has major importance for the antioxidant protection of lipids, under some conditions.

2.4.4. Formation of Endoperoxides

Product distribution can become quite complex in highly unsaturated lipids as it can involve intramolecular radical addition to double bonds to yield cyclized products.

In the autoxidation of arachidonic acid in the absence of fast H-atom donors (e.g., α-TOH), the peroxyl radical can undergo 5-*exo* cyclization occurring with a monomolecular rate constant of about 800 s^{-1} which outcompetes β-fragmentation with k_β of about 140 s^{-1} [1]. The result is the formation of an endoperoxide still bearing a C-centered radical (Figure 5), which undergoes another 5-*exo* cyclization, to form another alkyl radical, and further autoxidation steps to ultimately yield endoperoxide products or intermediates, which are stereoisomeric to some of the prostaglandins (e.g., PGG_2 and $PGF_2\alpha$) produced by cyclooxygenase (COX) enzyme [31], therefore generally named isoprostanes.

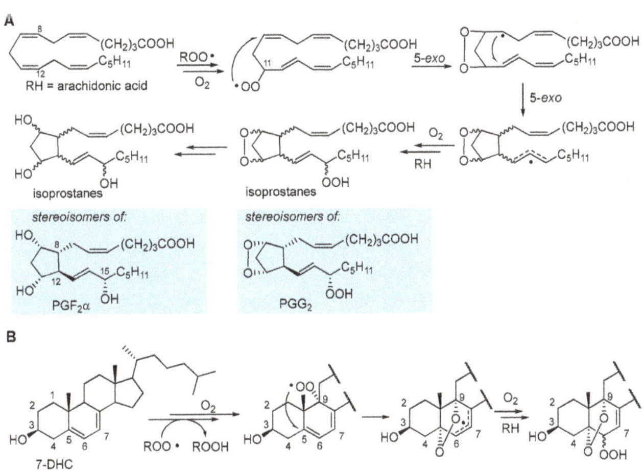

Figure 5. Formation of endoperoxides during the propagation of lipid peroxidation of arachidonic acid (**A**) and 7-dehydrocholesterol (**B**).

In 7-HDC, autoxidation proceeds both by ROO• addition to yield the epoxide (Figure 3) and by H-atom abstraction preferably from C9-H [31,38]. The resulting peroxyl radical was found to undergo 5-exo cyclization at C5 to yield a stabilized allyl radical that continues the autoxidation to afford a mixture of α-5,9-endoperoxides-α&β-6-hydroperoxide [38], as shown in Figure 5. In general, endoperoxide formation is observed when the unsaturated lipid geometry allows for 5-*exo* cyclization, possibly yielding an unstrained or resonance-stabilized radical.

2.5. Peroxidation of Intact Triglycerides and Phospholipids

Most of the knowledge and understanding of the chemistry and related kinetics of lipid peroxidation have been developed using isolated PUFAs and MUFAs and their simple monoesters (e.g., methyl linoleate) as model compounds, along with biologically relevant yet specific compounds, like cholesterol and some of its congeners, or it has been mutuated from knowledge on simpler hydrocarbons. However, on quantitative grounds, the majority of lipids in biological systems are phospholipids (e.g., in cell membranes) and triglycerides (e.g., in adipocytes and liporoteins). Unfortunately, only a limited number of studies have specifically addressed their peroxidation chemistry, owing to their complexity, showing, however, some distinctive features.

A pioneering study by Antunes et al. investigated the kinetics of peroxidation of phosphatidilcholine multilamellar liposomes in buffered water. They studied the liposomes of both 1-palmitoyl-2-linoleoyl-sn-glycero-3-phosphocholine (PLPC) and 1,2-dilinoleoyl-sn-glycero-3-phosphocholine (DLPC), containing, respectively, one and two oxidizable linoleic acid residues per phospholipid molecule [25]. For PLPC, where only one chain is really oxidizable, they found k_p = 16.6 $M^{-1}s^{-1}$ and $2k_t$ = 1.27 × 10^5 $M^{-1}s^{-1}$ at 37 °C, hence they were both significantly lower that those recorded for methyl linoleate in organic solution (Table 1). Interestingly, with DLPC, carrying two oxidizable chains per molecule, they were able to distinguish two rate constants k_p of 13.6 $M^{-1}s^{-1}$ and of 5.1 s^{-1} for intermolacular and intramolecular chain propagation, respectively, while chain termination gave the rate constant $2k_t$ = 1.02 × 10^5 $M^{-1}s^{-1}$ at 37 °C [25]. This clearly shows that the presence of two oxidizable chains in the same lipid molecules changes the kinetic behavior. Stemming from their results, Porter's group investigated, by the radical clock method, the kinetics of chain propagation for PC liposomes always containing palmitic acid esterified in position 1 of the gycerol and one PUFA in position 2: linoleic acid (C18:2; PLPC), used as reference, arachidonic acid (C20:4; PAPC), eicosapentaenoic acid (C20:5; PEPC) and docosahexanoic acid (C22:6; PDPC) [23]. The measured k_p values were proportional to the number of double bonds, as also seen for isolated PUFA in organic solution, but the absolute values were much lower (see Table 1), i.e., 35, 115, 145 and 172 $n^{-1}s^{-1}$, respectively, referring to the mole fraction of the oxidizable lipid in the lipsomes, instead of the molar concentration in solution [23]. Interestingly, they found a different kinetic behavior of the peroxyl radicals in C9 or in C13 in linoleoyl residue, indicating that oxidation in liposomes depends on the position of the formed peroxyl radical. While the above results refer to lipids having only one oxidizable PUFA residue, subsequent studies on 1,2-dilinoleoyl-glycero-3-phosphocholine (DLPC) and on tetralinoleoyl cardiolipin (L_4CL), a mitochondria-specific phospholipid carrying four linoleic acid residues, proved that arm-to-arm attack by peroxyl radical is relevant and it can form interarm peroxide (-O-O-) bridges by addition to double bonds (Figure 6), with consequences on the formation of toxic electrophilic fragmentation products like 4-hydroxynonenal (vide infra) [50].

The kinetics of peroxidation of triglycerides is more complex than that of the isolated fatty acids in solution, owing to arm-to-arm propagation processes, which are more relevant in dilute solution than at high concentration or in the bulk [51]. To date, the only complete kinetic characterization of an intact natural triglyceride is that of sunflower seed oil (SSO) recently reported by our group [24]. Each trigyceride molecule statistically contains 1.7 chains of linoleic acid and its oxidizability was found to be roughly 1.7-fold that of linoleic acid; however, to our surprise, the rate constant k_p was just slightly higher than that of linoleic acid (67 vs. 62 $M^{-1}s^{-1}$), hence the higher oxidizability is due to slower termination (see Table 1), possibly owing to steric impairment in Russel's mechanism [24].

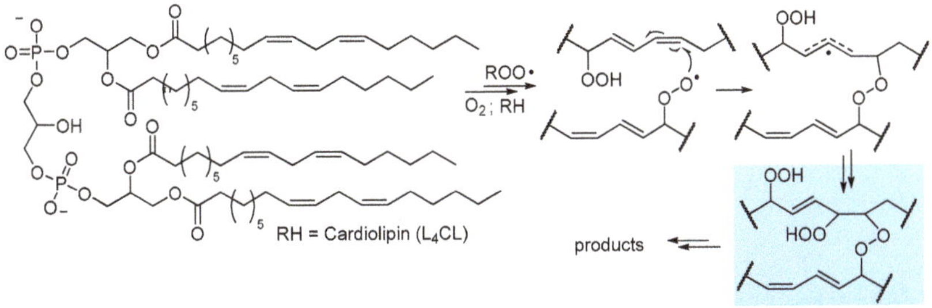

Figure 6. Formation of arm-to-arm peroxides in the peroxidation of mitochondrial cardiolipin.

Clearly, more data would be desirable for complex lipids like phospholipids and triglycerides, particularly in the light of their importance for the onsetting of antioxidant strategies in biological systems.

3. Secondary and Late Products of Lipid Peroxidation

Following the formation of primary oxidation products of LP, namely primary hydroperoxides and primary epoxides, formed directly in chain propagation, their further oxidation, followed or accompanied by other reactions, yields a multitude of secondary products. Several such products are electrophiles and can undergo other reactions, e.g., with proteins, and they have cell signaling functions. Hence, they are toxic or might have some active biological role and have been regarded as biomarkers of oxidative damage, prompting great attention and major efforts in the development of analytical techniques to detect them [16,52–61].

3.1. Formation of Electrophilic Carbonyl Compounds

The most regarded secondary product of LP, which is also among the most relevant LP biomarkers in biological systems, is 4-hydroxynonenal (4-HNE), formed as a late product of oxidation of linoleic acid [52,55,60]. It is an α-β unsaturated aldehyde with very high reactivity towards biological nucleophiles (e.g., amino acid residues and DNA bases), which grants both its role as a signaling molecule and its toxicity. Despite its importance, 4-HNE is not unique but a member of a broad family of short-chain aldehydes and ketones, often bearing conjugated unsaturation in the hydrocarbon chain, all characterized by the marked electrophilic character and all deriving from fragmentation of the hydrocarbon backbone of oxidized lipids; some of these carbonyl compounds are collected in Figure 7. Among them, two arguably stand out: 4-oxononenal (4-ONE), the oxidized form of 4-HNE, being orders of magnitude more reactive as an electrophile [6,58], and malondialdehyde (MDA), a dicarbonyl compound that is arguably the best-known marker of LP, owing to its facile detection (by spectrophotometry or HPLC) upon derivatization with thiobarbituric acid (the so-called TBARS assay) [52]. The mechanism by which 4-HNE forms from oxidized lipids has been highly debated, and several hypotheses have been proposed, which are not mutually exclusive and, possibly, might all be relevant to some extent.

The most classical mechanism is perhaps the one based on Hock rearrangement and fragmentation of the primary hydroperoxide in the presence of an acid catalyst [6,31]. The formed carbonyl compounds can then undergo further autoxidation to afford 4-HNE, as depicted in Figure 7A. Another well-established mechanism is based on Fenton-type cleavage of one hyroperoxyl group in a multiply oxidized linoleyl residue to afford an alkoxyl radical that undergoes fragmentation [6,55]. The reduction of the formed 4-hydroperoxynonenal affords 4-HNE (Figure 7B).

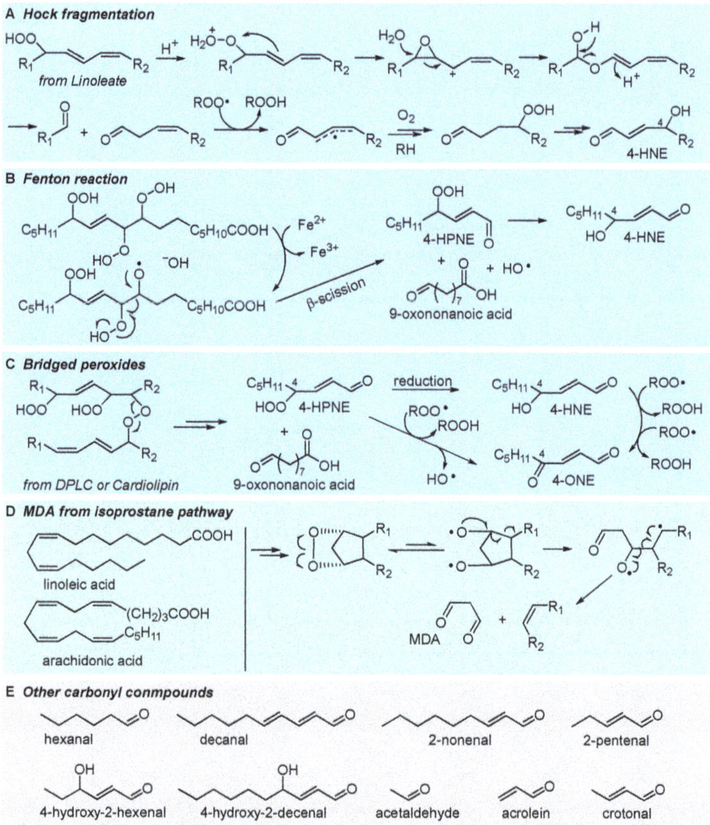

Figure 7. Formation of electrophiles as late products of LP: (**A**) 4-hydroxynonenal (4-HNE) from Hock rearrangment and fragmentation; (**B**) 4-HNE from Fenton reaction on dihydroperoxides; (**C**) 4-HNE and 4-oxononenal (4-ONE) from bridged peroxides; (**D**) malondialdehyde (MDA) from isoprostane pathway; (**E**) other common carbonyl compounds.

Perhaps the most relevant mechanistic proposal, by Porter's group, stems from the bridged peroxides formed by arm-to-arm propagation in LP of phospholipids, which we have discussed in Section 2.3. Breaking of the peroxide bridge affords 4-HNE and 4-ONE (as summarized in Figure 7C), respectively, by reduction of the -OOH group and by formal water elimination (actually H-abstraction from Cα and elimination of •OH radical from -OOH) [50]. Even more controversial is the mechanism of formation of MDA, which could follow different routes, including processes not necessarily involving LP [62]. One significant mechanism is along the pathway to isoprostanes (see Figure 5), actually in competition with their formation, which relies on the spontaneous or induced breaking of the endoperoxide [31], as illustrated in Figure 7D. Although the formation of endoperoxides via 5-*exo* peroxyl addition and 5-*exo* alkyl addition is normally described for arachidonic acid's LP [31,62], actually only two non-conjugated double bonds are necessary to form the dioxabicyclo[2.2.1]heptane structure, hence this mechanism can also apply to linoleic acid or other PUFAs.

Carbonyl compounds are also formed as secondary products of cholesterol hydroperoxides; for instance, Hock fragmentation and further oxidation of cholesterol 5-hydroperoxide yield secosterols [31], which have been implicated in several diseases [63–65].

3.2. Formation of Isoprostanes

Formation of isoprostanes, from sequential steps following the formation of the peroxyl radical in the peroxidation of arachidonic acid, is described in Section 2.3 and illustrated in Figure 5. Isoprostanes are stereoisomers of some of the otherwise identical prostaglandins, e.g., PGH_2, PGG_2 and $PGF_2\alpha$, formed (instead) with full stereochemical control by cyclooxygenase (COX 1 and 2 in humans), still from arachidonic acid, released by a phospholipase during the inflammatory cascade [38]. Owing to their high specificity, isoprostanes have gained importance as the most reliable markers of LP and oxidative stress in biological systems, particularly in relation to some conditions which they directly affect [4,64–68]. Among them, 8-epi-PGF2α (also called 8-isoprostane, 8-isoPGF2α, iPF2α-III and 15-F2t-IsoP), the epimer in C8 of $PGF_2\alpha$, is the best-established marker of oxidative stress in human studies, owing to its stability and selective urinary excretion, which makes its monitoring sensitive and conveneint [4,67]. The autoxidation of arachidonic acid occurs with preferential formation of 8-, 9-, 11- or 12-peroxy radicals; each of them generates one class of F2-IsoP [67,68] and each of the four classes consists of 16 stereoisomers, since the autoxidation and 5-exo ring closures proceed with no stereochemical control (Figure 5), resulting in a total of 64 F2-IsoPs that can be formed during peroxidation of arachidonic acid [67].

3.3. Interaction of LP Products with Amino Acids and Proteins

Carbonyl secondary products show high reactivity with cell nucleophiles, namely DNA and proteins [6,58]. The reaction with proteins is particularly noteworthy as it interferes with their biological functions in both ways and can express either cytoprotective or damaging roles [6,18]. The most important nucleophiles in peptides are thiols (e.g., the -SH of cycteine residues) and amines (e.g., the side chain -NH_2 of lysine, the inidazolyl residue in histidine, the guanidine residue in arginine) [69]. Taking conjugated alkenals like 4-HNE as a model, amino groups preferably undergo Schiff-base formation with the carbonyl, while other nucleophiles preferably undertake 1,4-Michael-type addition to the C=C; the subsequent reaction of 4-HNE with both an amine residue and a thiol residue of different proteins can result in protein cross-linking [6,58], as exemplified in Figure 8. Second-order rate constants for the reaction of 4-HNE and 4-ONE with model amino acid derivatives shows that 4-ONE is more reactive by two orders of magnitude with thiols (e.g., reaction with GSH at 23 °C, pH 7.4: k_2 is 1.33 and 150 $M^{-1}s^{-1}$, for 4-HNE and 4-ONE, respectively), while it is only about one order of magnude more reactive with other nucleophiles (e.g., reaction with N-acetylhistamine at 23 °C, pH 7.4: k_2 is 2.1×10^{-3} and 2.2×10^{-2} $M^{-1}s^{-1}$, for 4-HNE and 4-ONE. respectively) [69]. Protein alkylation is the main process responsible for the biological activity of LP secondary products (vide infra).

Interestingly, in parallel to the influence of lipid oxidation products in the structure and functions of proteins, oxidative damage to peptides and proteins, particularly concerning cysteine and methionine residues, would cause radical damage to lipids, altering their stucture and functions [70–73].

A main mechanism for peptide-to-lipid damage transfer was extensively investigated in Chatgilialoglu-Ferreri's group and consists of cis–trans isomerization of C=C double bonds in MUFA and PUFA, caused by reversible addition of thiyl radicals (RS•) and sulfhydryl radicals (HS•/$S^{-\bullet}$) generated from cystein, as exemplified in Figure 8E. Thiyl radicals undergo fast addition to cis C=C double bonds with a bimolecular rate constant of 1.6×10^5 $M^{-1}s^{-1}$ (for methyl oleate), but the adduct undergoes rapid fragmentation, which yields the trans isomer approximately 10-fold faster than the cis isomer (viz. 1.7×10^7 s^{-1} vs. 1.6×10^8 s^{-1} for methyl oleate), thereby converting the less stable cis isomer of the lipids into the stable trans isomer [71]. Although formally this process does not belong to LP, as the lipids have not changed their oxidation state, isomerization, occurring, for instance, in cell membrane phospholipids, alters membrane function similarly to LP (vide infra) by affecting its fluidity; additionally, the two processes are interconnected, and both are involved in the development and control of ferroptosis [74].

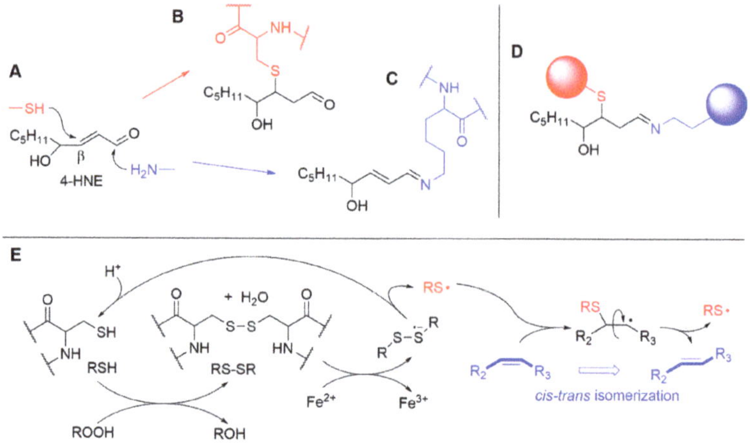

Figure 8. Examples of lipid–peptide interaction in peroxidation: (**A**) favorite nucleophilic attack sites in 4-HNE; (**B**) resulting Cys-HNE adduct; (**C**) Lys-HNE adduct; (**D**) protein cross-linking by 4-HNE; (**E**) lipid cis–trans isomerization caused by the reversible addition of Cys-derived thiyl radical.

4. Biological Consequences of Lipid Peroxidation

Lipid peroxidation, oxidative damage to other essential biomolecules, like proteins and DNA, and the generation in biological systems of oxidizing radicals not adequately balanced by antioxidant defenses have long been associated with different degenerative pathologies, particularly with cardiovascular diseases and cancer [1–3], and have been referred to as oxidative stress (OS), although the actual causal interplay has long escaped full rationalization. Understanding has enormously progressed in recent years and a number of valuable reviews have addressed it [4,6,10,14,15,18,57]. Therefore, we will limit this section to a brief overview, aimed at highlighting the potential of antioxidant strategies, presented in Section 5.

It is now recognized that LP is necessary to cell physiology, to maintain redox homeostasis, and it has essential signaling functions and protective roles. However, dysregulation causes alteration of cell metabolism and/or cell death and it is involved in a multitude of pathologic conditions.

4.1. LP and Membrane Integrity and Functions

Modification of cell membranes is the most direct biological consequence of LP. Accumulation of hydroperoxides in the phospholipid bilayer alters their physical–chemical character by increasing the polarity of phospholipids and by altering their aggregation and hence membrane structure. Studies on giant unilamellar vesicles (GUVs) as model systems showed that accumulation of hydroperoxides causes an increase in the area which is proportional to the extent of peroxidation, and it is accompanied by a decrease in thickness of the membrane. Additionally, changes in the liquid–gel transition temperature and phase separations in the lipid domain were observed [75]. Oxidation of cholesterol extends those changes but, alone, is insufficient to cause them. On extending the oxidation and releasing small-chain secondary products, permeability increases, and the integrity is compromised with creation of pores [75]. The mechanical properties of the membrane are significantly altered, with the stretching module decreasing linearly with the conversion into peroxidized lipids, from 200 mN m^{-1} for native lipids to 50 mN m^{-1} for fully oxidized bilayers [76]. Mechanical changes and increases in the surface area were judged to be the consequence of changes in phospholipid chain conformation, caused by the introduction of polar -OOH groups, with increased probability to settle at the lipid–water interface [75]. When changes to membrane structure are sufficiently extended to cause pores or altered permeability, the unavoidable consequence is cell death, which is one of the main effector strategies of ferroptosis (vide infra).

4.2. LP and Cell Signaling

ROS and LP electrophiles (LPEs) are two main families of players in non-enzymatic cell signaling; among them, LPEs have a prominent role due to generally higher persistency and ability to diffuse from the immediate surroundings of the generation site [57]. Their main signaling mechanism is by forming protein adducts. Due to the different location of the reacting amino acid residue and to the protein to which the amino acid pertains, this would result in modifying cell metabolism in different ways, both by activating and deactivating specific pathways [18]. At low levels (0.1–1 μM), 4-HNE produces adducts that promote the biosynthesis of antioxidant and detoxifying enzymes, thereby having a protective role. The main mechanism is the reaction with Cys residues of cytoplasmic Keap1 protein, which is part of the nuclear factor erythroid 2-related factor 2 (Nrf2) signaling system. A dimer of Keap1 protein normally holds the transcription factor Nrf2 in the cytoplasm, inhibiting its function. Upon alkylation by 4-HNE, a change in Keap1 conformation causes the release of Nrf2, which translocates into the nucleus and binds to Maf small protein. The resulting heterodimer binds DNA at antioxidant-responsive element (ARE) sites to upregulate genes coding for antioxidant enzymes, such as thioredoxin reductase, and for proteins that cause increased glutathione (GSH) levels [6,18]. However, at higher doses (50 μM), 4-HNE also creates adducts with histones, modifying protein expression at the DNA level, e.g., it blocks histone H2A acetylation, thereby impairing gene expression and contributing to the vulnerability of DNA to apoptosis. 4-HNE also affects the activity of NFκB transcription factor in opposite directions depending on the dose, with resulting anti-inflammatory of proinflammatory effects. The formation of MDA–protein adducts is associated with a proinflammatory reaction throughout the entire organism, via the activation of Th17 lymphocytes, which triggers autoimmune reactions [77]. Additionally, MDA leads to collagen cross-linking causing a loss of elasticity and disturbance in tissue remodeling, with systemic consequences, particularly on the blood vessel system [78]. The above are just examples of the complex pattern of influence of the metabolic machinery elicited by LPEs' signaling role.

Lipid hydroperoxides themselves are signaling factors and are key players in inflammatory processes. Besides being the primary products of LP formed in the propagation stage, they are also formed by direct oxidation of lipids by singlet oxygen (see: Section 2.1) as a form of defense from such highly reactive species and by enzymatic processes, most notably by lipoxygenases (LOXs), a family of dioxygenases converting lipids to their hydroperoxides that can further transform into other hormone-like signaling molecules (e.g., tromboxanes, leucotrienes, lipoxines and resolvines), which modulate the activity of the innate immune system (e.g., the action of macrophages) in both ways, exerting proinflammatory or anti-inflammatory roles [79]. Also, untransformed LOX-generated ROOHs primarily exert a signaling role towards the immune systems (typically proinflammatory), they directly alter membrane structure and permeability (see: Section 4.1) and might indirectly modify cell gene expression by changing the overall cell redox state [79]. LOX-generated hydroperoxides have also been attributed a key role in regulation of ferroptosis (see: Section 4.5) [80].

4.3. LP Association with Cancer and Apoptosis

Association between different markers of oxidative stress and cancer has been extensively reported in recent decades. Metabolic requirements of cancer cells are higher due to their rapid proliferation, which makes mitochondrial ROS generation higher than in non-transformed cells. Redox status and ROS are also of particular importance in tumor cell signaling associated with cancer progression. Therefore, compared to healthy non-malignant cells, cancer cells have altered redox homeostasis and higher levels of LP; in many types of tumor cells, these higher levels support their growth, proliferation and survival [81]. Therefore, higher levels of LP products and biomarkers are expected in association with cancer.

For instance, in a large population-based cohort study with 14 years of follow-up, it was found that a positive association exists between 8-isoprostane urinary levels and occurrence of lung cancer but not of other cancers. Actually, in smokers 8-isoprostane levels were inversely correlated with occurrence of prostate cancer, indicating a protective effect, possibly because higher OS levels associated with smoking, combined with testosterone, activate apoptosis, thereby protecting from cancer development in testosterone-sensitive tissues [68]. A recent study on the relationship between 4-HNE and prostate cancer based on metabolic profiling using LC-ESI-QTOF-MS and GC-EI-Q-MS revealed increased 4-HNE–protein adduct levels in the plasma of cancer patients while there were no 4-HNE–protein adducts in prostate carcinoma tissue [60]. Interestingly, while higher levels of LPEs are typically associated with cancer, some of them also promote defensive mechanisms.

A main defense mechanism against cancer development is apoptosis, a physiological mechanism of programmed cell death that can be triggered both by *death receptor* activation and by metabolic changes in the cell [64]. 4-HNE can modify the structure of the MDM2 protein, breaking down the MDM2–p53 complex. As a consequence, the p53 factor is activated and translocates to the nucleus, where it promotes the transcription of proapoptotic proteins, such as Bax and caspase-3 [66]. 4-HNE may also interact with other proteins; for example, with AKT kinase, which has inhibiting roles, thereby leading to a significant increase in apoptosis. 4-HNE and 4-hydroxyhexenal further promote apoptosis by other mechanisms, which have recently been reviewed [66].

4.4. LP and Neurological Disorders

The number of deaths associated with neurological disorders including neurodegenerative diseases like amyotrophic lateral sclerosis and Parkinson's, Alzheimer's and Huntington's diseases, as well as other diseases that involve neurodegenerative processes, like diabetes, have increased worldwide with time [82]. Additionally, these diseases have an important social impact as they compromise patients' quality of life. Although the causes of neurodegeneration are often unclear, OS and LP, along with inflammation, are recognized as prevailing molecular pathways leading to these diseases [82]. Some known inducers of neurodegeneration like arabin and methylmercury have been found to cause it through induction of LP and reduction of antioxidant defenses, or at least in parallel to inducing them [82,83], acrylamide induces neurodegeneration that can be reversed by antioxidant plant essential oils, while the link between retinal degeneration and OS is supported by the protective role of cytochrome b5 overexpression, which reduces lipid peroxidation [82]. Among the mechanisms linking LP to neurodegeneration, 4-HNE was identified as a trigger and its levels in an animal model increased following induction of neurodegenerative lesions [82].

In the animal model, both Parkinson's disease and Friedreich's ataxia were associated with dysregulated mitochondrial metabolism and increased lipid peroxidation, and they could be ameliorated with antioxidants. In Huntington's disease (HD), an increase in 4-HNE levels in different brain areas was also observed [82].

Alzheimer's disease (AD) is certainly the most investigated in connection with LP. In human studies, excessive ROS generation was reported under AD conditions, and it was found to be associated with β-amyloid aggregation, while, in turn, β-amyloid plaques produce ROS and LP, and increased OS biomarkers are found in association with AD [84]. Histological analysis revealed co-localization of LP secondary products and β-amyloid plaques in the brain, while there is a relationship with oxidized low-density lipoprotein (LDL) levels in AD [84]. Owing to this strong association, LP markers can be used both as diagnostic tools for AD and to monitor the efficacy of AD treatments [84]. Formation of Schiff bases of the cholesterol oxidation product 3-β-hydroxy-5-oxo-5,6-secocholestan-6-al and β-amyloid is known to be amyloidogenic and it was found that this modification, occurring in specific sites, has a different influence on aggregation kinetics, particularly adducts at Lys-16, obtained at physiological levels of β-amyloid, enabling both kinetic and thermodynamic aggregation that are sufficient to form neurotoxic lesions [65]. Based on the above, antioxidants are among the most promising therapeutic approaches for AD. Among them, metal chelators (acting as preventive antioxidants, vide infra), indirect antioxidants acting through the activation of the Nrf2 transcription factor, which induce de novo synthesis of antioxidant enzymes, and plant-based antioxidants like curcumin, resveratrol, capsaicin, epigallocatechin gallate (EGCG) from green tea and quercetin have been more extensively investigated, with favorable clinical outcomes [15]. Naringenin, a flavanone obtained mainly from grapefruit (*Citrus paradisi*) and other citrus fruits, possesses neuroprotective activity, along with anti-inflammatory effects, and it has beneficial effects on learning and memory in the AD model, through the mitigation of lipid peroxidation [15].

4.5. LP and Ferroptosis

Ferroptosis, first identified by Stockwell in 2012 [8], is the last of several forms of programmed cell death, differing from apoptosis, which have been described in recent decades. It owes its name to the initial attribution as iron-dependent oxidative cell death, promoted by a void in the antioxidant defenses based on Cys, which could be artificially induced by erastin, an inhibitor of cystine uptake working by irreversible blocking of the cystine/glutamate antiporter system (x_c^-), and it could be prevented by ferrostatin-1 [8], an iron-chelating aromatic amine that is instead unable to inhibit apoptosis. The discovery that, in mice, the knockout of selenoenzyme glutathione peroxidase 4 (GPX4)—a specific phospholipid hydroperoxide glutathione peroxidase that differs from other GPXs preferably reducing hydrogen peroxide—would cause cell death by inducing massive ferroptosis, which would be inhibited by liproxstatin-1, allowed recognizing the key role of GPX4 as a controller of ferroptosis [14,85]. The role of iron in ferroptosis appears linked to its ability to initiate LP via the Fenton reaction (Figure 1), by decomposing phospholipid hydroperoxides, which in turn could be the product of LP or formed enzymatically, e.g., by lipoxygenase (LOX) [11]. Although it was initially identified as a key player in the definition of ferroptosis, the actual necessity or dominance of iron in control of ferroptosis remains unclear [86]. Instead, the subsequent recognition that ferrostatin-1 and lipoxtatin-1 would act essentially as chain-breaking antioxidants (vide infra) in the inhibition of ferroptosis [12], and that other potent antioxidants like phenoxazines, diarylamines and nitroxides would also inhibit ferroptosis, points toward the prominent role of LP in ferroptosis, besides it being initiated by iron or by other processes [11,12,86]. Other antioxidants such as vitamin E [86] and vitamin K [87] were also found to inhibit ferroptosis. In other words, it is LP that drives the cell death [11] and the many degenerative diseases that are linked to ferroptosis. Regulation of LOX biosynthesis and of the LOX-mediated accumulation of hydroperoxides was identified as a key point in the control of ferroptosis [80]; however, its actual significance has lately been challenged by showing that, in different LOX-overexpressing cells, only some well-known LOX inhibitors were able to counteract ferroptosis and, incidentally, they were all effective chain-breaking antioxidants, likely acting through inhibition of LP [88]. Ferroptosis is related to cancer, which can be modulated by ferroptosis inducers or inhibitors [89]. Ferroptosis is also associated with neurodegenerative conditions such as AD, Parkinson's disease, Friedreich's ataxia and Huntington's disease [90], cardiovascular diseases and diseases of the urinary system [91]. Therefore, modulating ferroptosis in both directions, either by triggering it or by inhibiting it with antioxidants, might offer a potent tool in the therapy of many such diseases [90,91].

5. Antioxidants

Antioxidants are a very heterogeneous class of compounds, small molecules and enzymes that share the task of protecting oxidizable molecules or materials from oxidative transformation. In the biological context, the reference oxidative process is lipid peroxidation (LP); therefore, antioxidants are typically defined and discussed on the basis of their ability to prevent, slow down or block LP [3,21]. Based on where and how they interfere with the LP radical chain, antioxidants are classified as *preventive*, if they impair the initiation process, and *chain-breaking* if they block or slow down the propagation, while a new category, the *termination-enhancing* antioxidants, was recently introduced by our group, to include those molecules, such as some terpenes and terpenoids from essential oils, which act by favoring the radical-chain termination without actually impairing propagation [21,92,93].

5.1. Preventive Antioxidants

Antioxidants in this class are extremely heterogenous, ranging from small molecules to complex enzyme systems. They prevent the onsetting of LP by preventing the formation of the initial alkyl radical, which would give rise to the propagation cycle. Since triggering events can be different, their prevention occurs by a multitude of mechanisms, each addressing one specific mode of radical initiation. For instance, UV filters (sunscreens) might protect the skin by photoinitiated LP (i.e., preventing the action of photosensitizers) [21], but they would not affect radical initiation due to metal-catalyzed decomposition of peroxides and hydroperoxides (e.g., via the Fenton reaction, see Section 2.1). The latter is a prevailing mechanism of initiation, which explains the evolutionary development of enzyme systems like peroxidases aimed at clearing the biological medium of peroxides and hydroperoxides. Among them, catalase (CAT), which catalyzes the dismutation of H_2O_2 into water and molecular oxygen, and thiol peroxidases like GPX, which reduce hydroperoxides and hydrogen peroxide to alcohols or water, respectively, using a cysteine-derived thiol (e.g., the tripeptide glutathione) as the sacrificial reducing agent, can be classified as preventive antioxidants, since they "fuel down" the initiation process [21]. Along this line, superoxide dismutases (SODs), clearing superoxide radicals, and glutathione reductase (GR), which reduces oxidized glutathione using NADPH as the reducing agent, can also be assigned to the same class [21].

Metal chelators are also important members of this community. They block transition metal ions like iron and copper in a less redox-active form, allowing their indispensable presence in biological systems, yet neutralizing their ability to undergo Fenton-like chemistry. They can do so in case the two states participating in the redox cycle (e.g., Fe^{2+} and Fe^{3+}) form with the chelator complexes with sufficiently different stability. For instance, most chelators have higher affinity for Fe^{3+} than for Fe^{2+}, therefore chelated Fe^{3+} would have a reduction potential diminished by the difference in formation free energy ($\Delta\Delta G_{form} = \Delta G_{form}{}^{3+} - \Delta G_{form}{}^{2+}$) between Fe^{3+} and Fe^{2+} complexes, which would make the redox cycle thermodynamically not viable, blocking iron in the oxidized form. Examples of chelating agents of biological relevance are transferrin and ceruloplasmin [21].

5.2. Chain-Breaking Antioxidants

Chain-breaking antioxidants, also named radical trapping antioxidants (RTAs), are certainly the biggest class of small-molecule antioxidants in all fields of application, and *phenols* are the prototypical members [94,95]. In cells, α-tocopherol (α-TOH) is certainly the most important representative and the most potent lipophilic antioxidant, often used as the benchmark in antioxidant research [3]. The identification of the mechanism of its antioxidant action and of the relative reactivity of different tocopherols (α, β, γ, δ) highlighting the parallelism between vitamin E activity in vivo and the relative antioxidant activity [3], allowed understanding of the structure–activity relationship governing its properties and linked vitamin E to the knowledge developed with simpler synthetic phenolic antioxidants [3,21,94–97].

RTAs act by trapping alkylperoxyl radicals, thereby competing with chain propagation. In order to be effective, the reaction of RTAs with ROO• must be (much) faster than the rate of chain propagation, i.e., the rate at which peroxyl radicals attack the lipids (Equation (3)). Therefore, chain-breaking antioxidant activity works on purely kinetic bases, being a competition between two reactions. The rate constant of inhibition k_{inh} for trapping ROO• by the antioxidant (Equation (3)) is the key parameter deciding the effectiveness of an antioxidant and, since the antioxidant has normally much lower concentration that the oxidizable substrate it is called to protect (e.g., membrane lipids), $k_{inh} >> k_p$ is necessary. For instance, to protect linoleic acid residues ($k_p = 62\ M^{-1}s^{-1}$, see Table 1) with 1% antioxidant relative to lipids, $k_{inh} \geq 10^4\ M^{-1}s^{-1}$ is required [21].

$$V_{trapping} > V_{propagation} \Rightarrow k_{inh}[Antioxidant] > k_p[Substrate] \quad (3)$$

It should be noted that only the trapping of ROO• is relevant to afford antioxidant protection, as they are the sole chain-propagating species (with the exception of HOO•, see Section 2.2). In the case of phenols like α-TOH, the reaction with ROO• occurs by formal HAT from the phenolic -OH group, to afford a stabilized phenoxyl radical α-TO•, which is normally unable to propagate the radical chain. Instead, it is sufficiently long-lived to "wait" in solution to trap a second ROO•, normally by addition to the aromatic ring, as illustrated in Figure 9. Therefore, one molecule of (mono)phenolic antioxidant can break two radical chains, or it has a *stoichiometric factor n* = 2. This is the second most important parameter in quantifying antioxidant performance.

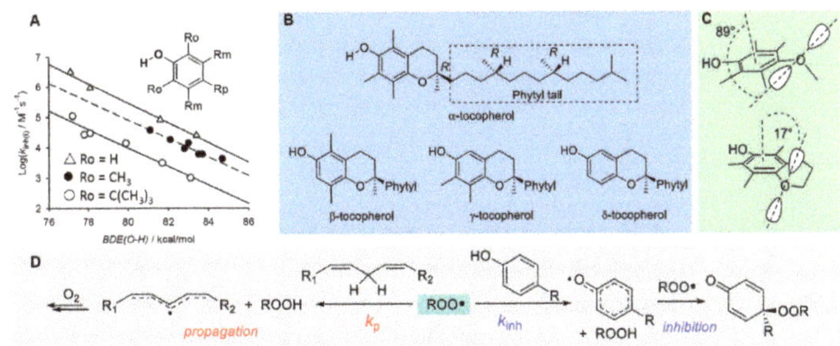

Figure 9. Linear free energy correlations between rate constant for ROO• trapping and BDE_{OH} for phenolic antioxidants bearing different ring substituents (**A**); structure of the main tocopherol congeners forming vitamin E (**B**); role of stereoelectronics in maximizing the ED contribution of -O- substituent included in the chroman ring or free to rotate (**C**); and representation of the kinetic competition involving chain-breaking antioxidants (**D**).

The rate of peroxidation during full inhibition by an antioxidant AH is given by Equation (4), where R_i is the rate of radical initiation.

$$-\frac{d[O_2]}{dt} = \frac{d[ROOH]}{dt} = \frac{k_p[RH]R_i}{nk_{inh}[AH]} + R_i \quad (4)$$

The rate constant k_{inh} depends on the BDE of the phenolic O-H being broken and of the ROO-H bond of newly formed hydroperoxide, which is worth about 88 kcal/mol [21,94]. Hence, phenols with lower BDE would give a more exothermic and faster reaction (BDE is 77.1 kcal/mol for α-TOH [94]). There exist linear free energy correlations between the BDE and k_{inh} for phenolic antioxidants, which also account for steric hindrance in *ortho* to the reactive -OH, as depicted in Figure 9 [21,94,95]. Substituents in the phenolic ring determine the reactivity with peroxyl radicals, according to their electronic properties: electron-donating groups (EDGs) decrease the bond dissociation enthalpy (BDE) of the phenolic O-H, making the HAT to peroxyl radicals faster, while electron-withdrawing groups (EWGs) have the opposite effects [21,94]. The effect is more marked when the substituent is in the *ortho* or *para* position, conjugated with the -OH group, while it is less important in the *meta* position and, in a first approximation, the effect of each substituent in decreasing the BDE is additive [94]. Stereoelectronic effects are also important; indeed, α-TOH has lower BDE and much higher k_{inh} than equivalently substituted phenols, due to the almost parallel alignment of one lone pair of the RO- substituent with respect to the axis of the aromatic π-system, forced by the geometric constraints in the chroman structure

(Figure 9). The performance of synthetic or natural monophenolic antioxidants, including currently popular compounds like bakuchiol [95] and curcumin derivatives [37], can be rationalized with the above concepts. The values of BDE, k_{inh} and n for some representative antioxidants are listed in Table 2.

5.2.1. Insertion of N(s) in the Phenolic Ring: 3-Pyridinols and 5-Pyrimidinols

Research aimed at the rational design of phenolic antioxidants that would be more effective than α-TOH, using the strategies outlined above, e.g., by inserting stronger ED substituents like the amino group and forcing its conjugation by geometrical constraints, eventually showed its limits. In parallel to decreasing the BDE and hence increasing the reactivity with ROO•, ED substituents also decrease the ionization potential (IP) of the molecule, to the point it directly reacts with molecular oxygen by electron transfer (ET) to yield the superoxide radical ($O_2^{-\bullet}$) (Figure 10). Therefore, these compounds are pro-oxidant and toxic, besides being terribly unstable under normal handling conditions [98]. In a joint effort with Pratt's group, we found that replacing one or two >CH moieties in the aromatic ring with nitrogens in non-conjugated positions (i.e., 3 and 5), to afford the corresponding 3-pyridinols and 5-pyrimidinols, would increase the BDE_{OH} by about 1.1 kcal/mole per nitrogen (ΔBDE of +1.1 kcal/mol and +2.5 kcal/mol for 3-pyridinols and 5-pyrimidinols, respectively), while the IP would increase 10-fold more (ΔIP of +11 kcal/mol and +24 kcal/mol for 5-pyrimidinol) [98,99]. In other words, insertion of heteroaromatic nitrogen(s) expands the thermodynamic gap between the two competing reaction pathways, HAT to ROO• and ET to O_2, while maintaining the same substituent effects known for phenols [99]. Two features favored this strategy in antioxidant design: (1) the slightly higher BDE_{OH} that would disfavor the reactivity can be compensated with stronger ED substituents, like the amino group (inaccessible in the phenolic series), without compromising the stability toward oxygen [99]; (2) 3-pyridinols and 5-pyrimidinols are more reactive toward ROO• than phenols having an identical BDE, owing to the intervention of polar effects stabilizing the transition state [100]. This led to the discovery of a wealth of potent RTAs (some examples in Table 2), many of which easily outperformed α-TOH despite the much simpler structure and easier synthetic accessibility [98–106], which also included the most potent chain-breaking antioxidant ever reported, able to quench peroxyl radicals at a diffusion-controlled rate (Pyr-7 in Figure 10) [101].

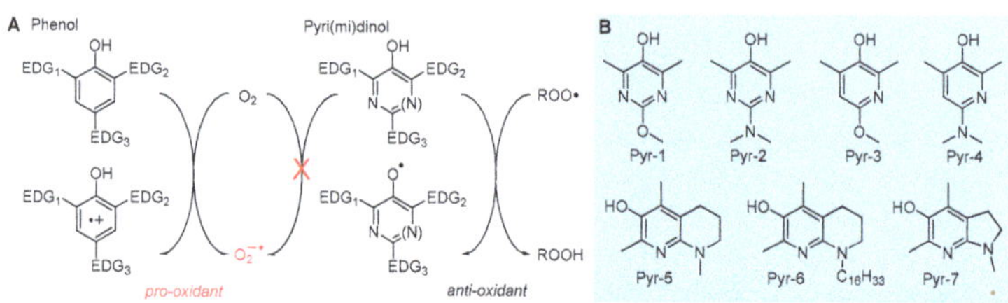

Figure 10. Different reactivity with oxygen and ROO• of electron-rich phenols and pyri(mi)dinols (**A**) and examples of 3-pyridinol and 5-pyrimidinol antioxidants (**B**).

5.2.2. Solvent and Medium Effects in Chain-Breaking Antioxidant Activity

The rate of radical trapping by phenols and most other RTAs is subject to solvent effects [107], which must be carefully considered in testing antioxidant activity [108] or in planning antioxidant strategies. In hydrogen-bond-accepting (HBA) solvents, k_{inh} is apparently decreased by as much as two orders of magnitude, depending on the actual HBA ability of the solvent (not its polarity) [107]. The effect does not depend on the radical reacting with the antioxidant (e.g., R•, RO•, ROO•) but on the ability of the antioxidant to act as a hydrogen-bond donor (HBD) to the solvent [109], since the effect is due to formation of an antioxidant–solvent complex in which the "active" -OH group (or, in general, the -XH group transferring the H-atom to radicals) is "blocked" by H-bonding to the solvent, and only the fraction of non-H-bonded antioxidant at equilibrium is available to react with radicals (Figure 11) [106,107]. Different antioxidants will have different sensitivity to solvent effects.

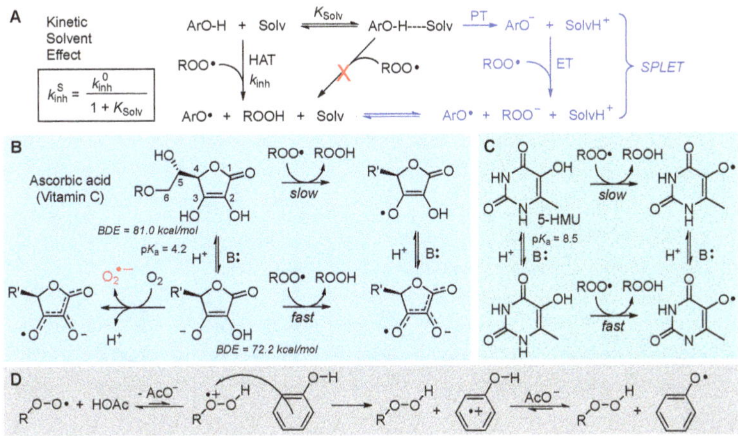

Figure 11. Solvent effects on the reactivity of antioxidants, showing the possible mechanistic change in protic solvents (**A**), pH-dependent antioxidant chemistry of ascorbic acid (**B**) and of 5-hydroxymethyluracil (**C**) and acid catalysis in the antioxidant activity of phenols in polar solvents (**D**).

The HBD ability of the antioxidant and the HBA ability of the solvent can be quantified by Abraham's solvatochromic parameters $\alpha_2{}^H$ and $\beta_2{}^H$, respectively [110,111], and solvent effects can be accounted for quantitatively by the Ingold–Snelgrove equation (Equation (5)), which allows predicting k_{inh} in any solvent ($k_{inh}{}^S$) once it is known in a non-H-bonding solvent ($k_{inh}{}^0$) [112].

$$Log\left(k_{inh}^{S}/M^{-1}s^{-1}\right) = Log\left(k_{inh}^{0}/M^{-1}s^{-1}\right) - 8.3 \times \alpha_2^H \beta_2^H \quad (5)$$

Triglycerides [24] and phospholipids [107] have strong HBA groups (i.e., the C=O and P=O), which visibly decrease the antioxidant protection by RTAs. In a heterogenous system of lipid particles in water, however, an additional factor affects the kinetics of antioxidant protection: the compartmentalization, which may render the rate of exchange of radicals and antioxidants among particles rate limiting. For instance, the value of k_{inh} for α-TOH measured in micelles of methyl linoleate and in phospholipid liposomes is, respectively, two orders of magnitude and three orders of magnitude lower than in homogenous non-H-bonding solution [21,107].

The decrease in the rate of formal HAT reaction to radicals in HBA solvents could favor other reaction mechanisms [113,114]. It has been shown that in protic solvents like alcohols, able to solvate anions, the reaction of acidic phenols (or other antioxidants) with radicals could take place via a mechanism named sequential proton loss electron transfer (SPLET), depicted in Figure 11, consisting in proton transfer (PT) to the solvent followed by ET from the electron-rich anion of the antioxidant to oxidizing radicals [113]. In water, this mechanism is favored in alkaline pH and can result in an increase in the rate of peroxyl radical trapping [114].

This change in the reaction mechanism does not occur only with phenols but controls the antioxidant activity of other biological antioxidants like 5-hydroxylmethyluracile (5-HMU) [115] and, most notably, of *ascorbic acid* and its derivatives [116,117], boosting their performance in the presence of a base (Figure 11). In the case of ascorbic acid, deprotonation also opens the way to its direct reaction with O_2 to yield superoxide radicals, accounting for its instability in aqueous solution. This is also a pro-oxidant process that partly counteracts the antioxidant behavior; as a consequence, the overall antioxidant performance of ascorbic acid largely depends on the experimental settings, and it can be significantly improved by inclusion in inert nanocarriers [117].

Opposite to the base effect outlined above, phenolic (and pyridinolic/pyrimidinolic) antioxidants have been found to undergo acid catalysis in their antioxidant activity. The behavior is only observed in polar solvents and can largely boost the reactivity, depending on the phenol. It is due to a change in reaction mechanism involving the partial protonation of the ROO• radical to form a highly oxidizing species that takes one electron from the phenol and forms the antioxidant's radical cation, which undergoes rapid acid–base exchange to afford the usual reaction products [118], as illustrated in Figure 11.

Table 2. Bond dissociation enthalpy (BDE), rate constants for ROO• radical trapping k_{inh} and stoichiometric factor n for representative antioxidants at 303 K in chlorobenzene.

Entry	Compound	BDE kcal/mol	k_{inh} $M^{-1}s^{-1}$	n	Ref.
1	α-Tocopherol	77.1	3.2×10^6	2.0	[3,21]
2	β-Tocopherol	-	1.3×10^6	2.0	[3,21]
3	γ-Tocopherol	-	1.4×10^6	2.0	[3,21]
4	δ-Tocopherol	-	4.4×10^5	2.0	[3,21]
5	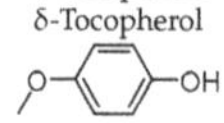	81.7	2.7×10^5	2.0	[21,94]
6	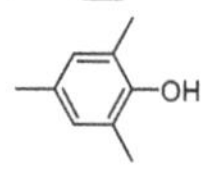	81.6	8.5×10^4	2.0	[21,94]
7	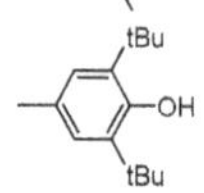	79.9	1.4×10^4	2.0	[21,94]
8	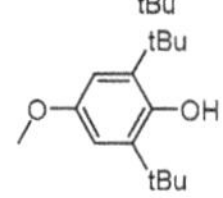	77.2	1.1×10^5	2.0	[21,94]
9	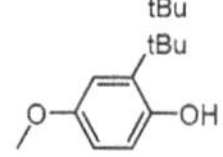	80.3	6.4×10^5	1.8	[21,94]
10	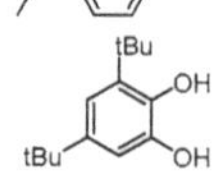	78.2	1.1×10^6	2.0	[94,119]
11	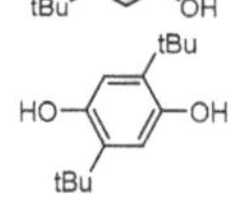	79.2	1.6×10^6 [1]	0.3	[120]
12	Quercetin	-	5.5×10^5	2.1	[119]
13	Pyr-1 [2]	81.4	2.1×10^5	2.0	[100]
14	Pyr-2 [2]	77.1	8.6×10^6	2.0	[100]
15	Pyr-3	78.9	4.4×10^5	2.1	[106]
16	Pyr-4	75.9	1.6×10^7	2.0	[101]
17	Pyr-5	75.2	8.8×10^7	1.3–2.0	[101]
18	Pyr-6	75.2	8.8×10^7	~2	[103]
19	Pyr-7	74.3	2.8×10^8	~2	[101]
20	α-Selenotocopherol	78.1	1.2×10^6	1.9	[121]
21	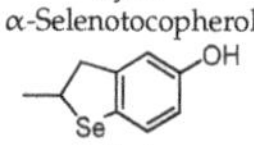	81,6	3.8×10^5	2.0	[122]
22	Te-1	-	9.2×10^6	0.4 [6]	[123]
23	Te-2	78.9	1.0×10^7	0.4 [6]	[124]
24	Te-3	-	1.6×10^6	0.3 [6]	[125]
25	Te-4	-	1.0×10^7	0.4 [6]	[125]
26	Phenoxazine [2,3]	76.1	2.9×10^7	5	[94]
27	Phenothiazine [2,4]	78.2	8.8×10^6	1.8	[94]
28	Dia-1	78.8	3.4×10^7	>2	[126]
29	Dia-2	79.0	3.7×10^7	>2	[126]
30	Ferrostatin-1 [5]	-	3.5×10^5	2.0	[11]
31	Liproxstatin-1 [5]	-	2.4×10^5	1.9	[11]

[1] k_{inh} is doubled by a statistical factor. [2] Data at 50 °C. [3] Values at 37 °C are: k_{inh} 4.1×10^7 $M^{-1}s^{-1}$, n = 2.3 from ref. [11]. [4] Values at 37 °C are: k_{inh} 8.0×10^6 $M^{-1}s^{-1}$, n = 2.1 from ref. [11]. [5] Data at 37 °C. [6] Without thiols.

5.2.3. Polyphenols and Flavonoids

Hydroquinone (1,4-dihydroxybenzene) and catechol (1,2-dihydroxybenzene) are the two key structural motifs in polyphenols and flavonoids. Even the simple unsubstituted molecules have high reactivity with ROO• radicals due to the strong ED character of each -OH group toward the other in the conjugated position [21,120]. Both hydroquinones and catechols are nominally able to trap two ROO• radicals by stepwise transfer of the two O-H. However, their relative position (*ortho* or *para*) deeply affects the antioxidant behavior [120], as shown in Figure 12. In hydroquinones (e.g., ubiquinol or coenzyme QH_2), the semiquinone radical formed upon transfer of the first H-atom has a very low BDE for the second O-H (~50 kcal/mol) and it is rather acidic (pKa ~ 4) [120]; therefore, via different mechanisms which have been discussed in detail [120] the semiquinone radical can react directly with O_2 to form a hydroperoxyl radical (HOO•) or superoxide radical anion ($O_2^{-\bullet}$) in lipidic medium or in water, respectively. This process, which parallels the generation of superoxide in the mitochondrial respiratory chain, has pro-oxidant action starting new oxidative chains; therefore, it partly counteracts (by shortening *n*) the otherwise very good antioxidant behavior of hydroquinones [120].

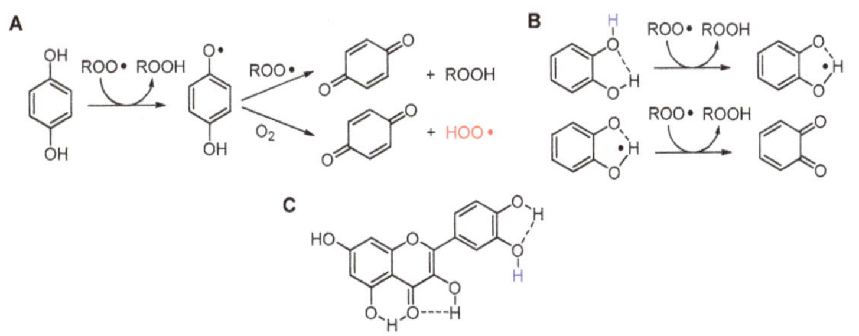

Figure 12. Radical reactivity of hydroquinone (**A**), of catechol (**B**) and structure of quercetin (**C**).

In catechols, the intramolecular H-bond makes one -OH unreactive, but it decreases the BDE of the unbound O-H, increasing its reactivity. The resulting semiquinone radical is stabilized by the intermolecular H-bonding, which prevents its reaction with O_2. Overall, even the unsubstituted catechol traps two peroxyl radicals with k_{inh} similar to δ-TOH, being an excellent antioxidant [119,127]. Not surprisingly, the catechol moiety is highly conserved in natural flavonoids (Figure 12) and, in a lipidic environment, their antioxidant activity is largely dictated by the catechol ring. Indeed, despite the large number of phenolic -OH groups in their structure, the reactivity of most polyphenols is modulated by the occurrence of intramolecular H-bonds [128]. As a consequence, flavonoids can trap more than two ROO• radicals, but with different rate constants k_{inh}, hence the trapping of the first two contributes most to the overall antioxidant performance [129]. In an aqueous environment, their reactivity is instead dictated by pH, and multiple mechanisms (PT-ET vs. HAT) can take place, as previously discussed in detail for quercetin [127].

5.2.4. Synergy among Antioxidants and Tocopherol-Mediated Peroxidation (TMP)

In nature, as in the protection of human-made materials, antioxidants are never used alone and cooperative effects among antioxidants are often the key to successful protection. The most relevant example in cells is certainly the synergic interplay between vitamin E and vitamin C. α-TOH is confined in the lipid bilayers and protects phospholipids from ROO• attack, being oxidized to an α-TO• radical, while water-soluble ascorbate reacts with α-TO• at the lipid–water interface, reducing it back to the starting α-TOH and extending its protective duration [130]. On doing so, ascorbate is oxidized to the corresponding radical; therefore, besides the recycling of α-TOH, the importance of this process resides in the "exporting of the unpaired electron" outside the lipid membrane, out of the reach of highly oxidizable lipid components. Indeed, it has been demonstrated by Bowry and

Ingold [131] that, in the autoxidation of human low-density lipoprotein (LDL), in the absence of ascorbate in the aqueous medium, the presence of α-TOH in the lipid core accelerates the peroxidation instead of blocking it—an apparently paradoxical behavior that they named tocopherol-mediated peroxidation (TMP)—due to the "excessive" resident time of α-TO• inside the lipid particle before it encounters a second ROO•, which gives time for the occurrence of even a slow and unfavored reaction of α-TO• with lipids, to (re)start the oxidative chain [131].

Interestingly, when pyridinol Pyr-6 was probed in place of α-TOH in the protection of LDL in the absence of vit C or co-antioxidants, it showed no sign of TMP, along with much higher antioxidant protection [103]. Cooperative effects similar to the vit E/vit C example also occur among phenolic and/or other antioxidants in homogenous solution, and their mechanism and efficiency have been thoroughly investigated [106,132].

5.2.5. Phenols Bearing Organochalcogen Substituents

The insertion of chalcogens heavier than oxygen (S, Se and Te) as substituents in the *ortho* or *para* position in phenolic (and pyridinolic) structures is an important strategy to afford novel antioxidants with a distinctive reactivity toward oxidizing radicals [121,133]. It has highlighted how intramolecular H-bonding can be used to afford antioxidants whose reactivity can predictably be modulated by conformational constraints and by the solvent [119]. Notable examples are lipoic acid adducts of natural catechols like hydroxytyrosol or ditocopheryl sulfides and disulfides [121].

Phenolic antioxidants cannot be recycled by thiols (unlike by ascorbate), which somewhat "wastes" the most abundant reducing source in biological systems. Looking for antioxidants that would boost this ability, we carried out a long cooperative project with Engman's group focused on Se- and Te-substituted phenols. All-*rac*-α-selenotocopherol, in which the chromanol -O- was replaced by -Se-, disappointingly showed slightly higher BDE_{OH} and slightly lower k_{inh} than the natural counterpart, with no additional property toward thiols [121]. However, some of its simplified congeners like Se-1 (Figure 13), besides being potent chain-breaking antioxidants, could be recycled by *N*-acetylcysteine (NAC) at the water–lipid interface in a biphasic lipid peroxidation model system [122].

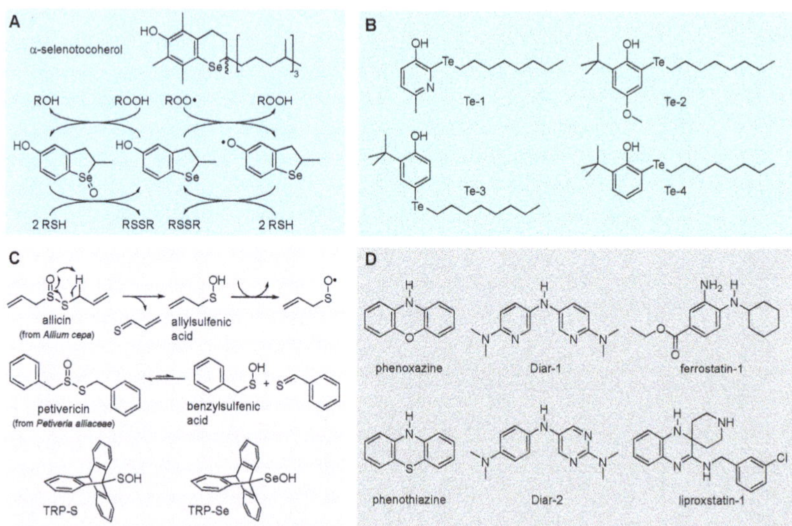

Figure 13. Examples of organochalcogen-containing phenolic antioxidants (**A**,**B**), sulfenic and selenenic acids (**C**) and diarylamine antioxidants (**D**).

Insertion of alkyltelluro (RTe-) substituents in the *ortho* or *para* position in the phenol or in the 3-pyridinol structure affords even more interesting antioxidants (Figure 13). These compounds can be recycled by thiols like cysteine derivatives both in homogenous lipid solution and heterogenous (aqueous biphasic) systems [123–125]. Additionally, even simple structures have a k_{inh} for trapping ROO• radicals much higher than reference α-TOH and their reactivity does not respond to the correlation with the BDE_{OH} reported in Figure 9 or to the solvent effect described in Figure 11. This proves a different reaction mechanism involving an oxygen-atom transfer from ROO• to RTe-, for which we refer to the original literature [125]. Interestingly, the efficiency of regeneration by thiols is much lower than 1, meaning that the same compounds could also be used as pro-oxidant agents in biological systems, acting through the depletion of Cys-derived thiols like glutathione [125]. This is particularly intriguing in the light of the role that this mechanism has in triggering ferroptosis [8].

5.2.6. Sulfenic and Selenenic Acids

Allium plants have long been regarded as a rich source of antioxidants, particularly owing to the distinctive content of sulfurated volatile components. Among them, thiosufinates like allicin and petivericin (Figure 13) have been demonstrated to be potent antioxidants [134,135]. This is due to the release of unstable sulfenic acids (RSOH), which would rapidly trap ROO• radicals to afford stabilized sulfenyl radicals (RSO•). Owing to their instability that makes their isolation prohibitive, the reactivity of sulfenic acids has remained largely unknown, despite their major relevance in biological systems, since they are involved in cellular redox homeostasis, being formed upon oxidation (e.g., by H_2O_2) and fragmentation of cystine. Only recently did the synthesis of stable tripticenesulfenic acid (TRP-S, Figure 13) shed light on their antioxidant chemistry [134]. With a BDE_{OH} as low as 71.9 kcal/mol, they trap ROO• radicals with k_{inh} in the range of 3×10^6 $M^{-1}s^{-1}$ from sterically hindered TRP-S to 3×10^7 $M^{-1}s^{-1}$ for benzylsulfenic acid from petivericin and up to ~1×10^8 $M^{-1}s^{-1}$ for allylsulfenic acid from allicin, i.e., they are among the most potent antioxidants in nature [135].

Their heavier homologues, the selenenic acids (RSeOH), are involved in the redox cycle of the GPX enzyme; however, they are even less stable than sulfenic acids. Synthesis of hindered tripticeneselenenic acid (TRP-Se, Figure 13) allowed clarifying their properties and reactivity, showing a higher BDE_{OH} (80.9 kcal/mol) compared to sulfenic acids, yet having unexpectedly high reactivity toward ROO•, i.e., k_{inh} = 1.7×10^5 $M^{-1}s^{-1}$ for sterically hindered TRP-Se (in PhCl at 30 °C) [136]. This is 18-fold lower than that of TRP-S despite the 9 kcal/mol higher BDE [136]. Unfortunately, the instability of these compounds hampers their use as antioxidants, unless they can be generated in situ from suitable precursors.

5.2.7. Aromatic Amines and Diarylamines as RTAs

Aromatic amines such as diarylamines are, after phenols, the main class of chain-breaking antioxidants. Owing to the higher thermal stability and lower reactivity compared to phenols, the most common substituted diphenylamines have mainly been used as antioxidants in high-temperature processes. However, their tricyclic analogues, such as phenoxazine and phenothiazine, have instead very high antioxidant activity at close to ambient temperature, owing to better conjugative stabilization of the aminyl radical formed upon quenching ROO• [21]. Insertion of heterocyclic nitrogens in the structure of diphenylamine, i.e., extending the same design strategy previously described for phenols (see Section 5.2.1), has created a large family of outstanding antioxidants, containing up to four heterocyclic nitrogens per molecule and a wealth of ED substituents, with large structural variability [98,126,137,138]. Many such molecules largely outperform reference α-TOH in terms of k_{inh}, in some cases having very low to negligible activation energy for ROO• trapping [126], implying negligible temperature dependence of the antioxidant activity [98]. Perhaps most notably, they outperform α-TOH in terms of the number of trapped ROO• radicals, having large stoichiometric factors that depend on temperature, due to a regeneration cycle involving the transient formation of the corresponding nitroxyl radicals [49].

Interestingly, recent work from Pratt's group has demonstrated that several antioxidants belonging to this class, namely diarylamines, phenoxazines and phenothiazines, including those displayed in Figure 13, are potent inhibitors of ferroptosis, able to match or surpass the performance of reference inhibitors liproxstatin-1 and ferrostatin-1 (Figure 13), which incidentally also belong to the class of aromatic amines [11].

5.2.8. Unconventional Antioxidant Mechanisms and HOO• as Co-Antioxidant

Sterically hindered dialkyl nitroxides like the 2,2,6,6-tetramethylpiperidin-1-oxyl radical (TEMPO) are sufficiently persistent to be manipulated like a "normal" molecule, which has allowed an enormous body of investigation. Among the many applications, a wealth of studies have shown a beneficial activity in a multitude of pathological conditions commonly associated with oxidative stress, which has suggested an antioxidant activity that has long escaped rationalization [139,140]. In water, TEMPO was demonstrated to behave as a SOD mimic, being able to decompose superoxide and peroxyl radicals via a catalytic cycle with the intermediation of the oxoammonium ion [141,142] (Figure 14). However, early knowledge indicated that nitroxides are unable to trap peroxyl radicals in a lipidic environment and have negligible antioxidant activity toward lipids.

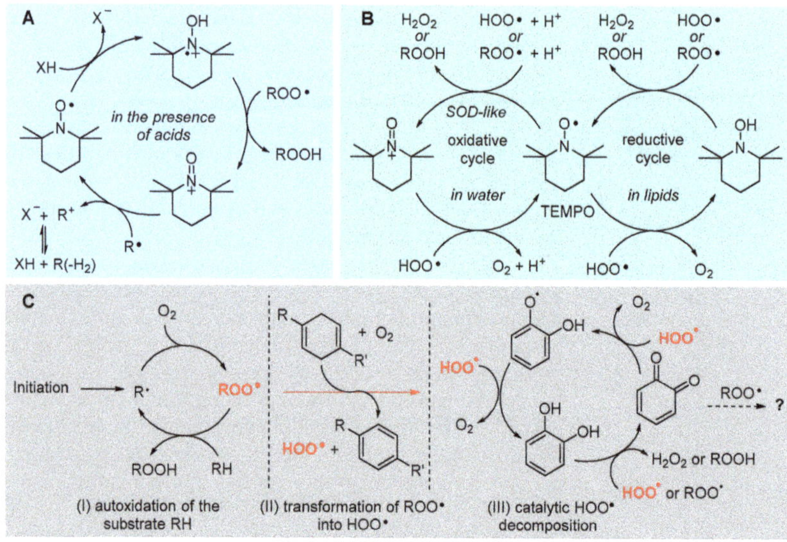

Figure 14. Non-conventional antioxidant mechanisms involving persistent nitroxides in acidic medium (**A**), or in the presence of HOO• (**B**), or involving *ortho*-quinones in the presence of HOO• (**C**).

We were surprised to find that they can indeed be excellent antioxidants in lipidic or organic media, but only in the presence of acids [143]. Of interest, depending on the strength of the acid, they could behave as stoichiometric antioxidants (trapping one radical per nitroxide) or they could be regenerated and work in a catalytic fashion if the acid is weak and forms a nucleophilic conjugated base, like acetic acid [144]. This chemistry, summarized in Figure 14, clearly helps explain their redox-related bioactivity, as carboxylic acids are abundant in biological systems, e.g., in proteins [144]. Even more surprising was the recent discovery of another catalytic antioxidant cycle of nitroxides, which is only expressed in lipidic or apolar environments, although it apparently resembles the SOD mimic behavior expressed only in water. The overall catalytic cycle occurs only in the presence of HOO• radicals (the neutral form of superoxide) and it does not involve the intermediacy of the oxoammonium ion from the nitroxide [145]. It was clarified, independently by our group and Pratt's group, that the cycle occurs by *reduction* of the nitroxide (>NO•) to the parent hydroxylamine (>NOH), which is then reoxidized to the nitroxide by HOO• or by ROO• [49,145] (Figure 14). Therefore, HOO•, normally regarded as an oxidizing chain-carrying radical, here works as sacrificial *reducing* agent or as a *co-antioxidant*. It can be formed spontaneously during the peroxidation of lipids (see Section 2.4.3), or a dedicated source like cyclohexadiene or γ-terpinene can be added along with the nitroxide, to form a co-antioxidant system [145]. Interestingly, this co-antioxidant system is among the most potent ever reported, largely outperforming α-TOH [145]. This chemistry is likely to have major relevance in the control of ferroptosis and some preliminary success in this direction has recently been reported [12].

Unfortunately, no nitroxide is currently approved for medicinal use; however, we were intrigued to find that *ortho* quinones, the oxidized "waste" of catechol-type polyphenolic antioxidants, can undergo a similar chemistry (Figure 14). In the presence of spontaneously generated HOO•, or of an added source, *ortho*-quinones (*o*-Q) (more efficiently than *para*) are stepwise reduced to the semiquinone radical (*o*-QH•) then to the catechol (*o*-QH2), which can then perform its antioxidant action trapping both HOO• and ROO•. Albeit not as efficient as that involving nitroxides, the synergic combination of *o*-Q/γ-terpinene or *o*-QH2/γ-terpinene has shown superior protection of highly oxidizable polyunsaturated lipids [146]. The kinetic aspects regulating this chemistry have been clarified and found to be key to the purported antioxidant behavior of melanin biopolymers [47]. Possibly, this chemistry would show its potential in the control of ferroptosis, an issue currently under investigation in our group.

5.3. Termination-Enhancing Antioxidants

Some hydrocarbons, aldehydes and other simple highly oxidizable molecules, such as many terpenic components of plant essential oils (e.g., γ-terpinene, limonene, linalool and citral), when subjected to some radical source, undergo autoxidation that is characterized by very fast chain termination [147]. As a consequence, if they are mixed with other highly oxidizable substrates, they will promptly co-oxidize but increase the overall rate of chain termination, thereby reducing the overall rate of autoxidation and slowing down the degradation of the oxidable substrate [147]. These compounds are unable to block the chain propagation, nonetheless they have antioxidant behavior, which we classified as termination-enhancing [147]. This mechanism, which is typical of many non-phenolic essential oil components is, however, less effective than the chain-breaking activity possessed, for instance, by phenolic essential oil components [148] and it largely depends on the experimental conditions. It has non-monotonic dependence on the concentration of the antioxidant and it can become pro-oxidant at a high concentration [147].

5.4. Indirect Antioxidants

This class is often used to group those mechanisms that cannot take place within a material or in a test tube but require a living cell or organism. Indeed, many molecules which might show antioxidant behavior in cells, e.g., by reducing the markers of lipid peroxidation or the related biological damage, do not trap peroxyl radicals at a significant rate and would be unable to directly protect lipids from peroxidation; however, they are able to enhance the antioxidant defenses in a living organism, e.g., by inducing the biosynthesis of antioxidant enzymes. Typical examples are isothiocyanates from brassica vegetables [149,150], but the list can be extended to many plant-derived compounds, such as essential oil components [150,151]. Indeed, even flavonoids like quercetin, having good direct (chain-breaking) antioxidant activity, can also boost the antioxidant enzymes in cells [152], therefore direct and indirect activity can co-exist and have different weight depending on the experimental settings. One main mechanism by which indirect activity is accomplished is via the Nrf2 signaling system [150], which has been illustrated in Section 4.2.

6. Conclusive Remarks and Future Perspective

Research in the broad area of lipid peroxidation and antioxidant protection has taken place during the past 70 years, with variable emphasis. In the biomedical area, the disappointing outcomes of early clinical studies aimed at using antioxidants as drugs in the treatment of those degenerative conditions lacking other curative pharmacological solutions had somewhat, gradually, curbed the enthusiasm. Lack of success might have been due to a combination of factors, such as incomplete understanding of the role of oxidative stress and lipid peroxidation in physiology and disease, incomplete understanding of the double-sided antioxidant–pro-oxidant behavior of some molecules in biological systems, a misjudgment in the strategy of using an antioxidant when a pro-oxidant would have been the more effective choice and, last but not least, the choosing of antioxidants that were not fit for purpose. The recognition of ferroptosis as a new LP-based form of programmed cell death has revived the area and enormously boosted the search for molecules able to address OS-related pathologies via the modulation of ferroptosis. Now, understanding of the interplay between LP and pathophysiology is more mature and a wealth of extremely potent new antioxidants have become available. Pyridinols, diarylamines and heavy organochalcogen-substituted phenols have underexplored potential in this regard. Additionally, non-conventional antioxidant systems based on the exploitation of the endogenous $HOO\bullet/O_2^{-\bullet}$ reducing system appear very promising in the modulation of ferroptosis, a research area we are currently set to address.

Funding: This research was funded by MUR (Rome, Italy), project FEROX, code number 20227XZKBY, call PRIN2022, and by the University of Bologna, grant RFO2022.

Conflicts of Interest: The authors declare no conflict of interest.

References

1. Porter, N.A. Mechanism for the Autoxidation of Polyunsaturated Lipids. *Acc. Chem. Res.* **1986**, *19*, 262–268. [CrossRef]
2. Ingold, K.U. Peroxyl radicals. *Acc. Chem. Res.* **1969**, *2*, 1–9. [CrossRef]
3. Burton, G.W.; Ingold, K.U. Vitamin E: Application of the Principles of Physical Organic Chemistry to the Exploration of Its Structure and Function. *Acc. Chem. Res.* **1986**, *19*, 194–201. [CrossRef]
4. Ito, F.; Sono, Y.; Ito, T. Measurement and Clinical Significance of Lipid Peroxidation as a Biomarker of Oxidative Stress: Oxidative Stress in Diabetes, Atherosclerosis, and Chronic Inflammation. *Antioxidants* **2019**, *8*, 72. [CrossRef]
5. Tarozzi, A.; Bartolini, M.; Piazzi, L.; Valgimigli, L.; Amorati, R.; Bolondi, C.; Djemil, A.; Mancini, F.; Andrisano, V.; Angela Rampa, A. From the dual function lead AP2238 to AP2469, a multi-target-directed ligand for the treatment of Alzheimer's disease. *Pharma. Res. Per.* **2014**, *2*, e00023. [CrossRef] [PubMed]
6. Foret, M.K.; Lincoln, R.; Do Cormo, S.; Cuello, A.C.; Cosa, G. Connecting the "Dots": From Free Radical Lipid Autoxidation to Cell Pathology and Disease. *Chem. Rev.* **2020**, *120*, 12757–12787. [CrossRef]
7. Canistro, D.; Boccia, C.; Falconi, R.; Bonamassa, B.; Valgimigli, L.; Vivarelli, F.; Soleti, A.; Genova, M.L.; Lenaz, G.; Sapone, A.; et al. Redox-Based Flagging of the Global Network of Oxidative Stress Greatly Promotes Longevity. *J. Gerontol. A Biol. Sci. Med. Sci.* **2015**, *70*, 936–943. [CrossRef]
8. Dixon, S.J.; Lemberg, K.M.; Lamprecht, M.R.; Skouta, R.; Zaitsev, E.M.; Gleason, C.E.; Patel, D.N.; Bauer, A.J.; Cantley, A.M.; Yang, W.S.; et al. Ferroptosis: An Iron Dependent Form of Nonapoptotic Cell Death. *Cell* **2012**, *149*, 1060–1072. [CrossRef]
9. von Krusenstiern, A.N.; Robson, R.N.; Qian, N.; Qiu, B.; Hu, F.; Reznik, E.; Smith, N.; Zandkarimi, F.; Estes, V.E.; Dupont, M.; et al. Identification of essential sites of lipid peroxidation in ferroptosis. *Nat. Chem. Biol.* **2023**, *19*, 719–730. [CrossRef]
10. Cheng, X.; Zhang, J.; Xiao, Y.; Wang, Z.; He, J.; Ke, M.; Liu, S.; Wang, Q.; Zhang, L. Mitochondrial Regulation of Ferroptosis in Cancer Therapy. *Int. J. Mol. Sci.* **2023**, *24*, 10037. [CrossRef]
11. Shah, R.; Margison, K.; Pratt, D.A. The Potency of Diarylamine Radical-Trapping Antioxidants as Inhibitors of Ferroptosis Underscores the Role of Autoxidation in the Mechanism of Cell Death. *ACS Chem. Biol.* **2017**, *12*, 2538–2545. [CrossRef]
12. Poon, J.-F.; Zilka, O.; Pratt, D.A. Potent Ferroptosis Inhibitors Can Catalyze the Cross-Dismutation of Phospholipid-Derived Peroxyl Radicals and Hydroperoxyl Radicals. *J. Am. Chem. Soc.* **2020**, *142*, 14331–14342. [CrossRef]
13. Conrad, M.; Lorenz, S.M.; Proneth, B. Targeting Ferroptosis: New Hope for As-Yet-Incurable Diseases. *Trends Mol. Med.* **2020**, *27*, 113–122. [CrossRef] [PubMed]
14. Ursini, F.; Maiorino, M. Lipid peroxidation and ferroptosis: The role of GSH and GPx4. *Free Radic. Biol. Med.* **2020**, *152*, 175–185. [CrossRef] [PubMed]
15. Angelova, P.R.; Esteras, N.; Abramov, A.Y. Mitochondria and lipid peroxidation in the mechanism of neurodegeneration: Finding ways for prevention. *Med. Res. Rev.* **2021**, *41*, 770–784. [CrossRef]

16. Gabbanini, S.; Matera, R.; Valvassori, A.; Valgimigli, L. Rapid liquid chromatography–tandem mass spectrometry analysis of 4-hydroxynonenal for the assessment of oxidative degradation and safety of vegetable oils. *Anal. Chim. Acta* **2015**, *869*, 50–58. [CrossRef]
17. Yang, S.; Takeuchi, M.; Friedrich, H.; van Duynhoven, J.P.M.; Hohlbein, J. Unravelling mechanisms of protein and lipid oxidation in mayonnaise at multiple length scales. *Food Chem.* **2023**, *402*, 134417. [CrossRef] [PubMed]
18. Gęgotek, A.; Skrzydlewska, A. Biological effect of protein modifications by lipid peroxidation products. *Chem. Phys. Lipids* **2019**, *221*, 46–52. [CrossRef] [PubMed]
19. Christensen, H.; Sehested, K. HO_2 and O_2 Radicals at Elevated Temperatures. *J. Phys. Chem.* **1988**, *92*, 3007–3011. [CrossRef]
20. Maillard, B.; Ingold, K.U.; Scaiano, J.C. Rate constants for the reactions of free radicals with oxygen in solution. *J. Am. Chem. Soc.* **1983**, *105*, 5095–5099. [CrossRef]
21. Valgimigli, L.; Pratt, D.A. Antioxidants in Chemistry and Biology. In *Encyclopedia of Radicals in Chemistry, Biology and Materials*; Chatgilialoglu, C., Studer, A., Eds.; John Wiley & Sons, Ltd.: Chirchester, UK, 2012; Volume 3, pp. 1623–1678. [CrossRef]
22. Howard, J.A.; Ingold, K.U. Absolute rate constants for hydrocarbon autoxidation. VI. Alkyl aromatic and olefinic hydrocarbons. *Can. J. Chem.* **1967**, *45*, 793–802. [CrossRef]
23. Xu, L.; Davis, T.A.; Porter, N.A. Rate Constants for Peroxidation of Polyunsaturated Fatty Acids and Sterols in Solution and in Liposomes. *J. Am. Chem. Soc.* **2009**, *131*, 13037–13044. [CrossRef]
24. Baschieri, A.; Pizzol, R.; Guo, Y.; Amorati, R.; Valgimigli, L. Calibration of Squalene, p-Cymene, and Sunflower Oil as Standard Oxidizable Substrates for Quantitative Antioxidant Testing. *J. Agric. Food Chem.* **2019**, *67*, 6902–6910. [CrossRef] [PubMed]
25. Antunes, F.; Barclay, L.R.C.; Vinqvist, M.R.; Pinto, R.E. Determination of propagation and termination rate constants by using an extension to the rotating-sector method: Application to PLPC and DLPC bilayers. *Int. J. Chem. Kinet.* **1998**, *30*, 753–767. [CrossRef]
26. Russell, G.A. Deuterium-Isotope Effects in the Autoxidation of Aralkyl Hydrocarbons. Mechanism of the Interaction of Peroxy Radicals1. *J. Am. Chem. Soc.* **1957**, *79*, 3871–3877. [CrossRef]
27. Miyamoto, S.; Martinez, G.R.; Medeiros, M.H.; Di Mascio, P. Singlet Molecular Oxygen Generated by Biological Hydroperoxides. *J. Photochem. Photobiol. B* **2014**, *139*, 24–33. [CrossRef]
28. Noguchi, N.; Nakada, A.; Itoh, Y.; Watanabe, A.; Niki, E. Formation of Active Oxygen Species and Lipid Peroxidation Induced by Hypochlorite. *Arch. Biochem. Biophys.* **2002**, *397*, 440–447. [CrossRef] [PubMed]
29. Miyamoto, S.; Martinez, G.R.; Rettori, D.; Augusto, O.; Medeiros, M.H.G.; Di Mascio, P. Linoleic acid hydroperoxide reacts with hypochlorous acid, generating peroxyl radical intermediates and singlet molecular oxygen. *Proc. Natl. Acad. Sci. USA* **2006**, *103*, 293–298. [CrossRef]
30. Miyamoto, S.; Nantes, I.L.; Faria, P.A.; Cunha, D.; Ronsein, G.E.; Medeiros, M.H.G.; Di Mascio, P. Cytochrome c-promoted cardiolipin oxidation generates singlet molecular oxygen. *Photochem. Photobiol. Sci.* **2012**, *11*, 1536–1546. [CrossRef]
31. Zielinski, Z.A.M.; Pratt, D.A. Lipid Peroxidation: Kinetics, Mechanisms, and Products. *J. Org. Chem.* **2017**, *82*, 2817–2825. [CrossRef]
32. Miranda-Apodaca, J.; Hananya, N.; Velázquez-Campoy, A.; Shabat, D.; Arellano, J.B. Emissive Enhancement of the Singlet Oxygen Chemiluminescence Probe after Binding to Bovine Serum Albumin. *Molecules* **2019**, *24*, 2422. [CrossRef]
33. Lissi, E.A.; Caceres, T.; Videla, L.A. Visible chemiluminescence from rat brain homogenates undergoing autoxidation. II Kinetics of theluminescence decay. *Free Radic. Biol. Med.* **1988**, *4*, 93–97. [CrossRef] [PubMed]
34. Timmins, G.S.; dos Santos, R.E.; Whitwood, A.C.; Catalani, L.H.; Di Mascio, P.; Gilbert, B.C.; Bechara, E.J.H. Lipid peroxidation-dependent chemiluminescence from the cyclization of alkylperoxyl radicals to dioxetane radical intermediates. *Chem. Res. Toxicol.* **1997**, *10*, 1090–1096. [CrossRef]
35. Fedorova, G.F.; Menshov, V.A.; Trofimov, A.V.; Vasil'Ev, R.F. Facile chemiluminescence assay for antioxidative properties of vegetable lipids: Fundamentals and illustrative examples. *Analyst* **2009**, *134*, 2128–2134. [CrossRef]
36. Vasil'ev, R.F.; Veprintsev, T.L.; Dolmatova, L.S.; Naumov, V.V.; Trofimov, A.V.; Tsaplev, Y.B. Kinetics of ethylbenzene oxy-chemiluminescence in the presence of antioxidants from tissues of the marine invertebrate *Eupentacta fraudatrix*: Estimating the concentration and reactivity of the natural antioxidants. *Kinet. Catal.* **2014**, *55*, 148–153. [CrossRef]
37. Kancheva, V.D.; Dettori, M.A.; Fabbri, D.; Alov, P.; Angelova, S.E.; Slavova-Kazakova, A.K.; Carta, P.; Menshov, V.A.; Yablonskaya, O.I.; Trofimov, A.V.; et al. Natural Chain-Breaking Antioxidants and Their Synthetic Analogs as Modulators of Oxidative Stress. *Antioxidants* **2021**, *10*, 624. [CrossRef]
38. Porter, N.A. A Perspective on Free Radical Autoxidation: The Physical Organic Chemistry of Polyunsaturated Fatty Acid and Sterol Peroxidation. *J. Org. Chem.* **2013**, *78*, 3511–3524. [CrossRef]
39. Pratt, D.A.; Tallman, K.A.; Porter, N.A. Free Radical Oxidation of Polyunsaturated Lipids: New Mechanistic Insights and the Development of Peroxyl Radical Clocks. *Acc. Chem. Res.* **2011**, *44*, 458–467. [CrossRef] [PubMed]
40. Howard, J.A.; Ingold, K.U. Absolute rate constants for hydrocarbon oxidation. XII. Rate constants for secondary peroxy radicals. *Can. J. Chem.* **1968**, *46*, 2661–2666. [CrossRef]
41. Maerker, G.; Haeberer, E.T.; Ault, W.C. Epoxidation of methyl linoleate. I. The question of positional selectivity in monoepoxidation. *J. Am. Oil Chem. Soc.* **1966**, *43*, 100–104. [CrossRef]
42. Aliwarga, T.; Raccor, B.S.; Lemaitre, R.N.; Sotoodehnia, N.; Gharib, S.A.; Xu, L.; Totah, R.A. Enzymatic and free radical formation of cis- and trans-epoxyeicosatrienoic acids in vitro and in vivo. *Free Radic. Biol. Med.* **2017**, *112*, 131–140. [CrossRef] [PubMed]

43. Zielinski, Z.A.M.; Pratt, D.A. H-Atom Abstraction vs Addition: Accounting for the Diverse Product Distribution in the Autoxidation of Cholesterol and Its Esters. *J. Am. Chem. Soc.* **2019**, *141*, 3037–3051. [CrossRef]
44. Do, Q.; Lee, D.D.; Dinh, A.N.; Seguin, R.P.; Zhang, R.; Xu, L. Development and Application of a Peroxyl Radical Clock Approach for Measuring Both Hydrogen-Atom Transfer and Peroxyl Radical Addition Rate Constants. *J. Org. Chem.* **2021**, *86*, 153–168. [CrossRef] [PubMed]
45. Denisov, E.T. The role of triplet repulsion in alkyl radical addition to a 7π-C-O bond and alkoxy radical addition to a π-C-C bond. *Kinet. Catal.* **2000**, *41*, 293–297. [CrossRef]
46. Degirmenci, I.; Coote, M.L. Comparison of Thiyl, Alkoxyl, and Alkyl Radical Addition to Double Bonds: The Unusual Contrasting Behavior of Sulfur and Oxygen Radical Chemistry. *J. Phys. Chem. A* **2016**, *120*, 1750–1755. [CrossRef]
47. Guo, Y.; Baschieri, A.; Mollica, F.; Valgimigli, L.; Cedrowski, J.; Litwinienko, G.; Amorati, R. Hydrogen Atom Transfer from HOO• to ortho-Quinones Explains the Antioxidant Activity of Polydopamine. *Angew. Chem. Int. Ed.* **2021**, *60*, 15220–15224. [CrossRef]
48. Foti, M.C.; Ingold, K.U. Mechanism of Inhibition of lipid peroxidation by γ-terpinene, an unusual and potentially useful hydrocarbon antioxidant. *J. Agric. Food Chem.* **2003**, *51*, 2758–2765. [CrossRef]
49. Harrison, K.A.; Haidasz, E.A.; Griesser, M.; Pratt, D.A. Inhibition of hydrocarbon autoxidation by nitroxide-catalyzed cross-dismutation of hydroperoxyl and alkylperoxyl radicals. *Chem. Sci.* **2018**, *9*, 6068–6079. [CrossRef]
50. Liu, W.; Porter, N.A.; Schneider, C.; Brash, A.R.; Yin, H. Formation of 4-Hydroxynonenal from Cardiolipin Oxidation: Intramolecular Peroxyl Radical Addition and Decomposition. *Free Radic. Biol. Med.* **2011**, *50*, 166–178. [CrossRef]
51. Bowry, V.W. Arm-to-Arm Autoxidation in a Triglyceride: Remote Group Reaction Kinetics. *J. Org. Chem.* **1994**, *59*, 2250–2252. [CrossRef]
52. Esterbauer, H.; Schaur, R.J.; Zollner, H. Chemistry and biochemistry of 4-hydroxynonenal, malonaldehyde and related aldehydes. *Free Radic. Biol. Med.* **1991**, *11*, 81–128. [CrossRef]
53. Ullery, J.C.; Marnett, L.J. Protein modification by oxidized phospholipids and hydrolytically released lipid electrophiles: Investigating cellular responses. *Biochim. Biophys. Acta Biomembr.* **2012**, *1818*, 2424–2435. [CrossRef]
54. Codreanu, S.G.; Ullery, J.C.; Zhu, J.; Tallman, K.A.; Beavers, W.N.; Porter, N.A.; Marnett, L.J.; Zhang, B.; Liebler, D.C. Alkylation Damage by Lipid Electrophiles Targets Functional Protein Systems. *Mol. Cell. Proteom.* **2014**, *13*, 849–859. [CrossRef] [PubMed]
55. Spickett, C.M. The lipidperoxidation product 4-hydroxy-2-nonenal: Advances in chemistry and analysis. *Redox Biol.* **2013**, *2013*, 145–152. [CrossRef] [PubMed]
56. Liang, X.; Qian, R.; Ou, Y.; Wang, D.; Lin, X.; Sun, C. Lipid peroxide-derived short-chain aldehydes promote programmed cell death in wheat roots under aluminum stress. *J. Hazard. Mater.* **2023**, *443*, 130142. [CrossRef]
57. Parvez, S.; Long, M.J.C.; Poganik, J.R.; Aye, Y. Redox Signaling by Reactive Electrophiles and Oxidants. *Chem. Rev.* **2018**, *118*, 8798–8888. [CrossRef] [PubMed]
58. Schopfer, F.J.; Cipollina, C.; Freeman, B.A. Formation and Signaling Actions of Electrophilic Lipids. *Chem. Rev.* **2011**, *111*, 5997–6021. [CrossRef]
59. Codreanu, S.G.; Liebler, D.C. Novel Approaches to identify protein adducts produced by lipid peroxidation. *Free Radic Res.* **2015**, *49*, 881–887. [CrossRef]
60. Perkovic, M.N.; Jaganjac, M.; Milkovic, L.; Horvat, T.; Rojo, D.; Zarkovic, K.; Coric, M.; Hudolin, T.; Waeg, G.; Orehovec, B.; et al. Relationship between 4-Hydroxynonenal (4-HNE) as Systemic Biomarker of Lipid Peroxidation and Metabolomic Profiling of Patients with Prostate Cancer. *Biomolecules* **2023**, *13*, 145. [CrossRef]
61. Riahi, Y.; Cohen, G.; Shamni, O.; Sasson, S. Signaling and cytotoxic functions of 4-hydroxyalkenals. *Am. J. Physiol. Endocrinol. Metab.* **2010**, *299*, E879–E886. [CrossRef]
62. Onyango, A.N.; Baba, N. New hypotheses on the pathways of formation of malondialdehyde and isofurans. *Free Radic. Biol. Med.* **2010**, *49*, 1594–1600. [CrossRef] [PubMed]
63. Takeuchi, C.; Galvé, R.; Nieva, J.; Witter, D.P.; Wentworth, A.D.; Troseth, R.P.; Lerner, R.A.; Wentworth, P. Proatherogenic Effects of the Cholesterol Ozonolysis Products, Atheronal-A and Atheronal-B. *Biochemistry* **2006**, *45*, 7162–7170. [CrossRef] [PubMed]
64. Nieva, J.; Song, B.-D.; Rogel, J.K.; Kujawara, D.; Altobel, L., III; Izharrudin, A.; Boldt, G.E.; Grover, R.K.; Wentworth, A.D.; Wentworth, P., Jr. Cholesterol Secosterol Aldehydes Induce Amyloidogenesis and Dysfunction of Wild-Type Tumor Protein p53. *Chem. Biol.* **2011**, *18*, 920–927. [CrossRef]
65. Usui, K.; Hulleman, J.D.; Paulsson, J.F.; Siegel, S.J.; Powers, E.T.; Kelly, J.W. Site-specific modification of Alzheimer's peptides by cholesterol oxidation products enhances aggregation energetics and neurotoxicity. *Proc. Natl. Acad. Sci. USA* **2009**, *106*, 18563–18568. [CrossRef] [PubMed]
66. Wójcik, P.; Žarkovi´, N.; Gegotek, A.; Skrzydlewska, E. Involvement of Metabolic Lipid Mediators in the Regulation of Apoptosis. *Biomolecules* **2020**, *10*, 402. [CrossRef]
67. Graille, M.; Wild, P.; Sauvain, J.-J.; Hemmendinger, M.; Guseva Canu, I.; Hopf, N.B. Urinary 8-isoprostane as a biomarker for oxidative stress. A systematic review and meta-analysis. *Toxicol. Lett.* **2020**, *328*, 19–27. [CrossRef]
68. Gao, X.; Brenner, H.; Holleczek, B.; Cuk, K.; Zhang, Y.; Anusruti, A.; Xuan, Y.; Xu, Y.; Schottker, B. Urinary 8-isoprostane levels and occurrence of lung, colorectal, prostate, breast and overall cancer: Results from a large, population-based cohort study with 14 years of follow-up. *Free Radic. Biol. Med.* **2018**, *123*, 20–26. [CrossRef]
69. Yang, J.; Tallman, K.A.; Porter, N.A.; Liebler, D.C. Quantitative Chemoproteomics for Site-Specific Analysis of Protein Alkylation by 4-Hydroxy-2-Nonenal in Cells. *Anal. Chem.* **2015**, *87*, 2535–2541. [CrossRef]

70. Ferreri, C.; Ferocino, A.; Batani, G.; Chatgilialoglu, C.; Randi, V.; Riontino, M.V.; Vetica, F.; Sansone, A. Plasmalogens: Free Radical Reactivity and Identification of Trans Isomers Relevant to Biological Membranes. *Biomolecules* **2023**, *13*, 730. [CrossRef]
71. Chatgilialoglu, C.; Ferreri, C.; Melchiorre, M.; Sansone, A.; Torreggiani, A. Lipid geometrical isomerism: From chemistry to biology and diagnostics. *Chem. Rev.* **2014**, *114*, 255–284. [CrossRef]
72. Torreggiani, A.; Tinti, A.; Jurasekova, Z.; Capdevila, M.; Saracino, M.; Foggia, M.D. Structural Lesions of Proteins Connected to Lipid Membrane Damages Caused by Radical Stress: Assessment by Biomimetic Systems and Raman Spectroscopy. *Biomolecules* **2019**, *9*, 794. [CrossRef] [PubMed]
73. Chatgilialoglu, C.; Ferreri, C. Reductive Stress of Sulfur-Containing Amino Acids within Proteins and Implication of Tandem Protein–Lipid Damage. *Int. J. Mol. Sci.* **2021**, *22*, 12863. [CrossRef]
74. Hirata, Y.; Ferreri, C.; Yamada, Y.; Inoue, A.; Sansone, A.; Vetica, F.; Suzuki, W.; Takano, S.; Noguchi, T.; Matsuzawa, A.; et al. Geometrical isomerization of arachidonic acid during lipid peroxidation interferes with ferroptosis. *Free Radic. Biol. Med.* **2023**, *204*, 374–384. [CrossRef]
75. Itri, R.; Junqueira, H.C.; Mertins, O.; Baptista, M.S. Membrane Changes under Oxidative Stress: The Impact of Oxidized Lipids. *Biophys. Rev.* **2014**, *6*, 47–61. [CrossRef]
76. Weber, G.; Charitat, T.; Baptista, M.S.; Uchoa, A.F.; Pavani, C.; Junqueira, H.C.; Guo, Y.; Baulin, V.A.; Itri, R.; Marques, C.M.; et al. Lipid Oxidation Induces Structural Changes in Biomimetic Membranes. *Soft Matter* **2014**, *10*, 4241–4247. [CrossRef] [PubMed]
77. Wang, G.; Wang, J.; Fan, X.; Ansari, G.A.S.; Khan, M.F. Protein adducts of malondialdehyde and 4-hydroxynonenal contribute to trichloroethene-mediated autoimmunity via activating Th17 cells: Dose–and time–response studies in female MRL +/+ mice. *Toxicology* **2012**, *292*, 113–122. [CrossRef] [PubMed]
78. Yamada, T.; Place, N.; Kosterina, N.; Östberg, T.; Zhang, S.J.; Grundtman, C.; Erlandsson-Harris, H.; Lundberg, I.E.; Glenmark, B.; Bruton, J.D.; et al. Impaired myofibrillar function in the soleus muscle of mice with collagen-induced arthritis. *Arthritis Rheumat.* **2009**, *60*, 3280–3289. [CrossRef] [PubMed]
79. Kuhn, H.; Banthiya, S.; van Leyen, K. Mammalian lipoxygenases and their biological relevance. *Biochim. Biophys. Acta* **2015**, *1851*, 308–330. [CrossRef]
80. Yang, W.S.; Kim, K.J.; Gaschler, M.M.; Patel, M.; Shchepinov, M.S.; Stockwell, B.R. Peroxidation of Polyunsaturated Fatty Acids by Lipoxygenases Drives Ferroptosis. *Proc. Natl. Acad. Sci. USA* **2016**, *113*, E4966–E4975. [CrossRef]
81. Kennedy, L.; Sandhu, J.K.; Harper, M.-E.; Cuperlovic-Culf, M. Role of Glutathione in Cancer: From Mechanisms to Therapies. *Biomolecules* **2020**, *10*, 1429. [CrossRef]
82. Peña-Bautista, C.; Vento, M.; Baquero, M.; Cháfer-Pericás, C. Lipid peroxidation in neurodegeneration. *Clin. Chim. Acta* **2019**, *497*, 178–188. [CrossRef] [PubMed]
83. da Santana, L.N.; Bittencourt, L.O.; Nascimento, P.C.; Fernandes, R.M.; Teixeira, F.B.; Fernandes, L.M.P.; Freitas Silva, M.C.; Nogueira, L.S.; Amado, L.L.; Crespo-Lopez, M.E.; et al. Low doses of methylmercury exposure during adulthood in rats display oxidative stress, neurodegeneration in the motor cortex and lead to impairment of motor skills. *J. Trace Elem. Med. Biol.* **2019**, *51*, 19–27. [CrossRef] [PubMed]
84. Peña-Bautista, C.; Baquero, M.; Vento, M.; Cháfer-Pericás, C. Free radicals in Alzheimer's disease: Lipid peroxidation biomarkers. *Clin. Chim. Acta* **2019**, *491*, 85–90. [CrossRef] [PubMed]
85. Friedmann Angeli, J.P.; Schneider, M.; Proneth, B.; Tyurina, Y.Y.; Tyurin, V.A.; Hammond, V.J.; Herbach, N.; Aichler, M.; Walch, A.; Eggenhofer, E.; et al. Inactivation of the ferroptosis regulator Gpx4 triggers acute renal failure in mice. *Nat. Cell Biol.* **2014**, *16*, 1180–1191. [CrossRef] [PubMed]
86. Conrad, M.; Pratt, D.A. The chemical basis of ferroptosis. *Nat. Chem. Biol.* **2019**, *15*, 1137–1147. [CrossRef]
87. Mishima, E.; Ito, J.; Wu, Z.; Nakamura, T.; Wahida, A.; Doll, S.; Tonnus, W.; Nepachalovich, P.; Eggenhofer, E.; Aldrovandi, M.; et al. A non-canonical vitamin K cycle is a potent ferroptosis suppressor. *Nature* **2022**, *608*, 778–783. [CrossRef]
88. Shah, R.; Shchepinov, M.S.; Pratt, D.A. Resolving the Role of Lipoxygenases in the Initiation and Execution of Ferroptosis. *ACS Cent. Sci.* **2018**, *4*, 387–396. [CrossRef]
89. Soula, M.; Weber, R.A.; Zilka, O.Z.; Alwaseem, H.; La, K.; Yen, F.; Molina, H.; Garcia-Bermudez, J.; Pratt, D.A.; Birsoy, K. Metabolic determinants of cancer cell sensitivity to canonical ferroptosis inducers. *Nat. Chem. Biol.* **2020**, *16*, 1351–1360. [CrossRef]
90. Friedmann Angeli, J.P.; Shah, R.; Derek, A.; Pratt, D.A.; Conrad, M. Ferroptosis Inhibition: Mechanisms and Opportunities. *Trends Pharmacol. Sci.* **2017**, *38*, 489–498. [CrossRef] [PubMed]
91. Zhang, L.; Jia, R.; Li, H.; Yu, H.; Ren, K.; Jia, S.; Li, Y.; Wang, Q. Insight into the Double-Edged Role of Ferroptosis in Disease. *Biomolecules* **2021**, *11*, 1790. [CrossRef]
92. Amorati, R.; Valgimigli, L. Methods to measure the antioxidant activity of phytochemicals and plant extracts. *J. Agric. Food Chem.* **2018**, *66*, 3324–3329. [CrossRef]
93. Amorati, R.; Foti, M.C.; Valgimigli, L. Antioxidant activity of essential oils. *J. Agric. Food Chem.* **2013**, *61*, 10835–10847. [CrossRef] [PubMed]
94. Lucarini, M.; Pedulli, G.F. Free radical intermediates in the inhibition of the autoxidation reaction. *Chem. Soc. Rev.* **2010**, *39*, 2106–2119. [CrossRef]
95. Cariola, A.; El Chami, M.; Granatieri, J.; Valgimigli, L. Anti-tyrosinase and antioxidant activity of meroterpene bakuchiol from *Psoralea corylifolia* (L.). *Food Chem.* **2023**, *405*, 134953. [CrossRef] [PubMed]
96. Helberg, J.; Pratt, D.A. Autoxidation vs. antioxidants—The fight for Forever. *Chem. Soc. Rev.* **2021**, *50*, 7343–7358. [CrossRef]

97. Ingold, K.U.; Pratt, D.A. Advances in Radical-Trapping Antioxidant Chemistry in the 21st Century: A Kinetics and Mechanisms Perspective. *Chem. Rev.* **2014**, *114*, 9022–9046. [CrossRef] [PubMed]
98. Valgimigli, L.; Pratt, D.A. Maximizing the Reactivity of Phenolic and Aminic Radical-Trapping Antioxidants: Just Add Nitrogen! *Acc. Chem. Res.* **2015**, *48*, 966–975. [CrossRef]
99. Pratt, D.A.; DiLabio, G.A.; Brigati, G.; Pedulli, G.F.; Valgimigli, L. 5-Pyrimidinols: Novel Chain-Breaking Antioxidants More Effective Than Phenols. *J. Am. Chem. Soc.* **2001**, *123*, 4625–4626. [CrossRef]
100. Valgimigli, L.; Brigati, G.; Pedulli, G.F.; DiLabio, G.A.; Mastragostino, M.; Arbizzani, C.; Pratt, D.A. The Effect of Ring Nitrogen Atoms on the Homolytic Reactivity of Phenolic Compounds: Understanding the Radical-Scavenging Ability of 5-Pyrimidinols. *Chem. Eur. J.* **2003**, *9*, 4997–5010. [CrossRef]
101. Wijtmans, M.; Pratt, D.A.; Valgimigli, L.; DiLabio, G.A.; Pedulli, G.F.; Porter, N.A. 6-Amino-3-Pyridinols: Towards Diffusion-Controlled Chain-Breaking Antioxidants. *Angew. Chem. Int. Ed.* **2003**, *42*, 4370–4373. [CrossRef]
102. Wijtmans, M.; Pratt, D.A.; Brinkhorst, J.; Serwa, R.; Valgimigli, L.; Pedulli, G.F.; Porter, N.A. Synthesis and Reactivity of Some 6-Substituted-2,4-Dimethyl-3-Pyridinols, a Novel Class of Chain-Breaking Antioxidants. *J. Org. Chem.* **2004**, *69*, 9215–9223. [CrossRef]
103. Kim, H.-Y.; Pratt, D.A.; Seal, J.R.; Wijtmans, M.; Porter, N.A. Lipid-Soluble 3-Pyridinol Antioxidants Spare α-Tocopherol and Do Not Efficiently Mediate Peroxidation of Cholesterol Esters in Human Low-Density Lipoprotein. *J. Med. Chem.* **2005**, *48*, 6787–6789. [CrossRef]
104. Nam, T.-G.; Nara, S.J.; Zagol-Ikapitte, I.; Cooper, T.; Valgimigli, L.; Oates, J.A.; Porter, N.A.; Boutaud, O.; Pratt, D.A. Pyridine and Pyrimidine Analogs of Acetaminophen as Inhibitors of Lipid Peroxidation and Cyclooxygenase and Lipoxygenase Catalysis. *Org. Biomol. Chem.* **2009**, *7*, 5103–5112. [CrossRef]
105. Serwa, R.; Nam, T.-g.; Valgimigli, L.; Culbertson, S.; Rector, C.L.; Jeong, B.-S.; Pratt, D.A.; Porter, N.A. Preparation and Investigation of Vitamin B 6-Derived Aminopyridinol Antioxidants. *Chem. Eur. J.* **2010**, *16*, 14106–14114. [CrossRef]
106. Valgimigli, L.; Bartolomei, D.; Amorati, R.; Haidasz, E.; Hanthorn, J.J.; Nara, S.J.; Brinkhorst, J.; Pratt, D.A. 3-Pyridinols and 5-Pyrimidinols: Tailor-Made for Use in Synergistic Radical-Trapping Co-antioxidant Systems. *Beilstein J. Org. Chem.* **2013**, *9*, 2781–2792. [CrossRef]
107. Valgimigli, L.; Ingold, K.U.; Lusztyk, J. Antioxidant activities of vitamin E analogues in water and a Kamlet-Taft β-value for water. *J. Am. Chem. Soc.* **1996**, *118*, 3545–3549. [CrossRef]
108. Valgimigli, L.; Ingold, K.U.; Lusztyk, J. Solvent effects on the reactivity and free spin distribution of 2,2-diphenyl-1-picrylhydrazyl radicals. *J. Org. Chem.* **1996**, *61*, 7947–7950. [CrossRef]
109. Lucarini, M.; Pedulli, G.F.; Valgimigli, L. Do Peroxyl Radicals Obey the Principle That Kinetic Solvent Effects on H-Atom Abstraction Are Independent of the Nature of the Abstracting Radical? *J. Org. Chem.* **1988**, *63*, 4497–449926. [CrossRef]
110. Abraham, M.H.; Grellier, P.L.; Prior, D.V.; Taft, R.W.; Morris, J.J.; Taylor, P.J.; Laurence, C.; Berthelot, M.; Doherty, R.M.; Kamlet, M.J.; et al. A general treatment of hydrogen bond complexation constants in tetrachloromethane. *J. Am. Chem. Soc.* **1988**, *110*, 8534–8536. [CrossRef]
111. Abraham, M.H.; Grellier, P.L.; Prior, D.V.; Morris, J.J.; Taylor, P.J. Hydrogen bonding. Part 10. A scale of solute hydrogen-bond basicity using log K values for complexation in tetrachloromethane. *J. Chem. Soc. Perkin Trans.* **1990**, *2*, 521–529. [CrossRef]
112. Snelgrove, D.W.; Lusztyk, J.; Banks, J.T.; Mulder, P.; Ingold, K.U. Kinetic Solvent Effects on Hydrogen-Atom Abstractions: Reliable, Quantitative Predictions via a Single Empirical Equation. *J. Am. Chem. Soc.* **2001**, *123*, 469–477. [CrossRef]
113. Litwinienko, G.; Ingold, K.U. Solvent Effects on the Rates and Mechanisms of Reaction of Phenols with Free Radicals. *Acc. Chem. Res.* **2007**, *40*, 222–230. [CrossRef] [PubMed]
114. Amorati, R.; Baschieri, A.; Morroni, G.; Gambino, R.; Valgimigli, L. Peroxyl Radical Reactions in Water Solution: A Gym for Proton-Coupled Electron-Transfer Theories. *Chem. Eur. J.* **2016**, *22*, 7924–7934. [CrossRef]
115. Amorati, R.; Valgimigli, L.; Pedulli, G.F.; Grabovskiy, S.A.; Kabalnova, N.N.; Chatgilialoglu, C. Base-promoted reaction of 5-hydroxyuracil derivatives with peroxyl radicals. *Org. Lett.* **2010**, *12*, 4130–4133. [CrossRef]
116. Amorati, R.; Pedulli, G.F.; Valgimigli, L. Kinetic and thermodynamic aspects of the chain-breaking antioxidant activity of ascorbic acid derivatives in non-aqueous media. *Org. Biomol. Chem.* **2011**, *9*, 3792–3800. [CrossRef] [PubMed]
117. Baschieri, A.; Amorati, R.; Benelli, T.; Mazzocchetti, L.; D'Angelo, E.; Valgimigli, L. Enhanced antioxidant activity under biomimetic settings of ascorbic acid included in halloysite nanotubes. *Antioxidants* **2019**, *8*, 30. [CrossRef] [PubMed]
118. Valgimigli, L.; Amorati, R.; Petrucci, S.; Pedulli, G.F.; Hu, D.; Hanthorn, J.J.; Pratt, D.A. Unexpected acid catalysis in reactions of peroxyl radicals with phenols. *Angew. Chem. Int. Ed.* **2009**, *48*, 8348–8351. [CrossRef]
119. Amorati, R.; Valgimigli, L.; Panzella, L.; Napolitano, A.; D'Ischia, M. 5-S-lipoylhydroxytyrosol, a multidefense antioxidant featuring a solvent-tunable peroxyl radical-scavenging 3-thio-1,2-dihydroxybenzene motif. *J. Org. Chem.* **2013**, *78*, 9857–9864. [CrossRef] [PubMed]
120. Valgimigli, L.; Amorati, R.; Fumo, M.G.; DiLabio, G.A.; Pedulli, G.F.; Ingold, K.U.; Pratt, D.A. The unusual reaction of semiquinone radicals with molecular oxygen. *J. Org. Chem.* **2008**, *73*, 1830–1841. [CrossRef] [PubMed]
121. Alfieri, M.L.; Panzella, L.; Amorati, R.; Cariola, A.; Valgimigli, L.; Napolitano, A. Role of Sulphur and Heavier Chalcogens on the Antioxidant Power and Bioactivity of Natural Phenolic Compounds. *Biomolecules* **2022**, *12*, 90. [CrossRef]
122. Kumar, S.; Johansson, H.; Engman, L.; Valgimigli, L.; Amorati, R.; Fumo, M.G.; Pedulli, G.F. Regenerable chain-breaking 2,3-dihydrobenzo[b]selenophene-5-ol antioxidants. *J. Org. Chem.* **2007**, *72*, 2583–2595. [CrossRef]

123. Kumar, S.; Johansson, H.; Kanda, T.; Engman, L.; Muller, T.; Jonsson, M.; Pedulli, G.F.; Petrucci, S.; Valgimigli, L. Catalytic chain-breaking pyridinol antioxidants. *Org. Lett.* **2008**, *10*, 4895–4898. [CrossRef]
124. Johansson, H.; Shanks, D.; Engman, L.; Amorati, R.; Pedulli, G.F.; Valgimigli, L. Long-lasting antioxidant protection: A regenerable BHA analogue. *J. Org. Chem.* **2010**, *75*, 7535–7541. [CrossRef] [PubMed]
125. Amorati, R.; Valgimigli, L.; Dinér, P.; Bakhtiari, K.; Saeedi, M.; Engman, L. Multi-faceted reactivity of alkyltellurophenols towards peroxyl radicals: Catalytic antioxidant versus thiol-depletion effect. *Chem. Eur. J.* **2013**, *19*, 7510–7522. [CrossRef]
126. Hanthorn, J.J.; Amorati, R.; Valgimigli, L.; Pratt, D.A. The reactivity of air-stable pyridine- and pyrimidine-containing diarylamine antioxidants. *J. Org. Chem.* **2012**, *77*, 6895–6907. [CrossRef]
127. Amorati, R.; Baschieri, A.; Cowden, A.; Valgimigli, L. The Antioxidant Activity of Quercetin in Water Solution. *Biomimetics* **2017**, *2*, 9. [CrossRef] [PubMed]
128. Lucarini, M.; Pedulli, G.F.; Valgimigli, L.; Amorati, R.; Minisci, F. Thermochemical and kinetic studies of a bisphenol antioxidant. *J. Org. Chem.* **2001**, *66*, 5456–546210. [CrossRef]
129. Matera, R.; Gabbanini, S.; Berretti, S.; Amorati, R.; De Nicola, G.R.; Iori, R.; Valgimigli, L. Acylated anthocyanins from sprouts of Raphanus sativus cv. Sango: Isolation, structure elucidation and antioxidant activity. *Food Chem.* **2015**, *166*, 397–406. [CrossRef] [PubMed]
130. Niki, E.; Saito, T.; Kawakami, A.; Kamiya, Y. Inhibition of oxidation of methyl linoleate in solution by vitamin E and vitamin C. *J. Biol. Chem.* **1984**, *259*, 4177–4182. [CrossRef]
131. Bowry, V.W.; Ingold, K.U. The Unexpected Role of Vitamin E (α-Tocopherol) in the Peroxidation of Human Low-Density Lipoprotein. *Acc. Chem. Res.* **1999**, *32*, 27–34. [CrossRef]
132. Valgimigli, L.; Lucarini, M.; Pedulli, G.F.; Ingold, K.U. Does β-carotene really protect vitamin E from oxidation? *J. Am. Chem. Soc.* **1997**, *119*, 8095–8096. [CrossRef]
133. Amorati, R.; Pedulli, G.F.; Valgimigli, L.; Johansson, H.; Engman, L. Organochalcogen substituents in phenolic antioxidants. *Org. Lett.* **2010**, *12*, 2326–2329. [CrossRef]
134. McGrath, A.J.; Garrett, G.E.; Valgimigli, L.; Pratt, D.A. The redox chemistry of sulfenic acids. *J. Am. Chem. Soc.* **2010**, *132*, 16759–16761. [CrossRef] [PubMed]
135. Amorati, R.; Lynett, P.T.; Valgimigli, L.; Pratt, D.A. The reaction of sulfenic acids with peroxyl radicals: Insights into the radical-trapping antioxidant activity of plant-derived thiosulfinates. *Chem. Eur. J.* **2012**, *18*, 6370–6379. [CrossRef]
136. Zielinski, Z.; Presseau, N.; Amorati, R.; Valgimigli, L.; Pratt, D.A. Redox chemistry of selenenic acids and the insight it brings on transition state geometry in the reactions of peroxyl radicals. *J. Am. Chem. Soc.* **2014**, *136*, 1570–1578. [CrossRef]
137. Hanthorn, J.J.; Valgimigli, L.; Pratt, D.A. Preparation of highly reactive pyridine- and pyrimidine-containing diarylamine antioxidants. *J. Org. Chem.* **2012**, *77*, 6908–6916. [CrossRef]
138. Shah, R.; Haidasz, E.A.; Valgimigli, L.; Pratt, D.A. Unprecedented inhibition of hydrocarbon autoxidation by diarylamine radical-trapping antioxidants. *J. Am. Chem. Soc.* **2015**, *137*, 2440–2443. [CrossRef] [PubMed]
139. Vaz, S.M.; Augusto, O. Reactive Oxygen Species Special Feature: Inhibition of myeloperoxidase-mediated protein nitration by tempol: Kinetics, mechanism, and implications. *Proc. Natl. Acad. Sci. USA* **2008**, *105*, 8191–8196. [CrossRef] [PubMed]
140. Bi, W.; Bi, Y.; Gao, X.; Li, P.; Hou, S.; Zhang, Y.; Bammert, C.; Jockusch, S.; Legalley, T.D.; Gibson, K.M.; et al. Indole-TEMPO conjugates alleviate ischemia-reperfusion injury via attenuation of oxidative stress and preservation of mitochondrial function. *Bioorg. Med. Chem.* **2017**, *5*, 2545–2568. [CrossRef]
141. Goldstein, S.; Merenyi, G.; Russo, A.; Samuni, A. The Role of Oxoammonium Cation in the SOD-Mimic Activity of Cyclic Nitroxides. *J. Am. Chem. Soc.* **2003**, *125*, 789–795. [CrossRef]
142. Goldstein, S.; Samuni, A. Kinetics and Mechanism of Peroxyl Radical Reactions with Nitroxides. *J. Phys. Chem. A* **2007**, *111*, 1066–1072. [CrossRef]
143. Amorati, R.; Pedulli, G.F.; Pratt, D.A.; Valgimigli, L. TEMPO reacts with oxygen-centered radicals under acidic conditions. *Chem. Commun.* **2010**, *46*, 5139–5141. [CrossRef] [PubMed]
144. Haidasz, E.A.; Meng, D.; Amorati, R.; Baschieri, A.; Ingold, K.U.; Valgimigli, L.; Pratt, D.A. Acid Is Key to the Radical-Trapping Antioxidant Activity of Nitroxides. *J. Am. Chem. Soc.* **2016**, *138*, 5290–5298. [CrossRef] [PubMed]
145. Baschieri, A.; Valgimigli, L.; Gabbanini, S.; DiLabio, G.A.; Romero-Montalvo, E.; Amorati, R. Extremely Fast Hydrogen Atom Transfer between Nitroxides and HOO· Radicals and Implication for Catalytic Coantioxidant Systems. *J. Am. Chem. Soc.* **2018**, *140*, 10354–10362. [CrossRef] [PubMed]
146. Guo, Y.; Baschieri, A.; Amorati, R.; Valgimigli, L. Synergic antioxidant activity of γ-terpinene with phenols and polyphenols enabled by hydroperoxyl radicals. *Food Chem.* **2021**, *345*, 128468. [CrossRef]
147. Baschieri, A.; Daci Ajvazi, M.; Folifack Tonfack, J.L.; Valgimigli, L.; Amorati, R. Explaining the antioxidant activity of some common non-phenolic components of essential oils. *Food Chem.* **2017**, *232*, 656–663. [CrossRef]
148. Guo, Y.; Pizzol, R.; Gabbanini, S.; Baschieri, A.; Amorati, R.; Valgimigli, L. Absolute Antioxidant Activity of Five Phenol-Rich Essential Oils. *Molecules* **2021**, *26*, 5237. [CrossRef]
149. Khoobchandani, M.; Ganesh, N.; Gabbanini, S.; Valgimigli, L.; Srivastava, M.M. Phytochemical potential of Eruca sativa for inhibition of melanoma tumor growth. *Fitoterapia* **2011**, *82*, 647–653. [CrossRef]
150. Cores, Á.; Piquero, M.; Villacampa, M.; León, R.; Menéndez, J.C. NRF2 Regulation Processes as a Source of Potential Drug Targets against Neurodegenerative Diseases. *Biomolecules* **2020**, *10*, 904. [CrossRef]

151. Matera, R.; Lucchi, E.; Valgimigli, L. Plant Essential Oils as Healthy Functional Ingredients of Nutraceuticals and Diet Supplements: A Review. *Molecules* **2023**, *28*, 901. [CrossRef]
152. Zamanian, M.Y.; Soltani, A.; Khodarahmi, Z.; Alameri, A.A.; Alwan, A.M.R.; Ramírez-Coronel, A.A.; Obaid, R.F.; Abosaooda, M.; Heidari, M.; Golmohammadi, M.; et al. Targeting Nrf2 signaling pathway by quercetin in the prevention and treatment of neurological disorders: An overview and update on new developments. *Fundam. Clin. Pharmacol.* **2023**, 1–15. [CrossRef] [PubMed]

biomolecules

Article

Relationship between 4-Hydroxynonenal (4-HNE) as Systemic Biomarker of Lipid Peroxidation and Metabolomic Profiling of Patients with Prostate Cancer

Matea Nikolac Perkovic [1], Morana Jaganjac [1], Lidija Milkovic [1], Tea Horvat [1], David Rojo [2], Kamelija Zarkovic [3], Marijana Ćorić [3], Tvrtko Hudolin [4], Georg Waeg [5], Biserka Orehovec [6] and Neven Zarkovic [1,*]

1 Division of Molecular Medicine, Ruder Boskovic Institute, HR-10000 Zagreb, Croatia
2 Centro de Metabolómica y Bioanálisis (CEMBIO), Facultad de Farmacia, Universidad CEU San Pablo, Campus Monteprincipe, ES-28003 Madrid, Spain
3 Division of Pathology, School of Medicine, University of Zagreb, University Hospital Centre Zagreb, Kispaticeva 12, HR-10000 Zagreb, Croatia
4 Division of Urology, School of Medicine, University of Zagreb, University Hospital Centre Zagreb, Kispaticeva 12, HR-10000 Zagreb, Croatia
5 Institute of Molecular Biosciences, Karl Franzens University, A-8010 Graz, Austria
6 Clinical Hospital Dubrava, HR-10000 Zagreb, Croatia
* Correspondence: zarkovic@irb.hr

Abstract: An oxidative degradation product of the polyunsaturated fatty acids, 4-hydroxynonenal (4-HNE), is of particular interest in cancer research due to its concentration-dependent pleiotropic activities affecting cellular antioxidants, metabolism, and growth control. Although an increase in oxidative stress and lipid peroxidation was already associated with prostate cancer progression a few decades ago, the knowledge of the involvement of 4-HNE in prostate cancer tumorigenesis is limited. This study investigated the appearance of 4-HNE-protein adducts in prostate cancer tissue by immunohistochemistry using a genuine 4-HNE monoclonal antibody. Plasma samples of the same patients and samples of the healthy controls were also analyzed for the presence of 4-HNE-protein adducts, followed by metabolic profiling using LC-ESI-QTOF-MS and GC-EI-Q-MS. Finally, the analysis of the metabolic pathways affected by 4-HNE was performed. The obtained results revealed the absence of 4-HNE-protein adducts in prostate carcinoma tissue but increased 4-HNE-protein levels in the plasma of these patients. Metabolomics revealed a positive association of different long-chain and medium-chain fatty acids with the presence of prostate cancer. Furthermore, while linoleic acid positively correlated with the levels of 4-HNE-protein adducts in the blood of healthy men, no correlation was obtained for cancer patients indicating altered lipid metabolism in this case. The metabolic pathway of unsaturated fatty acids biosynthesis emerged as significantly affected by 4-HNE. Overall, this is the first study linking 4-HNE adduction to plasma proteins with specific alterations in the plasma metabolome of prostate cancer patients. This study revealed that increased 4-HNE plasma protein adducts could modulate the unsaturated fatty acids biosynthesis pathway. It is yet to be determined if this is a direct result of 4-HNE or whether they are produced by the same underlying mechanisms. Further mechanistic studies are needed to grasp the biological significance of the observed changes in prostate cancer tumorigenesis.

Keywords: cancer; prostate carcinoma; lipid peroxidation; 4-hydroynonenal (4-HNE); metabolomics; GS-MS; LC-MS

Citation: Perkovic, M.N.; Jaganjac, M.; Milkovic, L.; Horvat, T.; Rojo, D.; Zarkovic, K.; Ćorić, M.; Hudolin, T.; Waeg, G.; Orehovec, B.; et al. Relationship between 4-Hydroxynonenal (4-HNE) as Systemic Biomarker of Lipid Peroxidation and Metabolomic Profiling of Patients with Prostate Cancer. *Biomolecules* **2023**, *13*, 145. https://doi.org/10.3390/biom13010145

Academic Editor: Chryssostomos Chatgilialoglu

Received: 30 November 2022
Revised: 29 December 2022
Accepted: 4 January 2023
Published: 10 January 2023

1. Introduction

Prostate cancer is the second most frequent malignancy in men worldwide, with more than 1,4 million new cases per year, causing 3.8% of all malignant deaths in 2020 [1]. It is a heterogeneous disease with incidence rates that vary across the world and increase with age. The etiology of prostate cancer is multifactorial and remains largely unknown compared to other common cancers. Some well-established risk factors include advanced age, positive family history, genetic factors, and African ancestry. Prostate cancer is a slow-growing cancer that may be asymptomatic at the early stage and often has an indolent course. Late-stage cancer can present with bone pain, urinary symptoms, and/or weight loss [2].

Most cases of prostate cancer are revealed using digital rectal examination, by diagnostic tests to determine prostate-specific antigen (PSA) levels, transrectal ultrasound (TRUS), and multiparametric MRI, while diagnosis is confirmed by tissue analysis obtained on image-guided transrectal biopsy. Almost all prostate cancers are histologically adenocarcinoma.

The International Society of Urological Pathology (ISUP) consensus conferences in 2005, 2014, and 2019, as well as the 2019 white paper by the Genitourinary Pathology Society (GUPS), report the introduction of computer-aided cancer grading using artificial intelligence [3]. Hence, a series of studies have shown that artificial intelligence-based algorithms can perform prostate cancer grading at the level of international experts in prostate pathology [4].

The tumorigenesis of prostate cancer is accompanied by excessive reactive oxygen species (ROS) production and impaired redox homeostasis. ROS have a dual role in tumorigenesis, and while the moderate elevation of ROS may promote anti-tumor effects, excessive ROS may support the development and progression of cancer [5,6]. Indeed, metastatic prostate cancer cells thrive under excessive ROS, and to achieve their addictive microenvironment, prostate cancer cells regulate their intracellular redox status and respiratory burst of neutrophils [7]. Unsaturated fatty acids are particularly sensitive to ROS-induced injury, and the abstraction of allylic hydrogen initiates lipid peroxidation, yielding reactive aldehydes, such are malondialdehyde, acrolein and 4-hydroxynonenal (4-HNE) [8]. Plasma malondialdehyde [9,10] and the presence of acrolein–protein conjugates in tumor tissues are associated with the progression of prostate carcinoma [11]. However, although the bioactive properties and biological significance of 4-HNE are well-recognized today [12], the possible involvement of 4-HNE in the pathogenesis of prostate cancer is limited, whereas our preliminary findings suggested that 4-HNE is not present in prostate carcinoma cells at all [13].

The biotransformation of free 4-HNE occurs within several minutes, with the maximal catabolism rate varying between tissues [14]. A number of primary and secondary metabolites have been identified, with the predominant urine metabolite being 1,4-dihydroxynonane mercapturic acid [15–17]. Moreover, the high reactivity of 4-HNE contributes to its removal via 4-HNE adduction to macromolecules. 4-HNE readily modifies diverse proteins, affecting their structure and function in both physiology and pathology [18–20], and is thus often studied in biological tissues and fluids.

Metabolic reprogramming is another hallmark of prostate cancer development and progression. Recent studies have evidenced marked changes in the plasma [21], serum [22], urine [23], and seminal plasma [24] metabolomes of prostate cancer patients. However, whether 4-HNE is linked to these perturbations is unknown. Therefore, in this study, we investigated the presence of 4-HNE-protein adducts in both plasma samples and tissues of prostate cancer patients to search for the possible link between 4-HNE and metabolome perturbation in order to better understand the potential role of 4-HNE in prostate cancer development and its association with metabolic alterations in prostate cancer patients.

2. Materials and Methods

2.1. Chemicals and Reagents

For metabolomic analyses, the following were used: acetonitrile (ACN) (LC-MS grade, Sigma-Aldrich, Steinheim, Germany), ethanol (EtOH) (LC-MS grade, Sigma-Aldrich, Steinheim, Germany), formic acid (FA) (MS grade, Sigma-Aldrich, Steinheim, Germany), heptane (Sigma-Aldrich, Steinheim, Germany), methanol (MeOH) (Sigma-Aldrich, Steinheim, Germany), O-methoxyamine hydrochloride (Sigma-Aldrich, Steinheim, Germany),

N,O-bis(trimethylsilyl) trifluoroacetamide (BSTFA) with 1% trimethylchlorosilane (TMCS) (Pierce Chemical Co., Rockford, IL, USA), and pyridine (Sigma-Aldrich, Steinheim, Germany). Ultrapure water was obtained from the MilliQ®plus 185 system (Millipore, Billerica, MA, USA). Tricosane (Sigma-Aldrich, Steinheim, Germany) and 4-chlorophenol (Sigma-Aldrich, Steinheim, Germany) were used as internal standards in the GC-MS analysis. The FAME mix (mix of fatty acid methyl esters; methyl caprylate, methyl caprate, methyl laurate, methyl myristate, methyl palmitate, methyl heptadecanoate, methyl oleate, methyl stearate, methyl eicosanoate, methyl docosanoate) for GC-MS analytical platform was purchased from Supelco (Bellefonte, PA, USA). Ammonium trifluoroacetate (TFA(NH4)), purine, and hexakis(1H,1H,3H-tetrafluoropropoxy)phosphazine (HP) from API-TOF reference mass solution kit (Agilent Technologies, Waldbronn, Germany) were diluted in 95:5 of ACN to water ratio and used as reference solution in the LC-MS analysis.

2.2. Subjects and Sample Collection

This study was carried out according to the approval of the Ethic Committee of the University Clinical Hospital Centre Zagreb (approval code 02/21 AG), while each patient signed the informed consent. All of the patients included in this study underwent open radical prostatectomy for localized prostate cancer based on prostate-specific antigen (PSA) values, digital rectal examination (clinical stage $\leq$ T2c), biopsy findings ($\leq$ Gleason grade group 3), and imaging modalities (computed tomography, bone scintigraphy, magnetic resonance). The blood samples of 30 patients were taken by venipuncture before surgery and collected into ethylenediaminetetraacetic acid (EDTA) tubes with BHT and were centrifuged at $3000\times g$ (4 °C) for 20 min to obtain the plasma. The control plasma samples were obtained from 41 fasted, healthy men following the same procedure as in the case of patients. The plasma samples were stored at −80 °C for subsequent analysis.

The surgically obtained tissue specimens were stored in formalin before further evaluation by pathologists, while patients were regularly followed up at the Urology Clinic at the University Hospital Centre Zagreb (approval code 02/21 AG).

2.3. Immunohistochemistry

The specimens of the tissue obtained by surgery were fixed in 10% buffered formalin immediately after resection, dehydrated in ethanol, and embedded in paraffin. The representative paraffin blocks of each tumor and surrounding mucosa were cut into three 5μm thin slices and examined by section staining with hematoxylin and eosin and immunohistochemistry using monoclonal antibody for 4-HNE-histidine obtained from the culture medium of the clone derived from a fusion of Sp2-Ag8 myeloma cells with B-cells of a BALBc mouse immunized by HNE-modified keyhole limpet hemocyane [25]. Dilutions of the antibody solution and appropriate reagents from the EnVision detection kit (K 8000, DAKO, Glostrup, Denmark) were used on a DAKO automated immunostainer. The 4-HNE-histidine antigens were localized using enzyme-labeled polymer conjugated to a secondary antibody with 3, 3,-diaminobenzidine as a chromogen and counterstained with hematoxylin (Kemika, Zagreb, Croatia).

The immunohistochemical investigation of the intensity and distribution of 4-HNE in the tumor and the surrounding non-tumorous mucosa was determined by expert pathologists. The presence of 4.HNE-protein adducts in carcinoma cells, stroma, and blood vessels was defined either as negative (0) if there was no immunohistochemical positivity observed or as positive (+) if the presence of 4-HNE-protein adducts was observed. Two pathologists diagnosed each specimen independently.

2.4. HNE-ELISA

A previously described in-house protocol [26,27] was used to measure the levels of 4-HNE-protein adducts in the plasma samples of the healthy controls (n = 41) and prostate cancer patients (n = 30). Briefly, 10 μL of standards/samples (adjusted to 10 mg/mL) per well of an ELISA plate (Nunc Immuno Maxisorp, Thermo Scientific, Nunc A/S,

4000 Roskilde, Denmark) was added into 100 μL of 0.05 M carbonate-binding buffer (pH 9.6; 0.015 M sodium carbonate, 0.035 M sodium bicarbonate) in triplicate and incubated for 5 h at 4 °C. Before the addition of the same monoclonal primary antibody as used for immunohistochemistry and overnight (ON) incubation at 4 °C, the wells were blocked with 5% fat-free dry milk in carbonate-binding buffer) for 3 h at room temperature (RT). The next day, after 30 min of endogenous peroxidase blocking, the wells were incubated with the secondary antibody for 1 h at RT, followed by the addition of the substrate solution and absorbance reading at 450/620 nm. Before each step, the wells were washed five times with a washing buffer. Concentrations of 4-HNE protein adducts were interpolated from the standard curve and expressed as pmol 4-HNE protein adducts/mg of proteins.

2.5. *Sample Preparation and Metabolite Extraction*

2.5.1. LC-MS Platform

On the day of analysis, the plasma samples (stored at −80 °C) were slowly defrosted on ice and vortex-mixed for 2 min. For metabolite extraction, 100 μL of each sample was mixed with cold MeOH:EtOH (1:1 ratio), vortex-mixed for 2 min and incubated on ice for 5 min. After centrifugation (16,000× *g*, 10 min, 4 °C), the remaining supernatant (200 μL) was transferred to crimp-top clear glass vials with a 300 μL insert for further LC-MS analysis.

2.5.2. GC-MS Platform

For the GC-MS analysis, the samples were aliquoted on the same day as the LC-MS analysis. The deproteinization was achieved by mixing the plasma samples (100 μL) with 300 μL of cold ACN. The samples were vortex-mixed for 2 min and incubated on ice for 5 min. The corresponding aliquot of each sample (100 μL) was centrifugated (16,000× *g*, 10 min, 4 °C), and 100 μL of the supernatant was transferred to the crimp-top clear glass vials with the insert, and subsequently, 20 μL of 4-chlorophenol (100 ppm, in ACN) was added. The samples were evaporated to dryness using a Speedvac Concentrator (Thermo Fisher Scientific, Waltham, MA, USA). Methoximation was performed by adding 10 μL of O-methoxyamine hydrochloride (15 mg/mL in pyridine). The samples were vigorously vortex-mixed for 5 min, followed by three cycles of ultrasonication (2 min) and vortex mixing (2 min). In the next step, the vials were incubated in darkness at room temperature for 16 h. The next day, 10 μL of BSTFA with 1% TMCS was added to each vial for silylation, and the samples were vortex-mixed for 5 min. Silylation was carried out at 70 °C for 1 h. After the samples cooled down to room temperature, 100 μL of tricosane (20 ppm in heptane) was added as an internal standard, and the samples were vortex-mixed for 2 min. Six blank samples (ACN to water ratio 3:1) were prepared in the same way as the plasma samples (including deproteinization and derivatization steps).

2.6. *Preparation of Quality Control Samples (QCs)*

The QCs were prepared separately for the LC-MS and GC-MS platforms. They are necessary to equilibrate/stabilize the system before analyzing the actual set of samples and, afterward, to monitor system stability and reproducibility during the analysis. Individual QCs were prepared by pooling and mixing equal volumes of each plasma sample (10 μL). They were processed in the same way as the actual samples following all of the previously described steps.

2.7. *Metabolomics Analysis*

To achieve a wider metabolite coverage, all of the samples were analyzed using two complementary analytical platforms, LC-ESI-QTOF-MS (abbreviated as LC-MS) and GC-EI-Q-MS (abbreviated as GC-MS).

2.7.1. Fingerprinting by LC-ESI-QTOF-MS

The liquid chromatography system, Agilent Technologies Series 1200 binary solvent delivery system (Agilent Technologies, Waldbronn, Germany), comprised of a binary pump, an integrated degasser, and an autosampler with a thermostat, coupled to an Agilent 6520 Accurate-Mass Q-TOF detector, was used to analyze the metabolic profile of the samples. For each sample, a volume of 10 µL has been injected into a reversed-phase column (Discovery®HS C18 HPLC Column, 515 cm × 2.1 mm, 3 µm; Supelco, Bellefonte, PA, USA) with a pre-column (Discovery®HS C18 HPLC Column, 2 cm × 2.1 mm, 3 µm; Supelco, Bellefonte, PA, USA), which was kept at 60 °C during the analysis. The elution conditions employed a flow rate of 0.6 mL/min with a gradient of solvent A (H2O with 0.1% FA) and solvent B (ACN with 0.1% FA). The analysis started with 25% of the mobile phase B and then increased to 95% of B in a time period of 35 min (0–35 min). The gradient then returned to the initial conditions in 1 min time (35–36 min), 25% of the mobile phase B, and these conditions were maintained until the end of the analysis (36–45 min). All of the samples were analyzed in both positive and negative ESI modes (full-scan ranging from 50 to 1000 m/z), with a scan rate of 1.02 scans/s. Two reference masses were continuously infused during the entire duration of the analysis to ensure a constant mass correction: 121.0509 (purine, detected m/z $[C_5H_4N_4+H]^+$) and 922.0098 (HP, detected m/z $[C_{18}H_{18}O_6N_3P_3F_{24}+H]^+$) for the positive mode, and 112.9855 (TFA(NH_4), detected m/z $[C_2O_2F_3(NH_4)\text{-}H]^-$) and 966.0007 (HP+FA, detected m/z $[C_{18}H_{18}O_6N_3P_3F_{24}+FA\text{-}H]^-$) for the negative mode.

Tandem mass spectrometry (MS/MS) was performed to facilitate the identification of significant metabolites using the same LC (Agilent 1200)-QTOF-MS (Agilent 6520) platform and the same chromatographic conditions as applied for the primary LC-MS analysis. The selected ions were targeted for fragmentation by collision-induced dissociation (CID) based on the previously determined accurate mass and retention time. Multiple collision energies (10 eV, 20 eV, and 40 eV) were used. The identity of the compounds was confirmed by comparing the fragmentation pattern of the selected ion with a public library of MS/MS spectra.

2.7.2. Fingerprinting by GC-EI-Q-MS

The Agilent 7890A gas chromatograph, with an autosampler (Agilent Technologies 7693), coupled to an inert MSD with Quadrupole (Agilent Technologies 5975) was used for the metabolomic fingerprinting of the plasma samples. For each derivatized sample, a volume of 2 µL was injected, with a split ratio of 1:10, into a Restek 20,782 deactivated glass–wool split liner. The compounds were separated using the GC-Column DB-5MS (length: 30 m, internal diameter: 0.25 mm, film thickness: 0.25 µm, packing: 95% dimethylpolysiloxane/5% diphenylpolysiloxane) with a pre-column (10 m J&W integrated with Agilent 122-5532G). The constant flow rate of the helium carrier gas was set to 1 mL/min, and the injector temperature was held at 250 °C. The temperature of the column oven was set at 60 °C (held for 1 min), with an increased rate of 10 °C/min until the temperature reached 325 °C. This temperature was maintained for up to 10 min before the injection of the next sample. The detector transfer line was set at 290 °C, while the filament source and quadrupole temperatures were set at 230 °C and 150 °C, respectively. The total analysis for each sample lasted 37.5 min. The electron ionization (EI) energy was set to 70 eV. The system collected the mass spectra in a mass range between 50 and 600 m/z at a rate of 2 spectra/s.

2.8. Data Treatment

2.8.1. LC-MS Data Treatment

The data treatment was done using Agilent MassHunter software tool (Agilent Technologies, Waldbronn, Germany). Firstly, the quality of the analysis for all of the analyzed samples and QCs was checked by inspecting and assessing the quality of the total ion chromatograms (TIC), checking the pressure curves to assess the stability of chromatographic conditions, and reviewing the stability of the reference masses' signal (using Agilent MassHunter Quantitative Analysis software, version B.07.00). All of the samples have passed all of the checkpoints. Subsequently, the raw data were imported into Agilent MassHunter Profinder software (version B.08.00) for deconvolution. The Molecular Feature Extraction (MFE) algorithm was used for deconvolution, creating a list of possible molecular features that matches a gaussian distribution of co-eluting ions related by charge-state, isotopic distribution and/or the presence of different adducts, and dimmers. Subsequently, a second deconvolution step was performed using the Recursive Feature Extraction (RFE) algorithm, which re-integrates the MFE results, improving the quality of the final features list. The resulting list of statistically significant accurate masses was annotated using the CEU Mass Mediator [28] search tool (maximum error mass $\pm$ 20 ppm) in order to assign tentative metabolite candidates. The matched compounds were identified using the accurate mass and by checking their isotopic pattern. Only the features with the highest score were kept for further identity confirmation. The biological role of each suggested compound was additionally evaluated to exclude the unrelated and impossible identification matches. Tandem mass spectrometry (MS/MS) was performed for the statistically significant annotated features in positive and negative ionization. The final identification of these compounds was performed by matching the fragmentation spectra as described by Naz and colleagues [29], using different databases such as HMDB [30], METLIN [31], KEGG [32], and LipidMaps [33]. For the identification of the compound, we considered proper retention time, accuracy mass (maximum error mass $\pm$ 20 ppm), and at least two MS/MS fragments.

2.8.2. GC-MS Data Treatment

The data treatment was done using Agilent MassHunter software tool (Agilent Technologies, Waldbronn, Germany). After inspecting and assessing the quality of the total ion chromatograms (TIC) for all of the analyzed samples, QCs, and blanks (using Agilent MassHunter Quantitative Analysis software, version B.07.00), and after checking the reproducibility of the signals of the internal standards (4-chlorophenol for derivatization and tricosane for analytical performance), all the samples were accepted. Then, the raw data files were imported into the Agilent MassHunter Unknowns Analysis software (version B.09.00) for deconvolution and identification using targeted libraries, Fiehn library, (version 2013, UC Davis, Davis, CA, USA) and the in-house CEMBIO spectral library for plasma samples. The identities of the compounds were assigned based on the retention time (RT) and mass spectra. The next step was an additional checking of the identified compounds and the non-identify features using the NIST library (National Institute of Standards and Technology, library 2.2 version 2014, Gaithersburg, MD, USA). The alignment of the obtained data was performed with the help of Agilent MassProfiler Professional software (version 13.0) and exported into Agilent MassHunter Quantitative Analysis (version B.09.00) for peak integration. The abundance of each compound in the obtained data matrix was normalized according to the IS (tricosane) abundance, and the blank subtraction was performed prior to statistical analysis.

2.9. *Statistical Analysis*

For both LC-MS and GC-MS, before the statistical analysis of the obtained data, the raw variables were filtered based on the criteria proposed by Godzien and colleagues [34]. The variables retained were (i) present in $\geq$80% of the QCs (with relative standard deviation (RSD) <30% in QC samples); or 9ii) present in <20% of the QCs, but present in $\geq$50% of the samples in a specific subject group. Both sets (i and ii) were subsequently modeled. To correct for the intra-batch effect, the analytical variations in QC samples were assessed, and the Quality Control-Robust Spline Correction (QC-RSC) algorithm was used when necessary, as suggested by Kuligowski and colleagues [35]. Support vector regression (QC-RSC) was performed using MATLAB (7.10.0.499, MathWorks, Natick, MA, USA) and the LIBSVM library [36] (software available at http://www.csie.ntu.edu.tw/~cjlin/libsvm).

After removing intra-batch effects, normalization was used to decrease the unwanted variations arising from errors in the sample preparation [37]. Auto Scaling (Unit Variance Scaling, UV) was used to normalize and scale metabolic signals based on the standard deviation of metabolomics data [38]. In the case of the GC-MS data, additionally, the abundance of all of the detected compounds was normalized by the signal of internal standard (tricosane) in each sample. The missing values in the data sets were replaced with zeros.

The SIMCA-P+ software (version 15.0.2.5959, Umetrics, Umea, Sweden) was used for multivariate statistical analyses, including building up Principal Component Analysis (PCA) models and Orthogonal PLS-DA (OPLS-DA) models. Based on OPLS-DA models, variable importance in the projection (VIP) values were obtained. Variables with VIP > 1.00 were considered significant. The MetaboAnalyst 5.0 software was used for metabolomics data pathway analysis [39,40].

Univariate statistical analysis was carried out using MATLAB (7.10.0.499, MathWorks, Natick, MA, USA) or GraphPad Prism software (GraphPad Software, San Diego, CA 92108, USA). The Shapiro–Wilk test was used to check for normal distribution in the data set. For the comparison of metabolite abundances between groups, Student's t-test or Mann–Whitney U test was performed, depending on the distribution of the data, followed by Benjamini–Hochberg (FDR, false discovery rate) post hoc correction for multiple comparisons. An unpaired t-test with Welch's correction was used to detect the differences in the levels of 4-HNE protein adducts between healthy controls and prostate cancer patients. The relationship between significantly altered metabolites and levels of plasma 4-HNE-modified proteins in both cancer patients and healthy subjects was tested using Spearman's correlation coefficient.

All of the tests were two-tailed, and $p < 0.050$ was considered significant. In the case of metabolomic data, metabolites with significantly different abundance between the two subject groups were determined by the combination of multivariate and univariate statistics.

Fold change was computed as the ratio between the mean metabolite abundance in the PC group cohort relative to the healthy control group. The percentage of change (%) was calculated as follows: [(average value in the PC group − average value in the control group)/(average value in the control group)] × 100, with positive values indicating increased abundance and negative values decreased abundance of specific metabolites in the PC group when compared to the healthy control group.

3. Results

This study included 30 patients with pathohistologically verified prostate cancer and 41 healthy male controls aiming to evaluate the impact of 4-HNE on plasma proteins and metabolome. The general characteristics of the prostate cancer patients are presented in Supplementary Table S1. The prostate-specific antigen (PSA) values before surgery ranged from 3.5 to 41 with a median value of 6.25, while the median of Gleason values of tumor differentiation was 7 (range 6–8) (Supplementary Table S1). The immunohistochemical appearance of 4-HNE-protein conjugates is shown in (Figure 1). The 4-HNE-modified proteins were not detected in any cancer cell, while the presence of 4-HNE-modified proteins was detected only in the stromal cells and blood vessels in cancer tissue of only one patient (patient no.1). The results of the immunohistochemical findings for each prostate cancer tissue are included in Supplementary Table S1. The comparison of 4-HNE-positive staining in cancer and adjacent tumor tissue also did not reveal any significant differences, i.e., 4-HNE was neither present in cancer nor in non-malignant surrounding prostate tissue. However, a significant increase in 4-HNE-modified proteins was observed in the plasma of patients with prostate cancer (Figure 1). Hence, the plasma samples of prostate cancer patients have significantly more 4-HNE-protein adducts compared to the healthy controls ($p < 0.0001$). The levels of 4-HNE-protein adducts did not correlate with the PSA values.

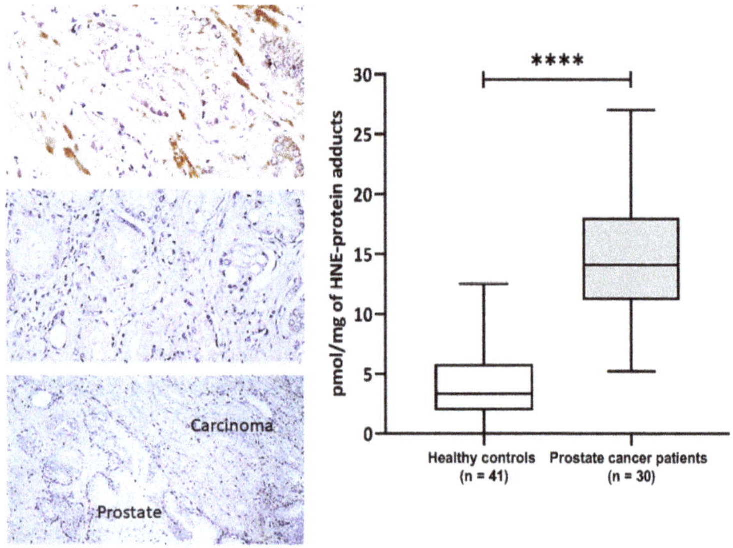

Figure 1. The 4-hydroxynonenal (4-HNE) modification of plasma and prostate cancer tissue proteins. Left—Immunohistochemistry of prostate carcinoma specimens obtained by monoclonal antibody specific for the 4-HNE-protein adducts visualized the 4-HNE presence by brown color (DAB staining), while negative cells were contrast-stained blue by hematoxylin. Photo on the top shows rare immunopositivity of the stromal cells observed only for one patient/cancer (200×), the middle photo shows the most usually observed absence of 4-HNE in cancer and in stromal cells (200×), while the lower photo shows the absence of 4-HNE both in cancer and in the adjacent prostate tissue (100×). Right—Plasma concentration of 4-HNE–protein adducts (pmol/mg of protein) in samples of healthy controls (n = 41) and prostate cancer patients (n = 30). Results are presented as a box and whisker plot. The line in the box represents the median, while the interquartile range (IQR) box represents the middle quartiles (the 75th minus the 25th percentile). The whiskers on either side of the IQR box represent the lowest and highest quartiles of the data. The ends of the whiskers represent the maximum and minimum of the data. **** $p < 0.0001$.

The fingerprinting of the plasma metabolome revealed distinct metabolic signatures of prostate cancer patients and healthy controls (Figure 2). To explore the differences in the overall metabolomic profiles of the prostate cancer patients and healthy control subjects, the PCA of all of the samples was used. The PCA showed a clear distribution between the two groups of subjects. Note that the PCA is a non-supervised multivariant model in which the tight cluster of the QCs validates the analytical performance and the clinical model, corroborating the biological differences between the clinical groups. Subsequently, a supervised OPLS-DA model and the VIP score were used in order to identify the metabolites that contributed the most to the difference between patients with prostate cancer and healthy subjects.

The PCA score plots revealed a clear and separate clustering between two groups of subjects (Figure 2). All of the OPLS-DA models were built from one predictive component and two orthogonal components (Figure 1). For all of the OPLS-DA models, the R2(cum) and Q2(cum) values exceeded 0.5, indicating the robustness of the models. This suggests that the models fit the data very well and have a good predictive ability. All three OPLS-DA models were evaluated with permutation analysis (Figure S1A). The permutation analyses strongly indicate that the original models are valid since the criteria for validity are satisfied (all permutated Q2-values to the left are lower than the original points to the right, and the blue regression line of the Q2-points intersects the vertical axis below zero) (Figure S1A). All of the permutated R2-values and Q2-values are lower than the original values on the right (Figure S1A). To identify strong and moderate outliers, we used Hoteling's T2 line plot (Figure S1B) and the DModX test (Figure S1C). In the case of GC-MS, an LC-MS(+), we detected one outlier in each data set (Figure S1B,C). In the case of GC-MS, the outlier was one participant from the healthy control group, and in the case of LC-MS(+), it was one participant with prostate cancer.

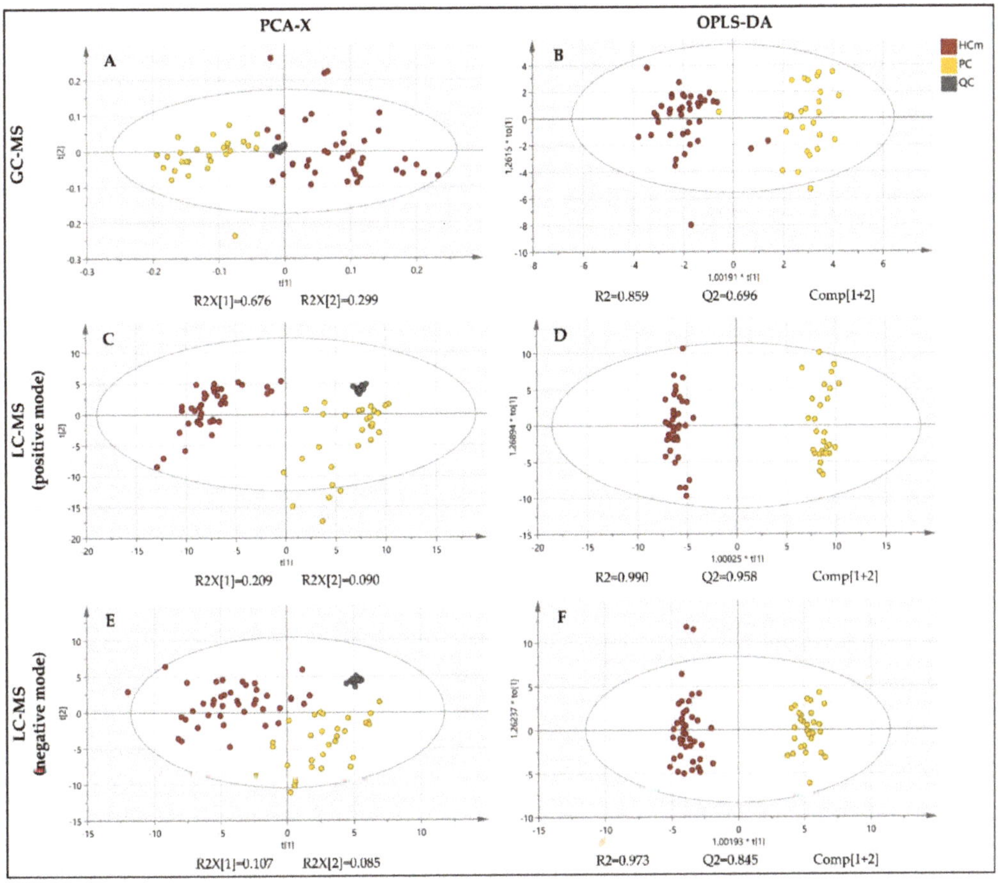

Figure 2. PCA and OPLS-DA score plot of the untargeted metabolomics analysis of plasma samples from patients diagnosed with prostate cancer (PC) and healthy male control subjects (HCm). Plots were obtained using SIMCA-P+ software (version 15.0.2.5959, Umetrics, Umea, Sweden). (**A**) PCA score plot for the GC-MS analysis; (**B**) OPLS-DA score plot fort the GC-MS analysis; (**C**) PCA score plot fort the LC-MS(+) analysis; (**D**) OPLS-DA score plot fort the LC-MS(+) analysis; (**E**) PCA score plot fort the LC-MS(−) analysis; (**F**) OPLS-DA score plot fort the LC-MS(−) analysis; QC = quality control. t[1]: the first principal component, t[2]: the second principal component, to[1]: the first orthogonal component, *: multiplication sign.

Using the GC-MS approach, a total of 18 metabolites were found to be significantly altered between patients with prostate cancer and healthy control subjects (Table 1). These differential metabolites were mainly fatty acyls, organic acids and their derivates, or different carbohydrates and carbohydrate conjugates. The list of metabolites with significantly different levels between the cancer patients and healthy men was determined by the combination of multivariate and univariate statistics, as presented in Table 1.

Table 1. Significantly altered metabolites, detected by GC-MS, between patients with prostate cancer and healthy control subjects.

Category	Compound	RT	%Δ	FC	log_2FC	pBH	VIP
Fatty Acyls	Caprylic acid (octanoic acid)	9.79	41.43	1.41	0.50	<0.0001	1.17
	Caproic acid (hexanoic acid)	7.06	−68.94	0.31	−1.69	<0.0001	2.04
	Lauric acid (dodecanoic acid)	14.75	69.22	1.69	0.76	0.002	1.00
	Palmitic acid (hexadecanoic acid)	18.87	92.58	1.93	0.95	<0.0001	1.06
	Stearic acid (octadecanoic acid)	20.69	80.19	1.80	0.85	0.002	1.02
	Palmitoleic acid (hexadecenoic acid)	18.68	181.07	2.81	1.49	<0.0001	1.00
	Linoleic acid (octadecadienoic acid)	20.41	140.74	2.41	1.27	<0.0001	1.02
	Oleic acid (octadecenoic acid)	20.46	177.32	2.77	1.47	<0.0001	1.07
Organic acids and derivatives	Lactic acid (2-hydroxypropanoic acid)	6.85	−56.24	0.44	−1.19	<0.0001	3.96
	2-hydroxybutyric acid	7.79	87.56	1.88	0.91	<0.0001	1.09
	3-hydroxybutyric acid	8.28	257.75	3.58	1.84	<0.0001	1.58
	Pyruvic acid (2-oxopropanoic acid)	6.70	−65.78	0.34	−1.55	<0.0001	2.04
	2-ketoisocaproic acid (ketoleucine)	8.54	33.61	1.34	0.42	0.049	1.03
Carbohydrates and carbohydrate conjugates	Glycerol	9.87	69.00	1.69	0.76	<0.0001	1.08
	Glyceric acid	10.65	30.93	1.31	0.39	0.002	1.00
	Mannose	17.22	19.80	1.20	0.26	0.002	1.02
	Galactose/glucose	17.55	17.57	1.18	0.23	0.002	1.66
Sterol Lipids	Cholesterol	27.57	30.47	1.30	0.38	0.003	1.04

%Δ, percentage of change; FC, fold change; pBH, Benjamini–Hochberg adjusted *p*-value; RT, retention time; VIP, variable importance in the projection.

Thus, the obtained results indicate increased levels of different fatty acids in patients with prostate cancer compared to the healthy controls, except for caproic acid, the level of which was significantly higher in the control subjects (Table 1). All carbohydrates were increased in prostate cancer, and the same trend was observed for cholesterol (Table 1). The analysis of the organic acids indicated elevated levels of 2-hydroxybutyric, 3-hydroxybutyric, and 2-ketoisocaproic acid and decreased levels of lactic and pyruvic acid in prostate cancer compared with healthy men (Table 1). It should be noted that the elevation of different fatty acids may indicate a catabolic shift of metabolism due to cancer.

The LC-MS metabolite profiling of the plasma samples from the prostate cancer patients resulted in large spectra of metabolites. To explore the differences in the overall metabolomic profiles of the prostate cancer patients and healthy control subjects, obtained by both positive and negative ionization modes, the PCA of all of the samples was used. The PCA showed a clear distribution between the two groups of subjects (Figure 2). A supervised OPLS-DA model and the VIP score were used to identify the metabolites that contributed the most to the difference between patients with prostate cancer and healthy subjects. The LC-MS followed an analogous data treatment, as described for GC-MS. The resulting list of compounds is presented in Table 2. The compounds that could not be identified according to their MS/MS spectra were excluded.

Table 2. Significantly altered metabolites, detected by LC-MS, between patients with prostate cancer and healthy control subjects.

Category	Compound	ESI Mode	*m/z*	RT	%Δ	FC	log_2FC	pBH	VIP
Fatty Acyls	Thapsic acid (hexadecanedioic acid)	-	285.2072	16.30	138.00	2.38	1.25	<0.0001	1.47
	Methylhexadecenoic acid	+	269.2461	18.70	86.68	1.87	0.90	<0.0001	1.05
	Palmitoleic acid	-	253.2176	27.60	88.09	1.88	0.91	<0.0001	1.14
	Linolenic acid (octadecatrienoic acid)	-	277.2174	25.90	91.04	1.91	0.93	<0.0001	1.12
	Oleic acid	-	281.2493	31.45	80.58	1.81	0.85	<0.0001	1.03
	Eicosapentaenoic acid	+	303.2313	25.54	252.02	3.52	1.82	<0.0001	1.19
	Docosapentaenoic acid	-	329.2488	28.63	103.96	2.04	1.03	<0.0001	1.19
	Decadienal	+	153.1267	12.45	−30.21	0.70	−0.52	<0.0001	0.28
	Octadecadienal (9,12)	+	265.2508	25.92	146.52	2.47	1.30	<0.0001	2.78
	Tetradecenoylcarnitine	+	370.2963	12.70	113.97	2.14	1.10	<0.0001	1.01
	9-hydroxyoctadecadienoic acid (9-HODE)	-	295.2278	18.71	388.27	4.88	2.29	<0.0001	3.97
Glycerolipids	MG(16:0)	+	331.2855	27.69	246.30	3.46	1.79	<0.0001	1.55
	MG(18:0)	+	359.3155	32.05	2966.39	30.66	4.94	<0.0001	6.00
	MG(18:2(9,12))	+	355.2844	25.33	238.69	3.39	1.76	<0.0001	1.23
Organic acids and derivatives	Arginine	+	175.1190	0.57	62.43	1.62	0.70	<0.0001	1.37
	Threonylhistidine	-	255.1121	1.79	91.05	1.91	0.93	<0.0001	1.57
	O-methoxycatechol-O-sulphate	-	203.0025	1.51	−71.64	0.28	−1.82	<0.0001	1.59
	Pyrocatechol sulfate	-	188.9878	1.19	−80.33	0.20	−2.35	<0.0001	3.47
Organoheterocyclic compounds	Biliverdin	+	583.2551	9.90	180.27	2.80	1.49	<0.0001	1.10
Prenol Lipids	Retinal	+	285.2222	25.55	754.02	8.54	3.09	<0.0001	2.91
Sterol Lipids	Hyodeoxycholic acid	-	391.2841	15.03	−83.22	0.17	−2.57	<0.0001	3.60
	Pregnenolone	+	317.2472	25.56	653.97	7.54	2.91	<0.0001	1.66

%Δ, percentage of change; FC, fold change; pBH, Benjamini–Hochberg adjusted *p*-value; RT, retention time; VIP, variable importance in the projection.

The metabolites that were shown to be significantly altered in the prostate cancer patients compared to the control subjects were mainly fatty acyls, glycerolipids, or organic acids and their derivates (Table 2). Notably, increased levels of fatty acyls, including different fatty acids, docosapentaenoic acid (docosanoid), 9-HODE (9-hydroxyoctadecadienoic acid), octadecadienal (fatty aldehyde), and fatty acid ester tetradecenoylcarnitine, were found in patients with prostate cancer when compared to healthy controls, except for one fatty aldehyde, decadienal, whose amount was significantly higher in the control subjects (Table 2).

The abundance of several monoacylglycerols was also higher in prostate cancer patients when compared to healthy subjects (Table 2). The levels of arginine, threonylhistidine, biliverdin, retinal, and pregnenolone were significantly increased in the plasma of cancer patients, while O-methoxycatechol-O-sulphate, pyrocatechol sulfate, and hyodeoxycholic acid had lower levels in prostate cancer when compared to the control group (Table 2).

The potential impact of elevated amounts of 4-HNE protein adducts in the plasma samples on plasma metabolome was further examined. Hence, correlation analysis of metabolites significantly altered in prostate cancer patients with levels of plasma 4-HNE-modified proteins in both cancer patients and healthy men samples is presented in Table 3.

Table 3. Correlation between altered metabolites and levels of plasma 4-HNE-protein adducts.

Compound	Healthy Controls			Prostate Cancer Patients		
	r	95% Confidence Interval	*p*	r	95% Confidence Interval	*p*
2-ketoisocaproic acid	0.146	−0.179 to 0.442	0.363	0.394	0.005 to 0.680	0.042 *
9-HODE	−0.014	−0.376 to 0.352	0.941	0.633	0.343 to 0.812	0.000 ***
Caproic acid (hexanoic acid)	−0.312	−0.571 to 0.005	0.047 *	0.179	−0.227 to 0.532	0.371
Eicosapentaenoic acid	0.120	−0.208 to 0.424	0.460	0.544	0.211 to 0.764	0.002 **
Hexadecanedioic acid	0.169	−0.160 to 0.464	0.297	−0.421	−0.684 to −0.060	0.021 *
Lactic acid (2-hydroxypropanoic acid)	−0.149	−0.444 to 0.176	0.353	0.316	−0.085 to 0.628	0.109
Linoleic acid (octadecadienoic acid)	0.357	0.042 to 0.608	0.024 *	−0.115	−0.484 to 0.288	0.567
Methyl hexadecanoic acid	0.127	−0.312 to 0.522	0.562	0.597	0.261 to 0.804	0.001 **
MG(18:2(9,12))	0.124	−0.204 to 0.427	0.446	0.554	0.231 to 0.767	0.002 **
Octadecadienal (9,12)	0.189	−0.244 to 0.559	0.378	0.523	0.175 to 0.755	0.004 **
Octadecenoic acid	0.150	−0.174 to 0.445	0.349	−0.330	−0.623 to 0.046	0.075
Palmitoleic acid (hexadecenoic acid)	0.198	−0.144 to 0.498	0.239	−0.509	−0.750 to −0.148	0.007 **
Pregnenolone	0.280	−0.044 to 0.551	0.080	0.629	0.330 to 0.813	0.000 ***
Pyruvic acid (2-oxopropanoic acid)	−0.393	−0.630 to −0.087	0.011 *	−0.183	−0.535 to 0.223	0.361
Retinal	0.333	−0.057 to 0.635	0.083	0.553	0.207 to 0.776	0.003 **
Stearic acid (octadecanoic acid)	0.466	0.168 to 0.687	0.003 **	−0.106	−0.476 to 0.297	0.601

r, Spearman correlation coefficient; Significance: * $p < 0.050$; ** $p < 0.010$; *** $p < 0.001$.

Plasma caproic acid, linoleic acid, stearic acid, and pyruvic acid showed significant correlations with the plasma levels of 4-HNE protein conjugates in healthy control samples. Stearic and linoleic acid correlated positively with the level of 4-HNE protein conjugates in healthy controls, although the strength of the correlation was moderate and weak, respectively. Contrary, in the same samples, a weak negative correlation was observed for caproic and pyruvic acid.

Only plasma linoleic acid, oleic acid, stearic acid, and pyruvic acid showed significant correlations with the plasma levels of 4-HNE protein conjugates in healthy control samples. For linoleic acid, oleic acid, and stearic acid, we observed a moderate positive correlation with the levels of 4-HNE protein conjugates, while pyruvic acid had a moderate negative correlation in the healthy controls. While all of these correlations were lost if prostate cancer was present, in the plasma samples of prostate cancer patients, new and significant correlations were observed between plasma 4-HNE protein conjugates and plasma metabolites 2-ketoisocaproic acid, 9-HODE, eicosapentaenoic acid, hexadecanedioic acid, methyl hexadecanoic acid, MG(18:2(9,12)), octadecadienal (9,12), palmitoleic acid, pregnenolone, and retinal.

A strong positive correlation between 4-HNE protein adducts and plasma metabolites was recorded for 9-HODE and pregnenolone, while a moderate positive correlation was observed for eicosapentaenoic acid, methyl hexadecanoic acid, MG(18:2(9.12)), and octadecadienal (9.12) and retinal. Furthermore, a significant but weak positive correlation was observed for 2-ketoisocaproic acid. On the contrary, a moderate negative correlation was found for 4-HNE and hexadecanedioic and palmitoleic acid.

Finally, to identify the pathways potentially affected by 4-HNE that could explain the observed changes, we further performed pathway analysis using MetaboAnalyst 5.0 to find that only the pathway of unsaturated fatty acids biosynthesis emerged as significant ($p = 0.000366$, FDR = 0.0308).

4. Discussion

The development and progression of cancer are closely associated with oxidative stress that may result in the peroxidation of lipids yielding 4-HNE. The presence of 4-HNE has been implicated in the tumorigenesis of diverse types of cancers, and nowadays, it is well recognized that the occurrence of 4-HNE in serum/plasma samples as well as in tumor and adjacent tumor tissue is tumor-type-specific [12]. However, this is the first study linking elevated 4-HNE-protein conjugates present in the plasma of prostate cancer samples with altered metabolome.

Interestingly, although a significant increase in 4-HNE levels was observed in the plasma samples of cancer patients, 4-HNE was not detected in the tumor tissue or in the adjacent normal tissue, unlike the previously revealed increase in 4-HNE in non-malignant cells surrounding different kinds of cancer tissue [8,41–43]. This could be attributed to induced transcription factor NF-E2-related factor 2 (Nrf2), which is activated in prostate cancer independently of ROS [44] and regulates 4-HNE metabolism and activity [45]. The removal of 4-HNE, especially its protein adducts, from cancer cells might be crucial for their survival because the proapoptotic effects of 4-HNE are associated with the aldehyde's binding to the cellular proteins, especially in cancer cells that are more sensitive to the toxic effect of 4-HNE than are normal, non-malignant cells [46–48].

Perhaps another reason why 4-HNE was detected neither in prostate carcinoma cells nor in non-malignant cells may be acrolein, another aldehydic product of lipid peroxidation, which can activate (similar to 4-HNE) the Nrf2 [49]. Namely, acrolein was found to be abundant in the malignant and stromal cells of prostate carcinoma tissues [11], as well as in benign and malignant colon neoplasms and in non-malignant tissue in the vicinity of tumors [50]. However, in the prostate tissue, acrolein could originate not only from oxidized lipids but also from protamines, such as spermine and spermidine, which are abundant in the prostate and are involved in the regulation of cellular proliferation and differentiation, again similar to 4-HNE [48,51]. Moreover, similar to 4-HNE, acrolein can be involved in inflammatory signaling and eventually even in spontaneous cancer regression [5,6,52–54]. Therefore, future studies should evaluate this option, too.

On the other hand, the absence of 4-HNE in prostate carcinoma tissue associated with the increase in 4-HNE in the blood of patients with prostate cancer might reflect the elimination of the aldehyde from the carcinoma cells into the blood or enhanced synthesis and release of 4-HNE from remote tissues, notably blood vessels, as was observed in the case of the visceral adipose tissue of people with metabolic syndrome and in COVID-19 patients due to oxidative vascular stress [55–57]. However, in both cases, the 4-HNE-protein adducts accumulated within the affected tissues, which was not the case with prostate carcinoma. Therefore, it is more likely that increased 4-HNE levels reflected systemic alteration of the lipid metabolism and oxidative stress in patients with prostate cancer, associated with the onset of cancer development. Since the control group in our study did not comprise patients but healthy men (mostly hospital crew volunteers) who did not have any records or symptoms of prostate or any other organ illness that would jeopardize their consideration as healthy persons, we did not collect their specific baseline characteristics. Therefore, further studies should consider deeper comparisons of the baseline characteristics of patients and control subjects to evaluate their possible differences and relevance for the findings obtained by metabolomics and specific 4-HNE-ELISA. Similarly, it would be interesting to repeat the analyses using the samples of blood of patients after a prolonged time after surgery, i.e., after complete recovery, to compare these with the results obtained before surgery. These are all goals of our further work.

The metabolomic profiling of the plasma samples obtained from patients with prostate cancer, performed using the combined GC-MS and LC-MS approach, revealed a significantly altered abundance of different fatty acids, organic acids, glycerolipids, carbohydrates, and sterol lipids in prostate cancer patients when compared to the healthy men controls. Using GC-MS, we detected and identified 40 compounds, out of which 18 were found to be differentially present in prostate cancer patients in comparison to the healthy control.

The LC-MS analysis, in positive and negative ionization mode, was also applied for the identification of metabolites that were significantly altered between two groups of subjects. Finally, we annotated 86 metabolites and confirmed the identity of 23 of them by using their exact mass, retention time, and specific MS/MS fragmentation patterns.

Our study suggests a positive association of different long-chain fatty acids (palmitoleic acid, linoleic acid, oleic acid, eicosapentaenoic acid (EPA), methylhexadecenoic acid, hexadecanedioic acid, docosapentaenoic acid (9-HODE) and medium-chain fatty acids (caprylic acid, lauric acid, palmitic acid, stearic acid) with a prostate cancer diagnosis. Only caproic acid levels were negatively related to prostate cancer diagnosis. Other studies have already suggested a possible relationship between blood fatty acid composition and the risk of prostate cancer, but the evidence is inconsistent. Our results, showing increased levels of many different fatty acids, along with higher amounts of plasma glycerol, emphasize the importance of altered lipid metabolism in the etiology of aggressive tumors.

The results of this study suggest a higher abundance of glycerol and glyceric acid, a glycerol oxidation product, in the plasma samples of patients diagnosed with prostate cancer, compared to healthy controls. Glyceric acid was found to be less abundant in metastatic breast cancer [58] and in the blood samples of individuals with advanced pancreatic cancer [59], which is not in line with our results that suggest higher levels of glyceric acid in patients with prostate cancer. Since we found the other carbohydrates and carbohydrate conjugates increased, one may assume that is a metabolic feature of prostate cancer presence. However, Huang and colleagues reported inverse associations between lethal prostate cancer risk and serum glycerol levels [60], so we should mention that all our patients survived the five years after surgery, which could be due to the proper therapy applied but also due to the less lethal character of the prostate cancer of our patients.

Furthermore, the level of different monoacylglycerols was found to be elevated in prostate cancer plasma samples. There is a lot of evidence supporting the role of monoacylglycerol lipase (MAGL) in tumorigenesis and metastasis. This lipolytic enzyme catalyzes the conversion of monoacylglycerides to free fatty acids and glycerol. The increased expression of MAGL was reported in diverse types of cancer, including prostate cancer [61]. Different fatty acids, glycerol and monoacylglycerols, which were found to be abundant in the prostate cancer plasma samples, could have a role as signaling molecules regulating cancer cell proliferation and aggressiveness.

A study by Crow and colleagues found a positive association between palmitic acid and prostate cancer risk and a positive association between α-linolenic acid and EPA with the risk of high-grade prostate cancer [62]. Non-targeted metabolomics analysis of the serum samples revealed 18 lipid or lipid-like compounds as potential biomarkers in the early diagnosis of prostate cancer [22]. These metabolites also included palmitic acid, which was found to be decreased in the prostate cancer samples, contrary to our results [22]. However, the above-mentioned study focused on metabolites that effectively discriminated prostate cancer patients from benign prostatic hyperplasia individuals. In the case of stearic acid, the lower proportion of this fatty acid was associated with a greater overall risk of prostate cancer [62], which is inconsistent with our results, indicating higher levels of this fatty acid in prostate cancer patients compared to healthy controls. The evidence regarding the association between dietary EPA and prostate cancer risk is inconclusive. Some studies reported a protective effect of EPA intake on prostate cancer [63–65]; however, others suggest a positive association between a higher intake of EPA and an increased risk of prostate cancer [66,67]. Circulating EPA levels have been positively associated with prostate cancer risk in a pooled analysis of individual data from prospective studies [68], and the dietary intake of EPA was associated with the risk of advanced or fatal prostate cancer in the NIH-American Association of Retired Persons (AARP) Diet and Health Study [69]. Elevated levels of 9-HODE and 13-HODE have been associated with increased prostate cancer mortality [70]. Interestingly, both 9-HODE and 13-HODE are synthesized from linoleic acid [71,72], a precursor of 4-HNE. The results showing a positive association of 9-HODE and 13-HODE serum concentration with increased risk for ovarian cancer are aligned with our results and highlight the importance of linoleic acid metabolites in the etiology of cancer [73].

A higher serum proportion of palmitoleic acid was associated with cancer mortality in the Swedish population [74], while oleic acid was found to promote the prostate cancer malignant phenotype of PC3 and DU-145 cells [75]. The intake of different short-chain fatty acids was shown to worsen prostate cancer survival [65], and higher intake of butyric acid, caproic acid, and the combined intake of different short-chain fatty acids (4:0–10:0) were associated with a higher risk of advanced prostate cancer [66].

We detected elevated levels of both 2-hydroxybutyric and 3-hydroxybutyric acid in patients diagnosed with prostate cancer. The 2-hydroxybutyric acid or α-hydroxybutyrate, which is produced by amino acid catabolism, was found to be elevated in the sera of colorectal cancer patients [76], and was suggested as a valuable biomarker for the early detection of colorectal cancer [77], as well as a potential early marker of insulin resistance and impaired glucose regulation, and was associated with elevated lipid oxidation and oxidative stress [78]. A meta-analysis of clinical metabolic profiling studies in 18 different cancer types, assessing blood, tissue, and urine samples, revealed that ketone bodies, which are produced during fatty acid metabolism, are metabolites most often found to be elevated in cancer compared to normal controls [79]. A ketone body 3-hydroxybutyric acid or β-hydroxybutyrate, which has been associated with tumor growth and cancer cachexia [80], was found to be upregulated in the blood samples from different cancer patients [79], and the same trend was detected in our study with patients diagnosed with prostate cancer.

Tumor metabolism is characterized by anaerobic glycolysis since cancer cells preferentially convert the glycolysis-induced pyruvate into lactic acid even in the presence of oxygen [59], resulting in elevated pericellular accumulation of organic acids, such as lactic acid, and low pyruvate status [81]. Our results also suggest a lower level of pyruvic acid in prostate cancer plasma samples if compared to healthy controls; however, contrary to our expectations, in our study, the levels of lactic acid were 56% lower in prostate cancer patients than in the healthy controls.

The blood level of glucose/galactose was elevated in prostate cancer patients compared with control subjects, confirming the positive associations between glucose and prostate cancer [82–84]. However, we also detected elevated levels of another hexose, mannose. Mannose, which is an isomer of glucose, was reported to have a potential anticancer effect on different tumors. Deng and colleagues demonstrated that mannose has an inhibitory effect on the proliferation and an apoptosis-promoting effect on prostate cancer cells in vitro [85]. The same trend was observed in mice [85]. However, high-throughput profiling of N-glycans in prostate cancer tissue samples revealed that high mannose glycans were more abundant in tumor regions than in non-tumor regions [86].

There has been a lot of evidence in the past decade supporting the role of cholesterol in prostate cancer and its progression [87–92]. Our results support the association between increased levels of blood cholesterol and prostate cancer diagnosis. Hypercholesterolemia could affect cancer cell proliferation [93,94], inflammation [95], lipid raft dynamics [96], and intertumoral steroidogenesis [97,98]. Different steroid hormones have a key role in the maintenance and progression of prostate cancer, and their effects are achieved through the androgen receptor [99]. The first step in steroidogenesis is the generation of the common steroid precursor pregnenolone from cholesterol. This common precursor of many steroids was found to induce the growth of LNCaP cells by binding to the mutated androgen receptor [100]. In our study, both cholesterol and pregnenolone levels were found to be much higher in patients with prostate cancer than in control subjects.

Hyodeoxycholic acid was found to be less abundant in patients with prostate cancer. This naturally occurring secondary bile acid is generated from lithocholic acid in the small intestine by the gut microbiota. Hyodeoxycholic acid is a weak liver X receptor α agonist with the capacity to decrease plasma glucose levels, improve plasma lipoprotein profiles, and increase the efficiency of intestinal cholesterol absorption [101]. Bile acids were primarily considered to promote carcinogenesis, but recently, more and more evidence suggests that lower concentrations of bile acids have potential anticancer activity, with an emphasis on ursodeoxycholic acid [102,103].

Alterations in amino acid metabolism are one of the hallmarks of cancer malignancy [104]. The 2-Ketoispcaproic acid, also known as α-ketoisocaproic acid or ketoleucine, is an abnormal metabolite associated with the incomplete breakdown of branched-chain amino acids. Ketoleucine was found to be increased in our group of patients diagnosed with prostate cancer, which is in accordance with the results from Jajin and colleagues, who detected increased 2-ketoisocaproic acid, α-ketoglutarate, and glutamine, along with decreased leucine and aspartate in medullary thyroid cancer [105]. In the last decade, the role of arginine in inflammation, cell activation, and growth has attracted a lot of attention due to the involvement of these processes in tumorigenesis [106,107]. The metabolism of arginine and the activity of arginase, the enzyme which hydrolyzes arginine to ornithine and urea, were found to be increased in prostate cancer and involved in differentiation [108–110]. In line with these studies, we also detected elevated levels of arginine in the plasma samples from prostate cancer patients. Dipeptide treonyhistidine was also found to be more abundant in the plasma of cancer patients than in healthy control subjects. The physical and functional properties of these dipeptides are different from those of single amino acids; however, their function in the human organism is not yet fully understood. Since our monoclonal antibody used in ELISA and immunohistochemistry is highly specific for the histidine adducts of 4-HNE [111], it might be possible that 4-HNE was, alongside 4-HNE-albumin and 4-HNE-LDL conjugates, also present in the plasma as a 4-HNE-treonyhistidine adduct. However, since there was no correlation between 4-HNE-His adducts determined by the ELISA and treonyhistidine levels determined by metabolomic analyses, we can assume that this dipeptide was not playing a major role in the formation of the 4-HNE-His adducts detected in the plasma of healthy controls and patients with prostate cancer. Different dipeptides were found to be elevated in the serum samples of prostate cancer patients [60], and recently, they have also attracted a lot of attention as potential disease biomarkers [112]. With respect to the derivatives of organic acids, it should be mentioned that levels of two phenylsulfates, O-methoxycatechol-O-sulphate, and pyrocatechol sulfate, were decreased significantly for 60–80% in the plasma of prostate cancer patients, but the biomedical relevance of these findings is still unknown.

Our results revealed elevated levels of biliverdin in patients diagnosed with prostate cancer. Biliverdin is a product of heme degradation by the heme oxygenase (HO) family of enzymes, so the increased level of biliverdin in cancer patients could be related to the altered activity of these enzymes. The HO-1 isoform was found to be expressed in a wide variety of cancers, including prostate cancer [113], and implicated in different biological processes which facilitate tumor progression and metastasis. The HO-1 possesses a dual role. It can prevent DNA damage and carcinogenesis in normal cells; however, in the late phase of tumorigenesis, the overexpression of HO-1 induces the proliferation and invasiveness of cancer cells [114–116].

Retinal belongs to the retinoid family, and it is the oxidized form of retinol. Retinoids have been shown to inhibit cell growth by inducing apoptosis and cellular differentiation, thus representing a potential approach to tumor therapy. However, the evidence of retinoids enhancing tumor growth [117] suggests a complicated and multifaceted role of retinoids in cancer. Our results indicate elevated levels of retinal in prostate cancer, which is partially in line with the findings suggesting the association of higher serum retinol with an elevated risk of prostate cancer [118].

In conclusion, our study revealed that the unsaturated fatty acids biosynthesis pathway could be modulated by elevated 4-HNE plasma protein adducts. This is a strong argument in favor of the 4-HNE analyses in clinical studies, as was suggested already years ago [119]. However, whether this is a direct effect of 4-HNE or they are simply induced by the same underlying mechanisms remains to be elucidated. Further mechanistic studies are needed to understand the biological relevance of the observed changes, which could also reveal with certainty the 4-HNE levels in the blood that might discriminate healthy from ill people, especially those suffering from cancer.

Supplementary Materials: The following supporting information can be downloaded at: https://www.mdpi.com/article/10.3390/biom13010145/s1, Table S1: General characteristics of prostate cancer patients and immunohistochemical appearance of 4-HNE in cancer tissue samples, Figure S1: Validity tests for OPLS-DA models. Plots were obtained using SIMCA-P+ software (version 15.0.2.5959, Umetrics, Umea, Sweden).

Author Contributions: Conceptualization, N.Z.; methodology, M.N.P., M.J., L.M., T.H. (Tea Horvat), D.R., M.Ć., T.H. (Tvrtko Hudolin), B.O. and G.W.; software, M.N.P., M.J. and L.M.; formal analysis, M.N.P., M.J., L.M., D.R. and K.Z.; resources, N.Z. and G.W.; data curation, M.N.P., M.J. and L.M.; writing—original draft preparation, M.N.P., M.J., L.M., K.Z., N.Z., M.Ć. and T.H.; writing—review and editing, M.N.P., M.J., L.M., K.Z. and N.Z.; visualization, M.N.P., L.M. and K.Z.; supervision, N.Z.; funding acquisition, N.Z. All authors have read and agreed to the published version of the manuscript.

Funding: This research was funded by offset project CRO_A-00033.

Institutional Review Board Statement: The study was conducted in accordance with the Declaration of Helsinki and approved by the Ethic Committee of the University Clinical Hospital Centre Zagreb (approval code 02/21 AG).

Informed Consent Statement: Informed consent was obtained from the patients included in the study.

Data Availability Statement: Not applicable.

Acknowledgments: This paper is dedicated to Nenad Grčević and Mira Šćukanec-Špoljar.

Conflicts of Interest: The authors declare no conflict of interest.

References

1. Sung, H.; Ferlay, J.; Siegel, R.L.; Laversanne, M.; Soerjomataram, I.; Jemal, A.; Bray, F. Global Cancer Statistics 2020: GLOBOCAN Estimates of Incidence and Mortality Worldwide for 36 Cancers in 185 Countries. *CA Cancer J. Clin.* **2021**, *71*, 209–249. [CrossRef] [PubMed]
2. Rawla, P. Epidemiology of Prostate Cancer. *World J. Oncol.* **2019**, *10*, 63–89. [CrossRef] [PubMed]
3. Netto, G.; Amin, M.; Kench, J.; Al, E. Tumours of the prostate. In *WHO Classification of Tumours: Urinary and Male Genital Tumours*; Srigley, J., Amin, M., Rubin, M., Tsuzuki, T., Eds.; International Agency for Research on Cancer: France, Lyon, 2022; p. 576.
4. Ström, P.; Kartasalo, K.; Olsson, H.; Solorzano, L.; Delahunt, B.; Berney, D.M.; Bostwick, D.G.; Evans, A.J.; Grignon, D.J.; Humphrey, P.A.; et al. Artificial intelligence for diagnosis and grading of prostate cancer in biopsies: A population-based, diagnostic study. *Lancet. Oncol.* **2020**, *21*, 222–232. [CrossRef] [PubMed]
5. Jaganjac, M.; Poljak-Blazi, M.; Zarkovic, K.; Schaur, R.J.; Zarkovic, N. The involvement of granulocytes in spontaneous regression of Walker 256 carcinoma. *Cancer Lett.* **2008**, *260*, 180–186. [CrossRef] [PubMed]
6. Žarković, N.; Jaganjac, M.; Žarković, K.; Gęgotek, A.; Skrzydlewska, E. Spontaneous Regression of Cancer: Revealing Granulocytes and Oxidative Stress as the Crucial Double-edge Sword. *Front. Biosci.* **2022**, *27*, 119. [CrossRef]
7. Costanzo-Garvey, D.L.; Case, A.J.; Watson, G.F.; Alsamraae, M.; Chatterjee, A.; Oberley-Deegan, R.E.; Dutta, S.; Abdalla, M.Y.; Kielian, T.; Lindsey, M.L.; et al. Prostate cancer addiction to oxidative stress defines sensitivity to anti-tumor neutrophils. *Clin. Exp. Metastasis* **2022**, *39*, 641–659. [CrossRef]
8. Jaganjac, M.; Cindrić, M.; Jakovčević, A.; Žarković, K.; Žarković, N. Lipid peroxidation in brain tumors. *Neurochem. Int.* **2021**, *149*, 105118. [CrossRef]
9. Hacer, İ.A.; Zeynep, A.A.; Can, Ö.; Riza, K.A.; Dildar, K.; Tülay, A. The effect of prostate cancer and antianrogenic therapy on lipid peroxidation and antioxidant systems. *Int. Urol. Nephrol.* **2003**, *36*, 57–62. [CrossRef]
10. Srivastava, D.S.L.; Mittal, R.D. Free radical injury and antioxidant status in patients with benign prostate hyperplasia and prostate cancer. *Indian J. Clin. Biochem.* **2005**, *20*, 162–165. [CrossRef]
11. Custovic, Z.; Zarkovic, K.; Cindric, M.; Cipak, A.; Jurkovic, I.; Sonicki, Z.; Uchida, K.; Zarkovic, N. Lipid peroxidation product acrolein as a predictive biomarker of prostate carcinoma relapse after radical surgery. *Free Radic. Res.* **2010**, *44*, 497–504. [CrossRef]
12. Jaganjac, M.; Milkovic, L.; Gegotek, A.; Cindric, M.; Zarkovic, K.; Skrzydlewska, E.; Zarkovic, N. The relevance of pathophysiological alterations in redox signaling of 4-hydroxynonenal for pharmacological therapies of major stress-associated diseases. *Free Radic. Biol. Med.* **2020**, *157*, 128–153. [CrossRef]
13. Zarkovic, K.; Jakovcevic, A.; Zarkovic, N. Contribution of the HNE-immunohistochemistry to modern pathological concepts of major human diseases. *Free Radic. Biol. Med.* **2017**, *111*, 110–126. [CrossRef]
14. Schaur, R.J.; Siems, W.; Bresgen, N.; Eckl, P.M. 4-Hydroxy-nonenal-A Bioactive Lipid Peroxidation Product. *Biomolecules* **2015**, *5*, 2247–2337. [CrossRef]

15. Peiro, G.; Alary, J.; Cravedi, J.-P.; Rathahao, E.; Steghens, J.-P.; Guéraud, F. Dihydroxynonene mercapturic acid, a urinary metabolite of 4-hydroxynonenal, as a biomarker of lipid peroxidation. *Biofactors* **2005**, *24*, 89–96. [CrossRef]
16. Pierre, F.; Peiro, G.; Taché, S.; Cross, A.J.; Bingham, S.A.; Gasc, N.; Gottardi, G.; Corpet, D.E.; Guéraud, F. New marker of colon cancer risk associated with heme intake: 1,4-dihydroxynonane mercapturic acid. *Cancer Epidemiol. Biomark. Prev.* **2006**, *15*, 2274–2279. [CrossRef]
17. Cherkas, A.; Golota, S.; Guéraud, F.; Abrahamovych, O.; Pichler, C.; Nersesyan, A.; Krupak, V.; Bugiichyk, V.; Yatskevych, O.; Pliatsko, M.; et al. A Helicobacter pylori-associated insulin resistance in asymptomatic sedentary young men does not correlate with inflammatory markers and urine levels of 8-iso-PGF(2)-α or 1,4-dihydroxynonane mercapturic acid. *Arch. Physiol. Biochem.* **2018**, *124*, 275–285. [CrossRef]
18. Al-Menhali, A.S.; Anderson, C.; Gourine, A.V.; Abramov, A.Y.; D'Souza, A.; Jaganjac, M. Proteomic Analysis of Cardiac Adaptation to Exercise by High Resolution Mass Spectrometry. *Front. Mol. Biosci.* **2021**, *8*, 723858. [CrossRef]
19. Gęgotek, A.; Domingues, P.; Wroński, A.; Ambrożewicz, E.; Skrzydlewska, E. The Proteomic Profile of Keratinocytes and Lymphocytes in Psoriatic Patients. *Proteom. Clin. Appl.* **2019**, *13*, e1800119. [CrossRef]
20. Hauck, A.K.; Zhou, T.; Upadhyay, A.; Sun, Y.; O'connor, M.B.; Chen, Y.; Bernlohr, D.A. Histone carbonylation is a redox-regulated epigenomic mark that accumulates with obesity and aging. *Antioxidants* **2020**, *9*, 1210. [CrossRef]
21. Wang, Y.; Jacobs, E.J.; Carter, B.D.; Gapstur, S.M.; Stevens, V.L. Plasma Metabolomic Profiles and Risk of Advanced and Fatal Prostate Cancer. *Eur. Urol. Oncol.* **2021**, *4*, 56–65. [CrossRef]
22. Xu, B.; Chen, Y.; Chen, X.; Gan, L.; Zhang, Y.; Feng, J.; Yu, L. Metabolomics Profiling Discriminates Prostate Cancer From Benign Prostatic Hyperplasia Within the Prostate-Specific Antigen Gray Zone. *Front. Oncol.* **2021**, *11*, 730638. [CrossRef] [PubMed]
23. Walz, S.; Wang, Q.; Zhao, X.; Hoene, M.; Häring, H.-U.; Hennenlotter, J.; Maas, M.; Peter, A.; Todenhöfer, T.; Stenzl, A.; et al. Comparison of the metabolome in urine prior and eight weeks after radical prostatectomy uncovers pathologic and molecular features of prostate cancer. *J. Pharm. Biomed. Anal.* **2021**, *205*, 114288. [CrossRef] [PubMed]
24. Falegan, O.S.; Jarvi, K.; Vogel, H.J.; Hyndman, M.E. Seminal plasma metabolomics reveals lysine and serine dysregulation as unique features distinguishing between prostate cancer tumors of Gleason grades 6 and 7. *Prostate* **2021**, *81*, 713–720. [CrossRef] [PubMed]
25. Zarkovic, K.; Juric, G.; Waeg, G.; Kolenc, D.; Zarkovic, N. Immunohistochemical appearance of HNE-protein conjugates in human astrocytomas. *BioFactors* **2005**, *24*, 33–40. [CrossRef] [PubMed]
26. Weber, D.; Milkovic, L.; Bennett, S.J.; Griffiths, H.R.; Zarkovic, N.; Grune, T. Measurement of HNE-protein adducts in human plasma and serum by ELISA-Comparison of two primary antibodies. *Redox Biol.* **2013**, *1*, 226–233. [CrossRef]
27. Perković, M.N.; Milković, L.; Uzun, S.; Mimica, N.; Pivac, N.; Waeg, G.; Žarković, N. Association of Lipid Peroxidation Product 4-Hydroxynonenal with Post-Traumatic Stress Disorder. *Biomolecules* **2021**, *11*, 1365. [CrossRef]
28. Gil de la Fuente, A.; Godzien, J.; Fernández López, M.; Rupérez, F.J.; Barbas, C.; Otero, A. Knowledge-based metabolite annotation tool: CEU Mass Mediator. *J. Pharm. Biomed. Anal.* **2018**, *154*, 138–149. [CrossRef]
29. Naz, S.; García, A.; Barbas, C. Multiplatform analytical methodology for metabolic fingerprinting of lung tissue. *Anal. Chem.* **2013**, *85*, 10941–10948. [CrossRef]
30. Wishart, D.S.; Feunang, Y.D.; Marcu, A.; Guo, A.C.; Liang, K.; Vázquez-Fresno, R.; Sajed, T.; Johnson, D.; Li, C.; Karu, N.; et al. HMDB 4.0: The human metabolome database for 2018. *Nucleic Acids Res.* **2018**, *46*, D608–D617. [CrossRef]
31. Smith, C.A.; O'Maille, G.; Want, E.J.; Qin, C.; Trauger, S.A.; Brandon, T.R.; Custodio, D.E.; Abagyan, R.; Siuzdak, G. METLIN: A metabolite mass spectral database. *Ther. Drug Monit.* **2005**, *27*, 747–751. [CrossRef]
32. Kanehisa, M.; Goto, S. KEGG: Kyoto encyclopedia of genes and genomes. *Nucleic Acids Res.* **2000**, *28*, 27–30. [CrossRef]
33. Fahy, E.; Sud, M.; Cotter, D.; Subramaniam, S. LIPID MAPS online tools for lipid research. *Nucleic Acids Res.* **2007**, *35*, W606–W612. [CrossRef]
34. Godzien, J.; Alonso-Herranz, V.; Barbas, C.; Armitage, E.G. Controlling the quality of metabolomics data: New strategies to get the best out of the QC sample. *Metabolomics* **2015**, *11*, 518–528. [CrossRef]
35. Kuligowski, J.; Sánchez-Illana, Á.; Sanjuán-Herráez, D.; Vento, M.; Quintás, G. Intra-batch effect correction in liquid chromatography-mass spectrometry using quality control samples and support vector regression (QC-SVRC). *Analyst* **2015**, *140*, 7810–7817. [CrossRef]
36. Chang, C.-C.; Lin, C.-J. LIBSVM: A library for support vector machines. *ACM Trans. Intell. Syst. Technol.* **2011**, *2*, 1–27. [CrossRef]
37. De Livera, A.M.; Dias, D.A.; De Souza, D.; Rupasinghe, T.; Pyke, J.; Tull, D.; Roessner, U.; McConville, M.; Speed, T.P. Normalizing and integrating metabolomics data. *Anal. Chem.* **2012**, *84*, 10768–10776. [CrossRef]
38. Gromski, P.S.; Xu, Y.; Hollywood, K.A.; Turner, M.L.; Goodacre, R. The influence of scaling metabolomics data on model classification accuracy. *Metabolomics* **2015**, *11*, 684–695. [CrossRef]
39. Pang, Z.; Chong, J.; Zhou, G.; de Lima Morais, D.A.; Chang, L.; Barrette, M.; Gauthier, C.; Jacques, P.-É.; Li, S.; Xia, J. MetaboAnalyst 5.0: Narrowing the gap between raw spectra and functional insights. *Nucleic Acids Res.* **2021**, *49*, W388–W396. [CrossRef]
40. Pang, Z.; Zhou, G.; Ewald, J.; Chang, L.; Hacariz, O.; Basu, N.; Xia, J. Using MetaboAnalyst 5.0 for LC–HRMS spectra processing, multi-omics integration and covariate adjustment of global metabolomics data. *Nat. Protoc.* **2022**, *17*, 1735–1761. [CrossRef]

41. Živković, N.P.; Petrovečki, M.; Lončarić, Č.T.; Nikolić, I.; Waeg, G.; Jaganjac, M.; Žarković, K.; Žarković, N. Positron emission tomography-computed tomography and 4-hydroxynonenal-histidine immunohistochemistry reveal differential onset of lipid peroxidation in primary lung cancer and in pulmonary metastasis of remote malignancies. *Redox Biol.* **2017**, *11*, 600–605. [CrossRef]
42. Zhong, H.; Xiao, M.; Zarkovic, K.; Zhu, M.; Sa, R.; Lu, J.; Tao, Y.; Chen, Q.; Xia, L.; Cheng, S.; et al. Mitochondrial control of apoptosis through modulation of cardiolipin oxidation in hepatocellular carcinoma: A novel link between oxidative stress and cancer. *Free Radic. Biol. Med.* **2017**, *102*, 67–76. [CrossRef] [PubMed]
43. Gęgotek, A.; Nikliński, J.; Žarković, N.; Žarković, K.; Waeg, G.; Łuczaj, W.; Charkiewicz, R.; Skrzydlewska, E. Lipid mediators involved in the oxidative stress and antioxidant defence of human lung cancer cells. *Redox Biol.* **2016**, *9*, 210–219. [CrossRef] [PubMed]
44. Bellezza, I.; Scarpelli, P.; Pizzo, S.V.; Grottelli, S.; Costanzi, E.; Minelli, A. ROS-independent Nrf2 activation in prostate cancer. *Oncotarget* **2017**, *8*, 67506–67518. [CrossRef] [PubMed]
45. Pettazzoni, P.; Ciamporcero, E.; Medana, C.; Pizzimenti, S.; Dal Bello, F.; Minero, V.G.; Toaldo, C.; Minelli, R.; Uchida, K.; Dianzani, M.U.; et al. Nuclear factor erythroid 2-related factor-2 activity controls 4-hydroxynonenal metabolism and activity in prostate cancer cells. *Free Radic. Biol. Med.* **2011**, *51*, 1610–1618. [CrossRef] [PubMed]
46. Sovic, A.; Borovic, S.; Loncaric, I.; Kreuzer, T.; Zarkovic, K.; Vukovic, T.; Wäg, G.; Hrascan, R.; Wintersteiger, R.; Klinger, R.; et al. The carcinostatic and proapoptotic potential of 4-hydroxynonenal in HeLa cells is associated with its conjugation to cellular proteins. *Anticancer Res.* **2001**, *21*, 1997–2004.
47. Sunjic, S.B.; Gasparovic, A.C.; Jaganjac, M.; Rechberger, G.; Meinitzer, A.; Grune, T.; Kohlwein, S.D.; Mihaljevic, B.; Zarkovic, N. Sensitivity of Osteosarcoma Cells to Concentration-Dependent Bioactivities of Lipid Peroxidation Product 4-Hydroxynonenal Depend on Their Level of Differentiation. *Cells* **2021**, *10*, 269. [CrossRef]
48. Sunjic, S.B.; Cipak, A.; Rabuzin, F.; Wildburger, R.; Zarkovic, N. The influence of 4-hydroxy-2-nonenal on proliferation, differentiation and apoptosis of human osteosarcoma cells. *Biofactors* **2005**, *24*, 141–148. [CrossRef]
49. Tirumalai, R.; Rajesh Kumar, T.; Mai, K.H.; Biswal, S. Acrolein causes transcriptional induction of phase II genes by activation of Nrf2 in human lung type II epithelial (A549) cells. *Toxicol. Lett.* **2002**, *132*, 27–36. [CrossRef]
50. Zarkovic, K.; Uchida, K.; Kolenc, D.; Hlupic, L.; Zarkovic, N. Tissue distribution of lipid peroxidation product acrolein in human colon carcinogenesis. *Free Radic. Res.* **2006**, *40*, 543–552. [CrossRef]
51. Tabor, C.W.; Tabor, H.; Bachrach, U. Identification of the aminoaldehydes produced by the oxidation of spermine and spermidine with purified plasma amine oxidase. *J. Biol. Chem.* **1964**, *239*, 2194–2203. [CrossRef]
52. Jaganjac, M.; Poljak-Blazi, M.; Schaur, R.J.; Zarkovic, K.; Borovic, S.; Cipak, A.; Cindric, M.; Uchida, K.; Waeg, G.; Zarkovic, N. Elevated neutrophil elastase and acrolein-protein adducts are associated with W256 regression. *Clin. Exp. Immunol.* **2012**, *170*, 178–185. [CrossRef]
53. Jaganjac, M.; Matijevic Glavan, T.; Zarkovic, N. The Role of Acrolein and NADPH Oxidase in the Granulocyte-Mediated Growth-Inhibition of Tumor Cells. *Cells* **2019**, *8*, 292. [CrossRef]
54. Bauer, G.; Zarkovic, N. Revealing mechanisms of selective, concentration-dependent potentials of 4-hydroxy-2-nonenal to induce apoptosis in cancer cells through inactivation of membrane-associated catalase. *Free Radic. Biol. Med.* **2015**, *81*, 128–144. [CrossRef]
55. Jaganjac, M.; Almuraikhy, S.; Al-Khelaifi, F.; Al-Jaber, M.; Bashah, M.; Mazloum, N.A.; Zarkovic, K.; Zarkovic, N.; Waeg, G.; Kafienah, W.; et al. Combined metformin and insulin treatment reverses metabolically impaired omental adipogenesis and accumulation of 4-hydroxynonenal in obese diabetic patients. *Redox Biol.* **2017**, *12*, 483–490. [CrossRef]
56. Zarkovic, N.; Jakovcevic, A.; Mataic, A.; Jaganjac, M.; Vukovic, T.; Waeg, G.; Zarkovic, K. Post-mortem Findings of Inflammatory Cells and the Association of 4-Hydroxynonenal with Systemic Vascular and Oxidative Stress in Lethal COVID-19. *Cells* **2022**, *11*, 444. [CrossRef]
57. Žarković, N.; Orehovec, B.; Baršić, B.; Tarle, M.; Kmet, M.; Lukšić, I.; Tatzber, F.; Wonisch, W.; Skrzydlewska, E.; Łuczaj, W. Lipidomics Revealed Plasma Phospholipid Profile Differences between Deceased and Recovered COVID-19 Patients. *Biomolecules* **2022**, *12*, 1488. [CrossRef]
58. Kim, H.-Y.; Lee, K.-M.; Kim, S.-H.; Kwon, Y.-J.; Chun, Y.-J.; Choi, H.-K. Comparative metabolic and lipidomic profiling of human breast cancer cells with different metastatic potentials. *Oncotarget* **2016**, *7*, 67111–67128. [CrossRef]
59. Ikeda, A.; Nishiumi, S.; Shinohara, M.; Yoshie, T.; Hatano, N.; Okuno, T.; Bamba, T.; Fukusaki, E.; Takenawa, T.; Azuma, T.; et al. Serum metabolomics as a novel diagnostic approach for gastrointestinal cancer. *Biomed. Chromatogr.* **2012**, *26*, 548–558. [CrossRef]
60. Huang, J.; Mondul, A.M.; Weinstein, S.J.; Derkach, A.; Moore, S.C.; Sampson, J.N.; Albanes, D. Prospective serum metabolomic profiling of lethal prostate cancer. *Int. J. Cancer* **2019**, *145*, 3231–3243. [CrossRef]
61. Nomura, D.K.; Lombardi, D.P.; Chang, J.W.; Niessen, S.; Ward, A.M.; Long, J.Z.; Hoover, H.H.; Cravatt, B.F. Monoacylglycerol Lipase Exerts Dual Control over Endocannabinoid and Fatty Acid Pathways to Support Prostate Cancer. *Chem. Biol.* **2011**, *18*, 846–856. [CrossRef]
62. Crowe, F.L.; Allen, N.E.; Appleby, P.N.; Overvad, K.; Aardestrup, I.V.; Johnsen, N.F.; Tjønneland, A.; Linseisen, J.; Kaaks, R.; Boeing, H.; et al. Fatty acid composition of plasma phospholipids and risk of prostate cancer in a case-control analysis nested within the European Prospective Investigation into Cancer and Nutrition. *Am. J. Clin. Nutr.* **2008**, *88*, 1353–1363. [CrossRef] [PubMed]

63. Leitzmann, M.F.; Stampfer, M.J.; Michaud, D.S.; Augustsson, K.; Colditz, G.C.; Willett, W.C.; Giovannucci, E.L. Dietary intake of n-3 and n-6 fatty acids and the risk of prostate cancer. *Am. J. Clin. Nutr.* **2004**, *80*, 204–216. [CrossRef] [PubMed]
64. Augustsson, K.; Michaud, D.S.; Rimm, E.B.; Leitzmann, M.F.; Stampfer, M.J.; Willett, W.C.; Giovannucci, E. A prospective study of intake of fish and marine fatty acids and prostate cancer. *Cancer Epidemiol. Biomark. Prev.* **2003**, *12*, 64–67.
65. Epstein, M.M.; Kasperzyk, J.L.; Mucci, L.A.; Giovannucci, E.; Price, A.; Wolk, A.; Håkansson, N.; Fall, K.; Andersson, S.-O.; Andrén, O. Dietary fatty acid intake and prostate cancer survival in Örebro County, Sweden. *Am. J. Epidemiol.* **2012**, *176*, 240–252. [CrossRef] [PubMed]
66. Perez-Cornago, A.; Huybrechts, I.; Appleby, P.N.; Schmidt, J.A.; Crowe, F.L.; Overvad, K.; Tjønneland, A.; Kühn, T.; Katzke, V.; Trichopoulou, A.; et al. Intake of individual fatty acids and risk of prostate cancer in the European prospective investigation into cancer and nutrition. *Int. J. Cancer* **2020**, *146*, 44–57. [CrossRef]
67. Wallström, P.; Bjartell, A.; Gullberg, B.; Olsson, H.; Wirfält, E. A prospective study on dietary fat and incidence of prostate cancer (Malmö, Sweden). *Cancer Causes Control* **2007**, *18*, 1107–1121. [CrossRef]
68. Crowe, F.L.; Appleby, P.N.; Travis, R.C.; Barnett, M.; Brasky, T.M.; Bueno-de-Mesquita, H.B.; Chajes, V.; Chavarro, J.E.; Chirlaque, M.-D.; English, D.R.; et al. Circulating fatty acids and prostate cancer risk: Individual participant meta-analysis of prospective studies. *J. Natl. Cancer Inst.* **2014**, *106*, dju240. [CrossRef]
69. Pelser, C.; Mondul, A.M.; Hollenbeck, A.R.; Park, Y. Dietary fat, fatty acids, and risk of prostate cancer in the NIH-AARP diet and health study. *Cancer Epidemiol. Biomark. Prev.* **2013**, *22*, 697–707. [CrossRef]
70. Huang, J.; Zhao, B.; Weinstein, S.J.; Albanes, D.; Mondul, A.M. Metabolomic profile of prostate cancer-specific survival among 1812 Finnish men. *BMC Med.* **2022**, *20*, 362. [CrossRef]
71. Kelavkar, U.P.; Nixon, J.B.; Cohen, C.; Dillehay, D.; Eling, T.E.; Badr, K.F. Overexpression of 15-lipoxygenase-1 in PC-3 human prostate cancer cells increases tumorigenesis. *Carcinogenesis* **2001**, *22*, 1765–1773. [CrossRef]
72. Kelavkar, U.P.; Hutzley, J.; McHugh, K.; Allen, K.G.D.; Parwani, A. Prostate tumor growth can be modulated by dietarily targeting the 15-lipoxygenase-1 and cyclooxygenase-2 enzymes. *Neoplasia* **2009**, *11*, 692–699. [CrossRef]
73. Hada, M.; Edin, M.L.; Hartge, P.; Lih, F.B.; Wentzensen, N.; Zeldin, D.C.; Trabert, B. Prediagnostic Serum Levels of Fatty Acid Metabolites and Risk of Ovarian Cancer in the Prostate, Lung, Colorectal, and Ovarian (PLCO) Cancer Screening Trial. *Cancer Epidemiol. Biomark. Prev.* **2019**, *28*, 189–197. [CrossRef]
74. Byberg, L.; Kilander, L.; Warensjö Lemming, E.; Michaëlsson, K.; Vessby, B. Cancer death is related to high palmitoleic acid in serum and to polymorphisms in the SCD-1 gene in healthy Swedish men. *Am. J. Clin. Nutr.* **2014**, *99*, 551–558. [CrossRef]
75. Liotti, A.; Cosimato, V.; Mirra, P.; Calì, G.; Conza, D.; Secondo, A.; Luongo, G.; Terracciano, D.; Formisano, P.; Beguinot, F.; et al. Oleic acid promotes prostate cancer malignant phenotype via the G protein-coupled receptor FFA1/GPR40. *J. Cell. Physiol.* **2018**, *233*, 7367–7378. [CrossRef]
76. Qiu, Y.; Cai, G.; Su, M.; Chen, T.; Zheng, X.; Xu, Y.; Ni, Y.; Zhao, A.; Xu, L.X.; Cai, S.; et al. Serum metabolite profiling of human colorectal cancer using GC-TOFMS and UPLC-QTOFMS. *J. Proteome Res.* **2009**, *8*, 4844–4850. [CrossRef]
77. Nishiumi, S.; Kobayashi, T.; Ikeda, A.; Yoshie, T.; Kibi, M.; Izumi, Y.; Okuno, T.; Hayashi, N.; Kawano, S.; Takenawa, T.; et al. A novel serum metabolomics-based diagnostic approach for colorectal cancer. *PLoS ONE* **2012**, *7*, e40459. [CrossRef]
78. Gall, W.E.; Beebe, K.; Lawton, K.A.; Adam, K.-P.; Mitchell, M.W.; Nakhle, P.J.; Ryals, J.A.; Milburn, M.V.; Nannipieri, M.; Camastra, S.; et al. Alpha-hydroxybutyrate is an early biomarker of insulin resistance and glucose intolerance in a nondiabetic population. *PLoS ONE* **2010**, *5*, e10883. [CrossRef]
79. Goveia, J.; Pircher, A.; Conradi, L.-C.; Kalucka, J.; Lagani, V.; Dewerchin, M.; Eelen, G.; DeBerardinis, R.J.; Wilson, I.D.; Carmeliet, P. Meta-analysis of clinical metabolic profiling studies in cancer: Challenges and opportunities. *EMBO Mol. Med.* **2016**, *8*, 1134–1142. [CrossRef]
80. Shukla, S.K.; Gebregiworgis, T.; Purohit, V.; Chaika, N.V.; Gunda, V.; Radhakrishnan, P.; Mehla, K.; Pipinos, I.I.; Powers, R.; Yu, F.; et al. Metabolic reprogramming induced by ketone bodies diminishes pancreatic cancer cachexia. *Cancer Metab.* **2014**, *2*, 18. [CrossRef]
81. Schornack, P.A.; Gillies, R.J. Contributions of cell metabolism and H+ diffusion to the acidic pH of tumors. *Neoplasia* **2003**, *5*, 135–145. [CrossRef]
82. Bensimon, L.; Yin, H.; Suissa, S.; Pollak, M.N.; Azoulay, L. Type 2 diabetes and the risk of mortality among patients with prostate cancer. *Cancer Causes Control* **2014**, *25*, 329–338. [CrossRef] [PubMed]
83. Cai, H.; Xu, Z.; Xu, T.; Yu, B.; Zou, Q. Diabetes mellitus is associated with elevated risk of mortality amongst patients with prostate cancer: A meta-analysis of 11 cohort studies. *Diabetes. Metab. Res. Rev.* **2015**, *31*, 336–343. [CrossRef] [PubMed]
84. Arthur, R.; Møller, H.; Garmo, H.; Häggström, C.; Holmberg, L.; Stattin, P.; Malmström, H.; Lambe, M.; Hammar, N.; Walldius, G.; et al. Serum glucose, triglycerides, and cholesterol in relation to prostate cancer death in the Swedish AMORIS study. *Cancer Causes Control* **2019**, *30*, 195–206. [CrossRef] [PubMed]
85. Deng, Y.-L.; Liu, R.; Cai, Z.-D.; Han, Z.-D.; Feng, Y.-F.; Cai, S.-H.; Chen, Q.-B.; Zhu, J.-G.; Zhong, W.-D. Mannose inhibits the growth of prostate cancer through a mitochondrial mechanism. *Asian J. Androl.* **2022**, *24*, 540–548. [CrossRef] [PubMed]
86. Conroy, L.R.; Stanback, A.E.; Young, L.E.A.; Clarke, H.A.; Austin, G.L.; Liu, J.; Allison, D.B.; Sun, R.C. In Situ Analysis of N-Linked Glycans as Potential Biomarkers of Clinical Course in Human Prostate Cancer. *Mol. Cancer Res.* **2021**, *19*, 1727–1738. [CrossRef]

87. Platz, E.A.; Till, C.; Goodman, P.J.; Parnes, H.L.; Figg, W.D.; Albanes, D.; Neuhouser, M.L.; Klein, E.A.; Thompson, I.M.J.; Kristal, A.R. Men with low serum cholesterol have a lower risk of high-grade prostate cancer in the placebo arm of the prostate cancer prevention trial. *Cancer Epidemiol. Biomark. Prev.* **2009**, *18*, 2807–2813. [CrossRef]
88. Mondul, A.M.; Clipp, S.L.; Helzlsouer, K.J.; Platz, E.A. Association between plasma total cholesterol concentration and incident prostate cancer in the CLUE II cohort. *Cancer Causes Control* **2010**, *21*, 61–68. [CrossRef]
89. Van Hemelrijck, M.; Walldius, G.; Jungner, I.; Hammar, N.; Garmo, H.; Binda, E.; Hayday, A.; Lambe, M.; Holmberg, L. Low levels of apolipoprotein A-I and HDL are associated with risk of prostate cancer in the Swedish AMORIS study. *Cancer Causes Control* **2011**, *22*, 1011–1019. [CrossRef]
90. Batty, G.D.; Kivimäki, M.; Clarke, R.; Davey Smith, G.; Shipley, M.J. Modifiable risk factors for prostate cancer mortality in London: Forty years of follow-up in the Whitehall study. *Cancer Causes Control* **2011**, *22*, 311–318. [CrossRef]
91. Farwell, W.R.; D'Avolio, L.W.; Scranton, R.E.; Lawler, E.V.; Gaziano, J.M. Statins and prostate cancer diagnosis and grade in a veterans population. *J. Natl. Cancer Inst.* **2011**, *103*, 885–892. [CrossRef]
92. Shafique, K.; McLoone, P.; Qureshi, K.; Leung, H.; Hart, C.; Morrison, D.S. Cholesterol and the risk of grade-specific prostate cancer incidence: Evidence from two large prospective cohort studies with up to 37 years' follow up. *BMC Cancer* **2012**, *12*, 25. [CrossRef]
93. Solomon, K.R.; Freeman, M.R. The complex interplay between cholesterol and prostate malignancy. *Urol. Clin. North Am.* **2011**, *38*, 243–259. [CrossRef]
94. Pelton, K.; Freeman, M.R.; Solomon, K.R. Cholesterol and prostate cancer. *Curr. Opin. Pharmacol.* **2012**, *12*, 751–759. [CrossRef]
95. Freeman, M.R.; Solomon, K.R. Cholesterol and benign prostate disease. *Differentiation* **2011**, *82*, 244–252. [CrossRef]
96. Zhuang, L.; Lin, J.; Lu, M.L.; Solomon, K.R.; Freeman, M.R. Cholesterol-rich lipid rafts mediate akt-regulated survival in prostate cancer cells. *Cancer Res.* **2002**, *62*, 2227–2231.
97. Locke, J.A.; Guns, E.S.; Lubik, A.A.; Adomat, H.H.; Hendy, S.C.; Wood, C.A.; Ettinger, S.L.; Gleave, M.E.; Nelson, C.C. Androgen levels increase by intratumoral de novo steroidogenesis during progression of castration-resistant prostate cancer. *Cancer Res.* **2008**, *68*, 6407–6415. [CrossRef]
98. Montgomery, R.B.; Mostaghel, E.A.; Vessella, R.; Hess, D.L.; Kalhorn, T.F.; Higano, C.S.; True, L.D.; Nelson, P.S. Maintenance of intratumoral androgens in metastatic prostate cancer: A mechanism for castration-resistant tumor growth. *Cancer Res.* **2008**, *68*, 4447–4454. [CrossRef]
99. Snaterse, G.; Visser, J.A.; Arlt, W.; Hofland, J. Circulating steroid hormone variations throughout different stages of prostate cancer. *Endocr. Relat. Cancer* **2017**, *24*, R403–R420. [CrossRef]
100. Grigoryev, D.N.; Long, B.J.; Njar, V.C.O.; Brodie, A.H.M. Pregnenolone stimulates LNCaP prostate cancer cell growth via the mutated androgen receptor. *J. Steroid Biochem. Mol. Biol.* **2000**, *75*, 1–10. [CrossRef]
101. Shih, D.M.; Shaposhnik, Z.; Meng, Y.; Rosales, M.; Wang, X.; Wu, J.; Ratiner, B.; Zadini, F.; Zadini, G.; Lusis, A.J. Hyodeoxycholic acid improves HDL function and inhibits atherosclerotic lesion formation in LDLR-knockout mice. *FASEB J. Off. Publ. Fed. Am. Soc. Exp. Biol.* **2013**, *27*, 3805–3817. [CrossRef]
102. Režen, T.; Rozman, D.; Kovács, T.; Kovács, P.; Sipos, A.; Bai, P.; Mikó, E. The role of bile acids in carcinogenesis. *Cell. Mol. Life Sci.* **2022**, *79*, 243. [CrossRef] [PubMed]
103. Fu, J.; Yu, M.; Xu, W.; Yu, S. Research Progress of Bile Acids in Cancer. *Front. Oncol.* **2021**, *11*, 778258. [CrossRef] [PubMed]
104. Wei, Z.; Liu, X.; Cheng, C.; Yu, W.; Yi, P. Metabolism of Amino Acids in Cancer. *Front. Cell Dev. Biol.* **2020**, *8*, 603837. [CrossRef] [PubMed]
105. Jajin, M.G.; Abooshahab, R.; Hooshmand, K.; Moradi, A.; Siadat, S.D.; Mirzazadeh, R.; Chegini, K.G.; Hedayati, M. Gas chromatography-mass spectrometry-based untargeted metabolomics reveals metabolic perturbations in medullary thyroid carcinoma. *Sci. Rep.* **2022**, *12*, 8397. [CrossRef] [PubMed]
106. Capuano, G.; Rigamonti, N.; Grioni, M.; Freschi, M.; Bellone, M. Modulators of arginine metabolism support cancer immunosurveillance. *BMC Immunol.* **2009**, *10*, 1. [CrossRef]
107. Wang, B.; Rong, X.; Palladino, E.N.D.; Wang, J.; Fogelman, A.M.; Martín, M.G.; Alrefai, W.A.; Ford, D.A.; Tontonoz, P. Phospholipid Remodeling and Cholesterol Availability Regulate Intestinal Stemness and Tumorigenesis. *Cell Stem Cell* **2018**, *22*, 206–220.e4. [CrossRef]
108. Bronte, V.; Kasic, T.; Gri, G.; Gallana, K.; Borsellino, G.; Marigo, I.; Battistini, L.; Iafrate, M.; Prayer-Galetti, T.; Pagano, F.; et al. Boosting antitumor responses of T lymphocytes infiltrating human prostate cancers. *J. Exp. Med.* **2005**, *201*, 1257–1268. [CrossRef]
109. Mumenthaler, S.M.; Yu, H.; Tze, S.; Cederbaum, S.D.; Pegg, A.E.; Seligson, D.B.; Grody, W.W. Expression of arginase II in prostate cancer. *Int. J. Oncol.* **2008**, *32*, 357–365. [CrossRef]
110. Gannon, P.O.; Godin-Ethier, J.; Hassler, M.; Delvoye, N.; Aversa, M.; Poisson, A.O.; Péant, B.; Alam Fahmy, M.; Saad, F.; Lapointe, R.; et al. Androgen-regulated expression of arginase 1, arginase 2 and interleukin-8 in human prostate cancer. *PLoS ONE* **2010**, *5*, e12107. [CrossRef]
111. Waeg, G.; Dimsity, G.; Esterbauer, H. Monoclonal antibodies for detection of 4-hydroxynonenal modified proteins. *Free Radic. Res.* **1996**, *25*, 149–159. [CrossRef]
112. Tang, Y.; Li, R.; Lin, G.; Li, L. PEP search in MyCompoundID: Detection and identification of dipeptides and tripeptides using dimethyl labeling and hydrophilic interaction liquid chromatography tandem mass spectrometry. *Anal. Chem.* **2014**, *86*, 3568–3574. [CrossRef]

113. Wegiel, B.; Gallo, D.; Csizmadia, E.; Harris, C.; Belcher, J.; Vercellotti, G.M.; Penacho, N.; Seth, P.; Sukhatme, V.; Ahmed, A.; et al. Carbon monoxide expedites metabolic exhaustion to inhibit tumor growth. *Cancer Res.* **2013**, *73*, 7009–7021. [CrossRef]
114. Sunamura, M.; Duda, D.G.; Ghattas, M.H.; Lozonschi, L.; Motoi, F.; Yamauchi, J.-I.; Matsuno, S.; Shibahara, S.; Abraham, N.G. Heme oxygenase-1 accelerates tumor angiogenesis of human pancreatic cancer. *Angiogenesis* **2003**, *6*, 15–24. [CrossRef]
115. Chen, G.G.; Liu, Z.M.; Vlantis, A.C.; Tse, G.M.K.; Leung, B.C.H.; van Hasselt, C.A. Heme oxygenase-1 protects against apoptosis induced by tumor necrosis factor-alpha and cycloheximide in papillary thyroid carcinoma cells. *J. Cell. Biochem.* **2004**, *92*, 1246–1256. [CrossRef]
116. Was, H.; Cichon, T.; Smolarczyk, R.; Rudnicka, D.; Stopa, M.; Chevalier, C.; Leger, J.J.; Lackowska, B.; Grochot, A.; Bojkowska, K.; et al. Overexpression of heme oxygenase-1 in murine melanoma: Increased proliferation and viability of tumor cells, decreased survival of mice. *Am. J. Pathol.* **2006**, *169*, 2181–2198. [CrossRef]
117. Matsumoto, T.; Mochizuki, W.; Nibe, Y.; Akiyama, S.; Matsumoto, Y.; Nozaki, K.; Fukuda, M.; Hayashi, A.; Mizutani, T.; Oshima, S.; et al. Retinol Promotes In Vitro Growth of Proximal Colon Organoids through a Retinoic Acid-Independent Mechanism. *PLoS ONE* **2016**, *11*, e0162049. [CrossRef]
118. Mondul, A.M.; Watters, J.L.; Männistö, S.; Weinstein, S.J.; Snyder, K.; Virtamo, J.; Albanes, D. Serum retinol and risk of prostate cancer. *Am. J. Epidemiol.* **2011**, *173*, 813–821. [CrossRef]
119. Frijhoff, J.; Winyard, P.G.; Zarkovic, N.; Davies, S.; Stocker, R.; Cheng, D.; Knight, A.; Taylor, E.L.; Oettrich, J.; Ruskovska, T.; et al. Clinical relevance of biomarkers of oxidative stress. *Antioxid. Redox Signal.* **2015**, *23*, 1144–1170. [CrossRef]

biomolecules

MDPI

Article

Plasmalogens: Free Radical Reactivity and Identification of Trans Isomers Relevant to Biological Membranes

Carla Ferreri [1,*], Alessandra Ferocino [1,†], Gessica Batani [1], Chryssostomos Chatgilialoglu [1,2], Vanda Randi [3], Maria Vittoria Riontino [3], Fabrizio Vetica [1,‡] and Anna Sansone [1]

1 Institute for Organic Synthesis and Photoreactivity (ISOF), National Research Council (CNR), Via P. Gobetti, 101, 40129 Bologna, Italy; alessandra.ferocino@kcl.ac.uk (A.F.); gessica.batani@isof.cnr.it (G.B.); chrys@isof.cnr.it (C.C.); fabrizio.vetica@uniroma1.it (F.V.)
2 Center for Advanced Technologies, Adam Mickiewicz University, 61-614 Poznan, Poland
3 Centro Regionale Sangue Regione Emilia Romagna (CRS-RER), Casa dei Donatori di Sangue, Via dell'Ospedale, 20, 40133 Bologna, Italy; vanda.randi@ausl.bologna.it (V.R.)
* Correspondence: carla.ferreri@isof.cnr.it
† Current address: Positron Emitting Radiotracer Laboratory, School of Biomedical Engineering and Imaging Sciences, King's College London, London SE1 7EH, UK.
‡ Current address: Department of Chemistry, Sapienza University of Rome, Piazzale Aldo Moro, 5, 00185 Rome, Italy.

Abstract: Plasmalogens are membrane phospholipids with two fatty acid hydrocarbon chains linked to L-glycerol, one containing a characteristic cis-vinyl ether function and the other one being a polyunsaturated fatty acid (PUFA) residue linked through an acyl function. All double bonds in these structures display the cis geometrical configuration due to desaturase enzymatic activity and they are known to be involved in the peroxidation process, whereas the reactivity through cis-trans double bond isomerization has not yet been identified. Using 1-(1Z-octadecenyl)-2-arachidonoyl-*sn*-glycero-3-phosphocholine (C18 plasm-20:4 PC) as a representative molecule, we showed that the cis-trans isomerization can occur at both plasmalogen unsaturated moieties, and the product has characteristic analytical signatures useful for omics applications. Using plasmalogen-containing liposomes and red blood cell (RBC) ghosts under biomimetic Fenton-like conditions, in the presence or absence of thiols, peroxidation, and isomerization processes were found to occur with different reaction outcomes due to the particular liposome compositions. These results allow gaining a full scenario of plasmalogen reactivity under free radical conditions. Moreover, clarification of the plasmalogen reactivity under acidic and alkaline conditions was carried out, identifying the best protocol for RBC membrane fatty acid analysis due to their plasmalogen content of 15–20%. These results are important for lipidomic applications and for achieving a full scenario of radical stress in living organisms.

Keywords: fatty acids; cis-trans isomerization; free radicals; trans plasmalogens; erythrocyte membrane; plasmalogen analysis; lipidomics

Citation: Ferreri, C.; Ferocino, A.; Batani, G.; Chatgilialoglu, C.; Randi, V.; Riontino, M.V.; Vetica, F.; Sansone, A. Plasmalogens: Free Radical Reactivity and Identification of Trans Isomers Relevant to Biological Membranes. *Biomolecules* **2023**, *13*, 730. https://doi.org/10.3390/biom13050730

Academic Editor: Robert V. Stahelin

Received: 17 March 2023
Revised: 18 April 2023
Accepted: 21 April 2023
Published: 24 April 2023

1. Introduction

Chemical and supramolecular properties of phospholipids are attracting interest for multidisciplinary applications, going from biologically relevant transformations [1–3], to nanobiotechnology, including liposome preparation and vesicle functions [4–6]. Plasmalogens (1-*O*-alk-1′-enyl 2-acyl glycerol phospholipids and glycolipids) are a class of phospholipids, the building blocks of cell membranes, with the general structure shown in Figure 1a. The vinyl-ether group in the *sn-1* position of L-glycerol is a peculiarity with respect to the ester function at both *sn*-1 and *sn*-2 positions, which characterizes the majority of membrane glycerophospholipids [7,8]. Recently, the orphan biosynthetic pathway relative to the formation of the cis vinyl function has been assigned to identify a delta-1 desaturase candidate gene [9,10]. Desaturase enzymes create fatty acid double bonds in a regio- and stereo-specifical manner, thus producing natural lipids as cis isomers with their biological effects depending upon such a configuration. In plasmalogens, the cis-vinyl ether function contributes to their behavior in biological structures; the cis geometry confers stronger intermolecular hydrogen bond interactions with contiguous phospholipids, influencing rigidity/fluidity, fission/fusion, and morphological transition [11].

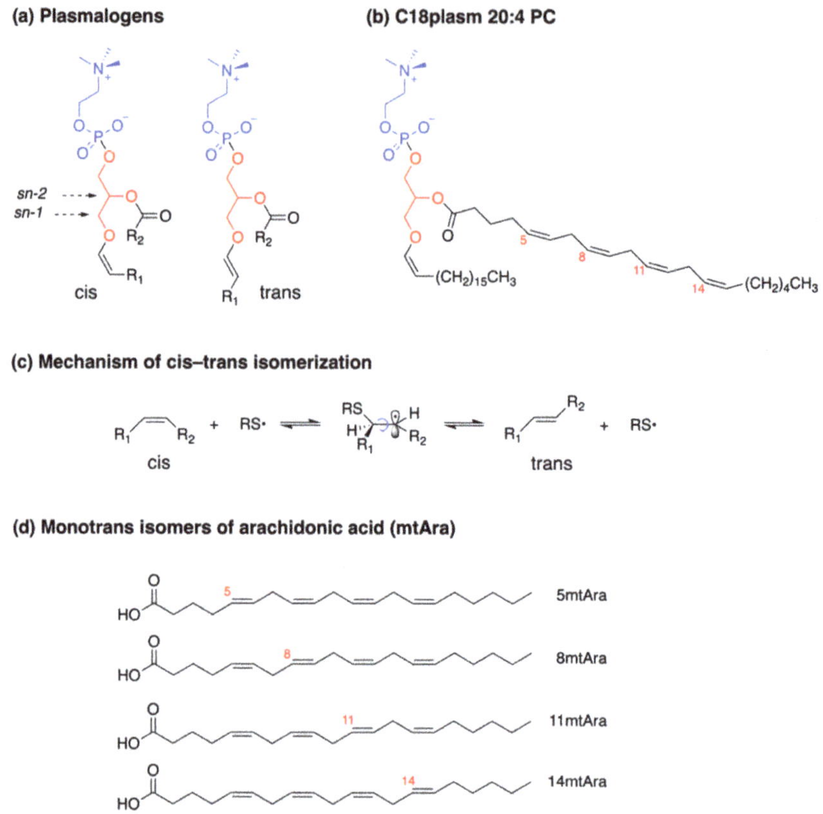

Figure 1. (**a**) General structure of plasmalogens with a fatty acid in position *sn*-2 and a vinyl ether-containing hydrocarbon chain in position *sn*-1. The natural structure has a cis-vinyl ether function. The trans-vinyl ether containing plasmalogen is a synthetically modified lipid. (**b**) 1-(1Z-octadecenyl)-2-arachidonoyl-*sn*-glycero-3-phosphocholine (acronym: C18 plasm 20:4-PC) with all cis double bonds. (**c**) Mechanism of thiyl radical catalyzed cis-trans double bond isomerization of unsaturated fatty acid moieties in lipids depicted as consecutive addition-elimination process. (**d**) The four monotrans isomers of arachidonic acid (mtAra) structures, as product mix obtained from the thiyl radical-catalyzed isomerization of Ara.

The plasmalogen content reaches up to 15–20% of the total phospholipid mass in red blood cells (RBC), and increases in tissues like the heart (32–50%), brain (20–50%), and spermatozoa (55%) [12–14].

From the chemical reactivity point of view, both the cis-vinyl ether and the allylic functions are known to react under oxidative conditions, thus playing a "sacrificial" role toward a series of attacks arriving at the membrane compartment, as reported for singlet oxygen [15] and iron-induced peroxidation [16]. This distinctive susceptibility is translationally applied in clinics, to evaluate diseases suffering from oxidative stress, such as Down syndrome [17], neurodegenerative and cardiometabolic disorders [18–20], cancer [13], or physiological processes, such as aging [21]. Moreover, the vinyl ether function is susceptible to acid-catalyzed hydrolysis, forming a lysophospholipid derivative and liberating the fatty acid chain in *sn*-1 position as fatty aldehydes, in the form of dimethylacetal (DMA) when MeOH is present [22]. In the case of infectious diseases, such in vivo hydrolysis exerts toxicity effects, due to the reduction of membrane plasmalogens with health consequences [23]. It is worth underlining that fatty acid-DMAs are known to be formed under HCl-MeOH conditions used as the conventional protocol in the RBC membrane lipidome analyses, in order to convert the fatty acid residues of phospholipids into their corresponding fatty acid methyl esters (FAME) for GC evaluation. Indeed, the GC quantification of DMA can be used to estimate the RBC plasmalogen content [24,25].

The unique formation of trans-vinyl ether-containing plasmalogen was reported in the synthetic route to 1-(1E-octadecenyl)-2–oleoyl-*sn*-glycero-3-phosphocholine (Figure 1a with R_1 as trans double bond and R_2 as oleic acid, 9cis-C18:1) [26]. No information on the trans plasmalogen formation in biologically related conditions is available so far.

On the other hand, the conversion of natural cis fatty acids into their corresponding trans fatty acid (TFA) isomers is known to be an endogenous process promoted by the in vivo formation of $RS^\bullet$ species from thiols, which donate an H atom during the repair of free radical damages. The cis-trans isomerization reaction occurs by the mechanism shown in Figure 1c [27]. For polyunsaturated fatty acids (PUFA) it occurs randomly on one of the double bonds of these structures, and the product is a mixture of mono-trans isomers, as shown in Figure 1d for Ara with its four monotrans Ara isomers (mtAra) [27]. The synthetic library of mtAra obtained by laboratory procedures has been successfully used as a molecular standard to prove and follow up the endogenous Ara isomerization in cells [28,29], animals [30,31], and humans [32,33].

We were interested to work on plasmalogens to investigate their chemical behavior under free radical conditions, in particular concerning the formation of the corresponding trans configurations at both cis double bonds of the structure. We chose the representative molecule reported in Figure 1b, i.e., 1-(1Z-octadecenyl)-2-arachidonoyl-*sn*-glycero-3-phosphocholine (acronym: C18 plasm 20:4-PC), containing the cis vinyl ether function in a C18 hydrocarbon chain at the *sn*-1 position and the omega-6 PUFA arachidonic acid (Ara, 5c,8c,12c,15c-20:4) esterified at the *sn*-2 position of L-glycerol. For the first time, we clarified the plasmalogen reactivity involving these two reactive sites of plasmalogens. The trans-plasmalogen structure, synthetically obtained by photolysis, was isolated and characterized by ^{1}H and ^{13}C mono- and bi-dimensional NMR, to get new information on peculiar spectral signatures for analytical applications. A comparative study of oxidation and isomerization of this plasmalogen provided new insights into chemical reactivity, to complete the scenario of lipid classes, previously performed on glycerophospholipids and cardiolipins [27,28,34,35]. Using unilamellar vesicles and RBC ghosts, we carried out biomimetic studies on the supramolecular assembly created by plasmalogens in a mix with other phospholipids. We also deepened the hydrolytic behavior in basic and acidic conditions as part of robust chemical protocols for the analyses of fatty acids in liposome and RBC experiments [34]. Generally, we aimed at reaching a better comprehension of the plasmalogen molecular reactivity to translate chemical information into powerful diagnostic tools for precision medicine [36].

2. Materials and Equipment

C18 plasm 20:4-PC (1-(1Z-octadecenyl)-2-arachidonoyl-*sn*-glycero-3-phosphocholine) and DMA:18:0 (18:0 dimethyl acetal) were produced by Avanti Lipids (Birmingham, AL, USA) and purchased from Merck (Darmstadt, Germany); chloroform, methanol, diethyl ether, n-hexane, acetonitrile, benzene (HPLC grade) were purchased from J. T. Baker (Phillipsburg, NJ, USA); ferrous ammonium sulfate $Fe(NH_4)_2(SO_4)_2 \times 6H_2O$ (Fe II AS) was from Carlo Erba, Milan, Italy; 2-mercaptoethanol, cis and trans FAME, Phosphate buffer, Hydrogen Peroxide, HCl/MeOH 0.5 M (1.5%; 1 mL/ampoule), BF_3/MeOH 14%, Tetramethylsilane (TMS) were purchased from Merck (Darmstadt, Germany); POPC (1-palmitoyl-2-oleoyl-*sn*-glycero-3-phosphocholine) and SAPC (1-stearoyl-2-arachidonoyl-*sn*-

glycero-3-phosphocholine) were purchased from Larodan (Solna, Sweden); benzene-d6 was purchased from CortecNet (Les Ulis, France); soybean lecithin was a gift from Lipoid GmbH (Ludwigshafen, Germany).

RBC was freshly isolated from anonymized blood samples of healthy volunteers received from the Department of Immunohematology and Transfusion Medicine (Metropolitan Area of Bologna, SIMT AMBO—Ethical Committee approval, #0083892).

NMR spectra were recorded at ambient temperature on a Varian 500 MHz spectrometer (Agilent, Cernusco sul Naviglio, Milan, Italy).

Hydrodynamic diameters of liposomes were measured using the dynamic light scattering (DLS) technique (Malvern Instruments Series NanoZS, Malvern Instruments, Malvern, UK) with a detection angle of 173°. All measurements were recorded at 25 °C.

Fatty acid methyl esters (FAME) were analyzed by comparison with authentic samples and chromatograms were examined and quantified by protocols that have been described previously [28,34]. 18:0 dimethyl acetal (DMA) was analyzed by GC (Agilent 6850, Agilent, Cernusco sul Naviglio, Milan, Italy), using the split mode (50:1) and a 60 m × 0.25 mm × 0.25 μm DB23 column (Agilent, Milan; Italy) with the same GC oven program and conditions used for FAME. DMA-18:0 was determined and quantified by its retention time and by multiple-point GC calibration curves of the standard reference (5 points) (LOQ: 0.0002 μg/mL; LOD: 0.0006 μg/mL; $R^2 = 0.9979$). Photolysis was performed as already described [28,34].

Liposome reactions were performed in an incubating orbital shaker (Argolab, Ski 4, Carpi, Italy), keeping the temperature at 37 °C.

Ag TLC preparation, TLC glass plates were pre-treated with 5% $AgNO_3$ solution in acetonitrile for 15 min and then dried at 120 °C for 1 h. Eluent for separation: $CHCl_3/MeOH/H_2O$ 4.5:2.3:0.2. The formation of a mixture of trans isomers after 4 min of UV-photolysis was monitored by Ag-TLC, using the cis plasmalogen as reference.

3. Methods

3.1. Transesterification Procedures of C18 Plasm-20:4 PC, FAME, Soybean Lecithin and RBC

A stock solution of 5 mg/5 mL of C18 plasm 20:4-PC (99% pure) in $CHCl_3$ was divided in aliquots of 0.25 mL in glass vials with Teflon cap (0.25 mg; 0.32 μmol) for each experiment. After the evaporation of $CHCl_3$, transesterification reactions were performed in the following conditions in triplicates:

(A) alkaline conditions: addition of 0.5 M KOH in MeOH (0.5 mL) to the plasmalogen, stirring of the reaction for 10 min, and quenching with a saturated solution of NaCl (0.5 mL); the FAME were extracted with n-hexane (4 × 1 mL), the organic solvent was dried on anhydrous Na_2SO_4, evaporated to dryness, and the residue was dissolved in n-hexane (50 μL) and analyzed by GC (1 μL) [37];

(B) acidic condition 1: addition of HCl/MeOH 0.5M (1.5%) (1 mL; fresh opened single-dose ampoule) to the plasmalogen and stirring of the reaction for 2 h at a controlled temperature of 100 °C in a thermostatic block [38,39]. The cooled methanolic reaction mixture was then extracted with n-hexane (4 × 1 mL). The organic layers were collected and evaporated to dryness. The residue containing FAME was dissolved in n–hexane (50 μL) and analyzed by GC (1 μL);

(C) acidic condition 2: addition of BF_3 in MeOH (14% wt/vol) (1mL) to the plasmalogen, stirring the reaction at a controlled temperature of 50 °C for 1 h in a thermostatic block [40]. After cooling at room temperature, the addition of 1 mL of a saturated solution of $NaHCO_3$ and n-hexane (4 × 1 mL) was performed followed by vortex-mixing; the hexane layers containing FAME were separated, collected, dried over anhydrous Na_2SO_4 and evaporated to dryness. The residue containing FAME was dissolved in n–hexane (50 μL) and analyzed by GC (1 μL).

In all FAME extraction steps, 100 μL of a stock solution of internal standard 17:0 methyl ester (1 mg/mL) was added to calculate the FAME recovery yield.

Table S1, Figures S1 and S2 (Supplementary Materials) show the results of plasmalogen transformation.

Alkaline (0.5 M KOH in MeOH) and acidic (HCl/MeOH 0.5M) procedures with the addition of C17:0 FAME, as above described, were also applied to evaluate FAME recovery yield of soybean lecithin (1 mg). The indicated experiment was performed in triplicates and the results are shown in Table S2.

The transesterification procedures were also tested with RBC. A single blood sample (40 mL), pooling 20 blood samples from different individuals, was created and 20 samples of 250 μL each (for a total of 5 mL) were taken to isolate RBC membranes and obtain membrane phospholipids, as described previously [32,33,41,42]. The 20 phospholipid extracts were divided into two groups to carry out alkaline (0.5 M KOH in MeOH) and acidic (HCl/MeOH 0.5M) transesterifications (n = 10 for each group). Only these two methodologies were chosen, excluding the acidic condition with BF_3/MeOH that has been shown to give an extensive PUFA deterioration in the C18 Plasm-20:4 PC transformation. FAME was obtained and examined by GC analysis as above described.

3.2. *Plasmalogen Cis-Trans Isomerization*

Cis-trans isomerization was performed by applying the previously described procedure for cardiolipins [34]. C18 plasm 20:4-PC (20 mg, 0.025 mmol) was dissolved in benzene (2 mL); the solution (12.6 mM) was transferred to a micro-photochemical reactor, degassed under N_2 for 20 min and, during the degassing, added with 0.5 equivalents of 2-mercaptoethanol in benzene (85 μL of a 146 mM stock solution; 6.3 mM). UV photo-irradiation was carried out for 4 min at a temperature of 22 ± 2 °C. Ag-TLC using $CHCl_3/MeOH/H_2O$ 4.5:2.3:0.2 as eluent evidenced three fractions (Rf: 0.7, 0.8, 0.86 corresponding to trans plasmalogen isomers) in comparison with starting material (Figure S3). The reaction mixture was evaporated to dryness and left under vacuum for 6 h to remove the residual thiol reagent. The crude (19 mg) was dissolved in 0.3 mL of deuterated benzene to carry out NMR spectra (^{1}H, ^{13}C, 2D HSQC) and compare them with the same NMR experiments run on the starting material.

^{1}H and ^{13}C NMR spectra were performed on the commercially available C18 plasm 20:4-PC in deuterated benzene, a solvent chosen to avoid hydrolysis of the vinyl ether which occurs in deuterated chloroform (Figures S4 and S5). Identification and attribution of the resonances are described in SI. 2D HSQC NMR (Figure S6a) was performed to assign proton and carbon resonances involved in double bonds, focusing specifically on those involved in vinyl ether moiety; Figure S6b reports magnification of the regions corresponding to vinyl ether group and arachidonic acid alkenyl carbons. ^{1}H and ^{13}C NMR spectra in deuterated benzene were carried out on the mixture of cis/trans C18 plasm 20:4-PC after photolysis (Figures S7 and S8). The assignment of resonances in the mixture is described in SI. 2D HSQC NMR of cis/trans plasmalogen mixture was performed (Figure S9a), to identify the vinyl ether group referred to as trans plasmalogen isomer (Figure S9b). An aliquot of the plasmalogen isomer mixture (0.5 mg) was also treated with KOH/MeOH (as the transesterification procedure) to obtain the fatty acid methyl ester (FAME) of arachidonic acid (ARA) moiety present at *sn*-2 position. GC analysis under known conditions separated and identified the four mono-trans isomers of the ARA residue (Figure S10).

To quantitate the progressive formation of trans isomers, photo-irradiations were performed on C18 plasm-20:4 PC (5 mg, 0.0063 mmol, 0.5 mL benzene) in presence of 0.5 equivalent of 2-mercaptoethanol, for 1, 2.5 and 4 min. At each time point, as previously described, the solvent was evaporated under vacuum and the residue dissolved in 150 μL of benzene-d6 containing 7.4 mM tetramethylsilane (TMS) as internal standard.

3.3. *Preparation of Liposomes*

Liposome preparation followed previously published procedures [28,34]. Briefly, a mixture of 1-palmitoyl-2-oleoyl-*sn*-glycero-3-phosphocholine (POPC) (85%) and C18 plasm 20:4-PC (15%) was dissolved in chloroform, evaporated to dryness to obtain a thin lipid film and left under vacuum for 1 h to eliminate traces of solvent. Then 1.3 mL of tridistilled water was added to reach a final lipid concentration of 10 mM and multilamellar vesicles (MLV) were formed by vigorous vortex stirring at 2200 rpm for 10 min. Large unilamellar vesicles were then prepared by extrusion technique (LUVET) using a filter of 200 nm diameter; the dimension was confirmed by Dynamic Light Scattering (DLS). The same procedure was applied for liposomes with 1-stearoyl-2-arachidonoyl-*sn*-glycero-3-phosphocholine (SAPC) 100%.

3.4. Preparation of RBC Ghosts

RBC was obtained from fresh EDTA-treated blood samples of healthy volunteers (see Material section) after separation from plasma as previously described [33,42]; briefly, 1 mL of whole blood was added with phosphate buffer (0.5 mL) and centrifuged at 5000 rpm for 5 min at 4 °C for two consecutive times. Plasma was removed and RBC lysis was then obtained by adding tridistilled H_2O followed by subsequent centrifugation at 15,000 rpm per 15 min at 4 °C 4 times to obtain a clear supernatant, which was then discarded; the erythrocyte membrane pellets were collected and used to prepare RBC ghosts as multilamellar vesicles (10 mM lipids as MLV suspended in water) following the above-described procedure for liposomes.

3.5. Isolation of Plasmalogens from RBC Membranes

RBC membrane pellets, obtained following the protocol described in Section 3.4, were extracted with a 2:1 $CHCl_3$/MeOH solution (4 times × 4 mL), according to the Folch method [43]. The organic layers were collected and dried over anhydrous Na_2SO_4, then evaporated under a vacuum to dryness. The lipid extracts were dissolved in 0.4 mL chloroform to isolate RBC plasmalogens by preparative TLC. The commercially available C18 plasm 20:4-PC was used as a standard reference. Eluent used for separation was $CHCl_3$/MeOH/H_2O 4.5:2.3:0.2 (adapted from ref [44]). The desired spot was scratched off, and silica gel was suspended in $CHCl_3$/MeOH/H_2O 4.5:2.3:0.2, stirred, and filtered; analytical TLC confirmed the purity of the plasmalogen fraction, estimated as 15% of the total lipid content.

3.6. Incubation Experiments under Oxidative Conditions

RBC ghosts or liposomes, prepared with SAPC and with 85:15 POPC/Plasmalogen mix (10 mM stock solution), were stored at 4 °C before use. Stock solutions of aqueous $Fe(NH_4)_2(SO_4)_2$ (1 mM), H_2O_2 (1 mM), and 2-mercaptoethanol (1 mM) were freshly prepared before incubations. To a 2 mL vial, the following reagents, at molarities similar to the biological ones, were added in sequence: $Fe(NH_4)_2(SO_4)_2 \times 6H_2O$ (10 μM), H_2O_2 (100 μM), liposome (1mM) and thiol (10 μM). Control samples containing only liposomes (1mM), in the absence of iron salts and thiols, were used to estimate the incubation effect (as blank). In all cases, a final reaction volume of 0.5 mL was reached. The oxidation reactions were performed in open vessels, keeping an incubation at 37 °C in an orbital shaker for 15 h, as described for cardiolipins [34], followed by work-up (extraction, transesterification, and GC analysis) as described in Sections 3.1 and 3.5.

3.7. Statistical Analysis

The results of the experiments were given as mean values ± SD and statistical analysis was performed using GraphPad Prism 8.0 software (GraphPad Software, Inc., San Diego, CA, USA). We used a non-parametric unpaired *t*-test two-tailed with 95% confidence interval.

4. Results and Discussion

4.1. Plasmalogen Cis-Trans Isomerization in Solution

The thiyl radical catalyzed isomerization of commercially available C18 plasm 20:4-PC (12.6 mM) was carried out in benzene by photolysis using 2-mercaptoethanol (6.3 mM) as the thiol [32,45]. We decided to perform the isomerization in an N_2-saturated benzene solution since we observed that degradation of the starting material is easier in chloroform than in benzene, likely due to the acidity of the former solvent. Irradiation in the photochemical reactor equipped with a mercury lamp (λ = 250–260 nm) was carried out for 4 min monitoring the formation of trans isomers by Ag-TLC individuating two new spots at higher Rf (Figure S3). Indeed, the Ag-assisted technique assigns the highest Rf to the compound with more trans double bonds [46]. At this stage we tried to separate the two plasmalogen isomer fractions; however, we did not succeed due to the sensitivity to hydrolysis during the work-up. The cis/trans plasmalogen mixture was used for mono and bi-dimensional NMR experiments, in order to individuate characteristic resonances of the trans isomer configuration of the vinyl ether moiety as well as of the trans double bonds of the Ara moiety. Figure 2 summarizes the results, highlighting the NMR regions of ^{1}H NMR (a) and ^{13}C NMR (b) spectra corresponding to the vinyl ether function OCH=CH in the *cis* and trans configurations: in the starting material, the red line identified the connections between the proton signal of OC<u>H</u>=CH at ppm 6.22 with its carbon at 145.66 ppm; while the blue line correlates the proton OCH=C<u>H</u> at 4.55 ppm with the carbon at 107.07 ppm. The integration of the trans-assigned regions gave an estimation of trans isomer formation at 32% yield. The trans vinyl ether assignments are corroborated by analogous assignments in the case of the previously cited synthetic plasmalogen [26].

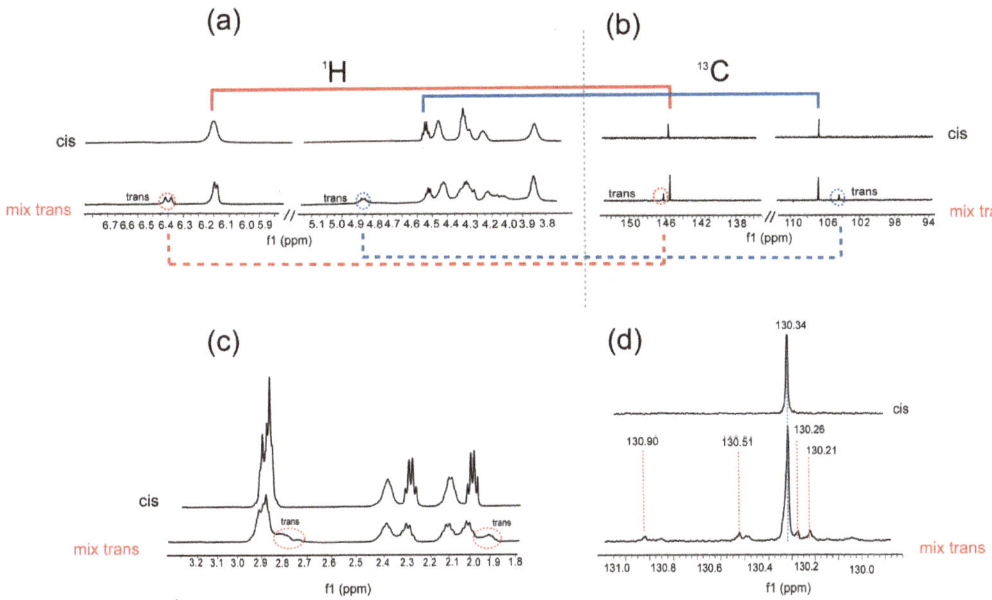

Figure 2. NMR spectra of the commercially available cis plasmalogen and the corresponding mixture of cis/ trans plasmalogen obtained after 4 min of UV photolysis; (**a**) ^{1}H NMR and (**b**) ^{13}C NMR regions relative to vinyl ether group; the red and blue lines give the H-C correlations of the OCH=CH group in cis configuration; red and blue dashed lines give the H-C correlations of the trans OCH=CH group; (**c**) region of bis-allylic and allylic proton signals; the red circle highlights the new signals at 2.78 and 1.97 ppm; (**d**) ^{13}C NMR spectral region relative to C15 resonance of cis plasmalogen and the corresponding cis/trans plasmalogen isomers, where red dashed lines indicate the mtAra isomers: 14-trans (130.90 ppm), 11-trans (130.51 ppm), 8-trans (130.26 ppm) and 5-trans (130.21 ppm) isomers [32].

In Figure 2c,d, the traces of the cis/trans isomer mixture show the two vinyl ether proton signals and the corresponding carbon atom resonances (red and blue dotted lines) of the trans vinyl ether-containing plasmalogen isomer. In Figure 2c,d, ^{1}H and ^{13}C NMR analyses are dedicated to the arachidonic acid moiety of the plasmalogen structure. The spectral region corresponding to allylic and bis-allylic hydrogen atoms is focused with two signals at 1.97 and 2.78 ppm, respectively, related to the mono-trans isomer of plasmalogen in comparison to the natural cis plasmalogen.

In Figure 2d the ^{13}C region corresponding to the C15 position of Ara residue is shown: the C-15 resonance of the cis isomer (130.34 ppm) is accompanied by four distinctive resonances at 130.90, 130.51, 130.26, 130.21 ppm, which belong to the mono-trans arachidonate isomers (mtAra) in position 14, 11, 8 and 5, respectively. This assignment is supported by previous data from the isomerization of 1-stearoyl-2-arachidonoyl-phosphatidyl choline (SAPC), and the four mono-trans isomers derived from the isomerization of Ara methyl ester (AraMe) [32,45].

^{1}H NMR experiments were also performed on the mixture of cis/trans plasmalogen after photo-irradiation at 1, 2.5, and 4 min of C18 plasm-20:4 PC (5 mg, 0.00625 mmol, 0.5 mL benzene) in presence of 0.5 equivalent of 2-mercaptoethanol; the above-described characteristic proton signals were followed up by ^{1}H NMR, at each time point of irradiation, dissolving in the crude reaction mixture in deuterated benzene 7.4 mM tetramethylsilane (TMS) as internal standard. This expedient allowed the calculation of the product yields, using the molarity of TMS standard to obtain those of cis/trans vinyl ether-containing compound, thus resulting in the yield of the trans plasmalogen (Table in Figure 3A). On the other hand, using the transesterification under alkaline conditions followed by quantitative GC analysis, the molarity of cis and mono-trans Ara isomers at the *sn*-2 position of plasmalogen was calculated (Figure S10). In Figure 3B the molarities expressed as relative quantitative percentages of the above-described analyses are reported in the function of the reaction time, showing the progressive formation of trans-plasmalogens (containing trans vinyl ether and mono-trans Ara isomers), reaching around 30% yield after 4 min photoirradiation, with 70% of the starting cis plasmalogen recovered in the mixture. Table S4 in Supplementary Materials gives details of the follow-up of the cis and trans contents in this experiment.

A

photolysis (min)	vinylether (C18) cis/trans (%) *	fatty acid (Ara) cis/trans (%) ¥
1	88.5/11.5	87.4/12.6
2.5	77.5/22.5	80.1/19.9
4	67.5/32.25	69.2/30.8

* obtained by ^{1}H NMR using TMS as reference (see Methods)
¥ obtained by GC analysis (see SI for details of monotrans isomers distribution)

B

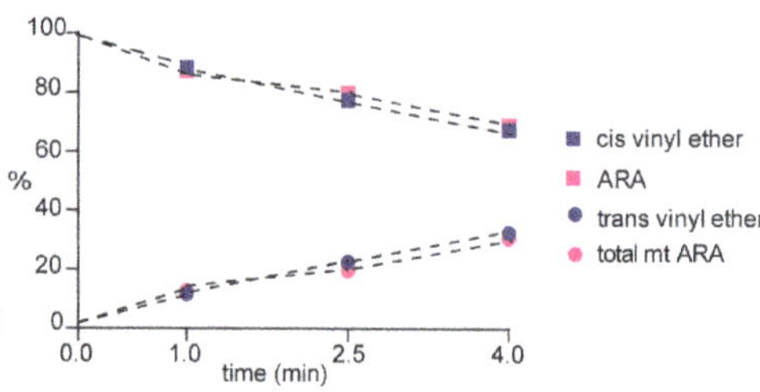

Figure 3. Follow-up of the isomerization of C18 plasm-20:4 PC (0.125 mM) performed in deuterated benzene in presence of 2-mercaptoethanol (0.5 equiv.) after 1, 2.5, and 4 min of photo-irradiation: (**A**) trans vinyl ether formation quantitatively determined by ^{1}H NMR, using tetramethylsilane (TMS) molarity (7.4 mM) as a reference, and mt Ara isomers obtained by quantitative GC analysis after transesterification of the reaction mixture at each time point; (**B**) graphical representation of the values reported in Table A for the plasmalogen photoirradiation with the formation of trans isomers and decrease of the starting cis material. See details in Table S4.

It is evident from Figure 3 that the vinyl ether reactivity in cis-trans isomerization is similar to the reactivity of Ara moiety that contains four C–C double bonds. The mechanism of the cis-trans isomerization of C-C double bond in MUFA methyl esters by thiyl radicals has been studied in detail [47–49], as well as for methyl linoleate as a model for PUFA isomerization [50]. The reversible addition of thiyl radical to cis fatty acids converts them to trans fatty acids in a catalytic process (see Figure 1c) and the cycle is interrupted by exothermic allylic or bisallylic abstraction [49,50]. Kinetic data are available for all elementary steps as well as the cycle lengths for individual fatty acids. On this basis, the reason for the high reactivity of cis vinyl ether for the isomerization reaction (Figure 4) with respect to cis double bond in unsaturated fatty acids can be attributed at least to two factors [51]: (i) higher rate constant of RS• addition due to the stabilization of adduct radical (see Figure 4), i.e., by the interaction of unpaired electron of thiyl radical with the lone pair electrons of oxygen, and (ii) the effect of the oxygen substituent increasing the pyramidality of the radical center and barriers to internal rotation about C^{α}–C^{β} and C^{α}–O, thus favoring conformational preference of intermediate radical towards the trans vinyl ether.

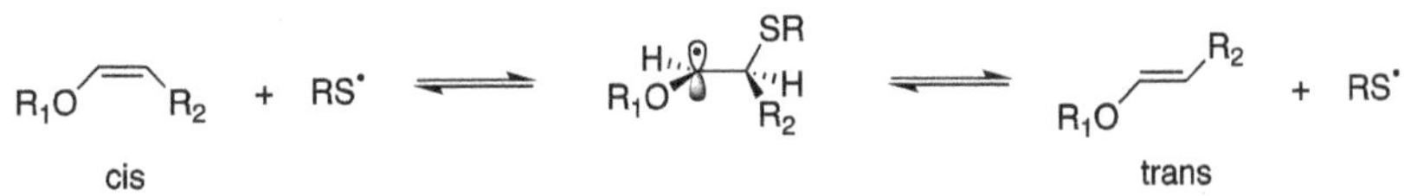

Figure 4. Proposed mechanism for the thiyl radical catalyzed cis-trans isomerization of vinyl ether moiety in plasmalogens.

4.2. Transesterification Procedures for Plasmalogen-Containing Samples

As explained in Section 1, plasmalogens present a typical vinyl ether bond, which is resistant to strong bases and reducing/oxidizing agents and is easily cleaved by acids, leading to an aldehyde function further transformed to dimethyl acetal (DMA) when the reaction is performed in methanol [23,52–54]. We performed a precise evaluation of the best transesterification condition for releasing the fatty acid in the *sn*-2 position as the corresponding FAME. This step is very important to treat biological samples containing plasmalogens, such as in the case of red blood cell (RBC) membranes. Indeed, the occurrence of vinyl ether hydrolysis results in the formation of a lysophospholipid derivative, affecting the behavior of the PUFA residue in *sn*-2 [23]. We first compared treatments of the commercially available C18 plasm 20:4-PC using alkaline conditions at room temperature (rt) using 0.5 M KOH/MeOH, and under two acidic conditions: 0.5 M (1.5%) HCl/MeOH at 100 °C and 14% BF_3/MeOH at 50 °C, as indicated in Section 3.1. Each experiment was performed in triplicate and results are reported in Table S1 (Supplementary Materials). Only alkaline transesterification (0.5 M KOH/MeOH) transformed the arachidonic acid residue into its methyl ester (Ara-Me) in quantitative yield, as ascertained by GC quantitative analysis using 17:0 (as an internal standard in the extraction phase) and by calibration of standard Ara-Me via multiple points GC calibration curves, as previously described [34]. No formation of DMA18.0 was detected. The methanolic acidic condition at a high temperature (100 °C) gave Ara-Me 92% yield and parallel formation of DMA C18:0 (52% yield). The acidic treatment with 14% BF_3/MeOH at 50 °C produced Ara-Me (58% yield) and DMA C18:0 (32.5% yield), showing that the two acidic conditions cause a consistent loss of information about the PUFA in the *sn*-2 position (Table S1). Figure S1 reports the GC chromatograms of the three transesterification reactions (A: KOH/MeOH at rt; B: HCl/MeOH at 100 °C; C: BF_3/MeOH at 50 °C); under acidic reactions, the presence of Ara-Me, DMA 18:0 and 17:0 (reference standard) is detected, together with a degradation product of DMA 18:0 which is known to be due to additional hydrolytic processes [52–54]. The formation of DMA18:0 was also confirmed by GC-MS (Figure S2) with diagnostic fragmentations at *m/z* 75 and 283 [54]. These results clarify the only condition to treat plasmalogen-containing specimens, i.e., KOH/MeOH at room temperature, whereas the results of samples that contain plasmalogens under acidic conditions and high temperatures are inexact [24,25,53,54] since the PUFA residues present in the *sn*-2 position are not transformed quantitatively. Additionally, the use of acidic conditions, besides the high temperatures (50 and 100 °C) and longer reaction times required, needs fast manipulation and hermetic vials to avoid irritations and water interference; in the case of BF_3 undesirable side reactions can also occur [55,56].

Alkaline and acidic transesterifications were also compared to examine the complexity of biological samples, such as soybean lecithin and RBC membrane pellet, taking into account that samples of biological origin and RBC often are reported under the condition of 0.5M HCl/MeOH at 100 °C [24,25,38,57]. In the lecithin experiment, performed in triplicate, the conversion of the linoleic acid residue into its methyl ester, calculated by calibration curve of the standard reference, was quantitative in basic condition (0.5 M KOH/MeOH at room temperature) and had a 93% yield in acidic condition (0.5 M HCl/MeOH, at 100 °C) (Table S2). It is worth noting that plasmalogens are not present in soybean lecithin. In RBC membranes, where a natural content of plasmalogen is around 15–20% of the total mass of phospholipids (see Experimental Section 3.5) [12–14,58], a membrane pellet was prepared (n = 10) to carry out the two transesterification reactions of the same sample, followed by work-up and analysis, as described in Section 3.1.

The results of the FAME composition of the same RBC membrane pellet treated by the two transesterification reaction conditions evidenced clear and statistically relevant differences, as reported in Table 1. This can be expected from the well-known chemical reactivity of lipid structures which contains amide-, ether, and ester bonds to link the fatty acid moiety [58–61]. In the acidic condition, it was seen the enrichment of SFA residues, explained by the transformation of SFA residues belonging to amide-containing lipids, such as sphingolipids, which need a such strong protocol to be converted [59–62]. Acidic conversion is the most frequently used method for automated protocols of blood lipidomic analysis for epidemiological studies [62,63]; however, it must be clear to the analyst that in this case sphingomyelins are transformed (rich in SFA), whereas plasmalogens, as well as other acid-sensitive PUFA-containing lipids, are not stable. On the other hand, under basic conditions, plasmalogens can release the ester-linked fatty acid moieties, which are mostly PUFA residues, as FAME, thus saving this important information about the presence of essential fatty acids. In Table 1 the different PUFA contents in the two conditions can be compared. It is worth noting that the results are obtained quantitatively (μg/μL) by parallel 17:0 and PUFA calibration, and then transformed into relative quantitative percentages of each element in the fatty acid cluster (13 elements). In nutritional studies and in evaluating the oxidative effects of diseases, using RBC membranes, as well as sperm lipids and other specimens rich in plasmalogens, our results demonstrate the importance of laboratory protocol choice affecting the resulting outcome. Another important point is to use both SFA and PUFA standards for the fatty acid calibration step, to ascertain that work-up conditions do not affect labile structures.

Table 1. Comparison of fatty acid methyl ester (FAME) percentages (% rel. quant.) obtained from the same RBC membrane pellet transformed by alkaline (0.5 M KOH/MeOH; room temperature) and by acidic (0.5 M HCl/MeOH—100 °C) transesterification conditions (n = 10 repetitions each).

FAME [1]	KOH/MeOH (% Rel. Quant.) Mean ± SD (n = 10)	HCl/MeOH (% Rel. Quant.) Mean ± SD (n = 10) [2]
16:0	23.3 ± 0.6	34.0 ± 3.2 ***
9cis-16:1	0.24 ± 0.03	0.18 ± 0.05 **
18:0	16.0 ± 1.3	21.8 ± 1.3 ***
9trans-18:1	0.01 ± 0.01	0.03 ± 0.01 ***
9cis-18:1	17.0 ± 0.5	13.8 ± 1.0 ***
11cis-18:1	1.2 ± 0.2	1.0 ± 0.1 ***
9cis,12cis-18:2	12.1±0.6	9.2 ± 0.4 ***
20:3 omega-6, DGLA	2.3 ± 0.1	1.5 ± 0.3 ***
20:4 omega-6, Ara	18.8 ± 0.4	13.3 ± 1.4 ***
5mtAra	0.02 ± 0.01	0.03 ± 0.01 *
20:5 omega-3, EPA	0.8 ± 0.1	0.6 ± 0.3 *
22:5 omega-3, DPA	2.7 ± 0.5	1.7 ± 0.4 ***

Table 1. *Cont.*

FAME [1]	KOH/MeOH (% Rel. Quant.) Mean ± SD (n = 10)	HCl/MeOH (% Rel. Quant.) Mean ± SD (n = 10) [2]
22:6 omega-3, DHA	5.4 ± 0.1	3.0 ± 0.5 ***
SFA	39.3 ± 1.3	55.8 ± 3.3 ***
MUFA	18.5 ± 0.7	14.9 ± 0.9 ***
PUFA	42.1 ± 0.7	29.3 ± 2.6 ***
omega-6	33.2 ± 0.7	24.0 ± 1.8 ***
omega-3	8.9 ± 0.4	5.2 ± 0.9 ***
TOT TRANS	0.03 ± 0.01	0.05 ± 0.02 *

[1] Identified by standard references and quantified as described in Materials and Methods. Values are obtained in µg/µL from the GC peak areas recognized and calibrated with standard references (corresponding to >97% of the total peaks of the chromatogram). Values are expressed in percentages relative to the sum of all the quantities of the recognized peaks (100%) ± Standard Deviation (S.D); [2] Statistical significance is estimated: * $p < 0.043$; ** $p \leq 0.0044$; *** $p \leq 0.0005$.

4.3. Plasmalogen-Containing Liposomes as Model for Radical Stress Conditions

Radical stress conditions were then studied in plasmalogen-containing liposomes using the alkaline transesterification procedure to follow up the two expected processes: isomerization with the formation of mtAra isomers of plasmalogen and peroxidation reaction of the PUFA evaluated by the loss of Ara (Section 3.6). We previously evidenced that cis-trans isomerization can occur during lipid peroxidation [28,35]; however, plasmalogen reactivity by combined isomerization-oxidation processes was never reported. This is a missing point in the whole scenario of the sensitivity of this lipid class to oxidative conditions, including the protective role of vinyl ether bonds against PUFA peroxidation [15,16,26]. As extensively explained in previous work on lipid isomerization [27–32,34,35], we first discovered the harmful consequence of thiyl radicals formed from thiols for the cis lipid double bonds (Figure 4). The biological significance of this reaction relies on the fact that thiyl radicals are produced from sulfur-containing compounds, such as cysteine, glutathione, and H_2S, as part of their antioxidant reactivity in cell metabolism [27,32]. We proved that trans fatty acid levels increase under biological stress conditions and nowadays these markers are considered in the scenario of pathological conditions [64]. The formation of free radicals from the antioxidant reaction is often under-evaluated in the whole scenario of biological damage repair. In our studies of the thiol involvement, we demonstrated that when the redox cycle of Fe^{2+}/Fe^{3+} is involved, such as in the case of bleomycin complex, both reactive radical species $OH^{\bullet}$ and $RS^{\bullet}$ are produced and trigger the reactivity shown in Figure 5, causing these species both PUFA peroxidation together with MUFA and PUFA double bond isomerization, respectively [34,35].

We used this knowledge to carry out a biomimetic experiment with plasmalogens using a 1 mM liposome suspension, composed of an 85:15 ratio of POPC and C18 plasm 20:4-PC, thus keeping the 15% plasmalogen content usually present in RBC membranes. The natural content of oxygen in the medium (open air) was left in all reactions, i.e., no degassing operation was performed, as previously reported [34]. We added the Fenton reagent (10 µM Fe^{2+} salt and 100 µM H_2O_2) and incubated at 37 °C for 15 h, as the time frame used for the previously described cardiolipin experiment, with or without 10 µM amphiphilic 2-mercaptoethanol (Section 3.6).

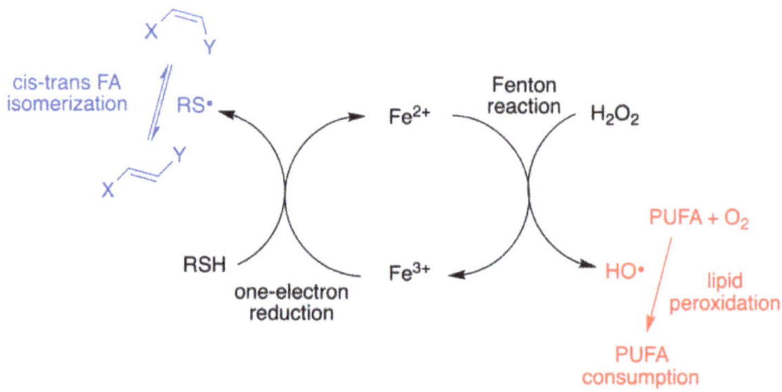

Figure 5. The isomerization and peroxidation processes occurring under Fenton-like conditions to double bond and PUFA moieties, respectively, in the presence of thiol RSH. The double bond can belong to the vinyl ether function or the unsaturated fatty acid chain.

In Table 2 the results of the two different suspensions as triplicates are reported, monitoring both peroxidation/isomerization of the Ara moiety of C18(plasm)20:4–PC (expected to give the four mono-trans isomers as depicted in Figure 1d) and isomerization for the oleic acid (9cis-18:1) moiety of POPC (i.e., expected to be transformed into elaidic acid, 9trans-C18:1 [27,28]). In the absence of thiol, no trans isomer was detected. The occurrence of peroxidation reaction was calculated from the recovery of Ara, quantitatively estimated using the calibration curve with Ara and the addition of an internal standard (C17:0 methyl ester). This is an indirect way to estimate Ara loss for peroxidation reaction, not using the direct measurement of peroxide formation [34,35]. In Table 2 the experiment without thiol showed no effect on the oleic acid moiety of POPC, whereas the Ara moiety of plasmalogen was reduced by 46%. In the presence of 10 μM 2-mercaptoethanol, as a source of thiyl radicals, the occurrence of both peroxidation and isomerization processes was evidenced for the oleic and arachidonic acids moieties, with the operative cycle shown in Figure 5.

Table 2. Contents of 9cis-18:1 (oleic acid) and Ara and their corresponding trans isomers in 1 mM Plasmalogen/POPC liposome suspensions (15:85) in the presence of 10 μM Fe^{2+}, 100 μM H_2O_2 incubated at 37 °C in the open air for 15 h with/without 10 μM 2-mercaptoethanol as a thiol [1].

Thiol (μM)	9cis-18:1 (%)	9trans-18:1 (%)	Ara (%)	Ara Loss (%)	mtAra (%)
0	99.9	-	54.0	46.0	n.d.
10	99.0	0.55	77.1	21.0	1.9

[1] Values are means ± standard deviation of n = 3 experiments under the same conditions; they are expressed as a relative percentage (% rel. quant.) of each fatty acid isomer with respect to the starting cis fatty acid residue, estimated by GC analysis using calibration with appropriate standard references and C17:0 as internal standard. The sd is not indicated since all experiments had sd < 0.1; n.d. = not detectable.

Finally, the model of RBC ghosts was examined as an additional biomimetic model to estimate the plasmalogen contribution to the exposition of biological membranes to radical stress. They were obtained following known procedures described in Section 3.4 [33,43]. RBC ghosts are membranes containing 15% plasmalogen in a mix with other phospholipids [58,65] and allow evaluating the behavior of oleic acid (9cis-18:1), linoleic acid (9cis,12cis-18:2, PUFA omega-6) and Ara residues. In this way, we could reach a satisfactory overview of reactivity environments to evaluate how plasmalogens behave toward peroxidation/isomerization. The results are listed in Table 3.

Table 3. Yields (% rel. quant.) of 18:1 (oleic acid), 18:2 (linoleic acid), and Ara and their corresponding monotrans (mt) isomers obtained after transesterification of RBC ghosts (1 mM) incubated at 37 °C in the open air for 15 h in the presence of 10 μM Fe^{2+}, 100 μM H_2O_2, with/without the addition of 10 μM 2-mercaptoethanol as a thiol.

Thiol (μM)	9cis-18:1	9trans-18:1	Ara	Ara Loss	mtAra	18:2	18:2 Loss	mt-18:2
0	99.0 ± 1.0	-	66.9 ± 1.1	33.1 ± 1.1	-	73.4 ± 2.1	26.5 ± 2.1	-
10	99.2 ± 0.1	0.7 ± 0.1	95.0 ± 0.2	2.3 ± 0.01	2.3 ± 0.2	96.1 ± 0.8	2.5 ± 0.6	1.4 ± 0.2

Values are means of n = 3 experiments under the same conditions expressed as a relative percentage (% rel. quant.) of each fatty acid isomer with respect to the starting cis fatty acid residue, estimated by GC analysis using calibration and 17:0 as internal standard; Yields of the three repetitions were found with errors <0.05%.

The most representative PUFA are linoleic acid (26% of loss) and Ara (loss of 33%) after incubation at 37 °C for 15 h (Table 3). Compared with 85:15 POPC: C18 plasm 20:4-PC, where the Ara loss is 46% (Table 2), PUFA reactivity to oxidation can be considered almost equivalent. It is remarkable that in RBC ghosts, containing approximately 18% MUFA and 42% PUFA residues, including 18% Ara (see Table 1 and [58]), Ara loss resulted to be similar to the 15/85 plasmalogen/POPC liposomes. When thiol was added, the scenario of peroxidation and isomerization reactivity showed almost complete protection of PUFA moieties (linoleic and arachidonic acids yields = ca 95%) with the appearance of trans isomers in a 0.7–2.3%. The behavior of RBC ghosts was completely different from the 85:15 POPC: C18 plasm 20:4-PC liposomes, which still showed an Ara loss of 21%.

In order to compare the arachidonic acid reactivity in lipids without the vinyl ether function, we also compared the behavior of liposomes made of 100% 1-stearoyl-2-arachidonoyl-*sn*-glycero-3-phosphocholine (SAPC) under oxidative conditions. The consumption of Ara moiety under Fenton conditions was found to be very high (73%, see Table S3), the highest among the liposome suspensions tested in this work. This can partly confirm the reported indication of oxidative protection given by the vinyl ether moiety of plasmalogens [15,16], at least reducing the PUFA loss comparing the data for liposomes containing 15:85 Plasmalogen/POPC, 100% SAPC, and RBC ghosts (Table 2, Table 3 and Table S3). In this experiment, it must be underlined that plasmalogens are a mix of structures having choline and ethanolamine as the polar head. This structural diversity should be considered part of the variables that can lead to a different molecular reactivity, which can be deepened in further studies. It must be also added that although the influence of the cholesterol in membranes is not the target of this work, the cholesterol content in RBC ghosts [65] is known to protect membrane phospholipids from oxidative damage [66].

5. Conclusions

The present work expands knowledge on the plasmalogen reactivity toward oxidative conditions completing the scenario of biological membrane transformations. The phenomenon of membrane lipid oxidation involves lipid bilayer properties and interactions between the lipid bilayer and integral membrane proteins [67], as well as the type of membrane lipids triggering oxidative reactivity, such as in the case of ferroptosis [68]. We demonstrated for the first time that plasmalogens display an additional reactivity linked to the cis double bonds at the vinyl ether function and the polyunsaturated fatty acid residue. Geometrical isomerization and formation of trans plasmalogens occur in the presence of thiyl radicals, generated from thiols in connection with the redox cycle Fe^{+2}/Fe^{+3}, which is considered a double-edged process in the scenario of cell death [69]. The design of appropriate models to estimate the potential of oxidative processes and antioxidant effects in biological membranes must take into account also the molecular composition and the plasmalogen contribution in tissues. We furnished clear proof that lipid composition with its supramolecular organization plays a crucial role in the overall protection of membranes from degradative processes, which can vary from tissue to tissue and from their plasmalogen contents.

Supplementary Materials: The following supporting information can be downloaded at: https://www.mdpi.com/article/10.3390/biom13050730/s1, Table S1: FAME yield (%) derived from the transesterification of C18 plasm 20:4-PC (0.32 mmol/mL) under three different conditions; Table S2: Transformation of linoleic acid (9cis,12cis-18:2) residues of soybean lecithin into fatty acid methyl ester (FAME); Table S3: SAPC liposome (1 mM) aqueous suspension in the presence of 10 μM Fe^{2+}, 100 μM H_2O_2; Table S4: Follow-up of the isomerization of C18 plasm-20:4 PC (0.125 mM); Figure S1: GC chromatograms of plasmalogen (C18 plasm 20:4-PC) transesterification performed in three conditions; Figure S2: GC/MS chromatogram showing 18:0 DMA and the side product after plasmalogen transesterification in acidic condition; Figure S3: Ag-TLC monitoring of the photoisomerization of plasmalogen; Figure S4: ^{1}H NMR spectrum of commercially available plasmalogen in C_6D_6; Figure S5: ^{13}C NMR spectrum of commercially available plasmalogen in C_6D_6; Figure S6a: HSQC 2D-NMR spectrum of commercially available plasmalogen in C_6D_6; Figure S6b: HSQC 2DNMR 4.1-6.4 ppm region containing the resonances of the vinyl ether and alkenyl moieties of all cis plasmalogen; Figure S7: ^{1}H NMR spectrum in C_6D_6 of the reaction mix of plasmalogen isomerization; Figure S8: ^{13}C NMR spectrum in C6D6 of the reaction mix of plasmalogen isomerization; Figure S9a: HSQC 2D NMR spectrum of the mixture cis-trans plasmalogen in C_6D_6; Figure S9b: HSQC 2D NMR 3.9–10 ppm enlargement; Figure S10: Enlargement of GC traces.

Author Contributions: Conceptualization, C.F.; methodology, C.C., A.S., F.V., V.R.; formal analysis, A.F., A.S., F.V., G.B.; investigation, A.S., F.V., G.B.; resources, C.F., V.R.; data curation, C.F., C.C., A.S.; writing—original draft preparation, C.F., A.S.; writing—review and editing, C.F., C.C., A.S., M.V.R.; visualization, C.F., C.C.; supervision, C.F., M.V.R.; project administration, C.F., V.R.; funding acquisition, C.F., V.R. All authors have read and agreed to the published version of the manuscript.

Funding: F.V. was supported for a 1-year grant by Lipinutragen srl; G.B. is a PhD student of the Department of Biotechnological, Biocomputational, Pharmaceutical and Pharmacological Sciences at the University of Bologna, in the frame of an industrial doctorate as an employee of Silfradent srl.

Institutional Review Board Statement: The study was approved by the Ethics Committee of the Metropolitan Area of Bologna, SIMT AMBO (protocol code #0083892).

Informed Consent Statement: Not applicable.

Data Availability Statement: The data presented in this study are available on request from the corresponding author. Data sets regarding humans subjects are not available due to privacy restrictions.

Acknowledgments: Authors thanks Paolo Dambruoso for some help in NMR techniques; authors thanks Andriani Chaidali for some laboratory activity during the work in the frame of her visit as Erasmus+ student from the Aristotle University of Thessaloniki, Greece under the supervision of C.F.; authors thanks Stefania Parenti of the CRC-RER for help in administrative matters.

Conflicts of Interest: The authors declare no conflict of interest.

References

1. Flor, A.C.; Kron, S.J. Lipid-derived reactive aldehydes link oxidative stress to cell senescence. *Cell Death Dis.* **2016**, *7*, e2366. [CrossRef] [PubMed]
2. Zhivaki, D.; Kagan, J.C. Innate immune detection of lipid oxidation as a threat assessment strategy. *Nat. Rev. Immunol.* **2022**, *22*, 322–330. [CrossRef] [PubMed]
3. Munir, R.; Lisec, J.; Swinnen, J.V.; Zaidi, N. Lipid metabolism in cancer cells under metabolic stress. *Br. J. Cancer* **2019**, *120*, 1090–1098. [CrossRef]
4. Editorial. Let's talk about lipid nanoparticles. *Nat. Rev. Mater.* **2021**, *6*, 99. [CrossRef]
5. Herrmann, I.K.; Wood, M.J.A.; Fuhrmann, G. Extracellular vesicles as a next-generation drug delivery platform. *Nat. Nanotechnol.* **2021**, *16*, 748–759. [CrossRef] [PubMed]
6. Skotland, T.; Sagini, K.; Sandvig, K.; Llorente, A. An emerging focus on lipids in extracellular vesicles. *Adv. Drug Deliv. Rev.* **2020**, *159*, 308–321. [CrossRef] [PubMed]
7. Dean, J.M.; Lodhi, I.J. Structural and functional roles of ether lipids. *Protein Cell* **2018**, *9*, 196–206. [CrossRef] [PubMed]
8. Koivuniemi, A. The biophysical properties of plasmalogens originating from their unique molecular architecture. *FEBS Lett.* **2017**, *591*, 2700–2713. [CrossRef] [PubMed]
9. Gallego-Garcia, A.; Monera-Girona, A.J.; Pajares-Martinez, E.; Bastida-Martinez, E.; Perez-Castano, R.; Iniesta, A.A.; Fontes, M.; Padmanabhan, S.; Elias-Arnanz, M. A bacterial light response reveals an orphan desaturase for human plasmalogen synthesis. *Science* **2019**, *366*, 128–132. [CrossRef]

10. Wainberg, M.; Kamber, R.A.; Balsubramani, A.; Meyers, R.M.; Sinnott-Armstrong, N.; Hornburg, D.; Jiang, L.H.; Chan, J.; Jian, R.Q.; Gu, M.X.; et al. A genome-wide atlas of co-essential modules assigns function to uncharacterized genes. *Nat. Genet.* **2021**, *53*, 638–649. [CrossRef]
11. Almsherqi, Z.A. Potential Role of Plasmalogens in the Modulation of Biomembrane Morphology. *Front. Cell Dev. Biol.* **2021**, *9*, 673917. [CrossRef] [PubMed]
12. Honsho, M.; Abe, Y.; Fujiki, Y. Plasmalogen biosynthesis is spatiotemporally regulated by sensing plasmalogens in the inner leaflet of plasma membranes. *Sci. Rep.* **2017**, *7*, 43936. [CrossRef]
13. Messias, M.C.F.; Mecatti, G.C.; Priolli, D.G.; Carvalho, P.D. Plasmalogen lipids: Functional mechanism and their involvement in gastrointestinal cancer. *Lipids Health Dis.* **2018**, *17*, 41. [CrossRef] [PubMed]
14. Braverman, N.E.; Moser, A.B. Functions of plasmalogen lipids in health and disease. *Biochim. Biophys. Acta-Mol. Basis Dis.* **2012**, *1822*, 1442–1452. [CrossRef]
15. Broniec, A.; Klosinski, R.; Pawlak, A.; Wrona-Krol, M.; Thompson, D.; Sarna, T. Interactions of plasmalogens and their diacyl analogs with singlet oxygen in selected model systems. *Free Radic. Biol. Med.* **2011**, *50*, 892–898. [CrossRef]
16. Sindelar, P.J.; Guan, Z.Z.; Dallner, G.; Ernster, L. The protective role of plasmalogens in iron-induced lipid peroxidation. *Free Radic. Biol. Med.* **1999**, *26*, 318–324. [CrossRef] [PubMed]
17. Bueno, A.A.; Brand, A.; Neville, M.M.; Lehane, C.; Brierley, N.; Crawford, M.A. Erythrocyte phospholipid molecular species and fatty acids of Down syndrome children compared with non-affected siblings. *Br. J. Nutr.* **2015**, *113*, 72–81. [CrossRef]
18. Dorninger, F.; Forss-Petter, S.; Berger, J. From peroxisomal disorders to common neurodegenerative diseases the role of ether phospholipids in the nervous system. *FEBS Lett.* **2017**, *591*, 2761–2788. [CrossRef]
19. Dorninger, F.; Forss-Petter, S.; Wimmer, I.; Berger, J. Plasmalogens, platelet-activating factor and beyond—Ether lipids in signaling and neurodegeneration. *Neurobiol. Dis.* **2020**, *145*, 105061. [CrossRef]
20. Paul, S.; Lancaster, G.I.; Meikle, P.J. Plasmalogens: A potential therapeutic target for neurodegenerative and cardiometabolic disease. *Prog. Lipid Res.* **2019**, *74*, 186–195. [CrossRef] [PubMed]
21. Maeba, R.; Maeda, T.; Kinoshita, M.; Takao, K.; Takenaka, H.; Kusano, J.; Yoshimura, N.; Takeoka, Y.; Yasuda, D.; Okazaki, T.; et al. Plasmalogens in human serum positively correlate with high-density lipoprotein and decrease with aging. *J. Atheroscler. Thromb.* **2007**, *14*, 12–18. [CrossRef] [PubMed]
22. Oberg, T.S.; Ward, R.E.; Steele, J.L.; Broadbent, J.R. Identification of plasmalogens in the cytoplasmic membrane of Bifidobacterium animalis subsp. lactis. *Appl. Environ. Microbiol.* **2012**, *78*, 880–884. [CrossRef]
23. Pike, D.P.; McGuffee, R.M.; Geerling, E.; Albert, C.J.; Hoft, D.F.; Shashaty, M.G.S.; Meyer, N.J.; Pinto, A.K.; Ford, D.A. Plasmalogen loss in sepsis and SARS-CoV-2 infection. *Front. Cell Dev. Biol.* **2022**, *10*, 912880. [CrossRef] [PubMed]
24. Acar, N.; Berdeaux, O.; Gregoire, S.; Cabaret, S.; Martine, L.; Gain, P.; Thuret, G.; Creuzot-Garcher, C.P.; Bron, A.M.; Bretillon, L. Lipid composition of the human eye: Are red blood cells a good mirror of retinal and optic nerve fatty acids? *PLoS ONE* **2012**, *7*, e35102. [CrossRef] [PubMed]
25. Pallot, C.; Mazzocco, J.; Meillon, C.; Semama, D.S.; Chantegret, C.; Ternoy, N.; Martin, D.; Donier, A.; Gregoire, S.; Creuzot-Garcher, C.P.; et al. Alteration of erythrocyte membrane polyunsaturated fatty acids in preterm newborns with retinopathy of prematurity. *Sci. Rep.* **2019**, *9*, 7930. [CrossRef] [PubMed]
26. Lankalapalli, R.S.; Eckelkamp, J.T.; Sircar, D.; Ford, D.A.; Subbaiah, P.V.; Bittman, R. Synthesis and antioxidant properties of an unnatural plasmalogen analogue bearing a trans O-vinyl ether linkage. *Org. Lett.* **2009**, *11*, 2784–2787. [CrossRef]
27. Chatgilialoglu, C.; Ferreri, C.; Melchiorre, M.; Sansone, A.; Torreggiani, A. Lipid geometrical isomerism: From chemistry to biology and diagnostics. *Chem. Rev.* **2014**, *114*, 255–284. [CrossRef]
28. Cort, A.; Ozben, T.; Sansone, A.; Barata-Vallejo, S.; Chatgilialoglu, C.; Ferreri, C. Bleomycin-induced trans lipid formation in cell membranes and in liposome models. *Org. Biomol. Chem.* **2015**, *13*, 1100–1105. [CrossRef]
29. Ferreri, C.; Kratzsch, S.; Brede, O.; Marciniak, B.; Chatgilialoglu, C. Trans lipids formation induced by thiols in human monocytic leukemia cells. *Free Radic. Biol. Med.* **2005**, *38*, 1180–1187. [CrossRef]
30. Zambonin, L.; Ferreri, C.; Cabrini, L.; Prata, C.; Chatgilialoglu, C.; Landi, L. Occurrence of trans fatty acids in rats fed a trans-free diet: A free radical-mediated formation? *Free Radic. Biol. Med.* **2006**, *40*, 1549–1556. [CrossRef]
31. Krokidis, M.G.; Prasinou, P.; Efthimiadou, E.K.; Boari, A.; Ferreri, C.; Chatgilialoglu, C. Effects of Aging and Disease Conditions in Brain of Tumor-Bearing Mice: Evaluation of Purine DNA Damages and Fatty Acid Pool Changes. *Biomolecules* **2022**, *12*, 1075. [CrossRef] [PubMed]
32. Ferreri, C.; Mennella, M.R.F.; Formisano, C.; Landi, L.; Chatgilialoglu, C. Arachidonate geometrical isomers generated by thiyl radicals: The relationship with trans lipids detected in biological samples. *Free Radic. Biol. Med.* **2002**, *33*, 1516–1526. [CrossRef] [PubMed]
33. Del Duca, E.; Sansone, A.; Sgrulletti, M.; Di Nolfo, F.; Chini, L.; Ferreri, C.; Moschese, V. Fatty-acid-based membrane lipidome profile of peanut allergy patients: An exploratory study of a lifelong health condition. *Int. J. Mol. Sci.* **2023**, *24*, 120. [CrossRef] [PubMed]
34. Vetica, F.; Sansone, A.; Meliota, C.; Batani, G.; Roberti, M.; Chatgilialoglu, C.; Ferreri, C. Free-radical-mediated formation of trans-cardiolipin isomers, analytical approaches for lipidomics and consequences of the structural organization of membranes. *Biomolecules* **2020**, *10*, 1189. [CrossRef]

35. Tartaro-Bujak, I.; Mihaljević, B.; Ferreri, C.; Chatgilialoglu, C. The influence of antioxidants in the thiyl radical induced lipid peroxidation and geometrical isomerization in micelles of linoleic acid. *Free Radic. Res.* **2016**, *50*, S18–S23. [CrossRef]
36. Zhou, Y.L.; Yu, N.; Zhao, J.; Xie, Z.M.; Yang, Z.N.; Tian, B. Advances in the biosynthetic pathways and application potential of plasmalogens in medicine. *Front. Cell Dev. Biol.* **2020**, *8*, 765. [CrossRef]
37. Kramer, J.F.K.; Fellner, V.; Dugan, M.E.R.; Sauer, F.D.; Mossoba, M.M.; Yurawecz, M.P. Evaluating acid and base catalysts in the methylation of milk and rumen fatty acids with special emphasis on conjugated dienes and total *trans* fatty acids. *Lipids* **1997**, *32*, 1219–1228. [CrossRef]
38. Dong, J.L.; Yu, L.S.H.; Xie, J.W. A simple and versatile method for the formation of acetals/ketals using trace conventional acids. *Acs Omega* **2018**, *3*, 4974–4985. [CrossRef]
39. Hanus, L.O.; Levitsky, D.O.; Shkrob, I.; Dembitsky, V.M. Plasmalogens, fatty acids and alkyl glyceryl ethers of marine and freshwater clams and mussels. *Food Chem.* **2009**, *116*, 491–498. [CrossRef]
40. Morrison, W.R.; Smith, L.M. Preparation of fatty acid methyl esters and dimethylacetals from lipids with boron fluoride-methanol. *J. Lipid Res.* **1964**, *5*, 600–608. [CrossRef]
41. Sansone, A.; Tolika, E.; Louka, M.; Sunda, V.; Deplano, S.; Melchiorre, M.; Anagnostopoulos, D.; Chatgilialoglu, C.; Formisano, C.; Di Micco, R.; et al. Hexadecenoic fatty acid isomers in human blood lipids and their relevance for the interpretation of lipidomic profiles. *PLoS ONE* **2016**, *11*, 152378. [CrossRef] [PubMed]
42. Sansone, A.; Melchiorre, M.; Chatgilialoglu, C.; Ferreri, C. Hexadecenoic fatty acid isomers: A chemical biology approach for human plasma biomarker development. *Chem. Res. Toxicol.* **2013**, *26*, 1703–1709. [CrossRef] [PubMed]
43. Folch, J.; Lees, M.; Sloane Stanley, G.H. A simple method for the isolation and purification of total lipides from animal tissues. *J. Biol. Chem.* **1957**, *226*, 497–509. [CrossRef] [PubMed]
44. Zhan, Y.Y.; Wang, L.; Liu, J.; Ma, K.L.; Liu, C.P.; Zhang, Y.; Zou, W. Choline Plasmalogens Isolated from Swine Liver Inhibit Hepatoma Cell Proliferation Associated with Caveolin-1/Akt Signaling. *PLoS ONE* **2013**, *8*, 77387. [CrossRef]
45. Ferreri, C.; Samadi, A.; Sassatelli, F.; Landi, L.; Chatgilialoglu, C. Regioselective cis-trans isomerization of arachidonic double bonds by thiyl radicals: The influence of phospholipid supramolecular organization. *J. Am. Chem. Soc.* **2004**, *126*, 1063–1072. [CrossRef]
46. Williams, C.M.; Mander, L.N. Chromatography with silver nitrate. *Tetrahedron* **2001**, *57*, 425–447. [CrossRef]
47. Chatgilialoglu, C.; Altieri, A.; Fischer, H. The kinetics of thiyl radical induced reactions of monounsaturated fatty acid esters. *J. Am. Chem. Soc.* **2002**, *124*, 12816–12823. [CrossRef]
48. Chatgilialoglu, C.; Samadi, A.; Guerra, M.; Fischer, H. The kinetics of Z/E isomerization of methyl oleate, catalyzed by photogenerated thiyl radicals. *ChemPhysChem* **2005**, *6*, 286–291. [CrossRef]
49. Chatgilialoglu, C.; Bowry, V.W. Why not trans? Inhibited radical isomerization cycles and coupling chains of lipids and alkenes with alkane-thiols. *J. Org. Chem.* **2018**, *83*, 9178–9189. [CrossRef]
50. Chatgilialoglu, C.; Ferreri, C.; Guerra, M.; Samadi, A.; Bowry, V.W. The reaction of thiyl radical with methyl linoleate: Completing the picture. *J. Am. Chem. Soc.* **2017**, *139*, 4704–4714. [CrossRef]
51. Chatgilialoglu, C.; Studer, A. (Eds.) *Encyclopedia of Radicals in Chemistry, Biology and Materials*; Wiley: Chichester, UK, 2012.
52. Fife, T.H. Vinyl ether hydrolysis. The facile general acid catalyzed conversion of 2-ethoxy-1-cyclopentene-1-carboxylic acid to cyclopentanone. *J. Am. Chem. Soc.* **1965**, *87*, 1084–1089. [CrossRef] [PubMed]
53. Alves, S.P.; Santos-Silva, J.; Cabrita, A.R.J.; Fonseca, A.J.M.; Bessa, R.J.B. Detailed dimethylacetal and fatty acid composition of rumen content from lambs fed lucerne or concentrate supplemented with soybean oil. *PLoS ONE* **2013**, *8*, 58386. [CrossRef] [PubMed]
54. See the LIPID MAPS library. Available online: https://www.lipidmaps.org/resources/lipidweb/lipidweb_html/ms/others/misclipids/index.htm (accessed on 18 April 2022).
55. Delmonte, P.; Belaunzaran, X.; Ridge, C.D.; Aldai, N.; Kramer, J.K.G. Separation and characterization of products from acidic methanolysis of plasmalogenic lipids by two-dimensional gas chromatography with online reduction. *J. Chromatogr. A* **2020**, *1619*, 460955. [CrossRef] [PubMed]
56. Otoki, Y.; Hennebelle, M.; Levitt, A.J.; Nakagawa, K.; Swardfager, W.; Taha, A.Y. Plasma Phosphatidylethanolamine and triacylglycerol fatty acid concentrations are altered in major depressive disorder patients with seasonal pattern. *Lipids* **2017**, *52*, 559–571. [CrossRef]
57. Aldai, N.; Murray, B.E.; Najera, A.I.; Troy, D.J.; Osoro, K. Derivatization of fatty acids and its application for conjugated linoleic acid studies in ruminant meat lipids. *J. Sci. Food Agric.* **2005**, *85*, 1073–1083. [CrossRef]
58. Hodson, L.; Skeaff, C.M.; Fielding, B.A. Fatty acid composition of adipose tissue and blood in humans and its use as a biomarker of dietary intake. *Prog. Lipid Res.* **2008**, *47*, 348–380. [CrossRef]
59. Hame, K.; Fujiwara, Y.; Yokoyama, K. Quantitative and qualitative method for sphingomyelin by LC-MS using two stable isotopically labeled sphingomyelin species. *J. Vis. Exp.* **2018**, *135*, e57293. [CrossRef]
60. Devle, H.; Naess-Andresen, C.F.; Stenstrom, Y.; Ekeberg, D. Rapid method for analysis of sphingomyelin by microwave derivatisation for gas chromatography-mass spectrometry. *Eur. J. Lipid Sci. Technol.* **2011**, *113*, 708–710. [CrossRef]
61. Takashima, S.; Toyoshi, K.; Itoh, T.; Kajiwara, N.; Honda, A.; Ohba, A.; Takemoto, S.; Yoshida, S.; Shimozawa, N. Detection of unusual very-long-chain fatty acid and ether lipid derivatives in the fibroblasts and plasma of patients with peroxisomal diseases using liquid chromatography-mass spectrometry. *Mol. Genet. Metab.* **2017**, *120*, 255–268. [CrossRef]

62. Wang, L.Y.; Summerhill, K.; Rodriguez-Canas, C.; Mather, I.; Patel, P.; Eiden, M.; Young, S.; Forouhi, N.G.; Koulman, A. Development and validation of a robust automated analysis of plasma phospholipid fatty acids for metabolic phenotyping of large epidemiological studies. *Genome Med.* **2013**, *5*, 39. [CrossRef]
63. Metherel, A.H.; Henao, J.J.A.; Ciobanu, F.; Taha, A.Y.; Stark, K.D. Microwave energy increases fatty acid methyl ester yield in human whole blood due to increased sphingomyelin transesterification. *Lipids* **2015**, *50*, 895–905. [CrossRef] [PubMed]
64. Hung, W.-L.; Hwang, L.S.; Shahidi, F.; Pan, M.-H.; Wang, Y.; Ho, C.-T. Endogenous formation of trans fatty acids: Health implications and potential dietary intervention. *J. Funct. Foods* **2016**, *25*, 14–24. [CrossRef]
65. Brosche, T.; Platt, D. Decrease of cholesterol concentration in human erythrocyte-ghosts in old-age. *Exp. Gerontol.* **1990**, *25*, 23–28. [CrossRef]
66. Zhang, X.X.; Barraza, K.M.; Beauchamp, J.L. Cholesterol provides nonsacrificial protection of membrane lipids from chemical damage at air-water interface. *Proc. Natl. Acad. Sci. USA* **2018**, *115*, 3255–3260. [CrossRef] [PubMed]
67. Elbaradei, A.; Wang, Z.K.; Malmstadt, N. Oxidation of membrane lipids alters the activity of the human serotonin 1A receptor. *Langmuir* **2022**, *38*, 6798–6807. [CrossRef]
68. Kagan, V.E.; Mao, G.W.; Qu, F.; Angeli, J.P.F.; Doll, S.; St Croix, C.; Dar, H.H.; Liu, B.; Tyurin, V.A.; Ritov, V.B.; et al. Oxidized arachidonic and adrenic PEs navigate cells to ferroptosis. *Nat. Chem. Biol.* **2017**, *13*, 81–90. [CrossRef]
69. Chen, X.; Yu, C.H.; Kang, R.; Tang, D.L. Iron metabolism in ferroptosis. *Front. Cell Dev. Biol.* **2020**, *8*, 590226. [CrossRef]

biomolecules

Review

Oxidative Stress in Healthy and Pathological Red Blood Cells

Florencia Orrico [1,2,3,†], Sandrine Laurance [4,†], Ana C. Lopez [1,2,3], Sophie D. Lefevre [4], Leonor Thomson [2,3], Matias N. Möller [1,3,‡] and Mariano A. Ostuni [4,*,‡]

1 Laboratorio de Fisicoquímica Biológica, Instituto de Química Biológica, Facultad de Ciencias, Universidad de la República, Montevideo 11400, Uruguay; forrico@fcien.edu.uy (F.O.); anclalop@fcien.edu.uy (A.C.L.); mmoller@fcien.edu.uy (M.N.M.)
2 Laboratorio de Enzimología, Instituto de Química Biológica, Facultad de Ciencias, Universidad de la República, Montevideo 11400, Uruguay; lthomson@fcien.edu.uy
3 Centro de Investigaciones Biomédicas (CEINBIO), Universidad de la República, Montevideo 11800, Uruguay
4 Université Paris Cité and Université des Antilles, UMR_S1134, BIGR, Inserm, F-75014 Paris, France; sandrine.laurance@inserm.fr (S.L.); sophie.lefevre@inserm.fr (S.D.L.)
* Correspondence: mariano.ostuni@inserm.fr
† Co-first author.
‡ Co-last author.

Abstract: Red cell diseases encompass a group of inherited or acquired erythrocyte disorders that affect the structure, function, or production of red blood cells (RBCs). These disorders can lead to various clinical manifestations, including anemia, hemolysis, inflammation, and impaired oxygen-carrying capacity. Oxidative stress, characterized by an imbalance between the production of reactive oxygen species (ROS) and the antioxidant defense mechanisms, plays a significant role in the pathophysiology of red cell diseases. In this review, we discuss the most relevant oxidant species involved in RBC damage, the enzymatic and low molecular weight antioxidant systems that protect RBCs against oxidative injury, and finally, the role of oxidative stress in different red cell diseases, including sickle cell disease, glucose 6-phosphate dehydrogenase deficiency, and pyruvate kinase deficiency, highlighting the underlying mechanisms leading to pathological RBC phenotypes.

Keywords: erythrocyte; reactive oxygen species; antioxidant; oxidative stress; sickle cell disease; glucose 6-phosphate dehydrogenase deficiency; pyruvate kinase deficiency

Citation: Orrico, F.; Laurance, S.; Lopez, A.C.; Lefevre, S.D.; Thomson, L.; Möller, M.N.; Ostuni, M.A. Oxidative Stress in Healthy and Pathological Red Blood Cells. *Biomolecules* **2023**, *13*, 1262. https://doi.org/10.3390/biom13081262

Academic Editor: Chryssostomos Chatgilialoglu

Received: 27 July 2023
Revised: 11 August 2023
Accepted: 16 August 2023
Published: 18 August 2023

1. Oxidative Stress in Healthy Red Blood Cells

Red blood cells (RBCs) are exposed to endogenous and exogenous oxidants, commonly referred to as reactive oxygen species (ROS) and reactive nitrogen species (RNS). These oxidants comprise a large group of molecules with different properties, including cellular sources and preferred molecular targets, which will be briefly discussed below, and have been discussed in more detail elsewhere [1,2].

One of the main mechanisms of endogenous oxidant production involves oxyhemoglobin (HbO_2). The autoxidation of HbO_2 occurs spontaneously at a low rate to yield superoxide ($O_2^{\bullet -}$) and methemoglobin (Hb-F^{III}, MetHb) [3]. Superoxide itself is a weak oxidant but can further react to make stronger oxidants. Superoxide can spontaneously dismutate to yield hydrogen peroxide (H_2O_2) and oxygen [4]. Hydrogen peroxide is a stronger oxidant that will react with thiols and metal centers [5]. Hydrogen peroxide reacts with HbO_2 to yield ferryl hemoglobin, which can oxidize other proteins and lipids [6]. Furthermore, in the presence of one electron reductant such as Fe^{II}, H_2O_2 can also generate hydroxyl radical ($HO^{\bullet}$), one of the strongest biological oxidants [4]. Hydroxyl radicals will react with most organic molecules at diffusion-controlled rates to yield organic radicals that can propagate oxidative damage [7].

RBCs will also be exposed to oxidants derived from endothelial and immune system cells, which generate nitric oxide ($NO^{\bullet}$), superoxide, peroxynitrite ($ONOO^-$), H_2O_2, and hypochlorous acid (HOCl). Nitric oxide is produced by the endothelial enzyme nitric oxide synthase (NOS3) as a signal molecule to induce vasodilation and by immune system cells at larger amounts by inducible NOS2, that result in the formation of more potent oxidizing species that can kill invading pathogens [8].

Because of its small size and hydrophobicity, $NO^\bullet$ can diffuse virtually unhindered through cellular membranes [9]. In the RBCs, $NO^\bullet$ will react with HbO_2 to give MetHb and nitrate and also with $O_2^{\bullet-}$ to give the powerful oxidant $ONOO^-$ [10,11]. Superoxide is also produced by NADPH oxidases in endothelial and immune system cells [12]. Although the ionic nature of superoxide limits its diffusion across cellular membranes, it has been observed to be transported by RBC band 3 bicarbonate/chloride exchange protein [13], and it can also protonate to the hydroperoxyl radical and diffuse across the lipid fraction of the membranes [14]. Peroxynitrite can use the same routes to diffuse into RBCs and cause intracellular oxidation, facilitated transport of the anion by band 3, and simple diffusion of the neutral peroxynitrous acid [15]. Peroxynitrite per se is a potent oxidizing molecule, but it rapidly reacts with CO_2 in solution to give the more potent oxidants nitrogen dioxide ($NO_2^\bullet$) and carbonate radical ($CO_3^{\bullet-}$) [16]. Nitrogen dioxide diffuses very rapidly across cellular membranes and reacts mainly with intracellular thiols but also with lipids in the membrane [1,17]. Carbonate radicals cannot diffuse across cellular membranes and will react preferentially with proteins [18].

Hydrogen peroxide produced by endothelial and immune-system cells directly or indirectly by NADPH oxidases also acts as a signaling molecule in the vascular system modulating endothelial cell proliferation and survival, for instance [19]. This exogenous H_2O_2 can rapidly diffuse across the RBC membrane to oxidize intracellular targets, through an aquaporin-independent pathway, likely by simple diffusion [2]. Hypochlorous acid is produced enzymatically by myeloperoxidase that is released by neutrophils and monocytes upon infection-related or inflammatory stimuli, using chloride and H_2O_2 as substrates [20]. Hypochlorous acid freely diffuses across the RBC membrane and oxidizes intracellular targets [21].

All these oxidants produced in and around the RBC can oxidize biomolecules in the RBC, but different oxidants have different preferential targets and lead to different types of damage, as will be discussed below.

1.1. Main Targets of Oxidants in RBCs

As mentioned above, both proteins and lipids can be damaged by oxidants in RBCs. The ultimate result of oxidative damage to RBCs is hemolysis, the loss of membrane integrity, and the release of hemoglobin and other intracellular proteins. Free hemoglobin is particularly toxic, and this is evident in several RBCs diseases [22].

The membrane of RBCs is composed of phospholipids, cholesterol, glycolipids, and proteins (some of them glycosylated). The polyunsaturated fatty acids (PUFA), 18% of the total fatty acids in RBCs [23], are the lipid components that are more susceptible to oxidation in a series of reactions that trigger lipoperoxidation. The first event is the abstraction of bis-allylic hydrogen from a polyunsaturated fatty acid to yield a lipid-derived radical, which rapidly reacts with molecular oxygen to yield a lipid-derived peroxyl radical ($LOO^\bullet$). This $LOO^\bullet$ can subsequently subtract hydrogen from a neighboring PUFA, and the reaction propagates as a chain reaction [24]. The further oxidation of LOOH yields reactive aldehydes such as hydroxynonenal and malondialdehyde. The lipid peroxidation propagation can be stopped by lipid-soluble antioxidants, such as α-tocopherol [24]. Also, a minor fraction of phospholipids (10–15%) is present as plasmalogens, which present a vinyl ether hydrocarbon group that has been associated with lipid antioxidant capacity in vitro [25,26].

Fresh normal RBCs do not contain products of lipid peroxidation, but increased membrane lipid peroxidation is evident in many RBCs diseases, such as thalassemia, unstable hemoglobin disease, and sickle cell disease [27]. RBC diseases are also accompanied by an increased susceptibility to lipid peroxidation. Additionally, lipid peroxidation products also increase during the storage of lipids for transfusion, especially when no leukoreduction is performed [28]. In many of these cases, the oxidation of lipids is associated with the oxidation of protein and crosslinking to cytoskeleton and membrane proteins [29]. Notably, much of the lipid oxidation is catalyzed by HbO_2 and is inhibited when hemoglobin is present as MetHb [27]. A possible explanation is that monomeric or unstable hemoglobin is associated with the RBC lipid membrane and catalyzes the formation of oxidizing radicals, such as $HO^•$, in situ, that cause lipid oxidation and protein cross-linking. In this line, hydrophobic oxidants like cumene hydroperoxide or tert-butyl hydroperoxide, which partition favorably in the membrane of RBCs, have been consistently observed to cause more damage than water-soluble H_2O_2 [30–32]. Recently, increased levels of the six-transmembrane epithelial antigen of prostate 3 (Steap3) protein in mice has been associated with increased lipid oxidation in RBCs and hemolysis [33]. The protein Steap3 reduces Fe^{3+} to Fe^{2+} after DMT1 transmembrane transport, and its deletion in mice causes iron deficiency anemia [34]. The mechanism of cellular damage is proposed to involve the reduction of free iron to Fe^{2+}, followed by a Fenton-type reaction with H_2O_2 to produce $HO^•$, that then oxidizes lipids and proteins [33].

The oxidation of hemoglobin is most easily recognized and has been found to be involved in many of the oxidative damage to RBCs. Some drugs, such as phenylhydrazine, can cause hemolytic anemia by means of hemoglobin oxidation that leads to the formation of hemichrome, a misfolded form of hemoglobin, its precipitation to form Heinz bodies, visible by microscopy, and then to hemolysis [35]. Similar effects have been observed in RBCs from mice deficient in peroxiredoxin2 (Prx2), an important antioxidant enzyme discussed below [36]. The oxidation of the cytoskeleton has also been observed during the storage of RBC for transfusion, associated with an increase in protein carbonyls and crosslinking [37]. Atomic force microscopy showed that spectrin filaments are altered during storage, forming thicker fibers and loss of connections [38].

1.2. Antioxidant Cellular Mechanisms in RBCs

Although the RBCs are exposed to large amounts of oxidants, both from endogenous and exogenous sources, they are well prepared to resist. Robust antioxidant defenses allow normal RBCs to survive 120 days in circulation. The main defenses against oxidant damage are provided by different enzymatic systems, aided by low molecular weight antioxidant and electron-rich molecules. The antioxidant defenses ultimately rely on the reducing power of NADPH, obtained from the oxidation of glucose by the pentose phosphate pathway.

Glucose is transported across the RBC membrane by the highly abundant GLUT1 transporter driven by the concentration gradient. In the cytosol, glucose is phosphorylated to glucose-6-phosphate (G6P) by hexokinase. A large fraction of glucose is used in glycolysis to produce the ATP necessary to keep important cellular functions working, such as the Na^+/K^+ ATPase and Ca^{2+} ATPase pumps, and NADH that is mostly used to reduce pyruvate to lactate and is also used to reduce MetHb to HbO_2 by methemoglobin reductase. A smaller fraction of the glucose is used in the pentose phosphate pathway to produce NADPH, which will provide the reducing power necessary to keep glutathione (GSH) and the thiol-dependent antioxidant enzymes reduced. NADPH is produced in the steps catalyzed by glucose-6-phosphate dehydrogenase (G6PDH) and 6-phosphogluconate dehydrogenase. The deficiency in G6PDH activity has important consequences in the capacity of RBCs to deal with oxidative stress and is recognized as a health problem that will be dealt with below.

Low molecular weight antioxidants include α-tocopherol in the plasma membrane to prevent lipid peroxidation, ascorbate, and GSH in the cytosol that can rapidly react with $NO_2^•$ or HOCl. Reduced ascorbate can reduce tocopheroxyl radicals in the membrane, and oxidized ascorbate can be reduced with GSH, forming a network of antioxidant power. It appears that the most important roles of low molecular weight antioxidants are to protect the lipids from oxidation, repair molecular radicals, and act as an additional reducing power reservoir [2].

RBCs also contain a large network of antioxidant proteins that can react and deactivate oxidant species very rapidly, preventing the damage of protein and lipids essential for RBC function. Some of them need the assistance of other proteins to function, some do not, but most of them act in a concerted fashion to decrease to a minimum the chances of damage to important molecules in RBCs. Given their relevance, a detailed discussion of the most important antioxidant proteins is given below.

- Superoxide dismutase

Superoxide dismutases (SODs) are antioxidant enzymes that catalyze the disproportionation of superoxide radicals and, therefore, constitute the first barrier against oxidative damage in cells. These enzymes contain metal atoms in their structure, based on which they can be classified into four groups, FeSOD, Cu, ZnSOD, MnSOD, and NiSOD [39]. Cu, ZnSOD (SOD1) is the isoform found in RBCs. It is present at an approximate concentration of 2 μM [40] and is associated to 95% of the entire pool of Cu^{2+} in the cell [41].

SOD1 is organized as a dimer, with identical active sites in each subunit. The Cu^{2+} atom is the redox-active metal that is sequentially reduced and oxidized by two superoxide molecules to yield one molecule of oxygen and one of H_2O_2 [42,43]. Zn^{2+}, on the other hand, is involved in structural stabilization and ensures correct coordination and reducing the potential of Cu^{2+} [44]. This process is very efficient, as the enzyme manages to increase the rate of the spontaneous reaction by four orders of magnitude, reaching practically diffusion-controlled values [43].

Kinetic model calculations estimate that if SOD1 were not present in RBCs, superoxide would increase its levels to values 100 times higher than the basal concentrations [3]. Experiments performed with SOD1 deficient mice show the consequences are a rise in MetHb and lipid peroxidation levels, accumulation of oxidized carbonic anhydrase, and, lastly, anemia due to a decrease in lifespan and oxidative damage of RBCs [45,46].

SOD1 is characterized by having disordered regions with undefined secondary or tertiary structures that can change rapidly in conformation. These regions appear to be often post-translationally modified and are proposed to have regulatory roles [47,48]. Furthermore, SOD1 was reported to be inactivated and more susceptible to proteolysis upon reaction with its product H_2O_2 [49,50]. However, given the abundance and reactivity of RBC peroxidases, this process is not likely to be of relevance in vivo. Other alterations in SOD1 activity are reported in circumstances that are often linked to oxidative stress, such as Alzheimer's disease and aging [51,52]. Because of this, it is often considered an interesting marker to study regarding oxidative stress.

- Glutathione peroxidase

Glutathione peroxidases (GPxs) are a family of enzymes that catalyze the reduction of hydroperoxides with varying substrate and electron donor specificities. The most abundant isoform in RBCs is GPx1, with an estimated concentration of 1.5 μM [53]. GPx1 is a homotetrameric enzyme with a selenium-containing active site [54]. Through its catalytic cycle, the enzyme is oxidized in order to reduce the peroxide and is later regenerated by GSH. The result is the formation of glutathione disulfide (GSSG), which in turn is reduced by glutathione reductase (GR) and NADPH [55].

Since its discovery, GPx1 has been associated with the protection of RBCs from oxidative damage, particularly inflicted by endogenously produced oxidants present in low concentrations [55,56]. It is proposed to act as a second barrier to H_2O_2, below Prx2, and to be especially implicated in the detoxification of small organic peroxides that could generate damage to the lipid membrane [3,53,57].

A second glutathione peroxidase with a similar role, GPx4, can also be found in RBCs, albeit 20 times less abundantly than GPx1 [58]. It is also a selenoenzyme, but it clearly differs in electron donor specificity, being able to be reduced by other thiols [59]. Furthermore, as it is a monomeric enzyme, it is allowed to metabolize more complex substrates, such as cholesterol and phospholipid hydroperoxides, even when they are bound to the membrane [59–61].

Although not frequent, deficiencies in GPx have been reported and subject the patients to a clinical profile similar to the observed G6PDH deficiency. Mainly, drug-induced hemolysis and hemoglobinuria have been described [62,63].

- Glutathione reductase

GR is a flavoprotein that belongs to the pyridine nucleotide-disulfide oxidoreductase family of enzymes. Structurally, it is a dimeric protein with specific binding sites to FAD, NADPH, and GSSG. Both these subunits are involved in the mechanism of catalysis, where a redox-active disulfide in the active site of the enzyme catalyzes the conversion of GSSG into two molecules of GSH. To start another cycle, this disulfide is later reduced by NADPH [64,65].

In physiological conditions, the concentration of oxidized GR is estimated to be very low since it is continuously being reduced [65,66]. GR is thus a very important part of the antioxidant system in RBCs, as it collaborates in maintaining GSH levels more than ten times higher than those of GSSG [67], in addition to helping in the regulation of the $NADP^+$/NADPH pool.

Given its role, the absence or decrease in GR activity entails various physiological complications. The causes behind such abnormalities vary and can be related or not to FAD metabolism, for example, an insufficient dietary intake of riboflavin or an inability to obtain FAD from riboflavin [68]. It has also been reported the existence of patients with inherited mutant variants of GR, where the altered enzyme inefficiently binds FAD or is eliminated due to problems in folding [69]. In these scenarios, as the capacity of RBCs to balance oxidative stress is impaired, the consequences are shown to be favism and drug-induced hemolysis, an increase in osmotic fragility, and a reduction of lifespan in RBCs, as well as spherocytosis [70,71].

- Peroxiredoxin 2

Peroxiredoxins (Prxs) are widely extended antioxidant proteins with thiol-dependent peroxidatic activity. Three different members of this family have been found in RBCs, namely Prx 1, Prx2, and Prx6 [72], although Prx2 is by far the most abundant. In fact, it represents the third most abundant protein in the RBCs, ranging between 240 and 410 μM [73,74].

Prx2 can metabolize peroxides of different natures, such as lipid hydroperoxides, peroxynitrite, and especially H_2O_2, with which it reacts at a very high rate [72,75]. Since it belongs to the typical 2-Cys type of peroxiredoxins, Prx2 forms homodimers with identical active sites in each subunit during catalysis. The catalytic cycle starts with the oxidation of one monomer peroxidatic cysteine to cysteine sulfenic acid upon reaction with the peroxide, followed by the formation of an intermolecular disulfide bond between this residue and a resolutive cysteine of a second monomer. The enzyme is later regenerated by thioredoxin (Trx), thioredoxin reductase (TR), and NADPH [75,76]. During this process, Prx2 can be hyperoxidized to cysteine sulfinic acid in its peroxidatic cysteine, which can only be reduced enzymatically by sulfiredoxin or further oxidized to an irreversible sulfonic form [77,78]. Probably regulated by its redox state, a dynamic equilibrium is established in vivo where the enzyme can exist as a dimer when oxidized or as a decamer when reduced or hyperoxidized. In addition, these decamers were shown to interact with each other and form greater molecular weight structures [76,79–81].

Due to its high abundance and reaction rates, Prx2 has been proposed to act as a first line of defense against endogenous H_2O_2 produced in RBCs, mainly by the autoxidation of hemoglobin [3,53,82]. Because of this, it is often considered a sensitive marker to oxidative stress. It can also act as a chaperone, binding to hemoglobin and preventing its aggregation due to oxidative damage [83–85]. In fact, studies report that deficiencies in Prx2 cause an increase in H_2O_2 and MetHb levels, alterations in cell morphology, formation of Heinz bodies, and hemolytic anemia [36,57]. Furthermore, Prx2 is also able to bind the cytosolic domain of band 3, so it could play a role in defending band 3 and other proteins in the membrane, as well as the lipidic fraction, from oxidants [85,86].

Apart from its peroxidase role, Prx2 has been previously associated with potassium transport in RBCs. The increase in calcium levels promotes the translocation of Prx2 to the membrane, where it can interact with the Gardos channel and thus alter potassium efflux and cell volume. However, the causes and mechanisms are not fully understood yet to this day [73,81]. More recently, a signaling role has also been proposed for peroxiredoxins, given their high reactivity with H_2O_2 and the ability to interact with other proteins. Post-translational modifications observed in Prx2, such as phosphorylation, proteolysis, and even the hyperoxidation of its peroxidatic cysteine, are involved in the regulation of its peroxidase activity and could affect its participation in redox signaling pathways [87–89].

- Thioredoxin

Thioredoxins (Trxs) are small antioxidant oxidoreductases. They can reduce a wide variety of protein disulfides since the active site surface of these enzymes can undergo chaperone-like conformational changes to accommodate different proteins [90]. Some targets include methionine sulfoxide reductases, sulfonucleotide reductases, and transcription factors, such as Ref-1 [91,92]. In RBCs, cytosolic Trx1 is particularly relevant for regenerating Prx2 and maintaining a reduced environment within the cell [75].

The catalytic mechanism of Trx1 depends on two cysteine residues to reduce its substrates. At first, a thiolate in Trx1 attacks the intermolecular disulfide bond that unites Prx2 dimers, forming a mixed disulfide. This disulfide is then disrupted by a second cysteine residue, obtaining a reduced Prx2 and an oxidized Trx1 that can be reduced by TR and NADPH [90]. There are three extra cysteine residues in the Trx1 structure, reported to be easily oxidized and implicated in the formation of molecular aggregates or oligomers and inactivation of the enzyme. These are suspected to have a physiological, possibly regulatory role [93–96].

- Thioredoxin reductase

TR, like GR, is a flavoprotein from the pyridine nucleotide-disulfide oxidoreductase family. Both these enzymes are highly similar in structure, as TR also acts as a homodimer with FAD and NADPH binding domains, as well as an interface domain [91,97]. However, they differ in their active sites because TR is a selenoprotein [98]. During the catalytic cycle, a selenenylsulfide bond is established between TR dimers after the reduction of the oxidized thiols in the substrate, which can be later reduced by a chained electron transfer from NADPH and FAD [97,99,100].

TR has been shown to have multiple possible substrates, among which are glutaredoxins and thioredoxins, hence collaborating with the antioxidant proteins in the RBCs. Furthermore, it can reduce smaller molecules that could react with H_2O_2 directly and is even proposed as an alternative system to reduce GSH in other organisms, so this could possibly occur in mammals as well [96,101,102]. There is also evidence that TR could play a role in the enzyme-mediated dehydroascorbic acid reduction in RBCs [103].

- Glutaredoxins

Glutaredoxins (Grxs) are a family of small and ubiquitous thiol-disulfide oxidoreductases. They have a considerable role in maintaining the sulfhydryl homeostasis in cells by reducing inter and intramolecular disulfides in proteins, as well as mixed protein-glutathione disulfides. Classified into two groups, Grxs can present either a dithiolic mechanism of catalysis or a monothiolic one. In the first case, two cysteine residues are involved, attacking the disulfide in the target protein and prompting its reduction while the enzyme gets oxidized. Grx is later regenerated by glutathione, with the formation of a glutathionylated intermediate. In the monothiol mechanism, the targets are glutathionylated proteins, and only one cysteine is needed for its reduction [104–106].

Red blood cells contain two isoforms of Grx, specifically Grx1 and Grx3. However, the dithiolic Grx1 is the most abundant, with reported concentrations of 4–8 μM [58]. It has been proven that Grx1 is capable of reducing oxidized forms of hemoglobin and membrane proteins, as well as regenerating metabolic enzymes phosphofructokinase, pyruvate kinase, and glyceraldehyde-3-phosphate dehydrogenase that are susceptible to inactivation by oxidative damage [107–109]. Nevertheless, glutathionylation of proteins is not exclusive to oxidative stress conditions, as it is also proposed to have a role in signaling pathways. In fact, Prx2, a candidate for the transduction of redox signals, can be glutathionylated in one or both of its active site cysteines and is deglutathionylated by Grx1 [110].

The role of Grx3 is not clear in mature RBCs. However, it could be particularly relevant during erythropoiesis since monothiolic Grxs, as is Grx3, are involved in iron homeostasis and the assembly of Fe-S clusters [111]. In other organisms, depletion of Grx3 during embryonic development affects the maturation of hemoglobin [112]. Grx5, also a monothiolic Grx, was shown to affect heme synthesis in erythroblasts, resulting in sideroblastic anemia in humans [113].

- Catalase

Catalases are enzymes that catalyze the dismutation of H_2O_2 into water and oxygen, thus protecting cells from oxidative damage. Human catalase organizes as a tetramer. It contains four heme groups and binds four molecules of NADPH, one in each subunit [114,115]. The tetramerization is essential for its activity, as it is proposed to allow the correct cycling of the enzyme and keep the iron of heme groups away from the protein surface, therefore avoiding the formation of hydroxyl radicals [114].

During catalysis, the elimination of H_2O_2 occurs in two sequential steps. First, ferric heme reduces one molecule of H_2O_2 to water, resulting in the formation of an oxyferryl species named compound I. In the second step, compound I is reduced using another molecule of peroxide, yielding water and oxygen and returning catalase to its basal redox state [114,116]. In a series of reactions involving superoxide and H_2O_2, compound I can be transformed into inactive forms of the enzyme (compounds II and III) [117]. NADPH, which stays in a reduced state when bound to catalase, is proposed to be there to prevent this inactivation by reacting with an intermediate species and blocking compound II accumulation [118].

RBCs are among the cells with the highest catalase activity in the organism, with an estimated subunit concentration of 11 μM [53]. Given the abundance and high reaction rates of Prx2, catalase is not very relevant in metabolizing endogenous peroxide in these cells [53]. However, it differs from other enzymes as it cannot be saturated with H_2O_2 [119]. This comes into play in circumstances where H_2O_2 levels are sufficiently high to react completely with all Prx2 and GPx available and deplete cellular NADPH, which is only slowly recovered [53]. This is often the case when experiments are performed in vitro with a relatively high concentration of H_2O_2 and low hematocrit, and has often led to believe that catalase is the main enzyme in the detoxification of exogenous H_2O_2 [53]. In line with this, experiments performed with mice RBCs deficient in catalase show they are more sensitive to high H_2O_2 concentrations, suffering an increase in hemoglobin oxidation [57].

- Considerations on the antioxidant systems in RBCs

RBCs are exposed to oxidants from endogenous sources and exogenous sources that can react with RBC lipids and proteins, leading to molecular damage and eventually to membrane rupture and hemolysis. To prevent these harmful effects, RBCs are well prepared with enzymatic and low molecular weight antioxidants (Figure 1). Actually, the enzymes Prx2, Gpx, and catalase appear to deal with most of the reactive oxygen and nitrogen species attacking RBCs. Low molecular weight antioxidants appear to be more important in preventing lipid oxidation (α-tocopherol/ascorbate/GSH system) and acting as a reducing power reservoir. At the same time, both antioxidant systems ultimately rely on the reducing power provided by NADPH from glucose in the pentose phosphate pathway.

Genome-wide association analysis has identified candidate genes that modify RBC susceptibility to osmotic hemolysis, including spectrin α-chain, ankyrin 1, aquaporin 1, and band 3, most of them involved with cytoskeleton stabilization or osmosis-driven water transport [120]. Other genes identified included hexokinase 1, the first enzyme in glucose metabolism, stress kinase MAPKAPK5, and the mechanosensitive calcium channel PIEZO1 [120]. Oxidative hemolysis is associated with changes in G6PDH, Grx, and GPx4, important proteins in redox metabolism and repair, and SEC14-like 4, a protein related to the transport of phospholipids and α-tocopherol [120]. Resistance to oxidative hemolysis was also associated with higher levels of GSH, a higher activity of the pentose phosphate pathway, and improved protein damage repair mechanisms [121].

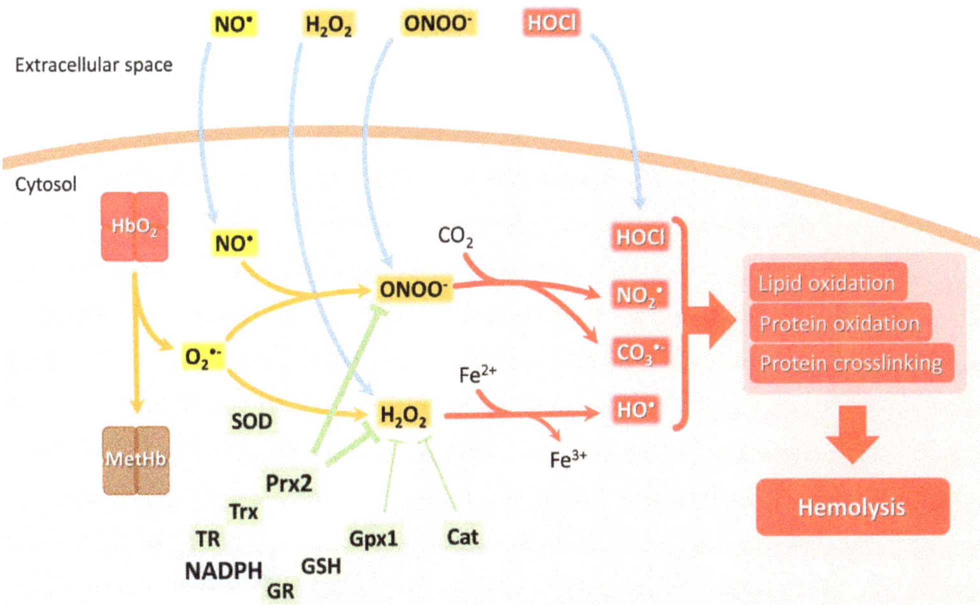

Figure 1. Scheme summarizing main endogenous and exogenous sources of ROS and RNS in the RBCs as well as the principal antioxidant actors. HbO_2: Oxyhemoglobin; MetHb: methemoglobin; $O_2^{\bullet-}$: superoxide anion; $NO^{\bullet}$: nitric oxide; H_2O_2: hydrogen peroxyde; ONOO-: peroxynitrite; $HO^{\bullet}$: hydroxyl radical; HOCl: hypochlorous acid; $CO_3^{\bullet-}$: carbonate radical; $NO_2^{\bullet}$: nitrogen dioxide radical; superoxide dismutase: SOD; catalase: Cat; glutathione reductase: GR; reduced glutathione: GSH, glutathione peroxidase 1: Gpx1; thioredoxin reductase: TR; thioredoxin: Trx; Peroxiredoxin 2: Prx2.

2. Oxidative Stress in Red Blood Cell Diseases

As developed above, RBCs' oxidative stress strongly depends on the balance among pathophysiological mechanisms producing ROS and enzymatic and non-enzymatic antioxidant systems. In this second part of the review, we aim to explore several RBC diseases where this balance is strongly altered both by increased ROS production or by diminished antioxidant capacity. The selection of RBC diseases is not exhaustive, but representative of those involving different oxidative stress causes.

2.1. Sickle Cell Disease

Sickle cell disease (SCD) is an inherited genetic disorder resulting in the production of an abnormal hemoglobin S (HbS) that undergoes deoxygenation-dependent polymerization [122,123]. The repeated cycles of HbS polymerization induce RBCs' shape distortion, cell rigidity, cell membrane alteration, and fragility, ultimately resulting in intravascular and extravascular hemolysis. Behind this primary described pathological mechanism, SCD pathophysiology appears to be more complex and involves an intricate network of molecular and cellular partners. In fact, in addition to HbS polymerization, an imbalance of the redox status is also observed in SCD due to an increase in the production of ROS and/or RNS conjugated to an impairment of the antioxidant systems (Figure 2 SCD panel). For example, peroxynitrite is involved in the oxidation and nitration of several intracellular targets (thiols, protein-membrane, lipids), leading to breakage of DNA, impairment of cell signaling and cell death (reviewed in [124,125]). In SCD, oxidative stress can arise from sickle RBCs and/or activated neutrophiles, platelets, and endothelial cells (ECs). Several erythroid and non-erythroid mechanisms have been described accounting for this pro-oxidant environment: (i) HbS autooxidation, (ii) heme and iron release, (iii) increased NADPH oxidase and endothelial xanthine oxidase (XO) activity, (iv) decreased $NO^{\bullet}$ bioavailability, (v) erythroid mitochondrial retention [126].

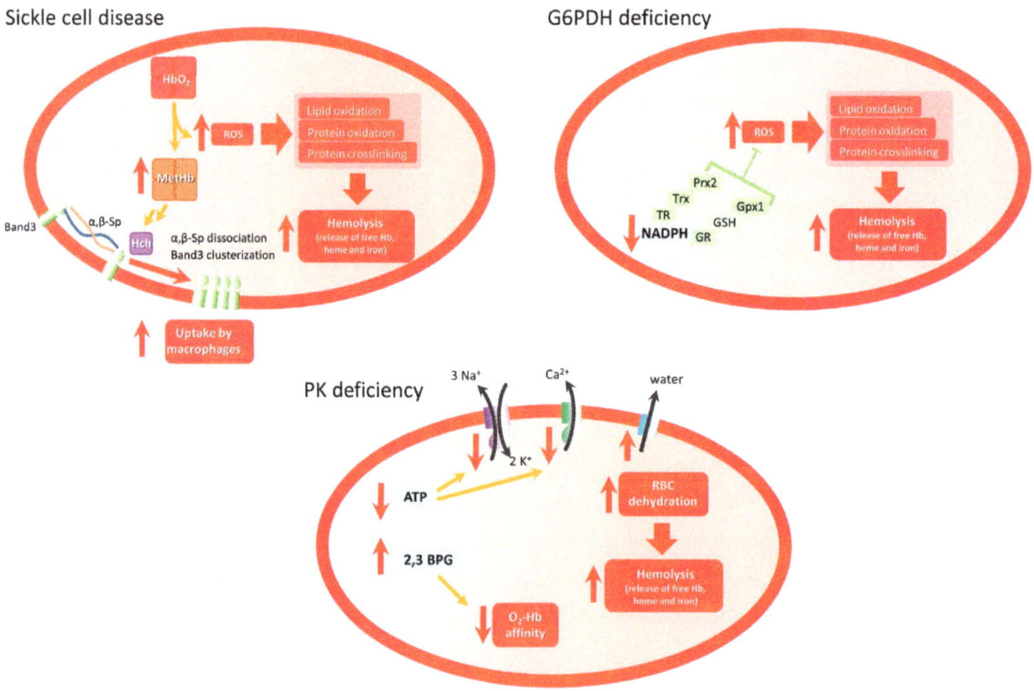

Figure 2. Scheme of main mechanism involved in oxidative stress and hemolytic clinical manifestations in Sickle Cell Disease, G6PDH deficiency, and PK deficiency. In SCD, highly unstable HbS will be converted in MetHb, favoring band 3 clustering and dissociation from membrane complexes, inducing membrane disorganization and membrane fragility. In G6PD, dramatic reduction of NADPH levels diminishes the antioxidant capacity of RBCs increasing ROS-induced hemolysis. In PKD, diminished ATP levels affect the functioning of membrane proteins such as Na^+/K^+ pump or PMCA pump, which will indirectly induce water efflux and RBC dehydration, incrementing RBC fragility and hemolysis.

- HbS autoxidation

HbS is very unstable and could easily undergo autoxidation in the presence of oxygen. The reaction leads to the production of MetHb that no longer binds oxygen and $O_2^{\bullet-}$ that dismutates to H_2O_2 [127–130]. This results in oxidative damage of the RBC membrane and lipid and protein oxidation, leading to hemolysis and release of toxic heme leading to the exacerbation of the pro-oxidative environment (see below). HbS-induced oxidative stress leads to post-translational modifications of hemoglobin (notably oxidation of Cys93 and ubiquitination of Lys96 and Lys145 of the β globin), phosphorylation of band 3, the most abundant protein of the RBC membrane, and ubiquitination of other erythroid proteins. ROS-induced band 3 modification induces its clusterization and dissociation from membrane/cytoskeleton complexes, leading to RBC membrane disorganization and potentially microparticle formation [131–133]. Several studies have highlighted the role of microparticles in several SCD complications, such as vaso-occlusion and kidney dysfunction [134–136].

- Hemolysis: heme and iron release

Repeated cycles of sickling/unsickling lead to the fragilization of the RBC membrane and thus to hemolysis that results in the release of extracellular hemoglobin, free heme, and free iron, all highly toxic to the vasculature by triggering vascular oxidative burden [22,137,138]. In fact, oxidative stress generated at the erythroid levels can affect not only RBCs but also neutrophils, monocytes, and endothelial cells. The released heme and ATP from hemolyzed RBCs will act as damage-associated molecular patterns (DAMPs), promoting the activation of endothelial cells, macrophages, and neutrophils through different cellular pathways involving several receptors such as P2X7, toll-like receptor 4 (TLR4) or other unidentified receptors. Those activation processes trigger the expression of adhesion molecules at the cell surface and also pro-inflammatory mediators resulting in the exacerbation of the pro-inflammatory and oxidant environment. These can ultimately lead to vaso-occlusion and other SCD complications. Heme promotes adhesion events and thus vaso-occlusion through the von Willebrand factor (vWF) release from endothelial granules, inter cellular adhesion molecule (ICAM-1), vascular cell adhesion molecule (VCAM-1), and P-selectin expression at the surface of the vessel wall. Thus, it promotes leukocyte recruitment to the vessel wall and leukocyte/sickle RBC interactions. In the SCD context, heme has also been shown to promote neutrophil activation and Neutrophil Extracellular Traps (NETs) that are composed of decondensed chromatin with cytoplasmic protein [139–141]. Those NETs can, in turn, contribute to the activation of the vascular system through the activation of the TLR4/TLR9 signaling pathways, thus exacerbating the oxidative environment.

- NADPH oxidase and XO activity

NADPH oxidase is one of the major enzymes responsible for the production of $O_2^{\bullet -}$ in leukocytes, RBCs, and endothelial cells. ROS produced by erythroid NADPH contributes to the exacerbation of erythroid dysfunction by exacerbating cell stiffness resulting in the increase of hemolysis [142]. Xanthine oxidase (XO) is also responsible for a large part of the production of $O_2^{\bullet -}$ and H_2O_2. Its activity is increased in SCD plasma, but its source remains to be clearly identified [143].

- $NO^{\bullet}$ bioavailability

$NO^{\bullet}$ plays an important role in vascular homeostasis and physiology. Notably, it acts on smooth muscle cells by regulating the vascular tone as a vasodilator, and on endothelial cells through downregulation of the expression of members of the selectin family, such as ICAM-1 and VCAM-1 [144]. $NO^{\bullet}$ could also inhibit platelet activation [145]. Interestingly, $NO^{\bullet}$ may also act on the RBC itself by modulating its deformability through, in part, the soluble guanylate cyclase (sGC) [146,147].

In SCD, hemolysis and consecutive free extracellular hemoglobin release lead to $NO^{\bullet}$ scavenging, thus decreasing its bioavailability in the circulation. Furthermore, the production and release of $O_2^{\bullet -}$ may participate in the decrease in $NO^{\bullet}$ through its reaction to form $ONOO^-$ [148]. Therefore, the decreased $NO^{\bullet}$ bioavailability in SCD, negatively affects vascular tone regulation and expression of adhesion proteins.

- Erythroid mitochondrial retention

Recently, Jagadeeswaran et al. showed that RBCs from SCD patients retain mitochondria [149], and this was confirmed by other groups [150–152]. However, the functionality of these mitochondria remains controversial. Some studies showed that they were still functional and that the mitochondrial retention was associated with high levels of ROS, but some of these observations have been made in SCD mice model or in a population of erythroid circulating cells that might also include reticulocytes, i.e., immature RBCs [149,151,152]. Another group did not detect any activity of these retained mitochondria in mature RBCs [150]. A clear link between mitochondrial retention and the increased oxidative stress in SCD remains to be fully determined as well as the mechanism leading to this mitochondrial retention.

It is well established now that oxidative stress plays an essential role in SCD pathophysiology and in complication occurrence. However, it appears that oxidative mechanisms are considerably complex as they involve not only the RBC as the primary pathological cell target but also vascular endothelial cells, monocytes, and neutrophils. The complexity is heightened by the intimate interplay between oxidative mechanisms and inflammation with the activation of innate immune cells and the production of pro-inflammatory mediators. A vicious circle sets in, exacerbating the pro-oxidative, pro-inflammatory, pro-coagulant, pro-adherent environment. This highly toxic milieu is deleterious in the short-term with the appearance of acute complications and also deleterious in the long-term, with end-organ damage. Consequently, new drugs targeting oxidative stress have been developed to counteract its detrimental consequences. To date, the main antioxidant therapy that has shown some benefits in SCD clinical trials is L-glutamine, an amino acid needed for the synthesis of nucleotides as NAD. Supplementing with L-glutamine could reduce the erythroid oxidative process and protect RBCs from oxidative damage. However, this treatment has shown limitations as some SCD patients did not tolerate the treatment, and it seems to fail to counteract anemia and hemolysis [153,154]. This observation means that the mechanism underlying pro-oxidative stress in SCD requires a lot more investigation in order to identify new potential therapeutical targets.

2.2. Glucose 6-Phosphate Dehydrogenase Deficiency

G6PDH catalyzes the first reaction in the pentose–phosphate pathway, oxidizing glucose-6-phosphate to 6-phosphogluconate and reducing NADP to NADPH, which is essential to provide reducing equivalents to several antioxidant systems [155–157], as discussed above.

The G6PDH deficiency (G6PD) is a chromosome X-linked highly polymorphic genetic disorder characterized by the reduced activity of the enzyme. Although most G6PD patients do not normally present clinical manifestations, RBCs from these patients present lower levels of NADPH and are more susceptible to oxidative stress (Figure 2 G6PD panel) induced by the action of drugs, anesthetics, infections, and metabolic disturbances [157–160], leading to hemolytic anemia and various health complications (reviewed in [156,157]).

G6PD is usually associated with favism, a hemolytic anemia syndrome induced by the ingestion of fava beans [161,162]. However, even though patients presenting fava bean intolerance carry some G6PD polymorphism, not all G6PD patients are intolerant to fava beans. Actually, different metabolites from fava beans, such as vicine and divicine, are highly oxidant and could induce hemolysis by depleting the antioxidant capacity of RBCs in a mechanism similar to that of synthetic drugs [163–165].

Interestingly, some of the drugs and compounds inducing hemolytic anemia in G6PD patients are not able to induce RBC hemolysis in vitro [163–165], supporting the hypothesis of other genetic factors contributing to the hemolytic phenotype [166]. Recently, Dinarelli et al. reported that RBCs from G6PD patients stored for 6–12 days were surprisingly less sensitive to hemolysis. Authors suggest that these aged RBCs presented a metabolic regulation leading to lower energy consumption and higher stress resistance [167]; however, this hypothesis should be further confirmed by studies including a higher number of patients. Moreover, these results are contradictory with those of Francis et al., which show that, after 42 days of storage, the quality of post-transfusion RBCs is significantly lower in G6PD patients compared with control subjects [168]. Infections, both from bacterial [169,170] or viral [171–173] origin, are also able to induce hemolytic anemia in G6PD patients, probably by inducing ROS production by circulating phagocytes.

Moreover, the severity of the clinical phenotype is patient-dependent. Looking for susceptibility factors, Tang et al. studied the metabolome changes in control or G6PD patients challenged by diamide-induced ROS production. They reported that diamide induced significant changes in RBC from G6PD patients leading to severe and irreversible loss of deformability [174].

Finally, as G6PD alters several cellular processes under oxidative stress and is frequently associated with anemia, it could be expected a deficient RBC maturation. However, in vitro differentiation of $CD34^+$ hematopoietic progenitor cells isolated from patients with different G6PD severity did not show any alteration in progenitor proliferation, nor differentiation or enucleation [175].

2.3. Pyruvate Kinase Deficiency

Pyruvate kinase (PK) is a critical enzyme in the glycolytic pathway, catalyzing the conversion of phosphoenolpyruvate (PEP) to pyruvate and generating ATP [176]. Pyruvate kinase deficiency (PKD) is an autosomal (chromosome 1q21) recessive genetic disorder that affects RBCs' ability to generate energy, leading to various degrees of hemolytic anemia [177–180].

Lacking mitochondria, mature RBCs' ATP production depends exclusively on glycolysis. Thus, impaired or reduced PK activity in PKD patients leads to a dramatic decrease in ATP levels, which are necessary for maintaining the cell's integrity and deformability [181,182]. Indeed, the main RBC membrane pumps controlling calcium, sodium, and potassium transport across the RBC membrane are P-type ATPase pumps, whose activity depends on ATP concentration. Reduced activity of these pumps leads to an altered ion balance that then induces water leak leading to RBC dehydration [183] (Figure 2 PKD panel). As a consequence, RBCs from PKD patients present altered membrane properties and become more susceptible to premature destruction at the spleen, leading to hemolytic anemia [184]. Other than the elimination of altered mature RBCs, PKD patients also present a diminished number of reticulocytes, which are most susceptible to low ATP levels.

Another consequence of PK deficiency is the accumulation of glycolytic intermediates such as 2,3-biphosphoglycerate (2,3-BPG), which diminishes the O_2-hemoglobin affinity favoring the tissue oxygenation that could partially compensate for anemia [185,186].

In a recent elegant article, Roy et al. developed a metabolomic approach to characterize some changes in metabolism pathways from PKD patients. They demonstrated that RBCs from PKD patients present higher levels of oxidative stress markers, such as polyamines, sulfur-containing compounds, and deaminated purines, correlated with increased pentose phosphate pathways metabolites. Moreover, these patients also showed higher levels of poly- and highly-unsaturated fatty acids and acyl carnitine [187].

3. Conclusions

During evolution, mammals' RBCs have lost nuclei, mitochondria, and other organelles, improving their efficiency in transporting and distributing oxygen and favoring the emergence of animal species with high energy demands.

However, the dark side of this process is the lack of transcriptional and translational tools to regulate oxidative stress. Indeed, RBCs need to "resist" during 120 days lifespan to the oxidative challenge. Oxidants derive both from endogenous (HbO_2 autoxidation) and exogenous sources (endothelial and white blood cells). Normal RBCs are prepared to deal with these oxidants with an efficient enzymatic and low molecular weight antioxidant system that acts in concert. These antioxidant defenses of the RBCs depend ultimately on the reducing power of NADPH, obtained in the pentose phosphate pathway.

Several RBC diseases are associated with exacerbated oxidative stress, either presenting increased production of ROS such as SCD or diminished antioxidant capacity (G6PD and PKD) (Figure 2). All of them are genetic diseases presenting a wide and patient-dependent spectrum of clinical manifestations. Another shared characteristic is that other than promising genetic therapy, there are still no curative treatments for these diseases. As patient susceptibility is probably related to individual differences in genetic and metabolic landscape, the upcoming challenge could be to take advantage of the highly improved genomic, proteomic, and metabolomic approaches to identify possible targets useful for personalized treatments.

Author Contributions: Conceptualization, M.N.M. and M.A.O.; writing—original draft preparation, F.O., S.L., A.C.L., S.D.L., L.T., M.N.M. and M.A.O.; writing—review and editing, F.O., S.L., M.N.M. and M.A.O.; supervision, M.N.M. and M.A.O.; project administration, M.N.M. and M.A.O. All authors have read and agreed to the published version of the manuscript.

Funding: This research was funded by CSIC I+D (2020-557), Universidad de la República, Uruguay to M.N.M., Fondo María Viñas (2019-155597), Agencia Nacional de Investigación e Innovación (ANII), Uruguay to L.T. FO y ACL received scholarships from Comisión Académica de Posgrados, Universidad de la República, Uruguay.

Institutional Review Board Statement: Not applicable.

Informed Consent Statement: Not applicable.

Data Availability Statement: Not applicable.

Acknowledgments: Authors thanks the ECOS Sud Program, Project U20S02, for Ph.D. internships and PI missions. We also thank Abdellah Nait and Corentine Chrysostome for their technical and administrative assistance, respectively.

Conflicts of Interest: The authors declare no conflict of interest.

References

1. Möller, M.N.; Cuevasanta, E.; Orrico, F.; Lopez, A.C.; Thomson, L.; Denicola, A. Diffusion and transport of reactive species across cell membranes. In *Bioactive Lipids in Health and Disease*; Trostchansky, A., Rubbo, H., Eds.; Springer: Cham, Switzerland, 2019; pp. 3–19.
2. Moller, M.N.; Orrico, F.; Villar, S.F.; Lopez, A.C.; Silva, N.; Donze, M.; Thomson, L.; Denicola, A. Oxidants and Antioxidants in the Redox Biochemistry of Human Red Blood Cells. *ACS Omega* **2023**, *8*, 147–168. [CrossRef]
3. Johnson, R.M.; Goyette, G., Jr.; Ravindranath, Y.; Ho, Y.-S. Hemoglobin autoxidation and regulation of endogenous H_2O_2 levels in erythrocytes. *Free Radic. Biol. Med.* **2005**, *39*, 1407–1417. [CrossRef]
4. Buettner, G.R. The pecking order of free radicals and antioxidants: Lipid peroxidation, α-tocopherol, and ascorbate. *Arch. Biochem. Biophys.* **1993**, *300*, 535–543. [CrossRef] [PubMed]
5. Winterbourn, C.C. The biological chemistry of hydrogen peroxide. In *Methods in Enzymology*; Elsevier: Amsterdam, The Netherlands, 2013; Volume 528, pp. 3–25.
6. Alayash, A.I.; Patel, R.P.; Cashon, R.E. Redox reactions of hemoglobin and myoglobin: Biological and toxicological implications. *Antioxid. Redox Signal.* **2001**, *3*, 313–327. [CrossRef] [PubMed]
7. Buxton, G.V.; Greenstock, C.L.; Helman, W.P.; Ross, A.B. Critical review of rate constants for reactions of hydrated electrons, hydrogen atoms and hydroxyl radicals (·OH/·O− in aqueous solution. *J. Phys. Chem. Ref. Data* **1988**, *17*, 513–886. [CrossRef]
8. Stuehr, D.J.; Haque, M.M. Nitric oxide synthase enzymology in the 20 years after the Nobel Prize. *Br. J. Pharmacol.* **2019**, *176*, 177–188. [CrossRef] [PubMed]
9. Möller, M.N.; Denicola, A. Diffusion of nitric oxide and oxygen in lipoproteins and membranes studied by pyrene fluorescence quenching. *Free Radic. Biol. Med.* **2018**, *128*, 137–143. [CrossRef] [PubMed]
10. Joshi, M.S.; Ferguson, T.B., Jr.; Han, T.H.; Hyduke, D.R.; Liao, J.C.; Rassaf, T.; Bryan, N.; Feelisch, M.; Lancaster, J.R., Jr. Nitric oxide is consumed, rather than conserved, by reaction with oxyhemoglobin under physiological conditions. *Proc. Natl. Acad. Sci. USA* **2002**, *99*, 10341–10346. [CrossRef]
11. Ferrer-Sueta, G.; Campolo, N.; Trujillo, M.; Bartesaghi, S.; Carballal, S.; Romero, N.; Alvarez, B.; Radi, R. Biochemistry of peroxynitrite and protein tyrosine nitration. *Chem. Rev.* **2018**, *118*, 1338–1408. [CrossRef]
12. Taylor, J.P.; Hubert, M.T. The role of NADPH oxidases in infectious and inflammatory diseases. *Redox Biol.* **2021**, *48*, 102159. [CrossRef]
13. Lynch, R.E.; Fridovich, I. Permeation of the erythrocyte stroma by superoxide radical. *J. Biol. Chem.* **1978**, *253*, 4697–4699. [CrossRef] [PubMed]
14. Gus'kova, R.A.; Ivanov, I.I.; Kol'tover, V.K.; Akhobadze, V.V.; Rubin, A.B. Permeability of bilayer lipid membranes for superoxide (O^{2-}) radicals. *Biochim. Biophys. Acta (BBA)-Biomembr.* **1984**, *778*, 579–585. [CrossRef]
15. Denicola, A.; Souza, J.M.; Radi, R. Diffusion of peroxynitrite across erythrocyte membranes. *Proc. Natl. Acad. Sci. USA* **1998**, *95*, 3566–3571. [CrossRef] [PubMed]

16. Denicola, A.; Freeman, B.A.; Trujillo, M.; Radi, R. Peroxynitrite reaction with carbon dioxide/bicarbonate: Kinetics and influence on peroxynitrite-mediated oxidations. *Arch. Biochem. Biophys.* **1996**, *333*, 49–58. [CrossRef]
17. Signorelli, S.; Möller, M.N.; Coitiño, E.L.; Denicola, A. Nitrogen dioxide solubility and permeation in lipid membranes. *Arch. Biochem. Biophys.* **2011**, *512*, 190–196. [CrossRef]
18. Romero, N.; Denicola, A.; Souza, J.M.; Radi, R. Diffusion of peroxynitrite in the presence of carbon dioxide. *Arch. Biochem. Biophys.* **1999**, *368*, 23–30. [CrossRef]
19. Breton-Romero, R.; Lamas, S. Hydrogen peroxide signaling in vascular endothelial cells. *Redox Biol.* **2014**, *2*, 529–534. [CrossRef]
20. Hawkins, C.L.; Davies, M.J. Role of myeloperoxidase and oxidant formation in the extracellular environment in inflammation-induced tissue damage. *Free Radic. Biol. Med.* **2021**, *172*, 633–651. [CrossRef]
21. Stacey, M.M.; Peskin, A.V.; Vissers, M.C.; Winterbourn, C.C. Chloramines and hypochlorous acid oxidize erythrocyte peroxiredoxin 2. *Free Radic. Biol. Med.* **2009**, *47*, 1468–1476. [CrossRef]
22. Schaer, D.J.; Buehler, P.W.; Alayash, A.I.; Belcher, J.D.; Vercellotti, G.M. Hemolysis and free hemoglobin revisited: Exploring hemoglobin and hemin scavengers as a novel class of therapeutic proteins. *Blood* **2013**, *121*, 1276–1284. [CrossRef]
23. Pacetti, D.; Gagliardi, R.; Balzano, M.; Frega, N.; Ojeda, M.; Borrero, M.; Ruiz, A.; Lucci, P. Changes in the fatty acid profile and phospholipid molecular species composition of human erythrocyte membranes after hybrid palm and extra virgin olive oil supplementation. *J. Agric. Food Chem.* **2016**, *64*, 5499–5507. [CrossRef]
24. Yin, H.; Xu, L.; Porter, N.A. Free radical lipid peroxidation: Mechanisms and analysis. *Chem. Rev.* **2011**, *111*, 5944–5972. [CrossRef] [PubMed]
25. Ferreri, C.; Ferocino, A.; Batani, G.; Chatgilialoglu, C.; Randi, V.; Riontino, M.V.; Vetica, F.; Sansone, A. Plasmalogens: Free Radical Reactivity and Identification of Trans Isomers Relevant to Biological Membranes. *Biomolecules* **2023**, *13*, 730. [CrossRef]
26. Sindelar, P.J.; Guan, Z.; Dallner, G.; Ernster, L. The protective role of plasmalogens in iron-induced lipid peroxidation. *Free Radic. Biol. Med.* **1999**, *26*, 318–324. [CrossRef] [PubMed]
27. Clemens, M.R.; Waller, H.D. Lipid peroxidation in erythrocytes. *Chem. Phys. Lipids* **1987**, *45*, 251–268. [CrossRef]
28. Fu, X.; Felcyn, J.R.; Odem-Davis, K.; Zimring, J.C. Bioactive lipids accumulate in stored red blood cells despite leukoreduction: A targeted metabolomics study. *Transfusion* **2016**, *56*, 2560–2570. [CrossRef]
29. Clemens, M.R.; Ruess, M.; Bursa, Z.; Waller, H.D. The relationship between lipid composition of red blood cells and their susceptibility to lipid peroxidation. *Free Radic. Res. Commun.* **1987**, *3*, 265–271. [CrossRef] [PubMed]
30. Hale, J.P.; Winlove, C.P.; Petrov, P.G. Effect of hydroperoxides on red blood cell membrane mechanical properties. *Biophys. J.* **2011**, *101*, 1921–1929. [CrossRef]
31. van den Berg, J.J.; den Kamp, J.A.O.; Lubin, B.H.; Roelofsen, B.; Kuypers, F.A. Kinetics and site specificity of hydroperoxide-induced oxidative damage in red blood cells. *Free Radic. Biol. Med.* **1992**, *12*, 487–498. [CrossRef]
32. Sinha, A.; Chu, T.T.; Dao, M.; Chandramohanadas, R. Single-cell evaluation of red blood cell bio-mechanical and nano-structural alterations upon chemically induced oxidative stress. *Sci. Rep.* **2015**, *5*, 9768. [CrossRef]
33. Howie, H.L.; Hay, A.M.; de Wolski, K.; Waterman, H.; Lebedev, J.; Fu, X.; Culp-Hill, R.; D'Alessandro, A.; Gorham, J.D.; Ranson, M.S. Differences in Steap3 expression are a mechanism of genetic variation of RBC storage and oxidative damage in mice. *Blood Adv.* **2019**, *3*, 2272–2285. [CrossRef] [PubMed]
34. Blanc, L.; Papoin, J.; Debnath, G.; Vidal, M.; Amson, R.; Telerman, A.; An, X.; Mohandas, N. Abnormal erythroid maturation leads to microcytic anemia in the TSAP6/Steap3 null mouse model. *Am. J. Hematol.* **2015**, *90*, 235–241. [CrossRef] [PubMed]
35. Waugh, S.M.; Low, P.S. Hemichrome binding to band 3: Nucleation of Heinz bodies on the erythrocyte membrane. *Biochemistry* **1985**, *24*, 34–39. [CrossRef]
36. Lee, T.-H.; Kim, S.-U.; Yu, S.-L.; Kim, S.H.; Park, D.S.; Moon, H.-B.; Dho, S.H.; Kwon, K.-S.; Kwon, H.J.; Han, Y.-H. Peroxiredoxin II is essential for sustaining life span of erythrocytes in mice. *Blood* **2003**, *101*, 5033–5038. [CrossRef] [PubMed]
37. Kriebardis, A.G.; Antonelou, M.H.; Stamoulis, K.E.; Economou-Petersen, E.; Margaritis, L.H.; Papassideri, I.S. Progressive oxidation of cytoskeletal proteins and accumulation of denatured hemoglobin in stored red cells. *J. Cell. Mol. Med.* **2007**, *11*, 148–155. [CrossRef]
38. Kozlova, E.; Chernysh, A.; Moroz, V.; Kozlov, A.; Sergunova, V.; Sherstyukova, E.; Gudkova, O. Two-step process of cytoskeletal structural damage during long-term storage of packed red blood cells. *Blood Transfus.* **2021**, *19*, 124–134.
39. Abreu, I.A.; Cabelli, D.E. Superoxide dismutases—A review of the metal-associated mechanistic variations. *Biochim. Biophys. Acta (BBA)-Proteins Proteom.* **2010**, *1804*, 263–274. [CrossRef]
40. Hartz, J.; Funakoshi, S.; Deutsch, H. The levels of superoxide dismutase and catalase in human tissues as determined immunochemically. *Clin. Chim. Acta* **1973**, *46*, 125–132. [CrossRef]
41. Gärtner, A.; Weser, U. Erythrocuprein (Cu_2Zn_2 superoxide dismutase) is the major copper protein of the red blood cell. *FEBS Lett.* **1983**, *155*, 15–18. [CrossRef]
42. Getzoff, E.D.; Tainer, J.A.; Weiner, P.K.; Kollman, P.A.; Richardson, J.S.; Richardson, D.C. Electrostatic recognition between superoxide and copper, zinc superoxide dismutase. *Nature* **1983**, *306*, 287–290. [CrossRef]
43. Fridovich, I. Superoxide radical and superoxide dismutases. *Annu. Rev. Biochem.* **1995**, *64*, 97–112. [CrossRef]
44. Nedd, S.; Redler, R.L.; Proctor, E.A.; Dokholyan, N.V.; Alexandrova, A.N. Cu, Zn-superoxide dismutase without Zn is folded but catalytically inactive. *J. Mol. Biol.* **2014**, *426*, 4112–4124. [CrossRef]

45. Iuchi, Y.; Okada, F.; Onuma, K.; Onoda, T.; Asao, H.; Kobayashi, M.; Fujii, J. Elevated oxidative stress in erythrocytes due to a SOD1 deficiency causes anaemia and triggers autoantibody production. *Biochem. J.* **2007**, *402*, 219–227. [CrossRef] [PubMed]
46. Homma, T.; Fujii, J. Oxidative Stress Caused by an SOD1 Deficiency Triggers the Accumulation of Oxidatively Modified Carbonic Anhydrase II in Erythrocytes. *React. Oxyg. Species* **2018**, *6*, 289–298. [CrossRef]
47. Eleutherio, E.C.A.; Magalhães, R.S.S.; de Araújo Brasil, A.; Neto, J.R.M.; de Holanda Paranhos, L. SOD1, more than just an antioxidant. *Arch. Biochem. Biophys.* **2021**, *697*, 108701. [CrossRef] [PubMed]
48. Wilcox, K.C.; Zhou, L.; Jordon, J.K.; Huang, Y.; Yu, Y.; Redler, R.L.; Chen, X.; Caplow, M.; Dokholyan, N.V. Modifications of superoxide dismutase (SOD1) in human erythrocytes: A possible role in amyotrophic lateral sclerosis. *J. Biol. Chem.* **2009**, *284*, 13940–13947. [CrossRef] [PubMed]
49. Hodgson, E.K.; Fridovich, I. Interaction of bovine erythrocyte superoxide dismutase with hydrogen peroxide. Inactivation of the enzyme. *Biochemistry* **1975**, *14*, 5294–5299. [CrossRef] [PubMed]
50. Salo, D.C.; Lin, S.W.; Pacifici, R.E.; Davies, K.J. Superoxide dismutase is preferentially degraded by a proteolytic system from red blood cells following oxidative modification by hydrogen peroxide. *Free Radic. Biol. Med.* **1988**, *5*, 335–339. [CrossRef]
51. Ceballos-Picot, I.; Trivier, J.-M.; Nicole, A.; Sinet, P.-M.; Thevenin, M. Age-correlated modifications of copper-zinc superoxide dismutase and glutathione-related enzyme activities in human erythrocytes. *Clin. Chem.* **1992**, *38*, 66–70. [CrossRef]
52. de Lustig, E.S.; Serra, J.A.; Kohan, S.; Canziani, G.A.; Famulari, A.L.; Dominguez, R.O. Copper-zinc superoxide dismutase activity in red blood cells and serum in demented patients and in aging. *J. Neurol. Sci.* **1993**, *115*, 18–25. [CrossRef]
53. Orrico, F.; Möller, M.N.; Cassina, A.; Denicola, A.; Thomson, L. Kinetic and stoichiometric constraints determine the pathway of H2O2 consumption by red blood cells. *Free Radic. Biol. Med.* **2018**, *121*, 231–239. [CrossRef]
54. Flohé, L.; Toppo, S.; Orian, L. The glutathione peroxidase family: Discoveries and mechanism. *Free Radic. Biol. Med.* **2022**, *187*, 113–122. [CrossRef] [PubMed]
55. Flohé, L.; Brand, I. Kinetics of glutathione peroxidase. *Biochim. Biophys. Acta (BBA)-Enzymol.* **1969**, *191*, 541–549. [CrossRef]
56. Mills, G.C. Hemoglobin catabolism: I. Glutathione peroxidase, an erythrocyte enzyme which protects hemoglobin from oxidative breakdown. *J. Biol. Chem.* **1957**, *229*, 189–197. [CrossRef]
57. Johnson, R.M.; Ho, Y.-S.; Yu, D.-Y.; Kuypers, F.A.; Ravindranath, Y.; Goyette, G.W. The effects of disruption of genes for peroxiredoxin-2, glutathione peroxidase-1, and catalase on erythrocyte oxidative metabolism. *Free Radic. Biol. Med.* **2010**, *48*, 519–525. [CrossRef] [PubMed]
58. Bryk, A.H.; Wiśniewski, J.R. Quantitative analysis of human red blood cell proteome. *J. Proteome Res.* **2017**, *16*, 2752–2761. [CrossRef]
59. Weaver, K.; Skouta, R. The selenoprotein glutathione peroxidase 4: From molecular mechanisms to novel therapeutic opportunities. *Biomedicines* **2022**, *10*, 891. [CrossRef] [PubMed]
60. Ursini, F.; Maiorino, M.; Gregolin, C. The selenoenzyme phospholipid hydroperoxide glutathione peroxidase. *Biochim. Biophys. Acta (BBA)-Gen. Subj.* **1985**, *839*, 62–70. [CrossRef]
61. Flohé, L.; Toppo, S.; Cozza, G.; Ursini, F. A comparison of thiol peroxidase mechanisms. *Antioxid. Redox Signal.* **2011**, *15*, 763–780. [CrossRef]
62. Necheles, T.F.; Rai, U.S.; Cameron, D. Congenital nonspherocytic hemolytic anemia associated with an unusual erythrocyte hexokinase abnormality. *J. Lab. Clin. Med.* **1970**, *76*, 593–602.
63. Gondo, H.; Ideguchi, H.; Hayashi, S.; Shibuya, T. Acute hemolysis in glutathione peroxidase deficiency. *Int. J. Hematol.* **1992**, *55*, 215–218. [PubMed]
64. Thieme, R.; Pai, E.F.; Schirmer, R.H.; Schulz, G.E. Three-dimensional structure of glutathione reductase at 2 A resolution. *J. Mol. Biol.* **1981**, *152*, 763–782. [CrossRef] [PubMed]
65. Deponte, M. Glutathione catalysis and the reaction mechanisms of glutathione-dependent enzymes. *Biochim. Biophys. Acta* **2013**, *1830*, 3217–3266. [CrossRef] [PubMed]
66. Worthington, D.J.; Rosemeyer, M.A. Glutathione reductase from human erythrocytes. Catalytic properties and aggregation. *Eur. J. Biochem.* **1976**, *67*, 231–238. [CrossRef] [PubMed]
67. Amen, F.; Machin, A.; Tourino, C.; Rodriguez, I.; Denicola, A.; Thomson, L. N-acetylcysteine improves the quality of red blood cells stored for transfusion. *Arch. Biochem. Biophys.* **2017**, *621*, 31–37. [CrossRef] [PubMed]
68. Beutler, E. Disorders due to enzyme defects in the red blood cell. *Adv. Metab. Disord.* **1972**, *60*, 131–160.
69. Kamerbeek, N.M.; van Zwieten, R.; de Boer, M.; Morren, G.; Vuil, H.; Bannink, N.; Lincke, C.; Dolman, K.M.; Becker, K.; Heiner Schirmer, R. Molecular basis of glutathione reductase deficiency in human blood cells. *Blood* **2007**, *109*, 3560–3566. [CrossRef]
70. Loos, H.; Roos, D.; Weening, R.; Houwerzijl, J. Familial deficiency of glutathione reductase in human blood cells. *Blood* **1976**, *48*, 53–62. [CrossRef]
71. Nakashima, K.; Yamauchi, K.; Miwa, S.; Fujimura, K.; Mizutani, A.; Kuramoto, A. Glutathione reductase deficiency in a kindred with hereditary spherocytosis. *Am. J. Hematol.* **1978**, *4*, 141–150. [CrossRef]
72. Cha, M.-K.; Yun, C.-H.; Kim, I.-H. Interaction of human thiol-specific antioxidant protein 1 with erythrocyte plasma membrane. *Biochemistry* **2000**, *39*, 6944–6950. [CrossRef]
73. Moore, R.B.; Mankad, M.V.; Shriver, S.K.; Mankad, V.N.; Plishker, G.A. Reconstitution of Ca(2+)-dependent K+ transport in erythrocyte membrane vesicles requires a cytoplasmic protein. *J. Biol. Chem.* **1991**, *266*, 18964–18968. [CrossRef]

74. Cho, C.S.; Kato, G.J.; Yang, S.H.; Bae, S.W.; Lee, J.S.; Gladwin, M.T.; Rhee, S.G. Hydroxyurea-induced expression of glutathione peroxidase 1 in red blood cells of individuals with sickle cell anemia. *Antioxid. Redox Signal* **2010**, *13*, 1–11. [CrossRef] [PubMed]
75. Manta, B.; Hugo, M.; Ortiz, C.; Ferrer-Sueta, G.; Trujillo, M.; Denicola, A. The peroxidase and peroxynitrite reductase activity of human erythrocyte peroxiredoxin 2. *Arch. Biochem. Biophys.* **2009**, *484*, 146–154. [CrossRef]
76. Schröder, E.; Littlechil, J.A.; Lebedev, A.A.; Errington, N.; Vagin, A.A.; Isupov, M.N. Crystal structure of decameric 2-Cys peroxiredoxin from human erythrocytes at 1.7 Å resolution. *Structure* **2000**, *8*, 605–615. [CrossRef] [PubMed]
77. Biteau, B.; Labarre, J.; Toledano, M.B. ATP-dependent reduction of cysteine-sulphinic acid by S. cerevisiae sulphiredoxin. *Nature* **2003**, *425*, 980–984. [CrossRef]
78. Jönsson, T.J.; Johnson, L.C.; Lowther, W.T. Structure of the sulphiredoxin–peroxiredoxin complex reveals an essential repair embrace. *Nature* **2008**, *451*, 98–101. [CrossRef]
79. Harris, J.R.; Schröder, E.; Isupov, M.N.; Scheffler, D.; Kristensen, P.; Littlechild, J.A.; Vagin, A.A.; Meissner, U. Comparison of the decameric structure of peroxiredoxin-II by transmission electron microscopy and X-ray crystallography. *Biochim. Biophys. Acta (BBA)-Protein Struct. Mol. Enzymol.* **2001**, *1547*, 221–234. [CrossRef]
80. Meissner, U.; Schröder, E.; Scheffler, D.; Martin, A.G.; Harris, J.R. Formation, TEM study and 3D reconstruction of the human erythrocyte peroxiredoxin-2 dodecahedral higher-order assembly. *Micron* **2007**, *38*, 29–39. [CrossRef]
81. Plishker, G.; Chevalier, D.; Seinsoth, L.; Moore, R. Calcium-activated potassium transport and high molecular weight forms of calpromotin. *J. Biol. Chem.* **1992**, *267*, 21839–21843. [CrossRef]
82. Low, F.M.; Hampton, M.B.; Peskin, A.V.; Winterbourn, C.C. Peroxiredoxin 2 functions as a noncatalytic scavenger of low-level hydrogen peroxide in the erythrocyte. *Blood* **2007**, *109*, 2611–2617. [CrossRef]
83. Han, Y.-H.; Kim, S.-U.; Kwon, T.-H.; Lee, D.-S.; Ha, H.-L.; Park, D.-S.; Woo, E.-J.; Lee, S.-H.; Kim, J.-M.; Chae, H.-B. Peroxiredoxin II is essential for preventing hemolytic anemia from oxidative stress through maintaining hemoglobin stability. *Biochem. Biophys. Res. Commun.* **2012**, *426*, 427–432. [CrossRef] [PubMed]
84. Ogasawara, Y.; Ohminato, T.; Nakamura, Y.; Ishii, K. Structural and functional analysis of native peroxiredoxin 2 in human red blood cells. *Int. J. Biochem. Cell Biol.* **2012**, *44*, 1072–1077. [CrossRef]
85. Bayer, S.B.; Low, F.M.; Hampton, M.B.; Winterbourn, C.C. Interactions between peroxiredoxin 2, hemichrome and the erythrocyte membrane. *Free Radic. Res.* **2016**, *50*, 1329–1339. [CrossRef] [PubMed]
86. Matte, A.; Bertoldi, M.; Mohandas, N.; An, X.; Bugatti, A.; Brunati, A.M.; Rusnati, M.; Tibaldi, E.; Siciliano, A.; Turrini, F. Membrane association of peroxiredoxin-2 in red cells is mediated by the N-terminal cytoplasmic domain of band 3. *Free Radic. Biol. Med.* **2013**, *55*, 27–35. [CrossRef] [PubMed]
87. Wood, Z.A.; Schröder, E.; Harris, J.R.; Poole, L.B. Structure, mechanism and regulation of peroxiredoxins. *Trends Biochem. Sci.* **2003**, *28*, 32–40. [CrossRef] [PubMed]
88. Randall, L.M.; Ferrer-Sueta, G.; Denicola, A. Peroxiredoxins as preferential targets in H2O2-induced signaling. In *Methods in Enzymology*; Elsevier: Amsterdam, The Netherlands, 2013; Volume 527, pp. 41–63.
89. Daneva, Z.; Marziano, C.; Ottolini, M.; Chen, Y.-L.; Baker, T.M.; Kuppusamy, M.; Zhang, A.; Ta, H.Q.; Reagan, C.E.; Mihalek, A.D. Caveolar peroxynitrite formation impairs endothelial TRPV4 channels and elevates pulmonary arterial pressure in pulmonary hypertension. *Proc. Natl. Acad. Sci. USA* **2021**, *118*, e2023130118. [CrossRef]
90. Holmgren, A. Thioredoxin structure and mechanism: Conformational changes on oxidation of the active-site sulfhydryls to a disulfide. *Structure* **1995**, *3*, 239–243. [CrossRef]
91. Lu, J.; Holmgren, A. The thioredoxin antioxidant system. *Free Radic. Biol. Med.* **2014**, *66*, 75–87. [CrossRef]
92. Palde, P.B.; Carroll, K.S. A universal entropy-driven mechanism for thioredoxin–target recognition. *Proc. Natl. Acad. Sci. USA* **2015**, *112*, 7960–7965. [CrossRef]
93. Holmgren, A. Thioredoxin. *Annu. Rev. Biochem.* **1985**, *54*, 237–271. [CrossRef]
94. Ren, X.; Bjoernstedt, M.; Shen, B.; Ericson, M.L.; Holmgren, A. Mutagenesis of structural half-cystine residues in human thioredoxin and effects on the regulation of activity by selenodiglutathione. *Biochemistry* **1993**, *32*, 9701–9708. [CrossRef] [PubMed]
95. Weichsel, A.; Gasdaska, J.R.; Powis, G.; Montfort, W.R. Crystal structures of reduced, oxidized, and mutated human thioredoxins: Evidence for a regulatory homodimer. *Structure* **1996**, *4*, 735–751. [CrossRef] [PubMed]
96. Holmgren, A. Antioxidant function of thioredoxin and glutaredoxin systems. *Antioxid. Redox Signal.* **2000**, *2*, 811–820. [CrossRef] [PubMed]
97. Zhong, L.; Arnér, E.S.; Holmgren, A. Structure and mechanism of mammalian thioredoxin reductase: The active site is a redox-active selenolthiol/selenenylsulfide formed from the conserved cysteine-selenocysteine sequence. *Proc. Natl. Acad. Sci. USA* **2000**, *97*, 5854–5859. [CrossRef] [PubMed]
98. Gladyshev, V.N.; Jeang, K.-T.; Stadtman, T.C. Selenocysteine, identified as the penultimate C-terminal residue in human T-cell thioredoxin reductase, corresponds to TGA in the human placental gene. *Proc. Natl. Acad. Sci. USA* **1996**, *93*, 6146–6151. [CrossRef]
99. Arscott, L.D.; Gromer, S.; Schirmer, R.H.; Becker, K.; Williams Jr, C.H. The mechanism of thioredoxin reductase from human placenta is similar to the mechanisms of lipoamide dehydrogenase and glutathione reductase and is distinct from the mechanism of thioredoxin reductase from *Escherichia coli*. *Proc. Natl. Acad. Sci. USA* **1997**, *94*, 3621–3626. [CrossRef]

100. Cheng, Q.; Sandalova, T.; Lindqvist, Y.; Arner, E.S. Crystal structure and catalysis of the selenoprotein thioredoxin reductase 1. *J. Biol. Chem.* **2009**, *284*, 3998–4008. [CrossRef]
101. Zhao, R.; Masayasu, H.; Holmgren, A. Ebselen: A substrate for human thioredoxin reductase strongly stimulating its hydroperoxide reductase activity and a superfast thioredoxin oxidant. *Proc. Natl. Acad. Sci. USA* **2002**, *99*, 8579–8584. [CrossRef]
102. Tan, S.-X.; Greetham, D.; Raeth, S.; Grant, C.M.; Dawes, I.W.; Perrone, G.G. The thioredoxin-thioredoxin reductase system can function in vivo as an alternative system to reduce oxidized glutathione in *Saccharomyces cerevisiae*. *J. Biol. Chem.* **2010**, *285*, 6118–6126. [CrossRef]
103. Mendiratta, S.; Qu, Z.-c.; May, J.M. Enzyme-dependent ascorbate recycling in human erythrocytes: Role of thioredoxin reductase. *Free Radic. Biol. Med.* **1998**, *25*, 221–228. [CrossRef]
104. Bushweller, J.H.; Aaslund, F.; Wuethrich, K.; Holmgren, A. Structural and functional characterization of the mutant Escherichia coli glutaredoxin (C14. fwdarw. S) and its mixed disulfide with glutathione. *Biochemistry* **1992**, *31*, 9288–9293. [CrossRef]
105. Fernandes, A.P.; Holmgren, A. Glutaredoxins: Glutathione-dependent redox enzymes with functions far beyond a simple thioredoxin backup system. *Antioxid. Redox Signal.* **2004**, *6*, 63–74. [CrossRef] [PubMed]
106. Lillig, C.H.; Berndt, C.; Holmgren, A. Glutaredoxin systems. *Biochim. Biophys. Acta (BBA)-Gen. Subj.* **2008**, *1780*, 1304–1317. [CrossRef] [PubMed]
107. Mieyal, J.J.; Starke, D.W.; Gravina, S.A.; Dothey, C.; Chung, J.S. Thioltransferase in human red blood cells: Purification and properties. *Biochemistry* **1991**, *30*, 6088–6097. [CrossRef] [PubMed]
108. Terada, T.; Oshida, T.; Nishimura, M.; Maeda, H.; Hara, T.; Hosomi, S.; Mizoguchi, T.; Nishihara, T. Study on human erythrocyte thioltransferase: Comparative characterization with bovine enzyme and its physiological role under oxidative stress. *J. Biochem.* **1992**, *111*, 688–692. [CrossRef]
109. Lind, C.; Gerdes, R.; Schuppe-Koistinen, I.; Cotgreave, I.A. Studies on the mechanism of oxidative modification of human glyceraldehyde-3-phosphate dehydrogenase by glutathione: Catalysis by glutaredoxin. *Biochem. Biophys. Res. Commun.* **1998**, *247*, 481–486. [CrossRef]
110. Peskin, A.V.; Pace, P.E.; Behring, J.B.; Paton, L.N.; Soethoudt, M.; Bachschmid, M.M.; Winterbourn, C.C. Glutathionylation of the active site cysteines of peroxiredoxin 2 and recycling by glutaredoxin. *J. Biol. Chem.* **2016**, *291*, 3053–3062. [CrossRef]
111. Hanschmann, E.-M.; Godoy, J.R.; Berndt, C.; Hudemann, C.; Lillig, C.H. Thioredoxins, glutaredoxins, and peroxiredoxins—Molecular mechanisms and health significance: From cofactors to antioxidants to redox signaling. *Antioxid. Redox Signal.* **2013**, *19*, 1539–1605. [CrossRef]
112. Haunhorst, P.; Hanschmann, E.-M.; Bräutigam, L.; Stehling, O.; Hoffmann, B.; Mühlenhoff, U.; Lill, R.; Berndt, C.; Lillig, C.H. Crucial function of vertebrate glutaredoxin 3 (PICOT) in iron homeostasis and hemoglobin maturation. *Mol. Biol. Cell* **2013**, *24*, 1895–1903. [CrossRef]
113. Ye, H.; Jeong, S.Y.; Ghosh, M.C.; Kovtunovych, G.; Silvestri, L.; Ortillo, D.; Uchida, N.; Tisdale, J.; Camaschella, C.; Rouault, T.A. Glutaredoxin 5 deficiency causes sideroblastic anemia by specifically impairing heme biosynthesis and depleting cytosolic iron in human erythroblasts. *J. Clin. Investig.* **2010**, *120*, 1749–1761. [CrossRef] [PubMed]
114. Putnam, C.D.; Arvai, A.S.; Bourne, Y.; Tainer, J.A. Active and inhibited human catalase structures: Ligand and NADPH binding and catalytic mechanism. *J. Mol. Biol.* **2000**, *296*, 295–309. [CrossRef] [PubMed]
115. Kirkman, H.N.; Gaetani, G.F. Catalase: A tetrameric enzyme with four tightly bound molecules of NADPH. *Proc. Natl. Acad. Sci. USA* **1984**, *81*, 4343–4347. [CrossRef] [PubMed]
116. Nicholls, P.; Fita, I.; Loewen, P.C. Enzymology and structure of catalases. *Adv. Inorg. Chem.* **2000**, *51*, 51–106. [CrossRef]
117. Kirkman, H.N.; Galiano, S.; Gaetani, G. The function of catalase-bound NADPH. *J. Biol. Chem.* **1987**, *262*, 660–666. [CrossRef] [PubMed]
118. Kirkman, H.N.; Rolfo, M.; Ferraris, A.M.; Gaetani, G.F. Mechanisms of protection of catalase by NADPH: Kinetics and stoichiometry. *J. Biol. Chem.* **1999**, *274*, 13908–13914. [CrossRef]
119. Aebi, H. [13] Catalase in vitro. In *Methods in Enzymology*; Elsevier: Amsterdam, The Netherlands, 1984; Volume 105, pp. 121–126.
120. Page, G.P.; Kanias, T.; Guo, Y.J.; Lanteri, M.C.; Zhang, X.; Mast, A.E.; Cable, R.G.; Spencer, B.R.; Kiss, J.E.; Fang, F. Multiple-ancestry genome-wide association study identifies 27 loci associated with measures of hemolysis following blood storage. *J. Clin. Investig.* **2021**, *131*, e146077. [CrossRef]
121. D'Alessandro, A.; Fu, X.; Kanias, T.; Reisz, J.A.; Culp-Hill, R.; Guo, Y.; Gladwin, M.T.; Page, G.; Kleinman, S.; Lanteri, M. Donor sex, age and ethnicity impact stored red blood cell antioxidant metabolism through mechanisms in part explained by glucose 6-phosphate dehydrogenase levels and activity. *Haematologica* **2021**, *106*, 1290. [CrossRef]
122. Stuart, M.J.; Nagel, R.L. Sickle-cell disease. *Lancet* **2004**, *364*, 1343–1360. [CrossRef]
123. Rees, D.C.; Williams, T.N.; Gladwin, M.T. Sickle-cell disease. *Lancet* **2010**, *376*, 2018–2031. [CrossRef]
124. Pacher, P.; Beckman, J.S.; Liaudet, L. Nitric oxide and peroxynitrite in health and disease. *Physiol. Rev.* **2007**, *87*, 315–424. [CrossRef]
125. Radi, R. Oxygen radicals, nitric oxide, and peroxynitrite: Redox pathways in molecular medicine. *Proc. Natl. Acad. Sci. USA* **2018**, *115*, 5839–5848. [CrossRef] [PubMed]
126. Vona, R.; Sposi, N.M.; Mattia, L.; Gambardella, L.; Straface, E.; Pietraforte, D. Sickle Cell Disease: Role of Oxidative Stress and Antioxidant Therapy. *Antioxidants* **2021**, *10*, 296. [CrossRef] [PubMed]

127. Hebbel, R.P.; Morgan, W.T.; Eaton, J.W.; Hedlund, B.E. Accelerated autoxidation and heme loss due to instability of sickle hemoglobin. *Proc. Natl. Acad. Sci. USA* **1988**, *85*, 237–241. [CrossRef] [PubMed]
128. Mohanty, J.G.; Nagababu, E.; Rifkind, J.M. Red blood cell oxidative stress impairs oxygen delivery and induces red blood cell aging. *Front. Physiol.* **2014**, *5*, 84. [CrossRef]
129. Sheng, K.; Shariff, M.; Hebbel, R.P. Comparative oxidation of hemoglobins A and S. *Blood* **1998**, *91*, 3467–3470. [CrossRef]
130. Umbreit, J. Methemoglobin—It's not just blue: A concise review. *Am. J. Hematol.* **2007**, *82*, 134–144. [CrossRef]
131. Welbourn, E.M.; Wilson, M.T.; Yusof, A.; Metodiev, M.V.; Cooper, C.E. The mechanism of formation, structure and physiological relevance of covalent hemoglobin attachment to the erythrocyte membrane. *Free Radic. Biol. Med.* **2017**, *103*, 95–106. [CrossRef]
132. Walder, J.A.; Chatterjee, R.; Steck, T.L.; Low, P.S.; Musso, G.F.; Kaiser, E.T.; Rogers, P.H.; Arnone, A. The interaction of hemoglobin with the cytoplasmic domain of band 3 of the human erythrocyte membrane. *J. Biol. Chem.* **1984**, *259*, 10238–10246. [CrossRef]
133. Jana, S.; Strader, M.B.; Meng, F.; Hicks, W.; Kassa, T.; Tarandovskiy, I.; De Paoli, S.; Simak, J.; Heaven, M.R.; Belcher, J.D.; et al. Hemoglobin oxidation-dependent reactions promote interactions with band 3 and oxidative changes in sickle cell-derived microparticles. *JCI Insight* **2018**, *3*, e120451. [CrossRef]
134. Camus, S.M.; Gausseres, B.; Bonnin, P.; Loufrani, L.; Grimaud, L.; Charue, D.; De Moraes, J.A.; Renard, J.M.; Tedgui, A.; Boulanger, C.M.; et al. Erythrocyte microparticles can induce kidney vaso-occlusions in a murine model of sickle cell disease. *Blood* **2012**, *120*, 5050–5058. [CrossRef]
135. Tharaux, P.L. Posttranslational modifications of sickle hemoglobin in microparticles may promote injury. *Kidney Int.* **2019**, *95*, 1289–1291. [CrossRef]
136. Camus, S.M.; De Moraes, J.A.; Bonnin, P.; Abbyad, P.; Le Jeune, S.; Lionnet, F.; Loufrani, L.; Grimaud, L.; Lambry, J.C.; Charue, D.; et al. Circulating cell membrane microparticles transfer heme to endothelial cells and trigger vasoocclusions in sickle cell disease. *Blood* **2015**, *125*, 3805–3814. [CrossRef] [PubMed]
137. Vinchi, F.; Sparla, R.; Passos, S.T.; Sharma, R.; Vance, S.Z.; Zreid, H.S.; Juaidi, H.; Manwani, D.; Yazdanbakhsh, K.; Nandi, V.; et al. Vasculo-toxic and pro-inflammatory action of unbound haemoglobin, haem and iron in transfusion-dependent patients with haemolytic anaemias. *Br. J. Haematol.* **2021**, *193*, 637–658. [CrossRef] [PubMed]
138. Woollard, K.J.; Sturgeon, S.; Chin-Dusting, J.P.; Salem, H.H.; Jackson, S.P. Erythrocyte hemolysis and hemoglobin oxidation promote ferric chloride-induced vascular injury. *J. Biol. Chem.* **2009**, *284*, 13110–13118. [CrossRef] [PubMed]
139. Graca-Souza, A.V.; Arruda, M.A.; de Freitas, M.S.; Barja-Fidalgo, C.; Oliveira, P.L. Neutrophil activation by heme: Implications for inflammatory processes. *Blood* **2002**, *99*, 4160–4165. [CrossRef]
140. Chen, G.; Zhang, D.; Fuchs, T.A.; Manwani, D.; Wagner, D.D.; Frenette, P.S. Heme-induced neutrophil extracellular traps contribute to the pathogenesis of sickle cell disease. *Blood* **2014**, *123*, 3818–3827. [CrossRef]
141. Schimmel, M.; Nur, E.; Biemond, B.J.; van Mierlo, G.J.; Solati, S.; Brandjes, D.P.; Otten, H.M.; Schnog, J.J.; Zeerleder, S.; Curama Study, G. Nucleosomes and neutrophil activation in sickle cell disease painful crisis. *Haematologica* **2013**, *98*, 1797–1803. [CrossRef]
142. George, A.; Pushkaran, S.; Konstantinidis, D.G.; Koochaki, S.; Malik, P.; Mohandas, N.; Zheng, Y.; Joiner, C.H.; Kalfa, T.A. Erythrocyte NADPH oxidase activity modulated by Rac GTPases, PKC, and plasma cytokines contributes to oxidative stress in sickle cell disease. *Blood* **2013**, *121*, 2099–2107. [CrossRef]
143. Aslan, M.; Ryan, T.M.; Adler, B.; Townes, T.M.; Parks, D.A.; Thompson, J.A.; Tousson, A.; Gladwin, M.T.; Patel, R.P.; Tarpey, M.M.; et al. Oxygen radical inhibition of nitric oxide-dependent vascular function in sickle cell disease. *Proc. Natl. Acad. Sci. USA* **2001**, *98*, 15215–15220. [CrossRef]
144. De Caterina, R.; Libby, P.; Peng, H.B.; Thannickal, V.J.; Rajavashisth, T.B.; Gimbrone, M.A., Jr.; Shin, W.S.; Liao, J.K. Nitric oxide decreases cytokine-induced endothelial activation. Nitric oxide selectively reduces endothelial expression of adhesion molecules and proinflammatory cytokines. *J. Clin. Invest.* **1995**, *96*, 60–68. [CrossRef]
145. Radomski, M.W.; Vallance, P.; Whitley, G.; Foxwell, N.; Moncada, S. Platelet adhesion to human vascular endothelium is modulated by constitutive and cytokine induced nitric oxide. *Cardiovasc. Res.* **1993**, *27*, 1380–1382. [CrossRef]
146. Simmonds, M.J.; Detterich, J.A.; Connes, P. Nitric oxide, vasodilation and the red blood cell. *Biorheology* **2014**, *51*, 121–134. [CrossRef] [PubMed]
147. Starzyk, D.; Korbut, R.; Gryglewski, R.J. The role of nitric oxide in regulation of deformability of red blood cells in acute phase of endotoxaemia in rats. *J. Physiol. Pharmacol.* **1997**, *48*, 731–735. [PubMed]
148. Beckman, J.S.; Beckman, T.W.; Chen, J.; Marshall, P.A.; Freeman, B.A. Apparent hydroxyl radical production by peroxynitrite: Implications for endothelial injury from nitric oxide and superoxide. *Proc. Natl. Acad. Sci. USA* **1990**, *87*, 1620–1624. [CrossRef] [PubMed]
149. Jagadeeswaran, R.; Vazquez, B.A.; Thiruppathi, M.; Ganesh, B.B.; Ibanez, V.; Cui, S.; Engel, J.D.; Diamond, A.M.; Molokie, R.E.; DeSimone, J.; et al. Pharmacological inhibition of LSD1 and mTOR reduces mitochondrial retention and associated ROS levels in the red blood cells of sickle cell disease. *Exp. Hematol.* **2017**, *50*, 46–52. [CrossRef]
150. Martino, S.; Arlet, J.B.; Odievre, M.H.; Jullien, V.; Moras, M.; Hattab, C.; Lefebvre, T.; Gouya, L.; Ostuni, M.A.; Lefevre, S.D.; et al. Deficient mitophagy pathways in sickle cell disease. *Br. J. Haematol.* **2021**, *193*, 988–993. [CrossRef]
151. Esperti, S.; Nader, E.; Stier, A.; Boisson, C.; Carin, R.; Marano, M.; Robert, M.; Martin, M.; Horand, F.; Cibiel, A.; et al. Increased retention of functional mitochondria in mature sickle red blood cells is associated with increased sickling tendency, hemolysis and oxidative stress. *Haematologica* 2023, *Epub ahead of print*. [CrossRef]

152. Moriconi, C.; Dzieciatkowska, M.; Roy, M.; D'Alessandro, A.; Roingeard, P.; Lee, J.Y.; Gibb, D.R.; Tredicine, M.; McGill, M.A.; Qiu, A.; et al. Retention of functional mitochondria in mature red blood cells from patients with sickle cell disease. *Br. J. Haematol.* **2022**, *198*, 574–586. [CrossRef]
153. Niihara, Y.; Miller, S.T.; Kanter, J.; Lanzkron, S.; Smith, W.R.; Hsu, L.L.; Gordeuk, V.R.; Viswanathan, K.; Sarnaik, S.; Osunkwo, I.; et al. A Phase 3 Trial of l-Glutamine in Sickle Cell Disease. *N. Engl. J. Med.* **2018**, *379*, 226–235. [CrossRef]
154. Cox, S.E.; Hart, E.; Kirkham, F.J.; Stotesbury, H. L-Glutamine in sickle cell disease. *Drugs Today* **2020**, *56*, 257–268. [CrossRef]
155. Berg, J.M.; Tymoczko, J.L.; Gatto, G.J.; Stryer, L. *Biochemistry*, 9th ed.; W.H. Freeman/Macmillan Learning: New York, NY, USA, 2019; p. xlii.
156. Cappellini, M.D.; Fiorelli, G. Glucose-6-phosphate dehydrogenase deficiency. *Lancet* **2008**, *371*, 64–74. [CrossRef] [PubMed]
157. Luzzatto, L.; Ally, M.; Notaro, R. Glucose-6-phosphate dehydrogenase deficiency. *Blood* **2020**, *136*, 1225–1240. [CrossRef] [PubMed]
158. Beutler, E. G6PD deficiency. *Blood* **1994**, *84*, 3613–3636. [CrossRef]
159. Gomez Gomez, S.; Ruano Santiago, M.; Rodriguez Morillo, A.; Perez Munoz, A.M.; Echevarria Moreno, M. Anesthetic management of glucose 6-phosphate dehydrogenase deficiency. *Rev. Esp. Anestesiol. Reanim.* **2023**, *70*, 235–239. [CrossRef]
160. Wilson, J. Rasburicase-induced methaemoglobinaemia and catastrophic oxidative haemolysis in undiagnosed G6PD deficiency. *Br. J. Haematol.* **2023**, *200*, 7. [CrossRef] [PubMed]
161. Arese, P.; Mannuzzu, L.; Turrini, F. Pathophysiology of favism. *Folia Haematol. Int. Mag. Klin. Morphol. Blutforsch.* **1989**, *116*, 745–752.
162. Luzzatto, L.; Arese, P. Favism and Glucose-6-Phosphate Dehydrogenase Deficiency. *N. Engl. J. Med.* **2018**, *378*, 60–71. [CrossRef]
163. Arese, P.; Bosia, A.; Naitana, A.; Gaetani, S.; D'Aquino, M.; Gaetani, G.F. Effect of divicine and isouramil on red cell metabolism in normal and G6PD-deficient (Mediterranean variant) subjects. Possible role in the genesis of favism. *Prog. Clin. Biol. Res.* **1981**, *55*, 725–746.
164. McMillan, D.C.; Jollow, D.J. Favism: Divicine hemotoxicity in the rat. *Toxicol. Sci.* **1999**, *51*, 310–316. [CrossRef]
165. McMillan, D.C.; Schey, K.L.; Meier, G.P.; Jollow, D.J. Chemical analysis and hemolytic activity of the fava bean aglycon divicine. *Chem. Res. Toxicol.* **1993**, *6*, 439–444. [CrossRef]
166. Stamatoyannopoulos, G.; Fraser, G.R.; Motulsky, A.C.; Fessas, P.; Akrivakis, A.; Papayannopoulou, T. On the familial predisposition to favism. *Am. J. Hum. Genet.* **1966**, *18*, 253–263. [PubMed]
167. Dinarelli, S.; Longo, G.; Germanova-Taneva, S.; Todinova, S.; Krumova, S.; Girasole, M. Surprising Structural and Functional Properties of Favism Erythrocytes Are Linked to Special Metabolic Regulation: A Cell Aging Study. *Int. J. Mol. Sci.* **2022**, *24*, 637. [CrossRef] [PubMed]
168. Francis, R.O.; D'Alessandro, A.; Eisenberger, A.; Soffing, M.; Yeh, R.; Coronel, E.; Sheikh, A.; Rapido, F.; La Carpia, F.; Reisz, J.A.; et al. Donor glucose-6-phosphate dehydrogenase deficiency decreases blood quality for transfusion. *J. Clin. Invest.* **2020**, *130*, 2270–2285. [CrossRef] [PubMed]
169. Pamuk, G.E.; Dogan Celik, A.; Uyanik, M.S. Brucellosis triggering hemolytic anemia in glucose-6-phosphate dehydrogenase deficiency. *Med. Princ. Pr.* **2009**, *18*, 329–331. [CrossRef]
170. Quereshy, F.A.; Gold, E.S.; Powers, M.P. Hemolytic anemia in a glucose-6-phosphate dehydrogenase-deficient patient triggered by a maxillofacial infection. *J. Oral. Maxillofac. Surg.* **2000**, *58*, 805–807. [CrossRef]
171. Araujo, T.; Katiyar, V.; Gonzales Zamora, J.A. Acute Retroviral Syndrome Presenting with Hemolytic Anemia Induced by G6PD Deficiency. *Trop. Med. Infect. Dis.* **2018**, *4*, 6. [CrossRef]
172. Meloni, T.; Forteleoni, G.; Porcu, A. Acute hemolytic anemia in two G6PD-deficient children with viral hepatitis. *Haematologica* **1988**, *73*, 397–399.
173. Oluboyede, O.A.; Ayoola, E.A. Glucose 6 phosphate dehydrogenase enzyme (G6PD) and viral hepatitis in Nigeria. *East. Afr. Med. J.* **1982**, *59*, 754–759.
174. Tang, H.Y.; Ho, H.Y.; Wu, P.R.; Chen, S.H.; Kuypers, F.A.; Cheng, M.L.; Chiu, D.T. Inability to maintain GSH pool in G6PD-deficient red cells causes futile AMPK activation and irreversible metabolic disturbance. *Antioxid. Redox Signal* **2015**, *22*, 744–759. [CrossRef]
175. Boonpeng, K.; Ketprasit, N.; Palasuwan, A.; Kulkeaw, K.; Palasuwan, D. Glucose-6-phosphate dehydrogenase is dispensable for human erythroid cell differentiation in vitro. *Exp. Hematol.* **2023**, *121*, 18–29.e2. [CrossRef]
176. Nelson, D.L.; Cox, M.M.; Lehninger, A.L. *Lehninger Principles of Biochemistry*, 7th ed.; W.H. Freeman and Company: New York, NY, USA; Macmillan Higher Education: Houndmills, UK, 2017; p. xxxiv.
177. Bianchi, P.; Fermo, E.; Lezon-Geyda, K.; van Beers, E.J.; Morton, H.D.; Barcellini, W.; Glader, B.; Chonat, S.; Ravindranath, Y.; Newburger, P.E.; et al. Genotype-phenotype correlation and molecular heterogeneity in pyruvate kinase deficiency. *Am. J. Hematol.* **2020**, *95*, 472–482. [CrossRef] [PubMed]
178. Luke, N.; Hillier, K.; Al-Samkari, H.; Grace, R.F. Updates and advances in pyruvate kinase deficiency. *Trends Mol. Med.* **2023**, *29*, 406–418. [CrossRef] [PubMed]
179. Svidnicki, M.; Santos, A.; Fernandez, J.A.A.; Yokoyama, A.P.H.; Magalhaes, I.Q.; Pinheiro, V.R.P.; Brandalise, S.R.; Silveira, P.A.A.; Costa, F.F.; Saad, S.T.O. Novel mutations associated with pyruvate kinase deficiency in Brazil. *Rev. Bras. Hematol. Hemoter.* **2018**, *40*, 5–11. [CrossRef]

180. Zanella, A.; Fermo, E.; Bianchi, P.; Chiarelli, L.R.; Valentini, G. Pyruvate kinase deficiency: The genotype-phenotype association. *Blood Rev.* **2007**, *21*, 217–231. [CrossRef]
181. Chapman, R.G.; Schaumburg, L. Glycolysis and glycolytic enzyme activity of aging red cells in man. Changes in hexokinase, aldolase, glyceraldehyde-3-phosphate dehydrogenase, pyruvate kinase and glutamic-oxalacetic transaminase. *Br. J. Haematol.* **1967**, *13*, 665–678. [CrossRef] [PubMed]
182. Delivoria-Papadopoulos, M.; Oski, F.A.; Gottlieb, A.J. Oxygen-hemoglobulin dissociation curves: Effect of inherited enzyme defects of the red cell. *Science* **1969**, *165*, 601–602. [CrossRef]
183. van Wijk, R.; van Solinge, W.W. The energy-less red blood cell is lost: Erythrocyte enzyme abnormalities of glycolysis. *Blood* **2005**, *106*, 4034–4042. [CrossRef]
184. Bowman, H.S.; Oski, F.A. Splenic macrophage interaction with red cells in pyruvate kinase deficiency and hereditary spherocytosis. *Vox Sang.* **1970**, *19*, 168–175. [CrossRef]
185. Lakomek, M.; Winkler, H.; Pekrun, A.; Kruger, N.; Sander, M.; Huppke, P.; Schroter, W. Erythrocyte pyruvate kinase deficiency. The influence of physiologically important metabolites on the function of normal and defective enzymes. *Enzym. Protein* **1994**, *48*, 149–163. [CrossRef]
186. Oski, F.A.; Marshall, B.E.; Cohen, P.J.; Sugerman, H.J.; Miller, L.D. The role of the left-shifted or right-shifted oxygen-hemoglobin equilibrium curve. *Ann. Intern. Med.* **1971**, *74*, 44–46. [CrossRef]
187. Roy, M.K.; Cendali, F.; Ooyama, G.; Gamboni, F.; Morton, H.; D'Alessandro, A. Red Blood Cell Metabolism in Pyruvate Kinase Deficient Patients. *Front. Physiol.* **2021**, *12*, 735543. [CrossRef] [PubMed]

biomolecules

MDPI

Review

Impact of Polyphenols on Inflammatory and Oxidative Stress Factors in Diabetes Mellitus: Nutritional Antioxidants and Their Application in Improving Antidiabetic Therapy

Michal Krawczyk †, Izabela Burzynska-Pedziwiatr †, Lucyna A. Wozniak and Malgorzata Bukowiecka-Matusiak *

Department of Structural Biology, Medical University of Lodz, 90-419 Lodz, Poland; michal.s.krawczyk@gmail.com (M.K.); izabela.burzynska-pedziwiatr@umed.lodz.pl (I.B.-P.); lucyna.wozniak@umed.lodz.pl (L.A.W.)

* Correspondence: malgorzata.bukowiecka-matusiak@umed.lodz.pl

† These authors contributed equally to this work.

Abstract: Diabetes mellitus is a chronic metabolic disorder characterized by hyperglycaemia and oxidative stress. Oxidative stress plays a crucial role in the development and progression of diabetes and its complications. Nutritional antioxidants derived from dietary sources have gained significant attention due to their potential to improve antidiabetic therapy. This review will delve into the world of polyphenols, investigating their origins in plants, metabolism in the human body, and relevance to the antioxidant mechanism in the context of improving antidiabetic therapy by attenuating oxidative stress, improving insulin sensitivity, and preserving β-cell function. The potential mechanisms of, clinical evidence for, and future perspectives on nutritional antioxidants as adjuvant therapy in diabetes management are discussed.

Keywords: oxidative stress; inflammation; polyphenols; diabetes

Citation: Krawczyk, M.; Burzynska-Pedziwiatr, I.; Wozniak, L.A.; Bukowiecka-Matusiak, M. Impact of Polyphenols on Inflammatory and Oxidative Stress Factors in Diabetes Mellitus: Nutritional Antioxidants and Their Application in Improving Antidiabetic Therapy. *Biomolecules* **2023**, *13*, 1402. https://doi.org/10.3390/biom13091402

Academic Editor: David E. Stec

Received: 3 August 2023
Revised: 10 September 2023
Accepted: 11 September 2023
Published: 17 September 2023

1. Introduction

Diabetes mellitus (DM) is a global health concern with increasing prevalence. It is a chronic metabolic disorder characterized by high blood glucose levels (hyperglycaemia) due to inadequate insulin production, impaired insulin action, or a combination of these [1]. Diabetes is classified into two types: type 1 and type 2. Diabetes can appear during pregnancy, under other conditions, such as drug or chemical toxicity, genetic disorders, endocrinopathies, and insulin receptor disorders, and in association with pancreatic exocrine diseases.

Hyperglycaemia occurs in type 1 diabetes mellitus (T1DM) due to a complex disease process in which genetic and environmental factors cause an autoimmune response that has yet to be fully understood [2]. The pancreatic cells within the islets of Langerhans are destroyed during this process, resulting in individuals with this condition relying primarily on exogenous insulin administration for survival. However, a subset has significant residual C-peptide production [3]. Type 1 diabetes is usually diagnosed in children and young adults, although it can develop at any age. It accounts for about 5–10% of all diabetes cases. Indeed, several major genetic determinants of T1DM account for only 40–50% of the familial clustering of this disorder [4]. Furthermore, there is a 6% annual increase in the risk of developing T1DM in developed countries, which remains unexplained but is thought to be caused by environmental triggers [5].

Type 2 diabetes mellitus (T2DM), also known as non-insulin-dependent diabetes or adult-onset diabetes, is characterized by insulin resistance and inadequate insulin secretion. In this condition, cells become less responsive to insulin, leading to elevated blood sugar levels. Type 2 diabetes is the most common form of diabetes, accounting for approximately 90–95% of all cases. It often develops in adults but has become more prevalent in children and adolescents due to lifestyle changes and obesity [6]. The link between the obesity and DM is related to adiposopathy, meaning the dysfunction in hormonal activity of adipose tissue promotes chronic inflammation, dysregulated glucose homeostasis, and impaired adipogenesis, leading to the accumulation of ectopic fat and insulin resistance [7]. Peripheral insulin resistance and the compensatory hypersecretion of insulin from pancreatic islets may precede a decline in islet secretory function in this form of the disease. Skeletal muscle, liver, and adipose tissue have the most prominently reduced insulin sensitivity due to these sites' specific glucose uptake and metabolism requirements. However, it is increasingly thought that in most subjects, a relative decrease in insulin secretion is the final event that leads to hyperglycaemia [8]. The rise in the incidence of type 2 diabetes, particularly in developing countries, corresponds to the trend of urbanization and lifestyle changes, perhaps most notably a "Western-style" diet with resulting obesity. This suggests that environmental factors play an essential role in this disease's pathogenesis, which also has a strong genetic component. It is still unlikely that genetic factors or aging alone can explain the dramatic rise in type 2 diabetes prevalence. It is unclear how increased caloric and dietary fat intake, reduced exercise, and an associated increase in body weight lead to type 2 diabetes.

Gestational diabetes mellitus (GDM) affects 2–10% of all pregnant women, depending on the type of applied diagnostic tests and investigated population [9]. There has been rapid growth recently in the incidence of GDM, both throughout the world and in Poland, caused by the obesity epidemic, the increase in mothers' ages, and a lack of physical activity [10]. Gestational diabetes leads to numerous severe complications in both mothers and their children [11]. Considering mothers, GDM is a significant risk factor for developing type 2 diabetes later in life; it is estimated that approx. 20–30% of women with gestational diabetes develop type 2 diabetes within 10 years of having a baby. The risk of T2DM is almost 10 times higher in women diagnosed with GDM than in women whose pregnancy is proceeding correctly [12]. In turn, children of women with GDM are at higher risk of developing macrosomia, neonatal hypoglycaemia, and jaundice complications. At a later age, the consequences may relate to the onset of metabolic syndrome, obesity, and diabetes [13].

In 2014, there were 422 million patients who were affected by diabetes, nearly twice as many as in 1980 [9]. According to projections, there will be 578 million diabetic patients by 2030 and 700 million by 2045 [14]. The disease reduces both a person's quality of life and life expectancy. It imposes a significant economic burden on the health care system globally and affects the lives of patients and their families.

Oxidative stress, resulting from an imbalance between reactive oxygen species (ROS) production and antioxidant defence, contributes to the pathogenesis of diabetes and its complications [15]. Nutritional antioxidants in various dietary sources have shown promise in reducing oxidative stress and improving glycaemic control in individuals with diabetes. These antioxidants can exert their effects by neutralizing ROS, modulating intracellular signalling pathways, and enhancing the antioxidant defence system [16]. An antioxidant-rich diet has been found to decrease oxidative markers and improve insulin sensitivity in T2DM individuals [17],which also has been supported by the results obtained by van der Schaft and coworkers [18].

The enzymatic antioxidant defence system is pivotal in neutralizing ROS and maintaining cellular redox homeostasis. Critical components of this system include superoxide dismutase (SOD), catalase, and glutathione peroxidase, along with sirtuins and peroxisome proliferator-activated receptor gamma (PPAR-gamma) [19]. The study conducted by Guzik et al. found that diabetes-related vessels produce significantly more superoxide from two main sources: first, the activity and protein levels of the vascular NAD(P)H oxidase system are increased; second, in diabetic vessels, the endothelium is a net contributor to the total vascular superoxide release rather than superoxide scavenging by NO production. This increased endothelial superoxide production appears to be caused by dysfunctional endothelial NO synthase (eNOS), which is mediated by tetrahydrobiopterin (BH_4) availability. Finally, Protein kinase C (PKC) signaling appears to be involved in these changes, at least in part [20].

A diverse group of organic compounds called polyphenols is abundant in plants. They have attracted much interest because of their potential health advantages, especially as antioxidants.

Numerous plant-based food sources, such as fruits, vegetables, whole grains, legumes, nuts, and seeds, are high in polyphenols. Berries (such as blueberries, strawberries, and raspberries), citrus fruits, apples, grapes, green tea, cocoa, and spices such as turmeric and cinnamon are notable sources of polyphenols. The specific polyphenol content varies among plant sources. Therefore, it is beneficial to incorporate a diverse range of fruits, vegetables, and other plant-based foods into one's diet [21].

This review outlines the recent progress in polyphenols' application as an anti-inflammatory and antioxidant factor in diabetes mellitus with particular emphasis on molecular mechanisms.

2. Damaging Effects Mediated by Oxygen Free Radicals

The term "reactive oxygen species" (ROS) encompasses molecules such as free radicals, neutral atoms, molecules, and ions [22,23]. Free radicals are molecules with one or more unpaired electrons on the atomic or molecular orbitals, and they are natural byproducts of oxygen metabolism in all aerobic organisms. Oxygen is the precursor of radicals in aerobic organisms. The molecule in the ground state is the diradical (•O=O•) consisting of two unpaired electrons on the two anti-bonding orbitals p*2p. The oxygen molecule can be excited with the change of spin of one electron on orbital p*2p, forming a singlet oxygen molecule ($^1O_2$). This molecule is highly reactive, can directly induce lipid peroxidation, DNA (particularly guanine), and protein oxidation (particularly histidine, tyrosine, and tryptophan), and participates in the synthesis of other ROS [24]. Free radicals are highly reactive species that can induce chain reactions resulting in new free radicals' synthesis. Singlet oxygen ($^1[O_2]$), hydrogen peroxide (H_2O_2), peroxide ion (O_2^-), and hypochlorous acid (HOCl) are nonradical ROS. ROS also comprise oxygen radicals, including the superoxide anion ($O_2\bullet^-$), hydroxyl radical (HO•), hydroxide radical ($HO_2\bullet$), peroxide radical (ROO•), and alkoxyl radical (RO•) [25–27].

ROS are formed in living organisms as a result of the action of various external environmental factors, including ionizing and ultraviolet radiation, ultrasound, or cigarette smoke (exogenous sources), as well as numerous intracellular biochemical processes occurring in mitochondria, peroxisomes, microsomes, or phagocytic leukocytes (endogenous sources) [27,28] (Figure 1).

The primary sources of ROS are mitochondria in the mitochondrial electron transport chain, where ATP, as a result of oxidative phosphorylation, is produced. About 1–3% of all electrons in the transport chain are "leaking" to generate superoxide which can be further reduced to $HO^\bullet$ and H_2O_2 instead of contributing to the reduction of oxygen to water. The mitochondrial respiratory chain (RC) is a significant site of ROS production in the cell. Therefore, it has been suggested that mitochondria are the prime targets for oxidative damage [25]. Several alternative pathways also lead to their formation: NADPH-oxidase (NOX), immune reactions, xanthine oxidase, arachidonic acid metabolism, etc. [29].

ROS are poisonous and pathogenic at uncontrolled high concentrations, but at physiological concentrations, ROS are mediators and regulators of many critical cellular processes, such as proliferation, differentiation, and apoptosis [30,31]. ROS can thus play a significant physiological role as secondary signalling messengers through protein kinases, transcription factors, and proinflammatory factors' genomic expression [32]. It has been found that some growth factors can induce ROS production in non-phagocytic cells by interacting with specific receptor proteins [33].

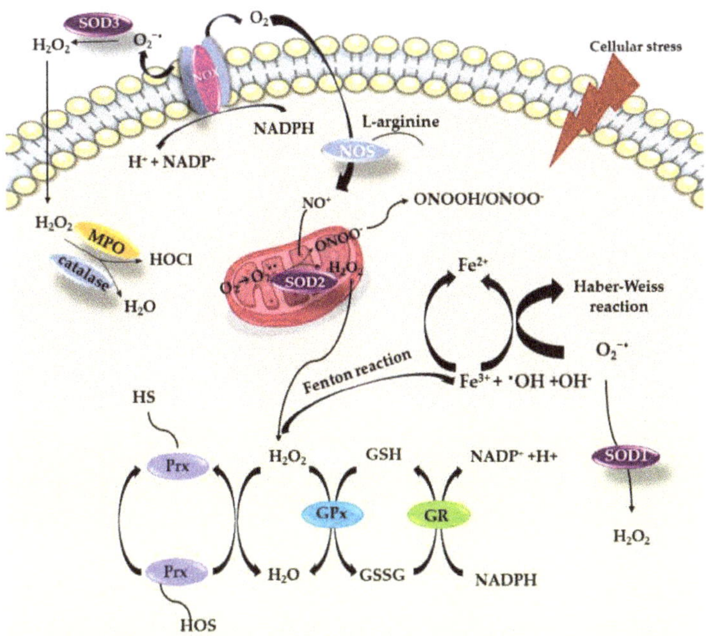

Figure 1. Sources of reactive oxygen species in the cell.

When ROS overwhelm the cellular antioxidant defence system, whether through an increase in ROS concentration or a decrease in the cellular antioxidant capacity, oxidative stress occurs. Oxidative stress results in direct or indirect ROS-mediated damage, including nucleic acids (DNA or RNA) [34], the oxidation of proteins [35], and the lipid peroxidation of polyunsaturated fatty acids (such as membrane phospholipids) [36]. These changes have been identified in carcinogenesis, neurodegeneration, atherosclerosis, diabetes, and ageing [37]. However, the involvement of ROS in the pathogenesis of diseases is not confined to macromolecular damage. There is increasing evidence that ROS signalling contributes to diseases.

2.1. Oxidative Damages of DNA

To date, more than 100 products have been identified from the oxidation of DNA. ROS reactions with DNA lead to a multitude of oxidative damage to macromolecules, which comprise (a) single- or double-strand breaks of DNA; (b) the structural modification of purine and pyrimidine bases; (c) the formation of DNA-protein binding and the formation of various types of adducts; (d) the introduction of basic sites; and (e) the cross-linking between DNA chains [38–40]. The consequences of damage to DNA may include the inhibition of replication and transcription or the presence of mutations leading to genomic instability. All these changes lead to disturbances in the normal functions of cells and the development of many diseases, including cancer and ageing [23,41].

Oxidative DNA damage encompasses mainly physicochemical changes in DNA, which can affect the interpretation and transmission of genetic information [42,43]. ROS (mainly HO•) in DNA can react with the nucleobases and deoxyribose, triggering significant oxidative reactions. This can lead to mutations, carcinogenesis, apoptosis, necrosis, and hereditary diseases [43]. DNA breaks occur due to nucleosome fragmentation (fundamental structures for the organization of DNA within chromosomes), thus initiating complications in forming DNA within chromatin. Chromatin plays an essential role in regulating gene transcription [44], and accordingly, modifications in its functional properties may result in errors leading to mutagenesis (Figure 2).

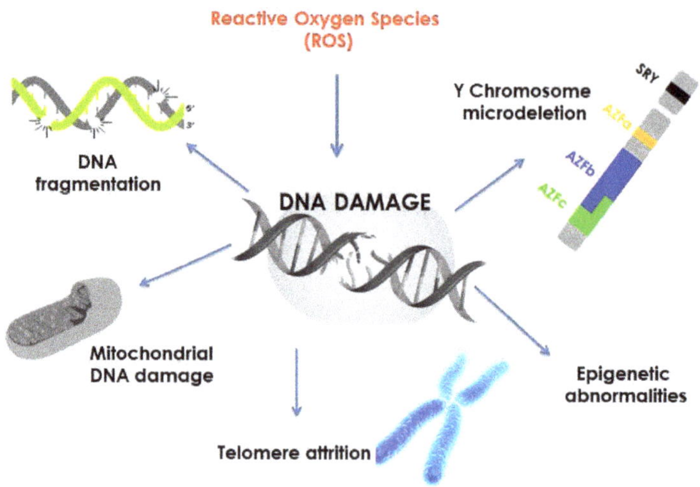

Figure 2. Types of ROS-mediated DNA damage based on Bui et al. [44].

The hydroxyl radical adds to purines, giving rise to C4-OH-, C5-OH-, and C8-OH-adduct radicals, while the addition to the pyrimidine leads to the formation of adducts at the C5-OH and C6-OH positions. 8-hydroxyguanine (8-OH-dG) has been the most extensively studied one so far. It is relatively quickly formed and is both mutagenic and carcinogenic. It is a promising biomarker of an organism's oxidative stress and a potential carcinogenesis biomarker [45]. It has been demonstrated that 8-OH-dG induces transversions of GC $\rightarrow$ TA, whose presence has been identified in several mutant oncogenes (e.g., the DNA mismatch repair protein—MSH2) and tumour suppressor genes (e.g., P53, BRCA1, APC) [46]. Other known products of oxidative DNA damage, which are also highly mutagenic, include 8-oxo-adenine, 5-hydroxyuracyl, thymine glycol, and uracil glycol [47].

Mitochondrial DNA (mtDNA) is particularly susceptible to oxidative damage since mitochondria have limited repair mechanisms through nucleotide excision. Moreover, mtDNA is not protected by histone proteins. The results of many studies have confirmed respiratory chain dysfunction in many cancer cells due to mutations in the genes encoding its enzymatic components [22,45].

DNA damage in cells leads to point mutations in particular regulatory genes that may activate oncogenes and inactivate tumour suppressor genes, thus initiating and developing cancer. However, specific and general repair mechanisms can repair DNA base modifications. In the case of extending oxidative damage to DNA that repair mechanisms cannot cope with, the cell goes into the process of programmed death called apoptosis.

Increased levels of 8-oxo-dG in the tissues of diabetic rats and the urine of patients with type 1 and type 2 diabetes were observed [48] with the levels being significantly higher in patients with albuminuria or other diabetic complications [49]. In patients with impaired fasting glucose in the prediabetic state, their levels of 8-OHdG increased compared to those of normoglycemic patients [50]. An increased concentration of 8-Oxo-dG and TG in familial combined hyperlipidaemic (FCH) subjects was detected compared to that of healthy controls. 8-oxo-dG was positively correlated with insulin and triglycerides and negatively with high-density lipoprotein cholesterol in FCH subjects [51]. According to Simone et al.'s study, the hyperglycaemia-induced redox-dependent activation of protein kinase—Akt—augmented the phosphorylation of the tuberin protein and, as a consequence, also downregulated human 8-oxoguanine-DNA glycosylase 1 (hOGG1), an enzyme engaged in the DNA base excision repair pathway (BER) [49].

Some studies have directly correlated DNA damage and HbA1c in T2DM individuals. When Xavier et al. [52] investigated peripheral blood mononuclear cells, individuals with poor glycaemic control (HbA1c > 7%) revealed higher DNA damage indicator levels compared to those of normoglycemic (HbA1c < 7%) individuals. In leukocytes from controlled T2DM patients, an elevated level of DNA damages was observed in comparison to that of healthy individuals [53,54]. However a positive correlation between antioxidant capacity and glucose or HbA1c levels in poorly controlled T2DM subjects was noted, which is consistent with the theory that their antioxidant capacity is improved to compensate for the excess of free radical species [54]. Other studies have examined blood lymphocytes obtained from T2DM patients, which exhibit increased DNA oxidative damage, a more heightened receptivity for mutagens, and insufficient DNA repair systems, further contributing to the development and progression of T2DM and an increased risk of cancer in those individuals [55].

Some literature data indicate a correlation between DNA damage and the risk of GDM development; however, the results are inconsistent. Patients with GDM or mild gestational hyperglycaemia may have higher oxidative DNA damage [56,57].

2.2. Lipid Peroxidation

Polyunsaturated fatty acids (PUFAs) present in membrane phospholipids are very susceptible to oxidation by ROS [58]. ROS, particularly hydroxyl and peroxyl radicals, can react with PUFAs and initiate lipid peroxidation chain reactions [59,60].

The first step of lipid peroxidation is an initiation reaction, which starts with the hydrogen abstraction from the methylene group of polyunsaturated fatty acids (in particular, linoleic and arachidonic acid), and, as a result, the alkyl radical ($R^•$) is formed. In the next propagation step, a radical $R^•$ reacts with oxygen to form peroxyl radicals ($ROO^•$) capable of abstracting a hydrogen atom from another polyunsaturated fatty acid and thus starting a chain reaction. As a result, fatty acid peroxides (ROOH) and another radical, $R^•$, are formed. The free radical chain reaction propagates until two free radicals conjugate each other to terminate the chain. The reaction can also terminate in the presence of a chain-breaking antioxidant such as vitamin C or vitamin E (α-tocopherol).

Lipid peroxidation products may be endoperoxides, epoxides, aldehydes, or dimers of fatty acids. Malondialdehyde (MDA) is the best-known end product of lipid peroxidation. It has been found that MDA has mutagenic properties by forming DNA adducts in bacterial and mammalian cells and is carcinogenic in rats [61]. In addition to MDA, in the process of lipid peroxidation, other α, β-unsaturated aldehydes and hydroxy aldehydes, including 4-hydroxyalkenals, 2-alkenals, hepta-2,4-dienal, 5-hydroxyoktanal, and 4-hydroxy-2-nonenal (HNE) are generated. HNE is mutagenic and highly toxic. MDA and HNE are good oxidative stress markers, and their presence has been confirmed, i.a., in cardiovascular diseases, diabetes, and Parkinson's and Alzheimer's diseases [62]. 4-HNE, which reacts with cellular proteins and DNA, giving stable adducts, is a signalling molecule that affects the regulation of several stress-sensitive transcription factors, such as Nrf2, activating protein-1 (AP-1), NF-κB, and peroxisome proliferator-activated receptors. Moreover, it is also involved in activating some stress response pathways, such as MAPK, epidermal growth factor receptor/Akt, and PKC pathways [59].

ROS-induced lipid peroxidation contributes to reactive carbonyl species (RCS) formation which plays a vital role in the development of various diseases, for example, diabetes. RCS may have cytotoxic and genotoxic properties depending on their concentration at the tissue or the systemic level. The augmented concentration of 4-HNE in serum, plasma, blood, urine, cells, and tissues may be linked to diabetic complications [63]. Many studies on type 1 and type 2 diabetes from various tissues for 4-HNE levels confirmed this concept [64,65]. In type 1 diabetic mice, about a 2.5-fold higher level of 4-HNE was detected compared to that of the controls [65].

A higher level of 4-HNE-modified proteins in the pancreatic beta cells of aged type 2 diabetic rats was observed compared to that of the controls, correlated with an increase in the fibrosis of pancreatic islets [66]. Toyokuni et al. revealed significantly higher levels of 4-HNE-modified albumin in the serum of type 2 diabetes patients [64].

MDA may be a promising biomarker in T2DM. Uncontrolled T2DM patients with an HbA1c level >7% exhibited significantly higher MDA levels than the control group. However, there was no statistically significant difference between T2DM-controlled patients (HbA1c level < 7%) and the control group. Moreover, a significant positive correlation between the MDA concentration and lipid parameters (except HDL-C) was observed, which may indicate the coexistence of atherogenic risk factors and oxidative stress [67]. Furthermore, the MDA levels were elevated compared to those of normoglycemic controls in T2DM patients without complications and with retinopathy [68]. Desco et al. also reported higher levels of plasma MDA in type 1 DM patients [69].

2.3. Oxidations of Proteins

The mediators of oxidative damage to proteins are $HO^{\bullet}$ and, to a lesser extent, $O^{\bullet}{}_2{}^-$ and H_2O_2. The reaction of $HO^{\bullet}$ with the protein backbone starts with the abstraction of the hydrogen atom from the α carbon atom of an amino acid, which leads to the creation of an alkyl radical which then reacts with oxygen to form a radical alkyl peroxide [70]. This product is then transformed into alkylperoxide. In the next step, alkylperoxide is converted into an alkoxyl radical which may be converted into a hydroxyl derivative of an amino acid or lead to the breaking of the polypeptide chain. In the case of oxygen deficiency, the formation of radical alkylperoxide is demanding, and alkyl radicals react with each other, forming cross-links between polypeptide chains.

Thiol groups of cysteine and methionine are particularly susceptible to ROS attacks resulting in the formation of both intramolecular (P1-SS-P1) and intermolecular (P1-SS-P2) disulfide bridges and oxidation products containing sulfenic group sulfinate or sulfonate, metionylosulfoxides, or metionylosulfonyl [71].

The aromatic amino acid residues are also highly susceptible to ROS attacks. It has been found that the oxidation of tyrosine residues formed 3,4- dihydroxyphenylalanine or 2,5-tyrosine and the oxidation of tryptophan residues formed formylkynurinine and kynurenine [72]. The markers of oxidative damage to proteins are carbonyl derivatives formed by the oxidation of amino acids containing free amino-, amido- or hydroxyl- groups by ROS. Carbonyl derivatives may react with the free amino groups of lysine residues, leading to cross-link formation in the same or another protein molecule. Modified and improperly folded protein chains are directed to the proteolytic pathways and eliminated through proteasome ubiquitination [72].

Advanced glycation end products (AGEs) are diverse compounds that may be formed in non-enzymatic reactions between glucose or other saccharides and proteins or lipids. Environmental factors, including cigarette smoke, high levels of refined and simple carbohydrates, hypercaloric diets, foods cooked at a high temperature, and a sedentary lifestyle, can initiate the production of AGEs and consequently damage cell lipids and proteins [67,73]. ROS may also catalyse the formation of AGEs and advance lipooxidation end products (ALEs). AGEs are synthesized in the non-enzymatic Maillard reaction of nonreducing sugars and amino groups in proteins, lipids, and nucleic acids, resulting in nonstable Schiff bases that undergo the Amadori rearrangement, leading to the formation of stable ketamine. These compounds may be transformed into more stable products (with peptides or proteins to form protein cross-links) in a reversible reaction. An alternative to AGEs is the polyol pathway in which the Heyns product is created, followed by Heyns rearrangement [74]. The AGEs of these pathways belong to the following groups: fructose-derived AGEs (Fru-AGE) and glucose-derived AGEs (Glu-AGE) (Figure 3)[75].

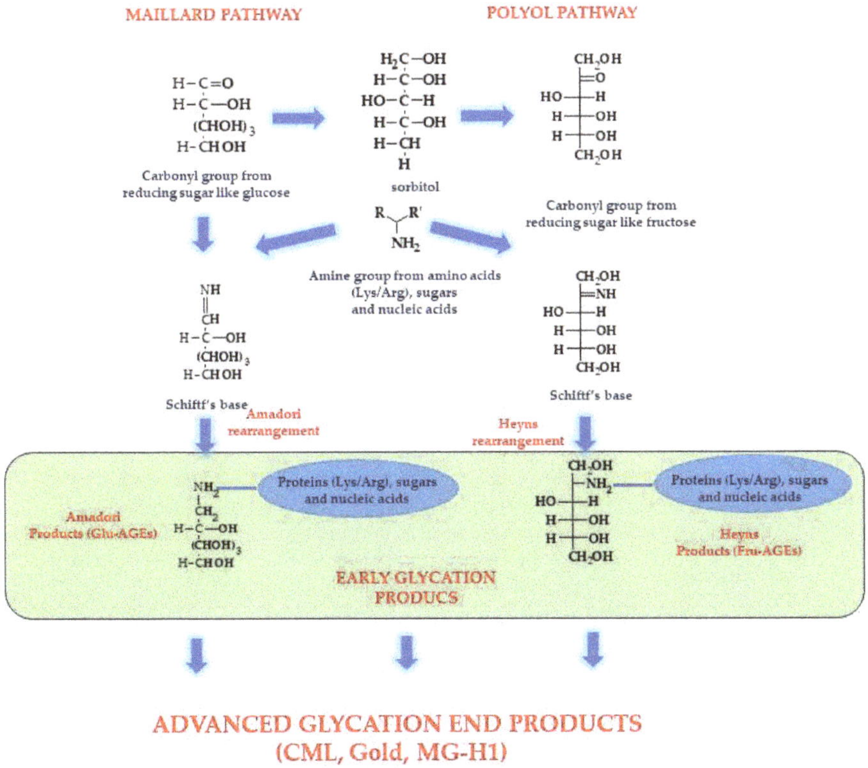

Figure 3. Pathways leading to advanced glycation end products (AGEs). Glu-AGE—glucose-derived AGE; Fru-AGE—fructose-derived AGE.

These compounds may also participate in oxidation, dehydration, or polymerization reactions to result in numerous other AGEs [76,77].

The oxidation of proteins is associated with the loss of biological activity and an increased susceptibility to aggregation and proteolysis. AGEs are involved in cell toxicity. AGEs can play a central role in the pathophysiology of different diseases, for example, diabetes. The AGE-induced pathophysiology of diabetes mellitus encompasses two mechanisms. AGEs may act directly by trapping and cross-linking proteins or indirectly by binding to the cell surface receptor.

AGEs can interact and modulate cell signalling through binding to various receptors, for example, Toll-like receptors, scavenger receptors, G-protein-coupled receptors, and pattern recognition receptors (RAGEs). The latter are specific receptors for AGEs, and, in homeostasis, they are expressed at basal levels.

The accumulation of a high amount of sugar compounds may contribute to increased levels of AGEs called glycotoxins and their interactions with human amino acids, peptides, and proteins. There is a hypothesis that carbohydrates from food can affect internal glucose levels and impair glucose pathways in tissues, which, consequently, can strongly and negatively influence human health [78]. Under pathological conditions or chronic inflammation, such as diabetes mellitus (DM), cardiovascular disease, Alzheimer's disease, cancer, and natural ageing, their expression is elevated. It has been reported that in diabetes, both ROS and AGE concentrations are increased [79]. Elevated concentrations of AGEs in the bloodstream, a consequence of hyperglycaemia in diabetes, may induce a signalling cascade through RAGE. This activation involves a variety of downstream effectors, including mitogen-activated protein kinase (MAPK), p38, stress-activated protein kinase/c-Jun N-terminal kinase (SAPK/JNK), Ras-mediated extracellular signal-regulated kinase (ERK1/2), and Janus kinase signal transducer and activator of transcription (JAK/STAT) pathway which will sequentially lead to sustained activation transcription factors, such as NF-κB, STAT3, HIF-1α, and AP-1 [74,80,81]. Mounting evidence suggests that AGEs/RAGE-induced signalling pathways encompassing NF-κB activation, inflammation, and ROS formation are directly linked to the pathogenesis of insulin resistance by increased IRS-1 serine phosphorylation and degradation, therefore blocking the insulin signalling pathway [82]. Hyperglycaemia resulting from elevated AGE levels in the pancreas may have a toxic effect on beta cells by triggering inflammatory cascades and oxidative stress [83,84]. This effect can be observed as upregulated RAGE expression in pancreatic islets [85]. It also confirmed that AGEs may be suitable biomarkers in gestational diabetes mellitus (GDM) by increasing the concentrations of TNF-α (tumour necrosis factor-alpha) and hs-CRP (high-sensitivity C-reactive protein), which are responsible for inflammatory reactions [86].

3. The Role of Oxidative Stress and Inflammation in Aetiology of Diabetes Mellitus

Pancreatic β-cell dysfunction and insulin resistance in hepatocytes, myocytes, and adipocytes significantly contribute to T2DM pathogenesis. The cellular and molecular mechanisms underlying these abnormalities are not fully understood. The imbalance between the production and elimination of reactive oxygen species (ROS) causes oxidative stress (OS), which stems from chronic low-grade inflammation and chronic hyperglycaemia resulting from insulin resistance, leading to abnormal cytokine overproduction and the activation of inflammatory signalling pathways in the above-mentioned insulin-sensitive tissues [87]. Hyperglycaemia linked to insulin resistance is the major pathogenetic factor of T2DM. Numerous pre-clinical and clinical studies investigate the influence of increased glycaemia on redox homeostasis and inflammatory responses. OS plays a pivotal role in the pathophysiology of diabetic complications related to lipid peroxidation, DNA damage, and mitochondrial dysfunction. Its involvement is visible in many other pathological conditions and age-related disorders. Ageing-related disorders are defined as the progressive loss of function in tissues subjected to OS and many other mechanisms [88]. The oxidative stress theory by many researchers is considered the primary background of ageing and age-related complications. Thus, maintaining the proper balance of the redox state is essential in oxidative stress-induced complications and insulin resistance prevention [89].

Some of the mechanisms described below underlying the development of T2DM and its complications in correlation to OS are presented in Figure 4. This chapter provides evidence on how increased glucose and insulin resistance affect ROS overproduction and the antioxidant defence system, leading to OS and immune system activation in diabetic individuals (Figure 4).

Prediabetes, an elevated blood glucose level not reaching the diabetes criteria (the range from 100 to 125 g/dL (5.6 mmol/L to 6.9 mmol/L) according to ADA criteria) [1], can predispose patients to T2DM due to hyperglycaemia-related OS, underlying ROS overproduction, and increased inflammation markers. Prolonged inflammation and OS overload can lead to impaired insulin secretion, insulin resistance, and further T2DM development [90]. Some OS- and inflammation-related changes are already present in the prediabetes phase; ongoing and progressing prediabetes-related hyperglycaemia enhances these alterations. Evolving OS leads to muscle insulin resistance, impaired pancreatic cell function, and insulin secretion. Some studies have revealed that using nicotinamide adenine dinucleotide phosphate (NADPH) oxidase inhibitors (Apocynin) causes the reduction of OS in a prediabetes state [91]. It has been shown that insulin resistance is associated with OS in non-diabetic individuals and those with increased risk factors for DM development, such as obesity or impaired fasting glucose. The involvement also of the gastrointestinal tract in glucose homeostasis is nowadays emphasised as an important factor in DM development and progression [92] which is why many ongoing studies investigate the influence of a polyphenol-rich diet and the incorporation of polyphenols into the gastrointestinal tract in other forms as well as their beneficial outcomes on DM management and treatment in the context of OS [93,94]. The Framingham Offspring Study on the association of oxidative stress, insulin resistance, and diabetes risk phenotypes revealed a positive correlation between insulin resistance and urinary 8-epi-prostaglandin F2α (8-epi-PGF2α),- an OS marker [95].

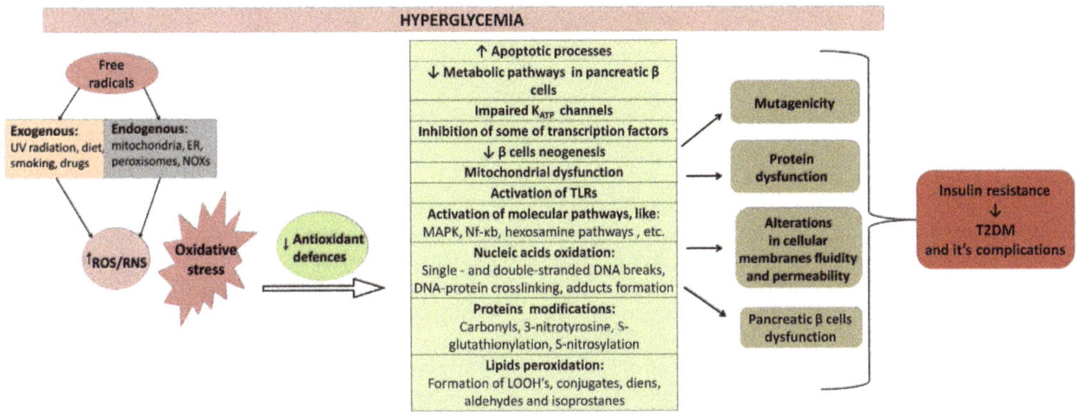

Figure 4. Molecular mechanisms correlating oxidative stress, its outcomes, and hyperglycaemia, which lead to insulin resistance and T2DM development. Abbreviations: TLRs—Toll-like receptors; Nf-κb—nuclear factor kappa b; MAPK—mitogen-activated kinases. Legend: ↓—decrease, ↑—increase.

Lipid peroxidation, DNA damage, and mitochondrial dysfunction are some of the examples of processes by which the OS executes a pivotal role in the pathophysiology of various complications of diabetes. Several experimental studies have proven that increased mitochondrial ROS production is associated with high glucose levels in leukocytes, endothelial cells, and adipocytes. Pancreatic β-cells are prone to glucose toxicity due to chronic OS in DM caused by the deficient expression of antioxidant enzymes compared to other tissues [96]. It has been proposed that excessive ROS production may cause the impairment of β-cells by the decreased expression and DNA-binding activity of some tissue-specific transcriptional factors, precisely pancreas duodenum homeobox-1 and V-maf musculoaponeurotic fibrosarcoma oncogene homolog A. This phenomenon causes reduced insulin gene expression, content, and secretion [97].

Acharya et al. [98] revealed the correlation between hyperglycaemia, oxidative stress, and β-cell dysfunction in newly diagnosed diabetic individuals when comparing the OS parameters at their time of diagnosis and eight weeks after antihyperglycemic treatment, thus finding a significantly increased OS at at the point of diagnosis with T2DM. Irrespective of the type of hypoglycaemic treatment used, lowered oxidative stress and improved β-cell function were observed.

Numerous ongoing and completed clinical trials have revealed the importance of anti-inflammatory constituents and their contribution to glucose homeostasis. A study investigating the glucose-lowering effect of Salsalate (a pro-drug of salicylate) demonstrated its effectiveness in lowering the HbA1c and fasting glucose levels of T2DM individuals [99].There is a growing body of evidence showing that low-grade chronic inflammation is one of the multiple factors underlying the pathophysiology of insulin resistance, DM, and its complications; furthermore, it is also clear that these processes might affect the insulin signalling transduction and beta cell function [61,100]. Janus kinase pathways (JNKs) induced by some cytokines further stimulate IRS-1 serine phosphorylation causing abnormalities in insulin signalling pathways [101]. Other inflammatory mediators, such as TNF-α and Nf-κb, have been reported to modulate IRS-1 serine phosphorylation and further affect the proper insulin signalling transduction.

In patients with T2DM, the increased activity of monocytes and macrophages and increased levels of proinflammatory mediators, such as CX3CL1 (fractalkine), CRP, TNF-α, IL-6 (interleukin-6), IL-1β, IL-18, MCP-1 (monocyte chemoattractant protein-1), resistin, PAI-1 (plasminogen activator inhibitor-1), E-selectin, and IFN-γ (interferon-gamma), have been reported, which indicates the potential benefits of incorporating agents of anti-inflammatory activity into therapeutic approaches in diabetes management. Additionally, Toll-like receptors (TLR) 2 and 4 were found to be suppressed in peripheral mononuclear cells in T2DM patients [102]. Guerrero-Romero et al. revealed a strong association between hyperglycaemia/poor glycaemic control and increased CRP levels in prediabetic and diabetic patients [103]. Besides the increased CRP levels, IL-6 was also found to be elevated in diabetic individuals [104]. However, it has been proven that an improvement in glycaemic control is accompanied by a reduction in the above-mentioned inflammatory markers and the TBARS (thiobarbituric acid-reactive substances) concentration—an OS marker [105].

The monocyte cell subset is essential among circulating immune cells considering diabetes' pathophysiology due to the increased levels of monocyte proinflammatory cytokines measured in T2DM individuals [106]. Thus, human monocytes have become the subject of many in vitro studies, revealing the upregulation of critical inflammatory genes such as cytokines—TNF-α and IL-1β, their receptors, and chemokines—MCP-1 in high-glucose conditions of culture [107]. The NF-κB regulates many hyperglycaemia-induced genes; thus, this transcription factor's importance in regulating high-glucose-induced immune responses has been broadly investigated. The CREB-binding protein (CBP) and p300, both displaying intrinsic HAT (histone acetyltransferase) activity, are coactivators for p65, the NF-κB component. The high-glucose-induced hyperacetylation of p65 and the inhibition of HDACs (histone deacetylases) have been reported to be stimulated by the CBP/p300, resulting in NF-κB activation and the further increased transcription of IL-6 and TNF-α in monocytes [108].The above-mentioned study by Yun et al. revealed curcumin's ability to decrease monocyte hyperglycaemia-induced cytokine production, probably via epigenetic changes in NF-κB, proving the direct action of curcumin on the inflammatory response. Since then, curcumin, a polyphenolic compound, has become a valuable subject in antidiabetic approaches.

Gonzalez et al. [109] reported the decreased monocyte CD33 expression and increased TNF-α production by human monocytes under hyperglycaemic conditions, thus making CD33—a member of the family of sialic acid-binding, immunoglobulin-like lectins—a subject of many studies related to the hyperglycaemia-induced inflammatory response. Gonzalez et al., using α-tocopherol, reported the reversal of the above-mentioned alternations, suggesting that hyperglycaemia-induced oxidative stress might cause CD33 downregulation. Oxidative stress and inflammation are two significant abnormalities underlying the development and progression of T2DM. They are promoted by chronic hyperglycaemia, a key factor for DM, thus indicating the reciprocal nature of their correlation. All of the above-mentioned findings strongly suggest that antioxidative compounds can effectively improve disturbed insulin secretion and action, thus positively influencing human glucose and oxidative homeostasis.

Enzymatic Antioxidant Defence System

Substances that can delay, prevent, or remove OS-related damages are classified as antioxidants and are categorized into two major groups: enzymatic and non-enzymatic antioxidants. Enzymatic antioxidants contain the enzymes that play a crucial role as the first line of defence against ROS, including superoxide dismutase (SOD; EC.1.15.1.1), catalase (CAT; EC 1.11.1.6), and glutathione peroxidase (GPx; EC 1.11.1.19). The nuclear erythroid 2-related factor-2 (Nrf2) is a transcription factor that regulates the expression of all the above-mentioned enzymes. Thus, Nrf2 maintains cellular redox balance by modulating the antioxidant response. In animal model studies, the beneficial effects of Nrf2 action on β-cell dysfunction and insulin resistance have been revealed, as well as on the development of diabetes-related micro- and macrovascular complications [110].

SOD is an enzyme that catalyses superoxide radicals' dismutation into hydrogen peroxide and molecular oxygen. It exists in different isoforms, including copper-zinc SOD (SOD1) which is present in the cytosol, manganese SOD (SOD2) which is found in the mitochondrial matrix, and extracellular SOD (SOD3). These isoforms are distributed in different cellular compartments and exhibit distinct regulatory mechanisms. SODs are critical in scavenging superoxide radicals and preventing their conversion into more reactive species [111].

Catalase is a heme-containing enzyme in peroxisomes that facilitates the decomposition of hydrogen peroxide into water and molecular oxygen. Catalase efficiently detoxifies hydrogen peroxide, thereby protecting cells from oxidative damage. Its activity is tightly regulated through transcriptional, post-transcriptional, and post-translational mechanisms. It is present in most cells, tissues, and organs and, at higher levels, in the liver and erythrocytes [112].

Glutathione peroxidases are a family of selenium-dependent enzymes that catalyse the reduction of hydrogen peroxide and lipid hydroperoxides using glutathione as a cofactor. These enzymes play a crucial role in maintaining the cellular redox balance and protecting against oxidative damage. Different isoforms of glutathione peroxidase are expressed in various tissues and exhibit distinct substrate specificities. Free thiol groups are oxidized to disulphide bonds during the process shown below, and reduced glutathione serves as an efficient electron donor [30] (Figure 5).

$$H_2O_2 + 2GSH \longrightarrow GS\text{-}SG + 2H_2O$$

Figure 5. General mechanism for obtaining glutathione disulphide. GSH—glutathione; GS-SG—glutathione disulphide.

Glutathione is a tripeptide (γ-L-glutamyl-L-cysteinylglycine) present in millimolar concentrations in cells. It acts as a major intracellular antioxidant and participates in redox reactions by working as a reducing agent and a cofactor for enzymes such as glutathione peroxidase. Glutathione levels are tightly regulated, and alterations in its concentration have been associated with oxidative-stress-related disorders [112].

4. Polyphenols as Nutritional Antioxidants

Like other secondary metabolites, polyphenols serve as plants' first line of defence. They are possibly the most significant non-nutrient bioactive groups in the human diet because plants constitute a primary nutritional component in human diets, and phenolics are more numerous than other phytochemicals in plant diets [113]. Among the several possible health advantages of dietary polyphenols, their capacity to inhibit oxidation stands out. They have consistently been demonstrated as powerful antioxidants, preventing oxidative damage and reducing inflammation. A diet high in phenolic-rich fruits, vegetables, and whole grains lowers one's risk of cancer, cardiovascular disease, chronic inflammation, and metabolic disorders [21]. The mechanisms of phenolic compounds' antioxidant and anti-inflammatory properties are linked with their capacity to scavenge free radicals, restore antioxidant enzyme activity, and regulate cytokine-induced inflammation.

In vitro, dietary phenolics are potent antioxidants, capable of neutralizing free radicals by donating an electron or a hydrogen atom to a wide range of reactive oxygen, nitrogen, and chlorine species, $^{\bullet}$, including $O_2^{\bullet -}$, OH peroxyl radicals $RO_2^{\bullet}$, hypochlorous acid (HOCl), and peroxynitrous acid (ONOOH) [114]. As efficient radical scavengers, polyphenols disrupt the propagation stage of the lipid autoxidation chain reactions or function as metal chelators to convert hydroperoxides or metal prooxidants into stable molecules. As metal chelators, phenolic compounds can directly block Fe^{3+}'s reduction by lowering the formation of reactive OH in the Fenton reaction [115]. Although phenolic acids and flavonoids have excellent radical scavenging action, their metal chelating potential and reducing power might differ depending on their structural characteristics.

Based on their chemical structure, the plant-derived antioxidants are classified as carotenoids, phenolics, alkaloids, nitrogen-containing compounds, and organosulfur compounds, along with the sterols, terpenes, fibre, and organoselenium compounds which are also distinguished in such a classification [113]. Flavonoids (such as flavonols, flavanols, flavones, and anthocyanins), phenolic acids (ellagic acid, caffeic acid), stilbenes (resveratrol), and lignans are just a few of the numerous substances that fall under the category of polyphenols. Every type of polyphenol has distinct antioxidant qualities and health advantages [116]. Synergistic interactions between polyphenols frequently result in combined effects that are more significant than the sum of their individual effects. These findings suggest that eating foods high in polyphenols may offer better antioxidant protection than taking supplements of isolated polyphenols [117].

Polyphenols have anti-inflammatory effects in addition to antioxidant ones. By lowering the production of proinflammatory molecules and fostering a more balanced immune response, they can help modulate inflammatory pathways. Polyphenols may aid in the prevention and treatment of inflammatory diseases by reducing chronic inflammation [118].

Additionally, polyphenols can affect the cellular signalling networks that control inflammation and oxidative stress. Specific enzymes, such as Nrf2 (nuclear factor erythroid 2-related factor 2), also govern antioxidant defence. Polyphenols can activate these enzymes, increasing the body's natural antioxidant capacity by modulating these pathways [119]. In addition to neutralizing free radicals in the aqueous environment, polyphenols shield lipids from oxidation. Lipid peroxidation can produce dangerous byproducts that harm cell membranes and exacerbate several diseases. The chain reaction of lipid peroxidation can be stopped by polyphenols, maintaining the structure and functionality of cellular membranes [120].

It is of high importance to also mention that numerous researchers are now making efforts to develop potentially beneficial and applicable synthetic derivatives of naturally occurring polyphenolic compounds that could be used as subjects for antidiabetic approaches. Compounds such as iminosugars and other sugar derivatives are some examples of subjects of ongoing studies and are being considered as promising for their potential applications as antidiabetic agents obtained by the chemical modification of plant-derived stem compounds [121].

5. Polyphenols Present in the Diet and Incidence of Various Diseases

Some studies indicate that the presence of polyphenols in one's diet can reduce the risk of developing diseases, such as cardiovascular diseases, some types of cancer, and neurodegenerative diseases [122,123]. In a randomized double-blind placebo-controlled trial study, after supplementation with polyphenolic extract (Hibiscus sabdariffa and Lippia citriodora) for over 84 days, type 1 hypertensive patients of both sexes exhibited a decrease in daytime systolic blood pressure (SBP) in comparison to a placebo group. However, their diastolic blood pressure remained at similar values as the baseline. The mechanism by which the polyphenol extracts decreased their blood pressure may be related to the modulation of different metabolic pathways and the activation of the AMPK pathway favouring lipolysis and, therefore, fat loss. Polyphenols present in HS and LC extracts may possess vasodilatory properties and the capacity to inhibit low-density lipoprotein oxidation as well as decrease the atherosclerotic process [124,125].

Xiao et al. demonstrated that green tea polyphenols alleviated the disorganized arterial wall and the increased intima-media thickness induced by a high-fat diet (HFD). The effect was also observed in the rats group fed with a standard diet. These results indicated that excess fat intakes could induce early vascular ageing (EVA) in young rats, and GTP could reverse such early vascular damage in HFD rats [126].

Talarid et al. revealed the hypertension- and glycaemia-lowering activity of polyphenols in grape-pomace-derived seasoning in subjects with a high level of cardiovascular risk. These results indicate that the reduction rate after the nutritional intervention (2 g of seasoning per day for 6 weeks) was relatively moderate. Grape pomace could also help in the control of cardiometabolic risk factors and metabolic syndrome (MetS), especially at the initial stages. [127].

Polyphenols regularly consumed in the diet may also exert anticancer properties by decreasing the growth and development of various types of cancer. Ávila-Gálvez et al. have attempted the metabolic profiling of isoflavones, lignans, and curcuminoids for the first time in breast tissues from breast cancer (BC) patients. Their results demonstrated that a mixture of polyphenolic metabolites reached the mammary tissues (MTs) from BC patients and suggest that some metabolites, especially the free curcumin occurring in MT, might exert anticancer activity after long-term exposure. The obtained results indicated that the consumption of high amounts of polyphenols could improve the glucuronidation reaction in terms of its saturation, which could allow some dietary polyphenols to reach systemic tissues in their free, much more bioactive form [128].

Polyphenols are often used to support the treatment of chronic non-communicable diseases such as diabetes and neurodegenerative diseases whose pathophysiology involves oxidative stress (OS) and inflammation [129]. Resveratrol—a polyphenol belonging to stilbenes—has proven antioxidative and anti-inflammatory activities that are attributed to its ability to activate sirtuin 1 (SIRT1), and, consequently, this activation stimulates AMP-dependent protein kinase (AMPK), which manages to improve biogenesis and mitochondrial function, increase insulin sensitivity, attenuate oxidative damage, and regulate metabolic homeostasis. In some studies on animal models with T2DM, resveratrol exerted antioxidant, anti-inflammatory, and even hypoglycaemic effects [130]. In this context, García-Martínez et al. demonstrated in a randomized clinical trial on adults with T2D that supplementation with resveratrol (with two doses: 1000 mg/day and 500 mg/day) resulted in an increase in the total antioxidant capacity, antioxidant gap, percentage of subjects without oxidant stress, and sirtuin 1 levels in comparison to those of the controls. Moreover, the observed antioxidant effect of resveratrol was more noticeable for the 1000 mg dose [131].

In the context of T2DM, Grabežet al. conducted a prospective, randomized, double-blind, placebo-controlled trial of the effect of polyphenols in pomegranate peel extract (PoPEx) on the outcomes of inflammatory factors and oxidative stress in patients with type 2 diabetes mellitus. Patients were randomly assigned to one of two groups: the first (n = 30) received PoPExin 250 mg capsules twice a day, while the second (n = 30) received placebo capsules twice a day. The plasma levels of inflammatory factors (IL-6, TNF-, and high-sensitivity C reactive protein (hsCRP)), oxidative stress biomarkers (thiobarbituric acid reactive substances (TBARS)), nitrites (NO_2^-), the superoxide anion radical ($O_2{\bullet}^-$), hydrogen peroxide (H_2O_2), the total antioxidant capacity (TAC), homocysteine, and the lipid profile were measured. The PoPEx therapy reduced inflammatory factors (IL-6, TNF-, hsCRP), oxidative stress biomarkers (TBARS, NO_2^-, $O_2{\bullet}^-$), and homocysteine, while increasing TAC. Furthermore, the PoPEx group showed a significant improvement in lipid profile. In the PoPEx group, there was a significant opposite relationship between the decreases in all the determined inflammatory markers and TAC [132].

Polyphenols belonging to various groups have also demonstrated neuroprotective and anti-ageing properties. Their activity in the brain encompasses the prevention of neuronal fatty acids' oxidation, a reduction in the damage caused by reactive oxygen and nitrogen species, and an improvement in neurocognition by facilitating de novo protein synthesis in key sites and neurogenesis in the dentate gyrus [133]. For example, Vina et al. found that patients with Alzheimer's disease supplemented with genistein at a dose of 120 mg/day for 12 months exhibited a significant improvement in two of the tests used (dichotomized direct TAVEC and dichotomized delayed Centil REY copy). In genistein-treated patients, the amyloid-beta deposition analysis revealed that they did not increase their uptake in the anterior cingulate gyrus after treatment as opposed to the placebo-treated group for which an increase was noted [134]. This study demonstrated that genistein may have a therapeutical role in and may have delayed the onset of Alzheimer's dementia in patients with prodromal Alzheimer's disease. In another study, Boespflug et al. showed that in older adults with cognitive impairment, blueberry powder (with an anthocyanin concentration near 15 mg) supplementation improved their neuronal responses, observed by magnetic resonance imaging (MRI) during memory challenges [135].

6. Metabolism of Polyphenols:

The bioavailability of polyphenols can be influenced by multiple factors such as the food matrix, processing methods, and individual variations in metabolism. Some polyphenols may have low bioavailability and thus a significantly low absorption or rapid metabolism. It needs to be emphasized that polyphenols present poor absorption in the small intestine, and their bioactivity in living organisms is also generally strictly correlated with the activity of the microbiota [136]. Nevertheless, even if their systemic concentration is relatively low, polyphenols can still exert local effects on the gastrointestinal tract, contributing to gut health and microbiota [137]. Polyphenols undergo various biotransformation processes, including phase I and phase II reactions. Phase I reactions involve the conversion of polyphenols by enzymes such as cytochrome P450, resulting in the formation of intermediate metabolites. These reactions include hydroxylation, demethylation, dehydroxylation, and oxidation. The resulting products are intermediate metabolites, which can exhibit different biological activities compared to the parent compounds. Phase II reactions include conjugation with endogenous compounds, such as glucuronic acid, sulphate, or methyl groups, leading to water-soluble metabolites that can be readily excreted. The gut microbiota also plays a crucial role in the metabolism of polyphenols, generating metabolites with potential biological activities. Polyphenols that escape metabolism in the upper gastrointestinal tract reach the large intestine, where they undergo extensive microbial metabolism by the gut microbiota. The gut microbiota can enzymatically convert polyphenols into smaller phenolic acids and other metabolites. These microbial metabolites can have unique biological activities and contribute to the overall health benefits associated with polyphenol consumption [138].

The daily consumption of polyphenols has been estimated to be 900 mg [139]. This level of intake is ten times that of vitamin C and one hundred times that of vitamin E or b-carotene [137]. Several factors, however, make an accurate calculation of polyphenol intake hard: (1) Polyphenols have an extensive variety of chemical structures. (2) Polyphenols are typically found in a wide range of foods. Some polyphenols, such as quercetin, can be found in a number of foods, whereas others, such as flavanones in citrus, are often specific to a particular species of plant or plant family. (3) The amount of each in a specific food can vary greatly as a result of a variety of factors. (4) There are no standard procedures for determining polyphenols in foods or methods of quantification. Catechins and proanthocyanidins are the most abundant flavonoids in the human diet, accounting for roughly three-quarters of the total flavonoids consumed. Phenolic acids make up a large portion of the polyphenols consumed by coffee drinkers. Estimates of total polyphenol intake are clearly affected by the number of polyphenol classes or subclasses included in the survey, particularly the inclusion of proanthocyanidins and phenolic acids. Few authors have included these two final polyphenol groups.

Several studies have investigated polyphenol intake in different populations by using 24-hour recalls or semi-quantitative food frequency questionnaires and polyphenols databases [140–143]. For example, the average daily intake of polyphenols in the Spanish diet is around 3000 mg/person/day [140], although the PREDIMED research of 7000 Spanish people found their total phenols intake to be 820 $\pm$ 323 mg per day, with flavonoids and phenolic acids being the most abundant components [142]. In France, the total polyphenol intake in 4000 adults is estimated at 1193 $\pm$ 510 mg/d [141] and in Poland, it is estimated at 1756.5 $\pm$ 695.8 mg/d [143].

Despite an increasing body of evidence for polyphenol bioavailability in the systemic circulation [144], little is known about their ability to reach the central nervous system (CNS). Animal studies show that polyphenols can cross the blood–brain barrier (BBB) and colocalize in tissues of the brain despite the route of administration. For example, naringenin was discovered in the brain after the injection [145], whereas epigallocatechin gallate [146], epicatechin [147], and anthocyanins [148,149] were identified after swallowing. Schaffer et al. demonstrated that polyphenols usually localize in the brain at levels below 1 nmol/g tissue [150]. Polyphenols and their degraded metabolites can also aid in the treatment of T2D. These components converted into secondary metabolites by certain gut microbiota have higher absorption and bioactivity than their precursors [151]. The flavonoid metabolites 3,4-dihydroxyphenylacetic acid and 3-hydroxyphenylpropionic acid which are derived from microbes may have anti-diabetic properties through encouraging the growth and functioning of pancreatic cells [152]. Furthermore, polyphenol microbial metabolites could control bile acid production, which impacts the metabolism of the host.

It is worth mentioning that the cellular environment and/or concentration of polyphenols is a factor with a diverse influence on the form of the displayed activity of polyphenols, either antioxidants or pro-oxidants [153]. Due to their pro-oxidant properties, polyphenols can have toxicological effects when consumed in high concentrations [154,155]. Their potential detrimental effects on health can be attributed to their capacity to activate transcription factors, such as NF-B, activating protein-1 (AP-1), tumour protein (p53), and Nrf2, as well as cause cell damage through the collapse of the mitochondrial membrane potential, favouring the propagation of reactive species and subsequently the inflammatory process [156].

In this regard, Moridani and colleagues evaluated the toxicity of flavonoids in isolated rat hepatocytes and in HeLa tumour cells. They discovered that epicatechin had the lowest toxicity (LD50 = 17,000 μmol) and that galangin had the highest toxicity after 2 h, with lipophilicity being the primary determinant of the cytotoxicity of the evaluated polyphenols [157]. According to a study by Chen et al. [158], genistein consumption at high concentrations (200 μmol) increased the production of reactive oxygen species (ROS) in a way that was dependent on 5-lipoxygenase, an essential enzyme in the biosynthesis of leukotrienes, which in turn plays a crucial role in the inflammatory response. Furthermore, high dietary doses of green tea polyphenols (0.5–10%) also worsen colitis and colon carcinogenesis in ICR mice and cause nephrotoxicity, hepatotoxicity, and the negative regulation of the expression of antioxidant enzymes and molecular chaperones [159]. Since phenolic compounds can adversely affect health depending on the concentration used, careful consideration must be given to their intake [156].

6.1. Mechanisms of Antidiabetic Activity of Phenolics

The inhibition of carbohydrate digestion by salivary and pancreatic α-amylase and α-glucosidase inhibition in the small intestine, glucose absorption inhibition, and insulin secretion stimulation and protective activity towards pancreatic β cells against glucotoxicity damage are indicated as potential mechanisms of the hypoglycaemic action of polyphenolic compounds. The effects of the suppression of gluconeogenesis in the liver and increase in glucose uptake in peripheral tissues on the modulation of intracellular signalling pathways are other proposed mechanisms of polyphenols' activity [160]. De Paulo Farias et al. [161] presented in their review the antidiabetic effects of dietary phenolic compounds and their mechanisms of action. It was discovered that these compounds have a high therapeutic potential for diabetes management, and their mode of action includes several mechanisms related to oxidative stress reduction, the inhibition of DPP-IV and enzymes involved in carbohydrate metabolism, a reduction in insulin resistance, and AGE formation, among others. Nevertheless, it is necessary to note that the anti-diabetic properties of phenolic compounds largely depend on factors such as food concentration, bioaccessibility, absorption, metabolism, and bioavailability, indicating the need for research into the relationship of these factors with phenolic bioactivity to outline strategies aimed at maximizing its effects on diabetes management. Moreover, care should be taken to establish safe doses, as these compounds can have negative health effects when consumed in large quantities.

Next sections of this paper are presenting some of the already revealed antidiabetic activities of most investigated phenolic compounds, the chemical structures of mentioned compounds are presented in Figure 6.

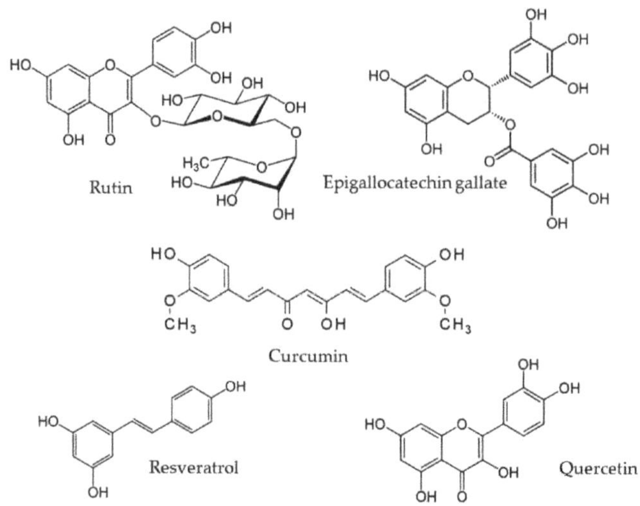

Figure 6. Structures of example polyphenols.

6.2. *Animal Studies on AntidiabeticEffects of Polyphenols*

Many studies have used diabetic streptozotocin (STZ) models or leptin-deficient *db/db* mice. Although in a mechanistic manner, some of the applied models seems to be more suitable for the investigation of T1DM than T2DM. Nonetheless, they provide helpful information on some of the numerous anti-diabetic properties of polyphenolic compounds.

Do et al. [162] studied resveratrol supplementation in *db/db* mice, and they concluded that this famous stilbene decreased blood glucose and HbA1c levels. Simultaneously, increased plasma and pancreatic insulin content was observed. It could suggest that resveratrol improves glucose tolerance and β-cell mass preservation [163].

Curcumin administration to rats fed a high-fat diet (HFD) and STZ resulted in lower fasting blood glucose levels, increased insulin levels, and insulin sensitivity [164]. Curcumin treatment reduced blood glucose and Hb1Ac levels in alloxan-induced diabetic rats and db/db mice [165]. Curcumin has also been shown to improve insulin resistance and glucose tolerance in db/db mice and HFD-fed mice [166], as well as to increase insulin levels in db/db mice. Curcumin supplementation did not, however, affect blood glucose levels in a similar study of db/db mice [167].

Studies on treating animals with green tea demonstrated that it could prevent hyperglycaemia, hyperinsulinemia, and insulin resistance in a high-fructose diet [168]. Additionally, it significantly reduced the amount of glucose in the blood of STZ-diabetic and *db/db* diabetic mice; nevertheless, it exhibited no effect on serum insulin levels [169]. However, some studies also demonstrated that particular green tea components showed inconsistent results. In the case of Epigallocatechin gallate (EGCG), its antidiabetic effects depend on the way of its administration. In STZ-treated mice, an intraperitoneal injection of EGCG decreased hyperglycaemia while preserving islet mass [170]. After 4 weeks, oral EGCG showed no effect on blood glucose levels in mice fed a HFD [171]. On the other hand, long-term oral dosing has been shown to reduce blood glucose in *db/db* mice, non-obese diabetic mice, and STZ rats [172].

Both rutin and its aglycone derivative compound, quercetin, also show interesting antidiabetic properties. STZ-rats treated orally with rutin showed decreased plasma glucose levels and increased insulin levels [173], and the C-peptide concentration was also increased [174]. Quercetin additionally protected β-cells against STZ-activated degeneration [175].

6.3. Human Studies on Antidiabetic Effects of Polyphenols

Ultimately, pre-clinical data have confirmed polyphenols as chemicals having metabolic regulation effects that may help to prevent or postpone the onset of type 2 diabetes, although human evidence is still lacking. There have only been a few meta-analyses of randomized controlled human studies examining the impact of polyphenols on diabetes biomarkers or incidence, and some of them have shown confusing results. For example, in one meta-analysis, biomarkers such as blood glucose, fasting insulin, or HbA1c were reported to be affected by a specific polyphenol but not another [176–178].

Many experiments with green tea and similar catechins have been conducted, but the findings have not been consistent. Zheng et al. [178] performed a meta-analysis to identify and quantify the effects of green tea catechins and green tea catechins' caffeine mixture on the glucose metabolism of adults. These studies showed that green tea catechins, with or without caffeine, substantially lowered fast blood glucose levels but had no effect on fast blood insulin, HbA1c, or the homeostatic model assessment of insulin resistance (HOMA-IR). A subgroup analysis revealed that the glucose-lowering impact was more pronounced when the follow-up period exceeded a median of 12 weeks. Green tea doses, intervention type (with or without caffeine), research quality, ethnicity, habitual caffeine intake, and participant health status did not appear to significantly impact the pooled mean differences in the fast blood glucose or fast blood insulin concentrations. Another meta-analysis of green tea's effect on individuals with T2DM or prediabetes revealed no effect on the fasting blood glucose, HbA1c, insulin, or HOMA-IR. However, these findings may be less credible due to the small sample numbers and the diverse quality of the research evaluated [179].

In turn, Rienks et al. [180] evaluated polyphenol exposure and risk of T2DM development and demonstrated that diets high in polyphenols, particularly flavonoids, can help avoid type 2 diabetes. Evidence of nonlinearity was discovered for most relationships, indicating a recommended level of consumption linked with the lowest risk of type 2 diabetes.

Human research on resveratrol has likewise had inconsistent results. Four weeks of resveratrol supplementation at a low dosage (10 mg/day) lowered HOMA-IR in T2DM patients but did not affect blood insulin levels or βß-cell function indicators [181]. In contrast, 12 weeks of a significantly larger dosage (3 g/day) in participants with T2DM did not improve HOMA-IR, despite a nonsignificant drop in HbA1c [182]. In contrast, Timmers et al. [183] discovered that 30 days of 150 mg/day resveratrol lowered levels of fasting plasma glucose, insulin, triglycerides, and HOMA-IR. Further follow-up trials have also failed to provide any significant differences. A 6-month experiment in T2DM patients exhibited no difference in levels of blood glucose, HbA1c, insulin, C-peptide, HOMA-IR, or FFA [184].

Studies on curcumin are much more promising. Some evidence has suggested that 12 weeks of therapy, including curcumin in participants with a high risk of T2DM development, caused decreased levels of plasma insulin with no effect on glucose levels [185]. Curcuminoid medication for three months lowered fasting blood glucose, HbA1c levels, and insulin resistance in individuals diagnosed with T2DM or prediabetes [186].

Ostadmohammadi et al. [187] conducted a systematic review to determine the effect of quercetin supplementation on glycaemic control among patients with metabolic syndrome (MetS) and related disorders. A meta-analysis found that quercetin supplementation did not affect glycaemic control in people with MetS and other associated illnesses. A subgroup study based on an 8-week-long trial with quercetin doses of 500 mg/day significantly lowered FPG levels. In addition, in trials with participants aged 45 years and quercetin doses of 500 mg/day, their insulin levels decreased considerably after quercetin supplementation.

More detailed data revealed in all described above studies are presented in the Table 1.

Table 1. Details of antidiabetic activities revealed in reported studies.

Investigated Phenolic Compounds and Their Origin	Experimental Design	Revealed Antidiabetic Activities	Ref.
Animal Model Studies In Vitro and In Vivo			
Resveratrol (3,5,4′-trihydroxy-trans-stilbene) Calss: Stilbenoid	C57BL/KsJ-db/db mice fed with a normal diet with RV (0.005% and 0.02%, *w*/*w*) or rosiglitazone (0.001%, *w*/*w*) for 6 weeks	↓blood glucose ↓plasma free fatty acid and triglyceride, ↓apo B/apo AI levels ↓HbA1c levels ↓hepatic gluconeogenic enzyme activity and hepatic glycogen ↓hepatic triglyceride content and p-IKK protein expression ↑plasma insulin levels ↑pancreatic insulin protein ↑skeletal muscle GLUT4 protein ↑plasma adiponectin levels ↑hepatic glycolytic gene expression and enzyme activity ↑skeletal muscle glycogen synthase protein expression ↑hepatic UCP and skeletal muscle UCP expression	[162]
	db/db and db/dm mice (non-diabetic control) were treated with or without RV (20 mg/kg BW daily) for 12 weeks	↑glucose tolerance at 2 h of OGTT in db/db mice ↑pancreas weight and β-cell mass ↓urinary 8-OHdG levels ↓percentage of islet nuclei that were positive for 8-OHdG immunostaining	[163]
Curcumin (1E,6E)-1,7-Bis(4-hydroxy-3-methoxyphenyl)hepta-1,6-diene-3,5-dione) Class: Curcuminoid	diabetic rats induced by high-fat diet plus STZ (30 mg/kg BW) were fed a diet containing 50, 150, or 250 mg/kg BW curcumin for 7 wks	↓plasma lipids and glucose ↓glucose and insulin tolerance ↓pyruvate dehydrogenase 4 and GS expression ↑2-deoxy-[(3)H]d-glucose uptake by L6 myotubes ↑phosphorylated AMPK, CD36, and carnitine palmitoyl transferase 1 expression ↑phosphorylated acetyl COA carboxylase in L6 myotubes	[164]
	male C57BL/KsJ-db/db mice and their age-matched lean non-diabetic db/+ mice, fed with or without curcumin (0.02%, *w*/*w*) for 6 weeks	↓blood glucose and HbA 1c levels, as well as body weight loss ↓glucose-6-phosphatase and phosphoenolpyruvate carboxykinase activities ↓hepatic activities of fatty acid synthase, beta-oxidation, 3-hydroxy-3-methylglutaryl coenzyme reductase, and acyl-CoA: cholesterol acyltransferase ↓plasma free fatty acid, cholesterol, and triglyceride concentrations ↑homeostasis model assessment of insulin resistance and glucose tolerance ↑plasma insulin level ↑hepatic glucokinase activity ↑hepatic glycogen and skeletal muscle lipoprotein lipase	[165]
	male C57BL/6J mice were fed either a normal diet or HFD.After 16 weeks, 10 HFD-fed mice were further treated with daily curcumin oral gavage (50 mg/kg BW)	↓glucose intolerance ↓HFD-induced elevations of MDA and ROS in the skeletal muscle ↑skeletal muscle content of Nrf2 and oxygenase-1	[166]
	db/db livers of 15-week-old mice treated with 0.75% curcumin mixed in their diet for 8 weeks	↑expression of AMPK and PPARγ ↓NF-κB protein levels	[167]

Table 1. *Cont.*

Investigated Phenolic Compounds and Their Origin	Experimental Design	Revealed Antidiabetic Activities	Ref.
Green tea polyphenols (mixture)	SpragueDawley rats fed with standard chow and deionized distilled water and a "green tea" group fed the same chow diet but with green tea instead of water (0.5 g of lyophilized green tea powder dissolved in 100 mL of deionized distilled water) for 12 weeks	↓fasting plasma levels of glucose, insulin, triglycerides, and free fatty acids ↑insulin-stimulated glucose uptake and insulin binding; adipocytes were significantly increased ↑basal and insulin-stimulated glucose uptake of adipocytes in vitro	[168]
	C57BLKS/J *db+/db+* mice and age-matched control C57BLKS/J +m/+m mice. Male ddY mice were singly injected with STZ (150 mg/kg, i.v.)and 4–6 weeks after the injection the samples were analyzed.The age-matched normal ddY mice were also used	↓blood glucose levels in diabetic db+/db+ mice and streptozotocin-diabetic mice 2–6 h after administration at 300 mg/kg ↔serum insulin level	[169]
Epigallocatechin gallate ((EGCG), (2R,3R)-3′,4′,5,5′,7-Pentahydroxyflavan-3-yl 3,4,5-trihydroxybenzoate) Class: Catechin	maleC57BL/KsJ mice aged 6 weeks induced with multiple low doses of streptozotocin (MLD-STZ, (40 mg/kg BW). EGCG (100 mg/kg/day) was administered with STZ for 5 days and then EGCG alone was administered for a further 5 days	↓blood glucose levels ↓iNOS expression ex vivo ↓decrease in islet mass induced by MLD-STZ	[170]
Epigallocatechin gallate continued	C57BL/6 male mice (6 weeks old) were fed with ND, HFD, or HFD with EGCG supplementation for 12 weeks. EGCG (50 mg/kg daily) was administered by gavage for 9–12 weeks	Inhibition of Caspase-1 activation and IL-1β secretion in mice bone marrow by suppressing NLRP3 inflammasome activation Inhibition of NLRP3-mediated ASC speckle formation and alleviated pyroptosis in BMDMs Improved high-fat-diet (HFD)-induced glucose tolerance	[171]
	male albino Wistar rats induced with by a single i.p. injection of 60mg/kg^{-1}STZ received EGCG 25mg/kg/day for 8 weeks 1 week after the induction of diabetes	Hypoglycaemic effect in diabetic rats Improved serum lipid profile Attenuation of increased MDA content and reduced activity of SOD in liver.	[172]
Rutin (3′,4′,5,7-Tetrahydroxy-3-[α-L-rhamnopyranosyl-(1→6)-β-D-glucopyranosyloxy] flavones) Class: Flavonoid glycoside	adult male SpragueDawley rats injected withSTZ i.p. (55mg/kg BW) after induction ofdiabetic neuropathy.Rutin (5mg/kg, 25mg/kg and 50mg/kg BW) was daily given to the diabetic rats for 2 weeks	Inhibition of mechanical hyperalgesia, thermal hyperalgesia, and coldallodynia ↑Na^+, K^+-ATPase activities insciatic nerves ↑hydrogen sulfide(H_2S) level, upregulated expression of nuclear factor-E2-related factor-2 (Nrf2), and heme oxygenase-1 (HO-1) in DRG ↓caspase-3 expression indorsal root ganglions ↓plasma glucose, attenuatedoxidative stress,andneuroinflammation Partial restoration ofnerve conductionvelocities in diabetic rats	[173]

Table 1. *Cont.*

Investigated Phenolic Compounds and Their Origin	Experimental Design	Revealed Antidiabetic Activities	Ref.
	male albino Wistar rats induced with i.p. injection of STZ (50 mg/kg BW).Rutin (25, 50, 100 mg/kg BW) was orally administered to normal and diabetic rats (1 mL/rat) using an intragastric tube for a period of 45 days	↑fasting plasma glucose, HbA1c, thiobarbituric acid reactive substances, and lipid hydroperoxides ↓blood insulin, C-peptide, total haemoglobin, protein levels, non-enzymic antioxidants: glutathione, vitamin C, vitamin E, and ceruloplasmin	[174]
Quercetin (3,3′,4′,5,7-Pentahydroxyflavone) Class: Falvonoid	Male adult albino Wistar rats induced by a single injection of STZ (45 mg/kg, i.p.). Diabetic rats were orally treated with sitagliptin (70 mg/kg BW), quercetin (50 mg/kg BW), or a combination of these daily for 3 weeks	↑increased SOD, GSH ↓NF-κB expression Normalized Islet number, β-cells' number, area, and perimeter alongside the restoration of the immunostaining intensity of β-cells.	[175]
		Human studies	
Green tea polyphenols (mixture)	A total of 17 trials comprising a total of 1133 subjects were included in the current meta-analysis	↓fasting blood glucose ↓Hb A1c ↓fasting blood insulin	[188]
Green tea catechins (mixture)	A total of 22 eligible randomized controlled trials with 1584 subjects were identified	↓fasting blood glucose ↔Hb A1c ↔fasting blood insulin ↔HOMA-IR	[189]
Green tea (mixture and extracts)	A total of six studies with 382 subjects were pooled into random-effects meta-analysis	↔I IOMA-IR ↔HbA1c ↔fasting blood insulin ↔fasting blood glucose	[176]
Polyphenols (mixture in supplements and food)	A total of 36 controlled randomized trials with 1954 subjects were included in 28 mg to 1.5 g of polyphenol mixture, supplemented for 0.7 to 12 months	↓HbA1c in T2DM individuals	[179]
Polyphenols (51 different compounds in Total)	A total of 18 studies investigated the association between polyphenols and type 2 diabetes	Evidence showing that diets rich in polyphenols, and particularly flavonoids, play a role in the prevention of type 2 diabetes.	[180]
Resveratrol (3,5,4′-trihydroxy-trans-stilbene) Calss: Stilbenoid	A total of 19 patients enrolled in the 4-week-long double-blind study were randomly assigned into two groups: an RV group receiving oral 2 × 5 mg RV and a control group receiving placebo	↓HOMA-IR ↓urinary ortho-tyrosine excretion ↑pAkt:Akt ratio in platelets	[181]
	Ten subjects with T2DM were randomized in a double-blind fashion to receive 3 g RV or placebo daily for 12 weeks	↑SIRT1 expression ↑pAMPK to AMPK expression ratio ↓average daily activity ↓step counts	[182]

Table 1. *Cont.*

Investigated Phenolic Compounds and Their Origin	Experimental Design	Revealed Antidiabetic Activities	Ref.
	A total of 11 healthy and obese men received placebo and 150 mg/day RV in a randomized double-blind crossover study for 30 days	↓sleeping and resting metabolic rate ↓intrahepatic lipid content ↓circulating glucose, triglycerides, and alanine-aminotransferase ↓inflammation markers ↓systolic blood presure ↑AMPK activity ↑SIRT1 and PGC-1α protein levels ↑citrate synthase activity ↑intramyocellular lipid levels Improved muscle mitochondrial respiration on a fatty acid-derived substrate and HOMA-IR index	[183]
Resveratrol continued	double-blind, randomized, placebo-controlled trial, with192 T2DM patients randomized to receive RV 500mg/day, 40mg/day, or placebo for 6months	↔weight, BMI, waist circumference ↔arterial blood pressure ↔fasting glucose, plasma insulin, C-peptide, free fatty acids, liver transaminases, uric acid, adiponectin, and interleukin-6 ↔HbA1c	[184]
Curcumin (1E,6E)-1,7-Bis(4-hydroxy-3-methoxyphenyl)hepta-1,6-diene-3,5-dione) Class: Curcuminoid	Plasma samples from 29 participants recruited for a randomised controlled trial with curcumin (180 mg/day) for 12 weeks were analysed	↓Levels of circulating GSK-3β and IAPP ↓insulin resistance	[190]
Curcumin + Zinc	A total of 84 subjects were randomized into curcumin (500 mg), zinc (30 mg), zinc and curcumin, and placebo groups for 90 days	↓BMI Improved fasting blood glucose, HbA1c, blood insulin, and HOMA-IR	[186]
Quercetin (3,3′,4′,5,7-Pentahydroxyflavone) Class: Falvonoid	A total of nine studies of 781 participants involved in meta-analysis	↔fasting plasma glucose, HOMA-IR, HbA1c ↓fasting blood glucose in studies with a duration of >8 weeks ↓insulin concentrations in studies that enrolled individuals aged <45 years	[187]

In conclusion, considering the significant attention and quantity of randomized controlled trials investigating the influence of polyphenols on metabolic- or diabetes-related outcomes, the results remain disputed. Inconsistent findings may be attributable to discrepancies in terms of the sample dose, time frame, health status of individuals, whether participants were taking other medications, and the small sample sizes frequently used. It needs to be emphasized that outcomes are not always comparable with a treatment's dose or time-dependent impact. The data show that polyphenol therapies may help decrease blood glucose and maybe improve insulin sensitivity, although more and better research is needed. Natural polyphenols have a favourable safety profile and are found in relatively high concentrations in the average human diet.

7. Conclusions

This review focused on the anti-diabetic properties of dietary phenolic compounds and their mechanisms of action. Nevertheless, it is essential to note that the final anti-diabetic activity of phenolic compounds is dependent on numerous factors, including their concentration in food, absorption, metabolism, and bioavailability, indicating the need for research into the association of these factors with phenolic bioactivity to outline strategies aimed at maximizing the impact on diabetes management. Therefore, given the importance of diabetes and the rising number of people suffering from it, there is a clear need for novel medicines that may lessen the detrimental effects of the condition while also ensuring the safety and well-being of the community. In this regard, polyphenols might be a possible option for controlling the course of this metabolic disease.

Presented in this review, numerous in vitro and in vivo studies have been selected from copious, currently ongoing, or recently completed studies and trials. They all indicate that a diet rich in polyphenolic compounds benefits glucose homeostasis through multiple and complex mechanisms of action in various human body organs such as the intestine, liver, muscle, adipocytes, and pancreatic β-cells. Polyphenol intake seems to be associated with a lower T2DM development and progression rate. The clinical studies that have been conducted have shown significant limitations and inconsistencies when considering dietary polyphenol supplementation. However, applying these compounds in glucose and insulin homeostasis management seems to be the right direction and a promising approach for further studies and research.

Due to the poor bioavailability of polyphenols, progress in micro- and nanoencapsulation methodologies of phenolic compounds could be applied as supportive supplements for diabetes treatment. The encapsulation of phenolic compounds may additionally cover the unpleasant flavour of some plant extracts, paving the way for the development of future functional foods, nutraceuticals, or orally administered polyphenol-based diabetes drugs.

Author Contributions: Conceptualization—M.B.-M.; validation, L.A.W. and M.B.-M.; writing—original draft preparation, I.B.-P., M.K., M.B.-M. and L.A.W.; writing—review and editing, M.B.-M., L.A.W.; supervision, L.A.W. All authors have read and agreed to the published version of the manuscript.

Funding: This research received no external funding.

Conflicts of Interest: The authors declare no conflict of interest.

References

1. American Diabetes Association (2). Classification and diagnosis of diabetes. *Diabetes Care* **2015**, *38*, S8–S16. [CrossRef] [PubMed]
2. Davies, J.L.; Kawaguchi, Y.; Bennett, S.T.; Copeman, J.B.; Cordell, H.J.; Pritchard, L.E.; Reed, P.W.; Gough, S.C.; Jenkins, S.C.; Palmer, S.M.; et al. A genome-wide search for human type 1 diabetes susceptibility genes. *Nature* **1994**, *371*, 130–136. [CrossRef] [PubMed]
3. Keenan, H.A.; Sun, J.K.; Levine, J.; Doria, A.; Aiello, L.P.; Eisenbarth, G.; Bonner-Weir, S.; King, G.L. Residual insulin production and pancreatic ß-cell turnover after 50 years of diabetes: Joslin Medalist Study. *Diabetes* **2010**, *59*, 2846–2853. [CrossRef] [PubMed]
4. Nejentsev, S.; Howson, J.M.; Walker, N.M.; Szeszko, J.; Field, S.F.; Stevens, H.E.; Reynolds, P.; Hardy, M.; King, E.; Masters, J.; et al. Localization of type 1 diabetes susceptibility to the MHC class I genes HLA-B and HLA-A. *Nature* **2007**, *450*, 887–892. [CrossRef] [PubMed]
5. Harjutsalo, V.; Sjöberg, L.; Tuomilehto, J. Time trends in the incidence of type 1 diabetes in Finnish children: A cohort study. *Lancet* **2008**, *371*, 1777–1782. [CrossRef]
6. Temneanu, O.R.; Trandafir, L.M.; Purcarea, M.R. Type 2 diabetes mellitus in children and adolescents: A relatively new clinical problem within pediatric practice. *J. Med. Life* **2016**, *9*, 235–239.
7. Artasensi, A.; Mazzolari, A.; Pedretti, A.; Vistoli, G.; Fumagalli, L. Obesity and Type 2 Diabetes: Adiposopathy as a Triggering Factor and Therapeutic Options. *Molecules* **2023**, *28*, 3094. [CrossRef]
8. Kahn, S.E.; Prigeon, R.L.; McCulloch, D.K.; Boyko, E.J.; Bergman, R.N.; Schwartz, M.W.; Neifing, J.L.; Ward, W.K.; Beard, J.C.; Palmer, J.P.; et al. Quantification of the relationship between insulin sensitivity and beta-cell function in human subjects. Evidence for a hyperbolic function. *Diabetes* **1993**, *42*, 1663–1672. [CrossRef]
9. World Health Organization Global Health Observatory. Available online: www.who.int: (accessed on 10 November 2021).
10. Ferrara, A. Increasing prevalence of gestational diabetes mellitus: A public health perspective. *Diabetes Care* **2007**, *30* (Suppl. S2), S141–S146. [CrossRef]
11. Plows, J.F.; Stanley, J.L.; Baker, P.N.; Reynolds, C.M.; Vickers, M.H. The Pathophysiology of Gestational Diabetes Mellitus. *Int. J. Mol. Sci.* **2018**, *19*, 3342. [CrossRef]
12. Li, Z.; Cheng, Y.; Wang, D.; Chen, H.; Ming, W.K.; Wang, Z. Incidence Rate of Type 2 Diabetes Mellitus after Gestational Diabetes Mellitus: A Systematic Review and Meta-Analysis of 170,139 Women. *J. Diabetes Res.* **2020**, *2020*, 3076463. [CrossRef] [PubMed]
13. Adane, A.A.; Mishra, G.D.; Tooth, L.R. Diabetes in Pregnancy and Childhood Cognitive Development: A Systematic Review. *Pediatrics* **2016**, *137*, e20154234. [CrossRef]

14. Saeedi, P.; Petersohn, I.; Salpea, P.; Malanda, B.; Karuranga, S.; Unwin, N.; Colagiuri, S.; Guariguata, L.; Motala, A.A.; Ogurtsova, K.; et al. Global and regional diabetes prevalence estimates for 2019 and projections for 2030 and 2045: Results from the International Diabetes Federation Diabetes Atlas, 9(th) edition. *Diabetes Res. Clin. Pract.* **2019**, *157*, 107843. [CrossRef]
15. Darenskaya, M.A.; Kolesnikova, L.I.; Kolesnikov, S.I. Oxidative Stress: Pathogenetic Role in Diabetes Mellitus and Its Complications and Therapeutic Approaches to Correction. *Bull. Exp. Biol. Med.* **2021**, *171*, 179–189. [CrossRef]
16. Deng, L.; Du, C.; Song, P.; Chen, T.; Rui, S.; Armstrong, D.G.; Deng, W. The Role of Oxidative Stress and Antioxidants in Diabetic Wound Healing. *Oxid. Med. Cell Longev.* **2021**, *2021*, 8852759. [CrossRef] [PubMed]
17. Ganjifrockwala, F.A.; Joseph, J.T.; George, G. Decreased total antioxidant levels and increased oxidative stress in South African type 2 diabetes mellitus patients. *J. Endocrinol. Metab. Diabetes South Afr.* **2017**, *22*, 21–25. [CrossRef]
18. van der Schaft, N.; Schoufour, J.D.; Nano, J.; Kiefte-de Jong, J.C.; Muka, T.; Sijbrands, E.J.G.; Ikram, M.A.; Franco, O.H.; Voortman, T. Dietary antioxidant capacity and risk of type 2 diabetes mellitus, prediabetes and insulin resistance: The Rotterdam Study. *Eur. J. Epidemiol.* **2019**, *34*, 853–861. [CrossRef]
19. Hassan, W.; Noreen, H.; Rehman, S.; Gul, S.; Kamal, M.A.; Kamdem, J.P.; Zaman, B.; da Rocha, J.B.T. Oxidative Stress and Antioxidant Potential of One Hundred Medicinal Plants. *Curr. Top. Med. Chem.* **2017**, *17*, 1336–1370. [CrossRef] [PubMed]
20. Guzik, T.J.; Mussa, S.; Gastaldi, D.; Sadowski, J.; Ratnatunga, C.; Pillai, R.; Channon, K.M. Mechanisms of Increased Vascular Superoxide Production in Human Diabetes Mellitus. *Circulation* **2002**, *105*, 1656–1662. [CrossRef] [PubMed]
21. Scalbert, A.; Manach, C.; Morand, C.; Remesy, C.; Jimenez, L. Dietary polyphenols and the prevention of diseases. *Crit. Rev. Food Sci. Nutr.* **2005**, *45*, 287–306. [CrossRef]
22. Valko, M.; Rhodes, C.J.; Moncol, J.; Izakovic, M.; Mazur, M. Free radicals, metals and antioxidants in oxidative stress-induced cancer. *Chem. -Biol. Interact.* **2006**, *160*, 1–40. [CrossRef] [PubMed]
23. Juan, C.A.; Pérez de la Lastra, J.M.; Plou, F.J.; Pérez-Lebeña, E. The Chemistry of Reactive Oxygen Species (ROS) Revisited: Outlining Their Role in Biological Macromolecules (DNA, Lipids and Proteins) and Induced Pathologies. *Int. J. Mol. Sci.* **2021**, *22*, 4642. [CrossRef]
24. Di Meo, S.; Venditti, P. Evolution of the Knowledge of Free Radicals and Other Oxidants. *Oxid. Med. Cell Longev.* **2020**, *2020*, 9829176. [CrossRef] [PubMed]
25. Valko, M.; Leibfritz, D.; Moncol, J.; Cronin, M.T.; Mazur, M.; Telser, J. Free radicals and antioxidants in normal physiological functions and human disease. *Int. J. Biochem. Cell Biol.* **2007**, *39*, 44–84. [CrossRef] [PubMed]
26. Jones, D.P. Radical-free biology of oxidative stress. *Am. J. Physiol. Cell Physiol.* **2008**, *295*, C849–C868. [CrossRef]
27. Halliwell, B. Reactive species and antioxidants. Redox biology is a fundamental theme of aerobic life. *Plant Physiol.* **2006**, *141*, 312–322. [CrossRef]
28. Kovacic, P.; Pozos, R.S.; Somanathan, R.; Shangari, N.; O'Brien, P.J. Mechanism of mitochondrial uncouplers, inhibitors, and toxins: Focus on electron transfer, free radicals, and structure-activity relationships. *Curr. Med. Chem.* **2005**, *12*, 2601–2623. [CrossRef]
29. Forrester, S.J.; Kikuchi, D.S.; Hernandes, M.S.; Xu, Q.; Griendling, K.K. Reactive Oxygen Species in Metabolic and Inflammatory Signaling. *Circ. Res.* **2018**, *122*, 877–902. [CrossRef]
30. Droge, W. Free radicals in the physiological control of cell function. *Physiol. Rev.* **2002**, *82*, 47–95. [CrossRef]
31. Kuksal, N.; Chalker, J.; Mailloux, R.J. Progress in understanding the molecular oxygen paradox—Function of mitochondrial reactive oxygen species in cell signaling. *Biol. Chem.* **2017**, *398*, 1209–1227. [CrossRef]
32. Sies, H. Oxidative Stress: Concept and Some Practical Aspects. *Antioxidants* **2020**, *9*, 852. [CrossRef] [PubMed]
33. Leonarduzzi, G.; Sottero, B.; Testa, G.; Biasi, F.; Poli, G. New insights into redox-modulated cell signaling. *Curr. Pharm. Des.* **2011**, *17*, 3994–4006. [CrossRef] [PubMed]
34. Yan, L.L.; Zaher, H.S. How do cells cope with RNA damage and its consequences? *J. Biol. Chem.* **2019**, *294*, 15158–15171. [CrossRef] [PubMed]
35. Hawkins, C.L.; Davies, M.J. Detection, identification, and quantification of oxidative protein modifications. *J. Biol. Chem.* **2019**, *294*, 19683–19708. [CrossRef] [PubMed]
36. Ito, F.; Sono, Y.; Ito, T. Measurement and Clinical Significance of Lipid Peroxidation as a Biomarker of Oxidative Stress: Oxidative Stress in Diabetes, Atherosclerosis, and Chronic Inflammation. *Antioxidants* **2019**, *8*, 72. [CrossRef]
37. Ray, P.D.; Huang, B.W.; Tsuji, Y. Reactive oxygen species (ROS) homeostasis and redox regulation in cellular signaling. *Cell. Signal.* **2012**, *24*, 981–990. [CrossRef]
38. Dahlmann, H.A.; Vaidyanathan, V.G.; Sturla, S.J. Investigating the biochemical impact of DNA damage with structure-based probes: Abasic sites, photodimers, alkylation adducts, and oxidative lesions. *Biochemistry* **2009**, *48*, 9347–9359. [CrossRef]
39. Thompson, P.S.; Cortez, D. New insights into abasic site repair and tolerance. *DNA Repair* **2020**, *90*, 102866. [CrossRef]
40. Dizdaroglu, M.; Jaruga, P. Mechanisms of free radical-induced damage to DNA. *Free Radic. Res.* **2012**, *46*, 382–419. [CrossRef]
41. Patel, J.; Baptiste, B.A.; Kim, E.; Hussain, M.; Croteau, D.L.; Bohr, V.A. DNA damage and mitochondria in cancer and aging. *Carcinogenesis* **2020**, *41*, 1625–1634. [CrossRef]
42. Auboeuf, D. Physicochemical Foundations of Life that Direct Evolution: Chance and Natural Selection are not Evolutionary Driving Forces. *Life* **2020**, *10*, 7. [CrossRef]
43. Srinivas, U.S.; Tan, B.W.Q.; Vellayappan, B.A.; Jeyasekharan, A.D. ROS and the DNA damage response in cancer. *Redox Biol.* **2019**, *25*, 101084. [CrossRef]

44. Bui, A.D.; Sharma, R.; Henkel, R.; Agarwal, A. Reactive oxygen species impact on sperm DNA and its role in male infertility. *Andrologia* **2018**, *50*, e13012. [CrossRef]
45. Valko, M.; Izakovic, M.; Mazur, M.; Rhodes, C.J.; Telser, J. Role of oxygen radicals in DNA damage and cancer incidence. *Mol. Cell Biochem.* **2004**, *266*, 37–56. [CrossRef]
46. D'Errico, M.; Parlanti, E.; Dogliotti, E. Mechanism of oxidative DNA damage repair and relevance to human pathology. *Mutat. Res.* **2008**, *659*, 4–14. [CrossRef] [PubMed]
47. Iyama, T.; Wilson, D.M., 3rd. DNA repair mechanisms in dividing and non-dividing cells. *DNA Repair* **2013**, *12*, 620–636. [CrossRef]
48. Dandona, P.; Thusu, K.; Cook, S.; Snyder, B.; Makowski, J.; Armstrong, D.; Nicotera, T. Oxidative damage to DNA in diabetes mellitus. *Lancet* **1996**, *347*, 444–445. [CrossRef] [PubMed]
49. Simone, S.; Gorin, Y.; Velagapudi, C.; Abboud, H.E.; Habib, S.L. Mechanism of oxidative DNA damage in diabetes: Tuberin inactivation and downregulation of DNA repair enzyme 8-oxo-7,8-dihydro-2′-deoxyguanosine-DNA glycosylase. *Diabetes* **2008**, *57*, 2626–2636. [CrossRef]
50. Al-Aubaidy, H.A.; Jelinek, H.F. Oxidative stress and triglycerides as predictors of subclinical atherosclerosis in prediabetes. *Redox Rep.* **2014**, *19*, 87–91. [CrossRef] [PubMed]
51. Ferri, J.; Martínez-Hervás, S.; Espinosa, O.; Fandos, M.; Pedro, T.; Real, J.T.; Chaves, F.J.; Cerdá, C.; Sáez, G.; Ascaso, J.F. 8-oxo-7,8-dihydro-2′-deoxyguanosine as a marker of DNA oxidative stress in individuals with combined familiar hyperlipidemia. *Med. Clin.* **2008**, *131*, 1–4. [CrossRef]
52. Xavier, D.J.; Takahashi, P.; Manoel-Caetano, F.S.; Foss-Freitas, M.C.; Foss, M.C.; Donadi, E.A.; Passos, G.A.; Sakamoto-Hojo, E.T. One-week intervention period led to improvements in glycemic control and reduction in DNA damage levels in patients with type 2 diabetes mellitus. *Diabetes Res. Clin. Pract.* **2014**, *105*, 356–363. [CrossRef]
53. Dinçer, Y.; Akçay, T.; Alademir, Z.; Ilkova, H. Assessment of DNA base oxidation and glutathione level in patients with type 2 diabetes. *Mutat. Res.* **2002**, *505*, 75–81. [CrossRef]
54. Lodovici, M.; Giovannelli, L.; Pitozzi, V.; Bigagli, E.; Bardini, G.; Rotella, C.M. Oxidative DNA damage and plasma antioxidant capacity in type 2 diabetic patients with good and poor glycaemic control. *Mutat. Res.* **2008**, *638*, 98–102. [CrossRef] [PubMed]
55. Blasiak, J.; Arabski, M.; Krupa, R.; Wozniak, K.; Zadrozny, M.; Kasznicki, J.; Zurawska, M.; Drzewoski, J. DNA damage and repair in type 2 diabetes mellitus. *Mutat. Res.* **2004**, *554*, 297–304. [CrossRef] [PubMed]
56. Gelaleti, R.B.; Damasceno, D.C.; Lima, P.H.; Salvadori, D.M.; Calderon, I.e.M.; Peraçoli, J.C.; Rudge, M.V. Oxidative DNA damage in diabetic and mild gestational hyperglycemic pregnant women. *Diabetol. Metab. Syndr.* **2015**, *7*, 1. [CrossRef] [PubMed]
57. Collins, A.R.; Raslová, K.; Somorovská, M.; Petrovská, H.; Ondrusová, A.; Vohnout, B.; Fábry, R.; Dusinská, M. DNA damage in diabetes: Correlation with a clinical marker. *Free Radic. Biol. Med.* **1998**, *25*, 373–377. [CrossRef]
58. Bukowiecka-Matusiak, M.; Burzynska-Pedziwiatr, I.; Szczesna, D.; Malgorzata, C.-K.; Fabian, A.; Wozniak, L.A. Erythrocyte membranes in metabolic and neurological diseases—Supplementation with fatty acids and membranes remodeling. In *Dietary Sugar, Salt and Fat in Human Health*; Preuss, H.G., Bagchi, D., Eds.; Elsevier: London, UK, 2020; pp. 505–528. [CrossRef]
59. Ayala, A.; Muñoz, M.F.; Argüelles, S. Lipid Peroxidation: Production, Metabolism, and Signaling Mechanisms of Malondialdehyde and 4-Hydroxy-2-Nonenal. *Oxidative Med. Cell. Longev.* **2014**, *2014*, 360438. [CrossRef]
60. Spiteller, G.; Afzal, M. The action of peroxyl radicals, powerful deleterious reagents, explains why neither cholesterol nor saturated fatty acids cause atherogenesis and age-related diseases. *Chemistry* **2014**, *20*, 14928–14945. [CrossRef]
61. Eizirik, D.; Colli, M.; Ortis, F.; Eizirik, D.L.; Colli, M.L.; Ortis, F. The role of inflammation in insulitis and β-cell loss in type 1 diabetes. *Nat. Rev. Endocrinol.* **2009**, *5*, 219–226. [CrossRef]
62. Peña-Bautista, C.; Vento, M.; Baquero, M.; Cháfer-Pericás, C. Lipid peroxidation in neurodegeneration. *Clin. Chim. Acta* **2019**, *497*, 178–188. [CrossRef]
63. Dham, D.; Roy, B.; Gowda, A.; Pan, G.; Sridhar, A.; Zeng, X.; Thandavarayan, R.A.; Palaniyandi, S.S. 4-Hydroxy-2-nonenal, a lipid peroxidation product, as a biomarker in diabetes and its complications: Challenges and opportunities. *Free Radic. Res.* **2021**, *55*, 547–561. [CrossRef]
64. Toyokuni, S.; Yamada, S.; Kashima, M.; Ihara, Y.; Yamada, Y.; Tanaka, T.; Hiai, H.; Seino, Y.; Uchida, K. Serum 4-hydroxy-2-nonenal-modified albumin is elevated in patients with type 2 diabetes mellitus. *Antioxid. Redox Signal.* **2000**, *2*, 681–685. [CrossRef]
65. Zhao, Y.; Song, W.; Wang, Z.; Jin, X.; Xu, J.; Bai, L.; Li, Y.; Cui, J.; Cai, L. Resveratrol attenuates testicular apoptosis in type 1 diabetic mice: Role of Akt-mediated Nrf2 activation and p62-dependent Keap1 degradation. *Redox Biol.* **2018**, *14*, 609–617. [CrossRef]
66. Ihara, Y.; Toyokuni, S.; Uchida, K.; Odaka, H.; Tanaka, T.; Ikeda, H.; Hiai, H.; Seino, Y.; Yamada, Y. Hyperglycemia causes oxidative stress in pancreatic beta-cells of GK rats, a model of type 2 diabetes. *Diabetes* **1999**, *48*, 927–932. [CrossRef]
67. Morsi, H.K.; Ismail, M.M.; Gaber, H.A.; Elbasmy, A.A. Macrophage Migration Inhibitory Factor and Malondialdehyde as Potential Predictors of Vascular Risk Complications in Type 2 Diabetes Mellitus: Cross-Sectional Case Control Study in Saudi Arabia. *Mediat. Inflamm.* **2016**, *2016*, 5797930. [CrossRef]
68. Khalili, F.; Vaisi-Raygani, A.; Shakiba, E.; Kohsari, M.; Dehbani, M.; Naseri, R.; Asadi, S.; Rahimi, Z.; Rahimi, M. Oxidative stress parameters and keap 1 variants in T2DM: Association with T2DM, diabetic neuropathy, diabetic retinopathy, and obesity. *J. Clin. Lab. Anal.* **2022**, *36*, e24163. [CrossRef] [PubMed]

69. Desco, M.C.; Asensi, M.; Márquez, R.; Martínez-Valls, J.; Vento, M.; Pallardó, F.V.; Sastre, J.; Viña, J. Xanthine oxidase is involved in free radical production in type 1 diabetes: Protection by allopurinol. *Diabetes* **2002**, *51*, 1118–1124. [CrossRef] [PubMed]
70. Caimi, G.; Hopps, E.; Noto, D.; Canino, B.; Montana, M.; Lucido, D.; Lo Presti, R.; Averna, M.R. Protein oxidation in a group of subjects with metabolic syndrome. *Diabetes Metab. Syndr.* **2013**, *7*, 38–41. [CrossRef] [PubMed]
71. Lo Conte, M.; Carroll, K.S. The redox biochemistry of protein sulfenylation and sulfinylation. *J. Biol. Chem.* **2013**, *288*, 26480–26488. [CrossRef]
72. Venturini, D.; Simao, A.N.; Dichi, I. Advanced oxidation protein products are more related to metabolic syndrome components than biomarkers of lipid peroxidation. *Nutr. Res.* **2015**, *35*, 759–765. [CrossRef]
73. Perrone, A.; Giovino, A.; Benny, J.; Martinelli, F. Advanced Glycation End Products (AGEs): Biochemistry, Signaling, Analytical Methods, and Epigenetic Effects. *Oxid. Med. Cell Longev.* **2020**, *2020*, 3818196. [CrossRef]
74. Twarda-Clapa, A.; Olczak, A.; Białkowska, A.M.; Koziołkiewicz, M. Advanced Glycation End-Products (AGEs): Formation, Chemistry, Classification, Receptors, and Diseases Related to AGEs. *Cells* **2022**, *11*, 1312. [CrossRef]
75. Wilmut, I.; Schnieke, A.; McWhir, J.; Kind, A.; Campbell, K. Viable Offspring Derived from Fetal And Adult Mammalian Cells. *Cloning Stem Cells* **2007**, *9*, 3–7. [CrossRef] [PubMed]
76. Perez-Burillo, S.; Rufian-Henares, J.A.; Pastoriza, S. Effect of home cooking on the antioxidant capacity of vegetables: Relationship with Maillard reaction indicators. *Food Res. Int.* **2019**, *121*, 514–523. [CrossRef] [PubMed]
77. Chuyen, N.V. Toxicity of the AGEs generated from the Maillard reaction: On the relationship of food-AGEs and biological-AGEs. *Mol. Nutr. Food Res.* **2006**, *50*, 1140–1149. [CrossRef] [PubMed]
78. Takeuchi, M.; Sakasai-Sakai, A.; Takata, T.; Takino, J.I.; Koriyama, Y.; Kikuchi, C.; Furukawa, A.; Nagamine, K.; Hori, T.; Matsunaga, T. Intracellular Toxic AGEs (TAGE) Triggers Numerous Types of Cell Damage. *Biomolecules* **2021**, *11*, 387. [CrossRef]
79. Frijhoff, J.; Winyard, P.G.; Zarkovic, N.; Davies, S.S.; Stocker, R.; Cheng, D.; Knight, A.R.; Taylor, E.L.; Oettrich, J.; Ruskovska, T.; et al. Clinical Relevance of Biomarkers of Oxidative Stress. *Antioxid. Redox Signal.* **2015**, *23*, 1144–1170. [CrossRef]
80. Yan, S.F.; Ramasamy, R.; Naka, Y.; Schmidt, A.M. Glycation, inflammation, and RAGE: A scaffold for the macrovascular complications of diabetes and beyond. *Circ. Res.* **2003**, *93*, 1159–1169. [CrossRef]
81. Sergi, D.; Boulestin, H.; Campbell, F.M.; Williams, L.M. The Role of Dietary Advanced Glycation End Products in Metabolic Dysfunction. *Mol. Nutr. Food Res.* **2021**, *65*, e1900934. [CrossRef]
82. Pinto-Junior, D.C.; Silva, K.S.; Michalani, M.L.; Yonamine, C.Y.; Esteves, J.V.; Fabre, N.T.; Thieme, K.; Catanozi, S.; Okamoto, M.M.; Seraphim, P.M.; et al. Advanced glycation end products-induced insulin resistance involves repression of skeletal muscle GLUT4 expression. *Sci. Rep.* **2018**, *8*, 8109. [CrossRef]
83. Le Bagge, S.; Fotheringham, A.K.; Leung, S.S.; Forbes, J.M. Targeting the receptor for advanced glycation end products (RAGE) in type 1 diabetes. *Med. Res. Rev.* **2020**, *40*, 1200–1219. [CrossRef]
84. Du, C.; Whiddett, R.O.; Buckle, I.; Chen, C.; Forbes, J.M.; Fotheringham, A.K. Advanced Glycation End Products and Inflammation in Type 1 Diabetes Development. *Cells* **2022**, *11*, 3503. [CrossRef]
85. Guan, S.S.; Sheu, M.L.; Yang, R.S.; Chan, D.C.; Wu, C.T.; Yang, T.H.; Chiang, C.K.; Liu, S.H. The pathological role of advanced glycation end products-downregulated heat shock protein 60 in islet β-cell hypertrophy and dysfunction. *Oncotarget* **2016**, *7*, 23072–23087. [CrossRef]
86. Li, S.; Yang, H. Relationship between advanced glycation end products and gestational diabetes mellitus. *J. Matern. Fetal Neonatal Med.* **2019**, *32*, 2783–2789. [CrossRef] [PubMed]
87. Wojcik, M.; Krawczyk, M.; Zielenlak, A.; Mac-Marcjanek, K.; Wozniak, L.A. Associations of high blood sugar with oxidative stress and inflammation in patients with type 2 diabetes. In *Dietary Sugar, Salt and Fat in Human Health*; Elsevier: Amsterdam, The Netherlands, 2020; pp. 305–323.
88. Liguori, I.; Russo, G.; Curcio, F.; Bulli, G.; Aran, L.; Della-Morte, D.; Gargiulo, G.; Testa, G.; Cacciatore, F.; Bonaduce, D.; et al. Oxidative stress, aging, and diseases. *Clin. Interv. Aging* **2018**, *13*, 757–772. [CrossRef] [PubMed]
89. Samuel, V.T.; Shulman, G.I. The pathogenesis of insulin resistance: Integrating signaling pathways and substrate flux. *J. Clin. Invest.* **2016**, *126*, 12–22. [CrossRef]
90. Luc, K.; Schramm-Luc, A.; Guzik, T.J.; Mikolajczyk, T.P. Oxidative stress and inflammatory markers in prediabetes and diabetes. *J. Physiol. Pharmacol. An. Off. J. Pol. Physiol. Soc.* **2019**, *70*, 809–824. [CrossRef]
91. Furukawa, S.; Fujita, T.; Shimabukuro, M.; Iwaki, M.; Yamada, Y.; Nakajima, Y.; Nakayama, O.; Makishima, M.; Matsuda, M.; Shimomura, I. Increased oxidative stress in obesity and its impact on metabolic syndrome. *J. Clin. Investig.* **2004**, *114*, 1752–1761. [CrossRef]
92. Wang, Y.; Alkhalidy, H.; Liu, D. The Emerging Role of Polyphenols in the Management of Type 2 Diabetes. *Molecules* **2021**, *26*, 703. [CrossRef] [PubMed]
93. Da Porto, A.; Cavarape, A.; Colussi, G.; Casarsa, V.; Catena, C.; Sechi, L.A. Polyphenols Rich Diets and Risk of Type 2 Diabetes. *Nutrients* **2021**, *13*, 1445. [CrossRef]
94. Naz, R.; Saqib, F.; Awadallah, S.; Wahid, M.; Latif, M.F.; Iqbal, I.; Mubarak, M.S. Food Polyphenols and Type II Diabetes Mellitus: Pharmacology and Mechanisms. *Molecules* **2023**, *28*, 3996. [CrossRef] [PubMed]
95. Meigs, J.B.; Larson, M.G.; Fox, C.S.; Keaney, J.F., Jr.; Vasan, R.S.; Benjamin, E.J. Association of oxidative stress, insulin resistance, and diabetes risk phenotypes: The Framingham Offspring Study. *Diabetes Care* **2007**, *30*, 2529–2535. [CrossRef] [PubMed]

96. Tiedge, M.; Lortz, S.; Drinkgern, J.; Lenzen, S. Relation between antioxidant enzyme gene expression and antioxidative defense status of insulin-producing cells. *Diabetes* **1997**, *46*, 1733–1742. [CrossRef] [PubMed]
97. Robertson, R.P. Oxidative stress and impaired insulin secretion in type 2 diabetes. *Curr. Opin. Pharmacol.* **2006**, *6*, 615–619. [CrossRef]
98. Acharya, J.D.; Pande, A.J.; Joshi, S.M.; Yajnik, C.S.; Ghaskadbi, S.S. Treatment of hyperglycaemia in newly diagnosed diabetic patients is associated with a reduction in oxidative stress and improvement in β-cell function. *Diabetes Metab. Res. Rev.* **2014**, *30*, 590–598. [CrossRef] [PubMed]
99. Goldfine, A.B.; Fonseca, V.; Jablonski, K.A.; Pyle, L.; Staten, M.A.; Shoelson, S.E. The Effects of Salsalate on Glycemic Control in Patients With Type 2 Diabetes. *Ann. Intern. Med.* **2010**, *152*, 346–357. [CrossRef]
100. Gupta, S.; Maratha, A.; Siednienko, J.; Natarajan, A.; Gajanayake, T.; Hoashi, S.; Miggin, S. Analysis of inflammatory cytokine and TLR expression levels in Type 2 Diabetes with complications. *Sci. Rep.* **2017**, *7*, 7633. [CrossRef]
101. Böni-Schnetzler, M.; Meier, D.T. Islet inflammation in type 2 diabetes. *Semin. Immunopathol.* **2019**, *41*, 501–513. [CrossRef]
102. Chaudhuri, A.; Ghanim, H.; Vora, M.; Sia, C.L.; Korzeniewski, K.; Dhindsa, S.; Makdissi, A.; Dandona, P. Exenatide exerts a potent antiinflammatory effect. *J. Clin. Endocrinol. Metab.* **2012**, *97*, 198–207. [CrossRef]
103. Guerrero-Romero, F.; Simental-Mendía, L.E.; Rodríguez-Morán, M. Association of C-reactive protein levels with fasting and postload glucose levels according to glucose tolerance status. *Arch. Med. Res.* **2014**, *45*, 70–75. [CrossRef]
104. Colak, A.; Akinci, B.; Diniz, G.; Turkon, H.; Ergonen, F.; Yalcin, H.; Coker, I. Postload hyperglycemia is associated with increased subclinical inflammation in patients with prediabetes. *Scand. J. Clin. Lab. Investig.* **2013**, *73*, 422–427. [CrossRef]
105. Arnalich, F.; Hernanz, A.; López-Maderuelo, D.; Peña, J.M.; Camacho, J.; Madero, R.; Vázquez, J.J.; Montiel, C. Enhanced acute-phase response and oxidative stress in older adults with type II diabetes. *Horm. Metab. Res.* **2000**, *32*, 407–412. [CrossRef] [PubMed]
106. Giulietti, A.; van Etten, E.; Overbergh, L.; Stoffels, K.; Bouillon, R.; Mathieu, C. Monocytes from type 2 diabetic patients have a pro-inflammatory profile. 1,25-Dihydroxyvitamin D(3) works as anti-inflammatory. *Diabetes Res. Clin. Pract.* **2007**, *77*, 47–57. [CrossRef]
107. Shanmugam, N.; Reddy, M.A.; Guha, M.; Natarajan, R. High glucose-induced expression of proinflammatory cytokine and chemokine genes in monocytic cells. *Diabetes* **2003**, *52*, 1256–1264. [CrossRef] [PubMed]
108. Yun, J.M.; Jialal, I.; Devaraj, S. Epigenetic regulation of high glucose-induced proinflammatory cytokine production in monocytes by curcumin. *J. Nutr. Biochem.* **2011**, *22*, 450–458. [CrossRef]
109. Gonzalez, Y.; Herrera, M.T.; Soldevila, G.; Garcia-Garcia, L.; Fabián, G.; Pérez-Armendariz, E.M.; Bobadilla, K.; Guzmán-Beltrán, S.; Sada, E.; Torres, M. High glucose concentrations induce TNF-α production through the down-regulation of CD33 in primary human monocytes. *BMC Immunol.* **2012**, *13*, 19. [CrossRef] [PubMed]
110. David, J.A.; Rifkin, W.J.; Rabbani, P.S.; Ceradini, D.J. The Nrf2/Keap1/ARE Pathway and Oxidative Stress as a Therapeutic Target in Type II Diabetes Mellitus. *J. Diabetes Res.* **2017**, *2017*, 4826724. [CrossRef]
111. Ghezzi, P.; Bonetto, V.; Fratelli, M. Thiol-disulfide balance: From the concept of oxidative stress to that of redox regulation. *Antioxid. Redox Signal.* **2005**, *7*, 964–972. [CrossRef]
112. Sung, C.C.; Hsu, Y.C.; Chen, C.C.; Lin, Y.F.; Wu, C.C. Oxidative stress and nucleic acid oxidation in patients with chronic kidney disease. *Oxid. Med. Cell Longev.* **2013**, *2013*, 301982. [CrossRef]
113. Bukowiecka-Matusiak, M.; Turek, I.A.; Woźniak, L.A. Natural phenolic antioxidants and their synthetic derivatives. In *Systems Biology of Free Radicals and Antioxidants*; Laher, I., Ed.; Springer: Berlin/Heidelberg, Germany, 2014; pp. 4047–4061. [CrossRef]
114. Rong, T. Antioxidant properties in vitro and in vivo: Realistic assessments of efficacy of plant extracts. *CAB Rev. Perspect. Agric. Vet. Sci. Nutr. Nat. Resour.* **2012**, *2012*, 7. [CrossRef]
115. Perron, N.R.; Brumaghim, J.L. A review of the antioxidant mechanisms of polyphenol compounds related to iron binding. *Cell Biochem. Biophys.* **2009**, *53*, 75–100. [CrossRef]
116. Dimitrios, B. Sources of natural phenolic antioxidants. *Trends Food Sci. Technol.* **2006**, *17*, 505–512. [CrossRef]
117. Fraga, C.G.; Croft, K.D.; Kennedy, D.O.; Tomás-Barberán, F.A. The effects of polyphenols and other bioactives on human health. *Food Funct.* **2019**, *10*, 514–528. [CrossRef] [PubMed]
118. Zhang, H.; Tsao, R. Dietary polyphenols, oxidative stress and antioxidant and anti-inflammatory effects. *Curr. Opin. Food Sci.* **2016**, *8*, 33–42. [CrossRef]
119. Nabavi, S.F.; Barber, A.J.; Spagnuolo, C.; Russo, G.L.; Daglia, M.; Nabavi, S.M.; Sobarzo-Sánchez, E. Nrf2 as molecular target for polyphenols: A novel therapeutic strategy in diabetic retinopathy. *Crit. Rev. Clin. Lab. Sci.* **2016**, *53*, 293–312. [CrossRef]
120. Ozgová, S.; Hermánek, J.; Gut, I. Different antioxidant effects of polyphenols on lipid peroxidation and hydroxyl radicals in the NADPH-, Fe-ascorbate- and Fe-microsomal systems. *Biochem. Pharmacol.* **2003**, *66*, 1127–1137. [CrossRef]
121. Tseng, P.S.; Ande, C.; Moremen, K.W.; Crich, D. Influence of Side Chain Conformation on the Activity of Glycosidase Inhibitors. *Angew. Chem. Int. Ed. Engl.* **2023**, *62*, e202217809. [CrossRef] [PubMed]
122. Rana, A.; Samtiya, M.; Dhewa, T.; Mishra, V.; Aluko, R.E. Health benefits of polyphenols: A concise review. *J. Food Biochem.* **2022**, *46*, e14264. [CrossRef]
123. Vauzour, D.; Rodriguez-Mateos, A.; Corona, G.; Oruna-Concha, M.J.; Spencer, J.P. Polyphenols and human health: Prevention of disease and mechanisms of action. *Nutrients* **2010**, *2*, 1106–1131. [CrossRef] [PubMed]

124. Marhuenda, J.; Pérez-Piñero, S.; Arcusa, R.; Victoria-Montesinos, D.; Cánovas, F.; Sánchez-Macarro, M.; García-Muñoz, A.M.; Querol-Calderón, M.; López-Román, F.J. A Randomized, Double-Blind, Placebo-Controlled Trial to Determine the Effectiveness of a Polyphenolic Extract (*Hibiscus sabdariffa* and *Lippia citriodora*) for Reducing Blood Pressure in Prehypertensive and Type 1 Hypertensive Subjects. *Molecules* **2021**, *26*, 1783. [CrossRef]
125. Hügel, H.M.; Jackson, N.; May, B.; Zhang, A.L.; Xue, C.C. Polyphenol protection and treatment of hypertension. *Phytomedicine* **2016**, *23*, 220–231. [CrossRef] [PubMed]
126. Xiao, X.T.; He, S.Q.; Wu, N.N.; Lin, X.C.; Zhao, J.; Tian, C. Green Tea Polyphenols Prevent Early Vascular Aging Induced by High-Fat Diet via Promoting Autophagy in Young Adult Rats. *Curr. Med. Sci.* **2022**, *42*, 981–990. [CrossRef]
127. Taladrid, D.; de Celis, M.; Belda, I.; Bartolomé, B.; Moreno-Arribas, M.V. Hypertension- and glycaemia-lowering effects of a grape-pomace-derived seasoning in high-cardiovascular risk and healthy subjects. Interplay with the gut microbiome. *Food Funct.* **2022**, *13*, 2068–2082. [CrossRef] [PubMed]
128. Ávila-Gálvez, M.; González-Sarrías, A.; Martínez-Díaz, F.; Abellán, B.; Martínez-Torrano, A.J.; Fernández-López, A.J.; Giménez-Bastida, J.A.; Espín, J.C. Disposition of Dietary Polyphenols in Breast Cancer Patients' Tumors, and Their Associated Anticancer Activity: The Particular Case of Curcumin. *Mol. Nutr. Food Res.* **2021**, *65*, e2100163. [CrossRef]
129. Pannu, N.; Bhatnagar, A. Resveratrol: From enhanced biosynthesis and bioavailability to multitargeting chronic diseases. *Biomed. Pharmacother.* **2019**, *109*, 2237–2251. [CrossRef] [PubMed]
130. Samuel, V.P.; Gupta, G.; Dahiya, R.; Jain, D.A.; Mishra, A.; Dua, K. Current Update on Preclinical and Clinical Studies of Resveratrol, a Naturally Occurring Phenolic Compound. *Crit. Rev. Eukaryot. Gene Expr.* **2019**, *29*, 529–537. [CrossRef]
131. García-Martínez, B.I.; Ruiz-Ramos, M.; Pedraza-Chaverri, J.; Santiago-Osorio, E.; Mendoza-Núñez, V.M. Effect of Resveratrol on Markers of Oxidative Stress and Sirtuin 1 in Elderly Adults with Type 2 Diabetes. *Int. J. Mol. Sci.* **2023**, *24*, 7422. [CrossRef]
132. Grabež, M.; Škrbić, R.; Stojiljković, M.P.; Vučić, V.; Rudić Grujić, V.; Jakovljević, V.; Djuric, D.M.; Suručić, R.; Šavikin, K.; Bigović, D.; et al. A prospective, randomized, double-blind, placebo-controlled trial of polyphenols on the outcomes of inflammatory factors and oxidative stress in patients with type 2 diabetes mellitus. *Rev. Cardiovasc. Med.* **2022**, *23*, 57. [CrossRef] [PubMed]
133. Devi, S.A.; Chamoli, A. Polyphenols as an Effective Therapeutic Intervention Against Cognitive Decline During Normal and Pathological Brain Aging. *Adv. Exp. Med. Biol.* **2020**, *1260*, 159–174. [CrossRef]
134. Viña, J.; Escudero, J.; Baquero, M.; Cebrián, M.; Carbonell-Asíns, J.A.; Muñoz, J.E.; Satorres, E.; Meléndez, J.C.; Ferrer-Rebolleda, J.; Cózar-Santiago, M.D.P.; et al. Genistein effect on cognition in prodromal Alzheimer's disease patients. The GENIAL clinical trial. *Alzheimers Res. Ther.* **2022**, *14*, 164. [CrossRef]
135. Boespflug, E.L.; Eliassen, J.C.; Dudley, J.A.; Shidler, M.D.; Kalt, W.; Summer, S.S.; Stein, A.L.; Stover, A.N.; Krikorian, R. Enhanced neural activation with blueberry supplementation in mild cognitive impairment. *Nutr. Neurosci.* **2018**, *21*, 297–305. [CrossRef]
136. Rubert, J.; Gatto, P.; Pancher, M.; Sidarovich, V.; Curti, C.; Mena, P.; Del Rio, D.; Quattrone, A.; Mattivi, F. A Screening of Native (Poly)phenols and Gut-Related Metabolites on 3D HCT116 Spheroids Reveals Gut Health Benefits of a Flavan-3-ol Metabolite. *Mol. Nutr. Food Res.* **2022**, *66*, e2101043. [CrossRef]
137. Scalbert, A.; Williamson, G. Dietary Intake and Bioavailability of Polyphenols. *J. Nutr.* **2000**, *130*, 2073S–2085S. [CrossRef]
138. Gupta, A.; Kagliwal, L.D.; Singhal, R.S. Biotransformation of polyphenols for improved bioavailability and processing stability. *Adv. Food Nutr. Res.* **2013**, *69*, 183–217. [CrossRef] [PubMed]
139. Ovaskainen, M.L.; Törrönen, R.; Koponen, J.M.; Sinkko, H.; Hellström, J.; Reinivuo, H.; Mattila, P. Dietary intake and major food sources of polyphenols in Finnish adults. *J. Nutr.* **2008**, *138*, 562–566. [CrossRef] [PubMed]
140. Saura-Calixto, F.; Serrano, J.; Goñi, I. Intake and bioaccessibility of total polyphenols in a whole diet. *Food Chem.* **2007**, *101*, 492–501. [CrossRef]
141. Pérez-Jiménez, J.; Fezeu, L.; Touvier, M.; Arnault, N.; Manach, C.; Hercberg, S.; Galan, P.; Scalbert, A. Dietary intake of 337 polyphenols in French adults. *Am. J. Clin. Nutr.* **2011**, *93*, 1220–1228. [CrossRef]
142. Tresserra-Rimbau, A.; Medina-Remón, A.; Pérez-Jiménez, J.; Martínez-González, M.A.; Covas, M.I.; Corella, D.; Salas-Salvadó, J.; Gómez-Gracia, E.; Lapetra, J.; Arós, F.; et al. Dietary intake and major food sources of polyphenols in a Spanish population at high cardiovascular risk: The PREDIMED study. *Nutr. Metab. Cardiovasc. Dis.* **2013**, *23*, 953–959. [CrossRef]
143. Grosso, G.; Stepaniak, U.; Topor-Mądry, R.; Szafraniec, K.; Pająk, A. Estimated dietary intake and major food sources of polyphenols in the Polish arm of the HAPIEE study. *Nutrition* **2014**, *30*, 1398–1403. [CrossRef]
144. Manach, C.; Scalbert, A.; Morand, C.; Rémésy, C.; Jiménez, L. Polyphenols: Food sources and bioavailability. *Am. J. Clin. Nutr.* **2004**, *79*, 727–747. [CrossRef]
145. Peng, H.W.; Cheng, F.C.; Huang, Y.T.; Chen, C.F.; Tsai, T.H. Determination of naringenin and its glucuronide conjugate in rat plasma and brain tissue by high-performance liquid chromatography. *J. Chromatogr. B Biomed. Sci. Appl.* **1998**, *714*, 369–374. [CrossRef]
146. Suganuma, M.; Okabe, S.; Oniyama, M.; Tada, Y.; Ito, H.; Fujiki, H. Wide distribution of [3H](-)-epigallocatechin gallate, a cancer preventive tea polyphenol, in mouse tissue. *Carcinogenesis* **1998**, *19*, 1771–1776. [CrossRef]
147. Abd El Mohsen, M.M.; Kuhnle, G.; Rechner, A.R.; Schroeter, H.; Rose, S.; Jenner, P.; Rice-Evans, C.A. Uptake and metabolism of epicatechin and its access to the brain after oral ingestion. *Free Radic. Biol. Med.* **2002**, *33*, 1693–1702. [CrossRef] [PubMed]
148. El Mohsen, M.A.; Marks, J.; Kuhnle, G.; Moore, K.; Debnam, E.; Kaila Srai, S.; Rice-Evans, C.; Spencer, J.P. Absorption, tissue distribution and excretion of pelargonidin and its metabolites following oral administration to rats. *Br. J. Nutr.* **2006**, *95*, 51–58. [CrossRef]

149. Talavéra, S.; Felgines, C.; Texier, O.; Besson, C.; Gil-Izquierdo, A.; Lamaison, J.L.; Rémésy, C. Anthocyanin metabolism in rats and their distribution to digestive area, kidney, and brain. *J. Agric. Food Chem.* **2005**, *53*, 3902–3908. [CrossRef] [PubMed]
150. Schaffer, S.; Halliwell, B. Do polyphenols enter the brain and does it matter? Some theoretical and practical considerations. *Genes Nutr.* **2012**, *7*, 99–109. [CrossRef]
151. Nie, Q.; Chen, H.; Hu, J.; Fan, S.; Nie, S. Dietary compounds and traditional Chinese medicine ameliorate type 2 diabetes by modulating gut microbiota. *Crit. Rev. Food Sci. Nutr.* **2019**, *59*, 848–863. [CrossRef] [PubMed]
152. Fernández-Millán, E.; Ramos, S.; Alvarez, C.; Bravo, L.; Goya, L.; Martín, M. Microbial phenolic metabolites improve glucose-stimulated insulin secretion and protect pancreatic beta cells against tert-butyl hydroperoxide-induced toxicity via ERKs and PKC pathways. *Food Chem. Toxicol.* **2014**, *66*, 245–253. [CrossRef]
153. Spissu, Y.; Gil, K.A.; Dore, A.; Sanna, G.; Palmieri, G.; Sanna, A.; Cossu, M.; Belhadj, F.; Gharbi, B.; Pinna, M.B.; et al. Anti- and Pro-Oxidant Activity of Polyphenols Extracts of Syrah and Chardonnay Grapevine Pomaces on Melanoma Cancer Cells. *Antioxidants* **2022**, *12*, 80. [CrossRef]
154. De Araújo, F.F.; de Paulo Farias, D.; Neri-Numa, I.A.; Pastore, G.M. Polyphenols and their applications: An approach in food chemistry and innovation potential. *Food Chem.* **2021**, *338*, 127535. [CrossRef]
155. De Paulo Farias, D.; Neri-Numa, I.A.; de Araújo, F.F.; Pastore, G.M. A critical review of some fruit trees from the Myrtaceae family as promising sources for food applications with functional claims. *Food Chem.* **2020**, *306*, 125630. [CrossRef] [PubMed]
156. Granato, D.; Mocan, A.; Câmara, J.S. Is a higher ingestion of phenolic compounds the best dietary strategy? A scientific opinion on the deleterious effects of polyphenols in vivo. *Trends Food Sci. Technol.* **2020**, *98*, 162–166. [CrossRef]
157. Moridani, M.Y.; Galati, G.; O'Brien, P.J. Comparative quantitative structure toxicity relationships for flavonoids evaluated in isolated rat hepatocytes and HeLa tumor cells. *Chem. -Biol. Interact.* **2002**, *139*, 251–264. [CrossRef]
158. Chen, W.; Lin, Y.C.; Ma, X.Y.; Jiang, Z.Y.; Lan, S.P. High concentrations of genistein exhibit pro-oxidant effects in primary muscle cells through mechanisms involving 5-lipoxygenase-mediated production of reactive oxygen species. *Food Chem. Toxicol.* **2014**, *67*, 72–79. [CrossRef] [PubMed]
159. Murakami, A. Dose-dependent functionality and toxicity of green tea polyphenols in experimental rodents. *Arch. Biochem. Biophys.* **2014**, *557*, 3–10. [CrossRef] [PubMed]
160. Dobson, C.C.; Mottawea, W.; Rodrigue, A.; Buzati Pereira, B.L.; Hammami, R.; Power, K.A.; Bordenave, N. Impact of molecular interactions with phenolic compounds on food polysaccharides functionality. *Adv. Food Nutr. Res.* **2019**, *90*, 135–181. [CrossRef]
161. de Paulo Farias, D.; de Araújo, F.F.; Neri-Numa, I.A.; Pastore, G.M. Antidiabetic potential of dietary polyphenols: A mechanistic review. *Food Res. Int.* **2021**, *145*, 110383. [CrossRef] [PubMed]
162. Do, G.M.; Jung, U.J.; Park, H.J.; Kwon, E.Y.; Jeon, S.M.; McGregor, R.A.; Choi, M.S. Resveratrol ameliorates diabetes-related metabolic changes via activation of AMP-activated protein kinase and its downstream targets in db/db mice. *Mol. Nutr. Food Res.* **2012**, *56*, 1282–1291. [CrossRef]
163. Lee, Y.E.; Kim, J.W.; Lee, E.M.; Ahn, Y.B.; Song, K.H.; Yoon, K.H.; Kim, H.W.; Park, C.W.; Li, G.; Liu, Z.; et al. Chronic resveratrol treatment protects pancreatic islets against oxidative stress in db/db mice. *PLoS ONE* **2012**, *7*, e50412. [CrossRef]
164. Na, L.X.; Zhang, Y.L.; Li, Y.; Liu, L.Y.; Li, R.; Kong, T.; Sun, C.H. Curcumin improves insulin resistance in skeletal muscle of rats. *Nutr. Metab. Cardiovasc. Dis.* **2011**, *21*, 526–533. [CrossRef]
165. Seo, K.I.; Choi, M.S.; Jung, U.J.; Kim, H.J.; Yeo, J.; Jeon, S.M.; Lee, M.K. Effect of curcumin supplementation on blood glucose, plasma insulin, and glucose homeostasis related enzyme activities in diabetic db/db mice. *Mol. Nutr. Food Res.* **2008**, *52*, 995–1004. [CrossRef] [PubMed]
166. He, H.J.; Wang, G.Y.; Gao, Y.; Ling, W.H.; Yu, Z.W.; Jin, T.R. Curcumin attenuates Nrf2 signaling defect, oxidative stress in muscle and glucose intolerance in high fat diet-fed mice. *World J. Diabetes* **2012**, *3*, 94–104. [CrossRef]
167. Jiménez-Flores, L.M.; López-Briones, S.; Macías-Cervantes, M.H.; Ramírez-Emiliano, J.; Pérez-Vázquez, V. A PPARγ, NF-κB and AMPK-dependent mechanism may be involved in the beneficial effects of curcumin in the diabetic db/db mice liver. *Molecules* **2014**, *19*, 8289–8302. [CrossRef] [PubMed]
168. Wu, L.Y.; Juan, C.C.; Ho, L.T.; Hsu, Y.P.; Hwang, L.S. Effect of green tea supplementation on insulin sensitivity in Sprague-Dawley rats. *J. Agric. Food Chem.* **2004**, *52*, 643–648. [CrossRef]
169. Tsuneki, H.; Ishizuka, M.; Terasawa, M.; Wu, J.B.; Sasaoka, T.; Kimura, I. Effect of green tea on blood glucose levels and serum proteomic patterns in diabetic (db/db) mice and on glucose metabolism in healthy humans. *BMC Pharmacol.* **2004**, *4*, 18. [CrossRef] [PubMed]
170. Song, E.K.; Hur, H.; Han, M.K. Epigallocatechin gallate prevents autoimmune diabetes induced by multiple low doses of streptozotocin in mice. *Arch. Pharm. Res.* **2003**, *26*, 559–563. [CrossRef]
171. Zhang, C.; Li, X.; Hu, X.; Xu, Q.; Zhang, Y.; Liu, H.; Diao, Y.; Zhang, X.; Li, L.; Yu, J.; et al. Epigallocatechin-3-gallate prevents inflammation and diabetes -Induced glucose tolerance through inhibition of NLRP3 inflammasome activation. *Int. Immunopharmacol.* **2021**, *93*, 107412. [CrossRef]
172. Roghani, M.; Baluchnejadmojarad, T. Hypoglycemic and hypolipidemic effect and antioxidant activity of chronic epigallocatechin-gallate in streptozotocin-diabetic rats. *Pathophysiology* **2010**, *17*, 55–59. [CrossRef]
173. Tian, R.; Yang, W.; Xue, Q.; Gao, L.; Huo, J.; Ren, D.; Chen, X. Rutin ameliorates diabetic neuropathy by lowering plasma glucose and decreasing oxidative stress via Nrf2 signaling pathway in rats. *Eur. J. Pharmacol.* **2016**, *771*, 84–92. [CrossRef]

174. Kamalakkannan, N.; Prince, P.S. Antihyperglycaemic and antioxidant effect of rutin, a polyphenolic flavonoid, in streptozotocin-induced diabetic wistar rats. *Basic. Clin. Pharmacol. Toxicol.* **2006**, *98*, 97–103. [CrossRef]
175. Eitah, H.E.; Maklad, Y.A.; Abdelkader, N.F.; Gamal El Din, A.A.; Badawi, M.A.; Kenawy, S.A. Modulating impacts of quercetin/sitagliptin combination on streptozotocin-induced diabetes mellitus in rats. *Toxicol. Appl. Pharmacol.* **2019**, *365*, 30–40. [CrossRef]
176. Palma-Duran, S.A.; Vlassopoulos, A.; Lean, M.; Govan, L.; Combet, E. Nutritional intervention and impact of polyphenol on glycohemoglobin (HbA1c) in non-diabetic and type 2 diabetic subjects: Systematic review and meta-analysis. *Crit. Rev. Food Sci. Nutr.* **2017**, *57*, 975–986. [CrossRef] [PubMed]
177. Liu, K.; Zhou, R.; Wang, B.; Chen, K.; Shi, L.-Y.; Zhu, J.-D.; Mi, M.-T. Effect of green tea on glucose control and insulin sensitivity: A meta-analysis of 17 randomized controlled trials. *Am. J. Clin. Nutr.* **2013**, *98*, 340–348. [CrossRef] [PubMed]
178. Zheng, X.-X.; Xu, Y.-L.; Li, S.-H.; Hui, R.; Wu, Y.-J.; Huang, X.-H. Effects of green tea catechins with or without caffeine on glycemic control in adults: A meta-analysis of randomized controlled trials. *Am. J. Clin. Nutr.* **2013**, *97*, 750–762. [CrossRef]
179. Yu, J.; Song, P.; Perry, R.; Penfold, C.; Cooper, A.R. The Effectiveness of Green Tea or Green Tea Extract on Insulin Resistance and Glycemic Control in Type 2 Diabetes Mellitus: A Meta-Analysis. *Diabetes Metab. J.* **2017**, *41*, 251–262. [CrossRef]
180. Rienks, J.; Barbaresko, J.; Oluwagbemigun, K.; Schmid, M.; Nöthlings, U. Polyphenol exposure and risk of type 2 diabetes: Dose-response meta-analyses and systematic review of prospective cohort studies. *Am. J. Clin. Nutr.* **2018**, *108*, 49–61. [CrossRef]
181. Brasnyó, P.; Molnár, G.A.; Mohás, M.; Markó, L.; Laczy, B.; Cseh, J.; Mikolás, E.; Szijártó, I.A.; Mérei, A.; Halmai, R.; et al. Resveratrol improves insulin sensitivity, reduces oxidative stress and activates the Akt pathway in type 2 diabetic patients. *Br. J. Nutr.* **2011**, *106*, 383–389. [CrossRef] [PubMed]
182. Goh, K.P.; Lee, H.Y.; Lau, D.P.; Supaat, W.; Chan, Y.H.; Koh, A.F. Effects of resveratrol in patients with type 2 diabetes mellitus on skeletal muscle SIRT1 expression and energy expenditure. *Int. J. Sport Nutr. Exerc. Metab.* **2014**, *24*, 2–13. [CrossRef]
183. Timmers, S.; Konings, E.; Bilet, L.; Houtkooper, R.H.; van de Weijer, T.; Goossens, G.H.; Hoeks, J.; van der Krieken, S.; Ryu, D.; Kersten, S.; et al. Calorie restriction-like effects of 30 days of resveratrol supplementation on energy metabolism and metabolic profile in obese humans. *Cell Metab.* **2011**, *14*, 612–622. [CrossRef]
184. Bo, S.; Ponzo, V.; Ciccone, G.; Evangelista, A.; Saba, F.; Goitre, I.; Procopio, M.; Pagano, G.F.; Cassader, M.; Gambino, R. Six months of resveratrol supplementation has no measurable effect in type 2 diabetic patients. A randomized, double blind, placebo-controlled trial. *Pharmacol. Res.* **2016**, *111*, 896–905. [CrossRef] [PubMed]
185. Thota, R.N.; Rosato, J.I.; Dias, C.B.; Burrows, T.L.; Martins, R.N.; Garg, M.L. Dietary Supplementation with Curcumin Reduce Circulating Levels of Glycogen Synthase Kinase-3β and Islet Amyloid Polypeptide in Adults with High Risk of Type 2 Diabetes and Alzheimer's Disease. *Nutrients* **2020**, *12*, 1032. [CrossRef] [PubMed]
186. Karandish, M.; Mozaffari-Khosravi, H.; Mohammadi, S.M.; Cheraghian, B.; Azhdari, M. The effect of curcumin and zinc co-supplementation on glycemic parameters in overweight or obese prediabetic subjects: A phase 2 randomized, placebo-controlled trial with a multi-arm, parallel-group design. *Phytother. Res.* **2021**, *35*, 4377–4387. [CrossRef]
187. Ostadmohammadi, V.; Milajerdi, A.; Ayati, E.; Kolahdooz, F.; Asemi, Z. Effects of quercetin supplementation on glycemic control among patients with metabolic syndrome and related disorders: A systematic review and meta-analysis of randomized controlled trials. *Phytother. Res.* **2019**, *33*, 1330–1340. [CrossRef] [PubMed]
188. Chen, C.R.; Liao, Y.W.; Wang, L.; Kuo, Y.H.; Liu, H.J.; Shih, W.L.; Cheng, H.L.; Chang, C.I. Cucurbitane triterpenoids from Momordica charantia and their cytoprotective activity in tert-butyl hydroperoxide-induced hepatotoxicity of HepG2 cells. *Chem. Pharm. Bull.* **2010**, *58*, 1639–1642. [CrossRef]
189. Chen, J.; Tian, R.; Qiu, M.; Lu, L.; Zheng, Y.; Zhang, Z. Trinorcucurbitane and cucurbitane triterpenoids from the roots of Momordica charantia. *Phytochemistry* **2008**, *69*, 1043–1048. [CrossRef]
190. Bennett, J.P., Jr.; Onyango, I.G. Energy, Entropy and Quantum Tunneling of Protons and Electrons in Brain Mitochondria: Relation to Mitochondrial Impairment in Aging-Related Human Brain Diseases and Therapeutic Measures. *Biomedicines* **2021**, *9*, 225. [CrossRef] [PubMed]

biomolecules

MDPI

Article

Melatonin as a Repairing Agent in Cadmium- and Free Fatty Acid-Induced Lipotoxicity

Anna Migni [1], Francesca Mancuso [2], Tiziano Baroni [2], Gabriele Di Sante [2], Mario Rende [2], Francesco Galli [1,*,†] and Desirée Bartolini [1,*,†]

1 Department of Pharmaceutical Sciences, University of Perugia, 06123 Perugia, Italy; annamigni4@gmail.com
2 Department of Medicine and Surgery, University of Perugia, 06123 Perugia, Italy; francesca.mancuso@unipg.it (F.M.); tiziano.baroni@unipg.it (T.B.); gabriele.disante@unipg.it (G.D.S.); mario.rende@unipg.it (M.R.)
* Correspondence: francesco.galli@unipg.it (F.G.); desiree.bartolini@unipg.it (D.B.); Tel.: +39-075585-7490 (F.G.); Tel.: +39-075585-7445 (D.B.)
† These authors contributed equally to this work.

Abstract: (1) Background: Cadmium (Cd) is a potentially toxic element with a long half-life in the human body (20–40 years). Cytotoxicity mechanisms of Cd include increased levels of oxidative stress and apoptotic signaling, and recent studies have suggested that these aspects of Cd toxicity contribute a role in the pathobiology of non-alcoholic fatty liver disease (NAFLD), a highly prevalent ailment associated with hepatic lipotoxicity and an increased generation of reactive oxygen species (ROS). In this study, Cd toxicity and its interplay with fatty acid (FA)-induced lipotoxicity have been studied in intestinal epithelium and liver cells; the cytoprotective function of melatonin (MLT) has been also evaluated. (2) Methods: human liver cells (HepaRG), primary murine hepatocytes and Caco-2 intestinal epithelial cells were exposed to $CdCl_2$ before and after induction of lipotoxicity with oleic acid (OA) and/or palmitic acid (PA), and in some experiments, FA was combined with MLT (50 nM) treatment. (3) Results: $CdCl_2$ toxicity was associated with ROS induction and reduced cell viability in both the hepatic and intestinal cells. Cd and FA synergized to induce lipid droplet formation and ROS production; the latter was higher for PA compared to OA in liver cells, resulting in a higher reduction in cell viability, especially in HepaRG and primary hepatocytes, whereas CACO-2 cells showed higher resistance to Cd/PA-induced lipotoxicity compared to liver cells. MLT showed significant protection against Cd toxicity either considered alone or combined with FFA-induced lipotoxicity in primary liver cells. (4) Conclusions: Cd and PA combine their pro-oxidant activity to induce lipotoxicity in cellular populations of the gut–liver axis. MLT can be used to lessen the synergistic effect of Cd-PA on cellular ROS formation.

Keywords: melatonin; cadmium; lipotoxicity

Citation: Migni, A.; Mancuso, F.; Baroni, T.; Di Sante, G.; Rende, M.; Galli, F.; Bartolini, D. Melatonin as a Repairing Agent in Cadmium- and Free Fatty Acid-Induced Lipotoxicity. *Biomolecules* **2023**, *13*, 1758. https://doi.org/10.3390/biom13121758

Academic Editor: Chryssostomos Chatgilialoglu

Received: 2 September 2023
Revised: 4 December 2023
Accepted: 5 December 2023
Published: 7 December 2023

1. Introduction

Cadmium (Cd) is a potentially toxic element (PTE) widely distributed in the environment due to anthropogenic activities (industrial processes, mining activities and the combustion of fossil fuels) [1]. Its persistence due to its long half-life and bioaccumulation in living organisms make it a significant public health concern [2]. Human exposure to Cd occurs via ingestion, inhalation and dermal contact [3]. Understanding the sources and exposure routes is crucial for evaluating the potential risks associated with Cd toxicity. Cd exerts its toxic effects through multiple mechanisms. It disrupts cellular homeostasis by interfering with essential biological processes, such as enzymatic activities, antioxidant defense systems and DNA repair mechanisms [4–7]. Additionally, Cd can induce oxidative stress, leading to the generation of reactive oxygen species (ROS) and subsequent damage to cellular components [8,9]. Cell- and tissue-specific mechanisms of toxicity have been described in different experimental models, with the kidney as a main target of the chronic effects of Cd toxicity and the liver of acute intoxication (reviewed in [10–12]).

In the liver cell, Cd exposure reduces cell viability, inducing ROS generation, lipid metabolism abnormalities, lipotoxicity and inflammatory gene activation [13–21]. Moreover, fat accumulation in the liver, a condition that can present as either physiological or pathological, increases the tissue-specific accumulation of Cd [22] and may interfere with the stress response and detoxification mechanisms that protect the liver from Cd toxicity (reviewed in [10]). Recent studies in rats demonstrated that Cd modifies the hepatic lipidome inducing de novo lipogenesis and lipoprotein metabolism alterations compatible with the insulin resistance phenotype of metabolic syndrome and non-alcoholic fatty liver disease (NAFLD) [23]. Again, the chronic exposure to Cd in mice fed a high-fat diet accelerates the development of non-alcoholic steatohepatitis (NASH) [21]. NASH is now a main cause of liver transplantation in developed countries and approximately one-fourth of the general population is at risk of developing this form of chronic liver disease [24]. NAFLD pathogenesis depends on a multiple-hit process in which lipotoxicity, an excess of cellular ROS and the activation of inflammatory genes are the earliest instigators of damage and activation of death programs of the liver cells, as well as of other cellular elements, that may contribute to harm the liver health [16,17], including intestinal epithelial cells that represent a natural barrier of the gastrointestinal tract and a system to preserve the liver homeostasis and function. To our knowledge, the effects of lipotoxicity on intestinal epithelial cells remains poorly characterized.

These premises support the hypothesis that Cd may synergize with free fatty acids (FFA) to induce cellular lipotoxicity. The hypothesis was explored in this study using HepaRG human liver cells and CACO-2 intestinal epithelial cells exposed to Cd, and FFA oleic acid (OA) and palmitic acid (PA) in different combinations. Lipotoxicity was investigated, assessing cellular lipid accumulation, cytotoxicity parameters, ROS production and the levels of the inflammatory indicators' interleukin-6 (IL-6) and LOX-5. Moreover, we tested the study hypothesis using melatonin (MLT), a hormonal regulator of the circadian rhythm, endowed with pharmacological properties as an antioxidant and cytoprotective agent [25–30]. Previous studies have demonstrated that MLT provides cytoprotective effects in different models of Cd toxicity [31,32], and preliminary data from our group supported the efficacy of this molecule as a repairing agent in liver cells during combinatorial treatments of Cd and FFA [33], and the pharmacological utilization of MLT was demonstrated to prevent lipotoxicity in experimental models of NAFLD as well as in randomized clinical trials (see [34–36] and references therein).

2. Materials and Methods

2.1. Chemicals

Melatonin (M5250, Sigma-Aldrich, St. Louis, MI, USA), oleic Acid (BioReagent from Sigma-Aldrich, St. Louis, MI, USA; product number: O1383), and palmitic Acid (P0500, Sigma-Aldrich, St. Louis, MI, USA); cadmium chloride ($CdCl_2$; 202908, Sigma-Aldrich, St. Louis, MI, USA); PD 98059 (P215; Sigma-Aldrich, St. Louis, MI, USA); 2′,7′-Dichlorofluorescein (35848; Sigma-Aldrich, St. Louis, MI, USA); Oil Red O (O0625; Sigma-Aldrich, St. Louis, MI, USA); MTT (M2003, Sigma-Aldrich, St. Louis, MI, USA); hydrocortisone 21-hemisuccinate (H2882, Supelco, Munich, Germany); insulin (I6634; Sigma-Aldrich); impermeable thiol-reactive AlexaFluor 488 C5 maleimide (AFM, Thermo Fisher Scientific, Waltham, MA, USA).

2.2. Cell Cultures and Treatments

2.2.1. Cell Lines and Cell Culture Conditions

HepaRG cells (Thermo Scientific, MA, USA) were immortalized human cells with metabolic and morphological similarities to primary hepatocytes. These cells were maintained in William's E medium (Sigma-Aldrich) containing 1% GlutaMAX (Invitrogen, Carlsbad, CA, USA), 10% fetal bovine serum (FBS, GIBCO, Life Technologies, Carlsbad,

CA, USA), 5 μg/mL insulin (Sigma-Aldrich), and 50 μM hydrocortisone 21-hemisuccinate (Supelco). Experiments were performed between cell passages 18–22.

For 2D differentiation, undifferentiated HepaRG were seeded in gelatin-coated plates and cultured in complete William's E medium for 48 h until complete fusion. Then, the medium was supplemented with 1.7% (*v*/*v*) DMSO for a further 18 days to allow the undifferentiated HepaRG to mature hepatocytes. We changed the medium every 3 days and captured the optical microscope images to monitor the differentiation stage using a EVOSTM XL Core imaging system for qualitative analysis (Thermo Fisher Scientific).

CACO-2 cells are colon carcinoma cells that grow in monolayers often used to mimic the intestinal epithelial barrier. These were cultured in Dulbecco's Modified Eagle Medium (DMEM, Thermo Scientific) with the addition of 10% FBS, 1% non-essential ammino acids (NEAA; Euroclone, Milan, Italy) and 1% GlutaMAX (Invitrogen, Carlsbad, CA, USA).

The cell lines were maintained in culture in 96-well plates with a density of 1×10^5 cell/well or in 6-well plates with a density of 2×10^5 cells/m^{-2} in an incubator at 37 °C with a humidified atmosphere of 5% CO_2.

2.2.2. Primary Hepatocyte Isolation and Culture

Hepatocytes were isolated from 4–6-month-old C57BL/6 J mice purchased from Charles River Laboratories (Calco, Italy). The mice were housed in a pathogen-free facility at the University of Perugia (Perugia, Italy) under controlled light and temperature conditions and treated according to European Community guidelines. The experiments were authorized by the Italian Ministry of Health (protocol number 605/2019-PR). Hepatocyte isolation was performed by two-step perfusion using Liver Perfusion and Liver Digest Media (Life Technologies, Pleasanton, CA, USA) followed by separation with 50% Percoll (GE Healthcare Life Sciences, Pittsburgh, PA, USA) density gradient. Once the liver was washed by perfusion, hepatocytes were dissociated by collagenase, separated from other cells and cultured [37]. The purity of live hepatocytes was routinely monitored by trypan blue exclusion (≥90%). Hepatocytes were cultured in HepatoZYME SFM medium (Gibco). Experimental assays were performed 72 h after hepatocyte plating.

2.2.3. Cell Treatments

In a preliminary series of experiments, the toxicity of Cd was investigated in intestinal and hepatic cells using Experimental Protocol 1 (EP1) of Figure 1. The cells were treated from 4 to 96 h with different concentrations of $CdCl_2$ (from 10 to 200 μM). These treatments were performed in the presence or absence of the cytoprotective agent MLT (50 nM). In control experiments of EP1, the cells were treated with vehicles of Cd and/or MLT, namely, 0.001% (*v*/*v*) H_2O or ethanol (EtOH), respectively. The possible interactive effect between Cd toxicity and FFA-induced lipotoxicity hypothesized in this study was investigated using the EP2 of Figure 1. CACO-2 and liver cells were plated in the 96-well black and transparent color plates (1×10^5 cell/well), and then the cells were pre-treated for 24 h with different concentrations of $CdCl_2$ (20 μM and 50 μM for CACO-2 cells; 50 μM and 100 μM for liver cells) and then treated for 48 h with the FFA oleic acid (OA, 200 μM) and/or palmitic acid (PA, 200 μM), or in combination (OA + PA; 200 μM each), in the presence or absence of MLT. OA was dissolved in DMSO and PA was prepared as reported in [38].

For MAPK-ERK1/2 signaling assay, HepaRG and CACO-2 cells were pre-treated for 1 h with ERK inhibitor (PD98059) 50 μM and then treated for 3 h with $CdCl_2$ 50 μM or 20 μM, respectively, in the presence or absence of MLT 50 nM.

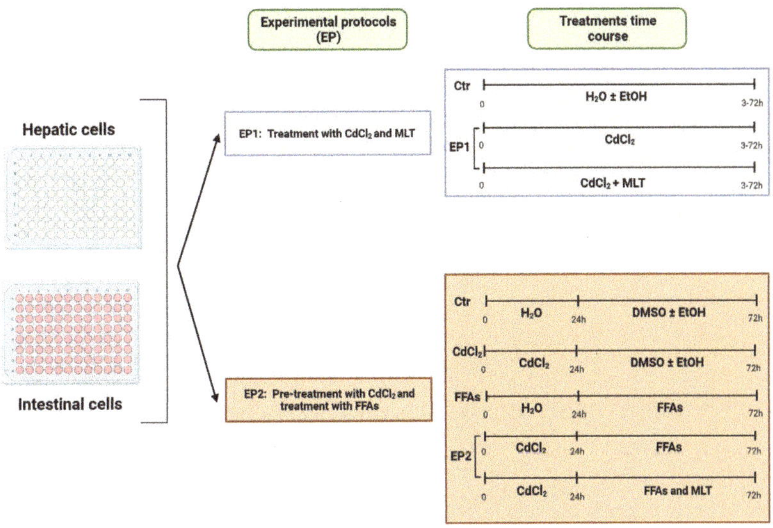

Figure 1. Experimental protocols (EP). The scheme shows the two-arm EP used in this study; namely, EP1 was designed to assess $CdCl_2$ cytotoxicity effects as studied by the cell viability test and the activity of survival and stress MAPKs. EP2 was set to explore the hypothesis that Cd can interact with FFA in inducing lipotoxicity and ROS generation in liver and intestinal cells. In EP2, the cells were pre-treated with different concentrations of $CdCl_2$ for 24 h and then treated for 48 h with 200 μM final concentration of the FFA oleic acid (OA) or palmitic acid (PA) or a combination of both (OA + PA). Effects of Cd and FFA were studied either in the presence or absence of the cytoprotective molecule MLT (50 nM) and against control experiments run with the vehicles of these treatments. Further details on experimental conditions are reported in the text.

2.3. Cell Viability Assay

Cell toxicity of $CdCl_2$ and FFAs and the cytoprotective effect of MLT were determined by MTT test (Sigma-Aldrich) in 96-well plates with a seeding density of 10,000 cells/cm^2, as reported in [39]. Briefly, after different times of incubation (3 to 72 h), the cell medium was replaced with fresh medium containing the MTT solution (1:10 *v*/*v*) and the cells were incubated for 2.5 h at 37 °C. An MTT solubilization solution (0.1 N HCl containing 10% vol/vol Triton X-100 in anhydrous isopropanol) was used to dissolve the formazan crystals formed during the incubation by the activity of mitochondrial dehydrogenases of viable cells, and a microplate reader (DTX880 Multimode Detector, Beckman Coulter, Brea, CA, USA) was used to measure the absorbance of the cell culture medium at 570 nm. The results were expressed as percentage of the optical density (OD) observed in control cells.

2.4. Apoptosis Assay by Flow Cytometry

Each group of cells was treated with Trypsin-EDTA solution (Sigma-Aldrich) to create a single-cell suspension, centrifuged (1500 rpm, 5 min) and washed twice with PBS solution, resuspended with 500 μL of binding buffer, then mixed sequentially with 5 μL of FITC-labeled Annexin V and 5 μL of PI stain, and incubated for 10 min at room temperature, and then immediately, cell apoptosis was measured using BD Accuri C6 Plus System (BD Biosciences, San Jose, CA, USA). An Annexin V-FITC/PI Apoptosis Kit was purchased from Elabscience (Cat.No.: E-CK-A211).

2.5. Intracellular and Extracellular ROS Analysis

Intracellular ROS were measured by the oxidative conversion of the intracellular probe 2′,7′-dichlorofluorescein-diacetate (DCFH-DA; Sigma-Aldrich) to the fluorescent derivative 2′,7′-dichlorofluorescein (DCF). Briefly, the cells were treated with 50 μM dichlorofluorescein diacetate (DCFH-DA) in PBS and incubated in the dark for 30 min at 37 °C to allow the probe to enter the cell and react with cellular ROS after enzymatic hydrolysis by intracellular esterases to its derivative dichloro-fluorescein (DCFH). This is a non-fluorescent derivative that, other than losing the ability to cross the membrane, is reactive to cellular ROS to produce a highly fluorescent form of the probe, i.e., the oxidized DCFH derivative. After washing with PBS to remove the non-internalized DCFH-DA, the fluorescence of the cells was recorded using a microplate reader set with excitation λ of 485 nm and emission λ of 535 nm. The obtained fluorescence (FL) was normalized against the absorbance (Abs) of the MTT test (FL/Abs), and the results were expressed as the percentage of the control cell value (identified as CTL).

Extracellular H_2O_2 was determined using a colorimetric assay kit (Elabscience, Houston, TX, USA) adapted to a 96-well microplate reader. A total of 100 µL of Reagent 1 (buffer solution of Elabscience kit) was pipetted in each well and incubated at 37 °C for 10 min. Then, 10 µL of the cell supernatant or of the negative and positive control test (namely, bi-distilled water and 60 mmol/L H_2O_2, respectively) were added, followed by 100 µL of Reagent 2 (ammonium molybdate); the absorbance of the solution was recorded at 405 nm as optical density units (OD) in a DTX880 Multimode Detector microplate reader (Beckman Coulter). Hydrogen peroxide concentrations were calculated using the following formula: $[H_2O_2]$ (mmol/L) = $(\Delta A1/\Delta A2) \times c \times f$.

ΔA1: ODSample—ODBlank;
ΔA2: ODStandard—ODBlank;
c: H_2O_2 concentration of the standard sample = 60 mmol/L;
f: dilution factor of sample.

2.6. Oil Red O (ORO) Assay

Cellular lipids were measured by Oil Red O (ORO) staining according to the procedure described in [38]. Briefly, CACO-2 and HepaRG cells were fixed with 10% neutral buffer formalin (Leica) for 30 min; the fixed cells were thus washed twice with sterile bi-distilled water and then incubated with 60% isopropanol for 5 min. The cells were stained with ORO solution for 2–5 min and then washed four times with sterile bi-distilled water before staining with hematoxylin solution (Sigma-Aldrich, St. Louis, MI, USA) for 1 min. The stained cells were washed again with sterile bi-distilled water and then were assessed by optical microscopy using a EVOSTM XL Core imaging system for qualitative analysis (Thermo Fisher). Furthermore, to quantify the cellular content of ORO, the cell pellet was incubated for 10 min with 100 µL of isopropanol, and the absorbance of the extract was assessed at 510 nm using a multiplate reader monochromator (TECAN, Männedorf, Switzerland).

2.7. Analysis of Cell-Surface Thiols by Flow Cytometry

After treatments, HepaRG cells were recovered and washed with PBS twice, then incubated with 10 µM AlexaFluor 488 C5 maleimide (AFM) probe in PBS for 30 min at 37 °C. After having been washed again in PBS, the cells were analyzed by flow cytometry using BD Accuri C6 Plus System (BD Biosciences, San Jose, CA, USA). Dead cells were gated out by staining with Propidium Iodide (PI, Thermo Fisher Scientific, Waltham, MA, USA), and 1×10^4 living cells was analyzed. The results were expressed as mean fluorescence of the AFM probe.

2.8. Total Protein Extraction and Quantification

The cells were plated in 6-well plates, and proteins were extracted with 100 µL of cell lysis buffer (Cell Signaling Technology Inc., Danvers, MA, USA) and protease inhibitor cocktail (Pierce, Thermo Fisher Scientific Inc., Waltham, MA, USA). The cell lysates were maintained in ice for 40 min before centrifugation at 14,000 rpm for 20 min at 4 °C, and the supernatants containing the total protein extract were collected and stored at −80 °C before determination. Protein concentrations were measured by bicinchoninic acid assay (BCA assay, Pierce, Thermo Fisher Scientific Inc., Waltham, MA, USA). A total of 200 µL of a solution 1:50 (*v*/*v*) of reagent A (bicinchoninc acid, sodium bicarbonate, sodium tartrate and sodium carbonate in 0.1 N NaOH, 11.25 final pH) and reagent B (4% w/v of $CuSO_4 \cdot 5 H_2O$ in water) were pipetted in a 96-well plate together with 10 μL of a sample or bovine serum albumin (BSA) at different concentration that was used as external standard for assay calibration. After 30 min of incubation at 37 °C in the dark, the absorbance of the samples and analytical standard was recorded at 570 nm using a microplate reader (DTX880 Multimode Detector, Beckman Coulter, Brea, CA, USA).

2.9. SDS-PAGE and Immunoblotting

Cell proteins (10–30 μg) were separated by 10–12% sodium dodecyl sulfate–polyacrylamide gel electrophoresis (SDS–PAGE) and then transferred to a nitrocellulose membrane (Millipore, Billerica, MA, USA) for immunoblot analysis. Membranes were blocked for 2 h at room temperature with 5% skim milk (Sigma-Aldrich) in Tris-buffered saline and 0.1% Tween20 (TBST) and then incubated overnight at 4 °C with the primary antibodies (Cell Signaling Technology, Danvers, MA, USA) that included the following: phospho-p44/42 MAPK (ERK1/2) rabbit monoclonal antibody (mAb) (1:1000), pp42/44 MAPK (Erk1/2) rabbit mAb (1:1000), phospho-SAPK/JNK rabbit mAb (1:1000), SAPK/JNK rabbit mAb (1:1000), GAPDH rabbit mAb (1:1000), phospho-p38 rabbit mAb (1:1000), p38 rabbit mAb (1:1000), Catalase rabbit mAb (1:1000), SOD-2 rabbit mAb (1:1000), 5-LOX rabbit mAb (1:1000) and α-Tubulin rabbit mAb (1:1000). The day after the incubation, the membranes were washed 3 times with TBST 0.1% and incubated with anti-rabbit or anti-mouse IgG (1:2000) horseradish peroxidase-linked secondary antibodies (Cell Signaling Technology Inc., Danvers, MA, USA). Protein bands were detected using an ECL Clarity Max (BioRad, Hercules, CA, USA), and band quantification was performed with a Gel-Pro Analyzer.

2.10. IL-6 Analysis

IL-6 levels in cell culture media of liver cells were detected using a commercial IL-6 ELISA kit (cat. no. BMS603-2) from eBioscience (Thermo Fisher Scientific, Inc., MA, USA) following the manufacturer's instructions.

2.11. Statistical Analyses

Statistical comparisons were performed using one-way ANOVA test. Data were expressed as mean ± SD of 3 independent experiments. The probability of error accepted for significant differences was $p < 0.05$ (* or #), and highly significant differences were identified for $p < 0.01$ (** or ##). Data analysis and graphical presentation were performed with GraphPad Prism 9 (Version 9.0.2).

3. Results

3.1. $CdCl_2$ Toxicity and Cytoprotective Effect of MLT

$CdCl_2$ toxicity in the liver and intestinal cell models utilized in this study, namely, primary murine hepatocytes, HepaRG cells before and after differentiation to hepatocyte-like cells, and CACO-2 intestinal epithelial cells, were investigated using the treatment protocol EP1 (Figure 1). MTT test data showed a concentration- and time-dependent toxicity of $CdCl_2$ in all of the cell models (Supplementary Figures S1 and S2), with a significant reduction in cell viability levels after 24 h of treatment with Cd concentrations > 20 μM. Cytofluorimetry data demonstrated that this toxicity effect of Cd, assessed at 48 h, is associated with a concentration-dependent induction of apoptosis and necrotic cell death in both the HepaRG and CACO-2 cell lines (Figures 2A and 2B, respectively). However, Cd toxicity in CACO-2 cells was characterized by higher levels of late apoptosis and necrosis compared to HepaRG cells (Figure 2). These cell death data indicate that CACO-2 cells are more prone than HepaRG cells to Cd toxicity.

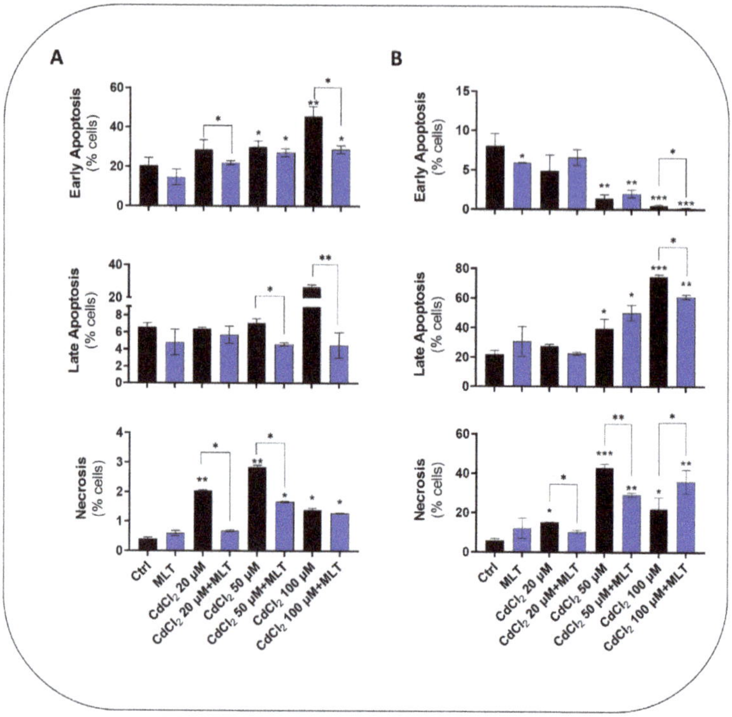

Figure 2. Cell death levels in HepaRG (**A**) and CACO-2 (**B**) cells treated with $CdCl_2$ and MLT. Apoptosis assay was studied by flow cytometry analysis after 48 h of treatment with different concentrations of $CdCl_2$ (as shown in the chart) and 50 nM MLT using EP1 of Figure 1. Apoptosis and necrosis levels were determined by staining the cells by annexin V-FITC and propidium iodide (PI). Data were mean ± SD of three independent experiments. One-way ANOVA test: * $p < 0,05$; ** $p < 0.001$; *** $p < 0.0001$ (control vs. all treatments) and ($CdCl_2$ vs. $CdCl_2$ + MLT).

MLT produced a significant cytoprotective effect, improving cell viability data of both the differentiated and undifferentiated form of HepaRG cells (Supplementary Figure S1A–D and Supplementary Figure S2, respectively), as well as in CACO-2 cells (Supplementary Figure S1E–H) exposed to $CdCl_2$ toxicity. Cell viability was also improved in primary murine hepatocytes, in which MLT exploited a protective effect both during co-treatment (Supplementary Figure S1I–L) and after treatment with Cd (Supplementary Figure S1M,N). Apoptosis and necrosis data confirmed that this cytoprotective activity of MLT depends on a significant reduction in cell death levels in both the two cell lines (Figure 2).

The expression and activity of the survival protein kinase MAPK-ERK1/2 and the stress-activated protein kinases (SAPKs) p38 and JNK were investigated in HepaRG cells exposed to Cd toxicity, since these control cell survival and death pathways during the exposure to cellular stressors also mediating the MLT cytoprotective function [40]. Cd toxicity in HepaRG cells was associated with a significant increase in phosphorylation levels of all protein kinases, namely, MAPK-ERK1/2 (Figure 3A), JNK (Figure 3B) and p38 (Figure 3C), indicating the role of these kinases in the Cd toxicity and its effects on the balance between survival and death pathways of the liver cell.

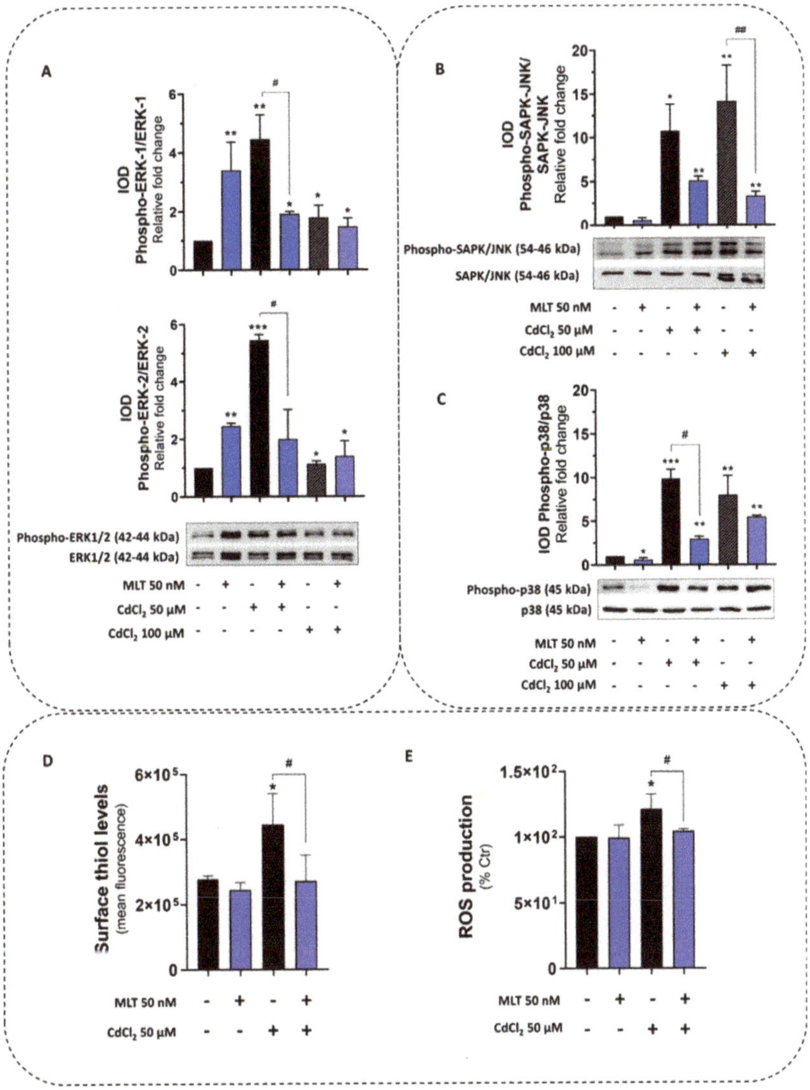

Figure 3. Protein kinase activity and redox parameters of HepaRG cells treated with $CdCl_2$ and MLT. (**A**) MAPK-ERK1-2, (**B**) SAPK/JNK and (**C**) p38-MAPK activity was studied by immunoblot assessing phosphorylation and the native form of the proteins. (**D**) Cell-surface thiols and (**E**) cellular ROS were assessed by FACS-scan analysis. HepaRG cells were exposed for 3 h to $CdCl_2$ (50 and 100 μM) and MLT (50 nM) that were studied as both separate treatments and in co-treatment mode (EP1). One-way ANOVA test: * $p < 0.05$; ** $p < 0.01$; *** $p < 0.001$ (control vs. all treatments). # $p < 0.05$; ## $p < 0.01$ ($CdCl_2$ vs. $CdCl_2$ + MLT).

MLT was confirmed to behave as a potent ERK1/2 agonist [41] by the stimulation of both the protein expression and phosphorylation of MAPK in HepaRG cells (Figure 3A), which is a specific effect as demonstrated by the utilization of the pharmacological inhibitor PD98059 (Supplementary Figure S3). However, during Cd exposure, MLT counteracted MAPK-ERK1/2 and SAPKs (p38 and JNK) activation of this liver cell line (Figure 3A–C), also restoring stress response parameters affected by the activity of these kinases, including the redox indicators cellular surface thiols and ROS (Figures 3D and 3E, respectively), the antioxidant enzymes catalase (CAT) and superoxide dismutase 2 (SOD2) (Supplementary Figure S4A–D), and the inflammatory enzyme 5-LOX (Supplementary Figure S5). The MLT activation effect on the antioxidant enzymes CAT and SOD2 was also confirmed in intestinal cells (Supplementary Figure S4C,D).

3.2. *Effect of $CdCl_2$ and FFA Treatments on Cellular Lipid Accumulation*

The effects of Cd and FFA on lipid levels of the different cell models were studied using the lipid probe ORO in EP2 of Figure 1.

In undifferentiated HepaRG cells, both Cd and FFA, when investigated as separate and independent treatments, were found to induce lipid accumulation in droplets (see light microscopy images of Figure 4A and spectrophotometric data of Figure 4B). This lipid accumulation was associated with a marked reduction in the cell mass revealed by light microscopy images (Figure 4A), demonstrating induction of lipotoxicity by these treatments that was concentration-dependent with respect to $CdCl_2$ in this human liver cell line. When $CdCl_2$ and FFA treatments were combined, these synergized to exacerbate their lipid accumulation and cell mass reduction effects with PA, which was confirmed to induce a stronger interaction with $CdCl_2$ compared to OA. Rather, the latter FFA species appeared to mitigate the lipid accumulation effect of PA in the OA + PA treatment (Figure 4B) but not its lipotoxicity effect, as assessed by the number of cells remaining in the light microscopy fields of Figure 4A.

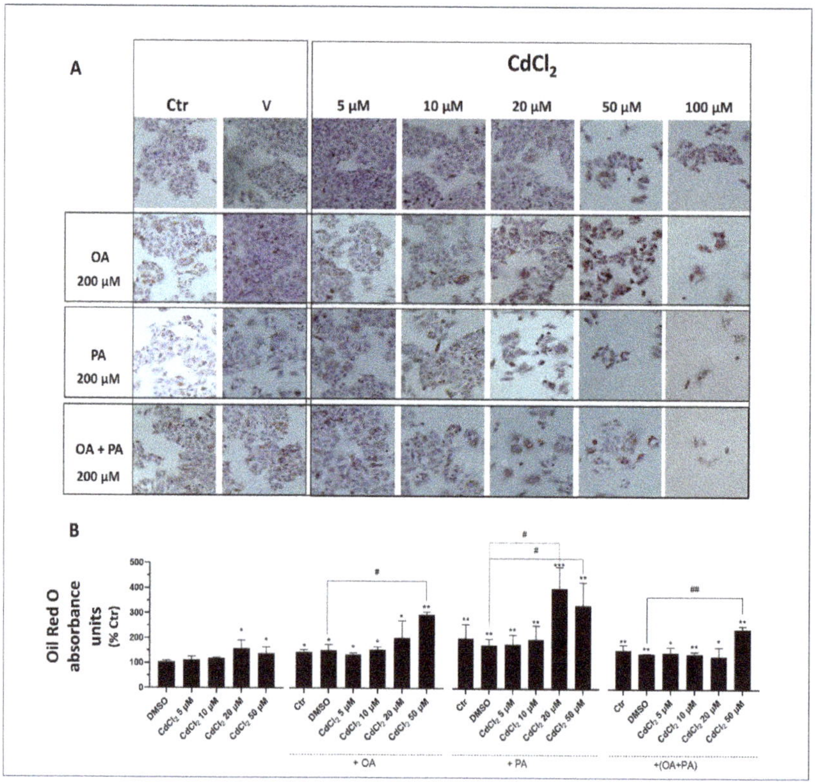

Figure 4. Effect of $CdCl_2$ and FFAs on cellular lipid accumulation in undifferentiated human liver HepaRG cells. Cell treatments were performed with the Experimental Protocol EP2 described in

detail in Figure 1 and in the section "Methods". Briefly, the cells were studied after a 24 h pre-treatment with $CdCl_2$ and 48 h treatment with FFAs (200 μM final concentration each). (**A**) Hematoxylin and Oil Red O (ORO) were used to stain liver cells and lipid droplets, respectively (40× magnification). (**B**) Quantification of cellular lipids by spectrophotometric determination of ORO absorbance at 510 nm. The FFAs used for the treatments were oleic acid (OA), palmitic acid (PA) and their combination (OA + PA). One-way ANOVA test: * $p < 0.05$; ** $p < 0.01$; *** $p < 0.001$ (control test vs. all treatments); # $p < 0.05$; ## $p < 0.01$ (FFAs vs. all treatments).

The concentration-dependent effect of Cd on neutral lipid accumulation was also confirmed in primary mouse hepatocytes studied by spectrophotometric analysis of cellular ORO (Figure 5).

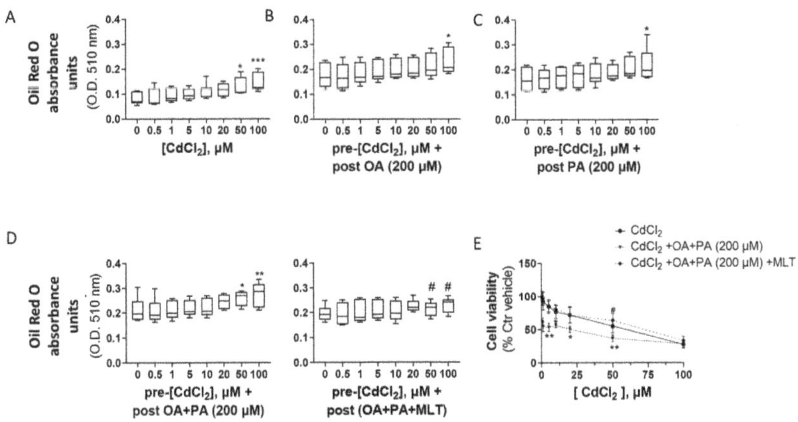

Figure 5. Effect $CdCl_2$, FFA and MLT on cellular lipids (**A–D**) and cell viability levels of primary murine hepatocytes. Cellular lipids were assessed by ORO staining and spectrophotometric analysis at 510 nm of cellular extracts, as described in detail in the section "Materials and Methods". (**A**) Lipids were assessed after 24 h pre-treatment with $CdCl_2$ and 48 h treatment with different FFAs (200 μM final concentration each), namely, OA (**B**), PA (**C**) and OA + PA tested alone or in combination with the cytoprotective agent MLT (50 nM) (**D**). One-way ANOVA test: * $p < 0.05$; ** $p < 0.01$, *** $p < 0.001$ (control test vs. all treatments); # $p < 0.05$ (FFAs vs. FFAs/MLT). Cell viability was measured by MTT assay as described in detail in the section "Materials and Methods" (**E**). The cells were pre-treated for 24 h with increasing concentrations of $CdCl_2$ and then treated with FFAs (200 μM each) for 48 h in the presence or absence of the cytoprotective agent MLT (50 nm) (**E**). One-way ANOVA test: * $p < 0.05$; ** $p < 0.01$ ($CdCl_2$ + OA + PA vs. all treatments); # $p < 0.05$ ($CdCl_2$ + OA + PA vs. $CdCl_2$ + OA + PA + MLT).

Light microscopy images and spectrophotometric data of CACO-2 cells (Figure 6) demonstrate that this cell line is less susceptible to the lipid accumulation effects of Cd and FFA compared to liver cells (Figures 4 and 5).

3.3. Cell Death Levels in Cd- and FFA-Induced Lipotoxicity, and MLT Cytoprotective Effect

Cell viability (Supplementary Figure S6) and cell death data (Figure 7) demonstrated the higher toxicity of PA compared to OA in both the HepaRG and CACO-2 cell lines; rather, OA appears to protect these cells form PA toxicity in combinatorial treatments (i.e., OA + PA). When combined with Cd exposure in the EP2 of Figure 1, FFA increased their toxicity with a higher reduction in viability and increased cell death levels in both the two cell lines, suggesting synergizing effects of Cd and FFA in inducing lipotoxicity; cell viability data confirmed this synergy of effects also in primary murine hepatocytes (Figure 5E).

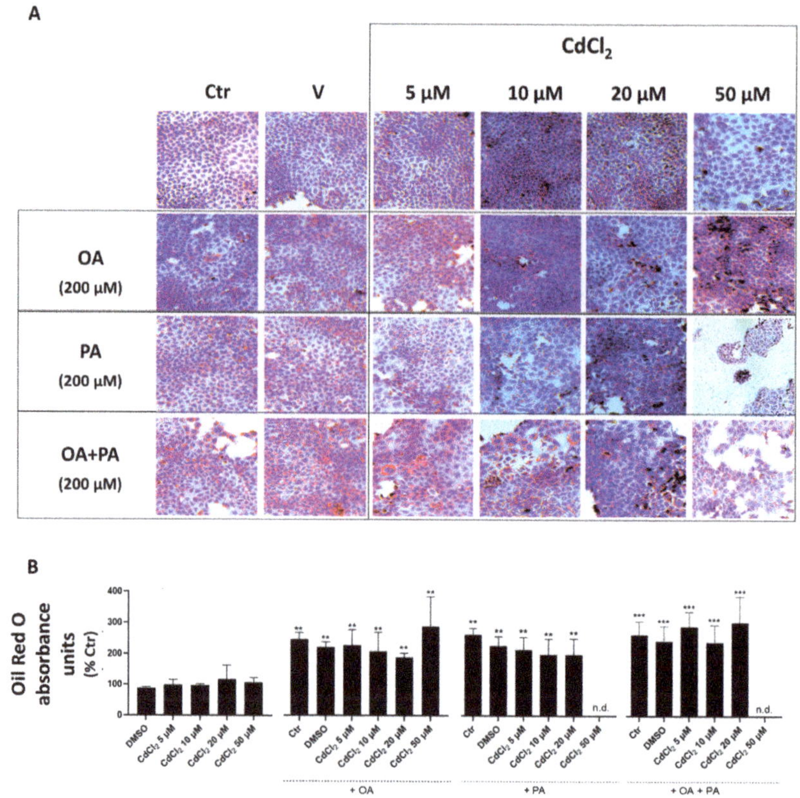

Figure 6. Effect of $CdCl_2$ and FFAs on the accumulation of cellular lipids in CACO-2 intestinal epithelial cells. Cell treatments were as described in the legend of Figure 7 (Experimental Protocol EP2, Figure 1). (**A**) Hematoxylin and Oil Red O (ORO) were used to stain liver cells and lipid droplets, respectively (40× magnification). (**B**) Quantification of cellular lipids by spectrophotometric determination of ORO absorbance at 510 nm. The FFAs used for the treatments were oleic acid (OA), palmitic acid (PA) and a combination of the two (OA + PA). One-way ANOVA test: ** $p < 0.01$; *** $p < 0.001$ (control test vs. all treatments). n.d. (non-detectable).

MLT lessened the cell death induction effect of the different combinations of Cd and FFA in HepaRG and CACO-2 cells (Figure 7A–D), whereas the cell viability reduction effect was not significantly affected (Supplementary Figure S6), indicating different sensitivity of these cell toxicity assays. However, MLT was found to protect the mouse primary hepatocytes from the cell viability reduction effect of Cd exposure combined with OA + PA treatment (Figure 5E), and this effect of MLT was in good agreement with that of lipid accumulation of these liver cells (Figure 5D).

3.4. *Intracellular and Extracellular ROS*

An increased production of cellular ROS is considered a causal indicator of the cytotoxicity process of both the Cd exposure and FFA-induced lipotoxicity. When ROS were studied in HepaRG cells (Figure 8A), the exposure to Cd or FFA increased their levels, with PA generating the most potent ROS generating response; important enough was the observation that Cd synergized with FFA to induce this oxygen activation effect in these cells, which is consistent with other indicators of lipotoxicity shown earlier in Sections 3.2 and 3.3. FFA synergized their ROS production effect with Cd also in CACO-2 intestinal cells (Figure 8B), and MLT was very effective in reducing this combined effect of Cd and FFA treatment in both of the two cell lines (Figures 8A and 8B, respectively).

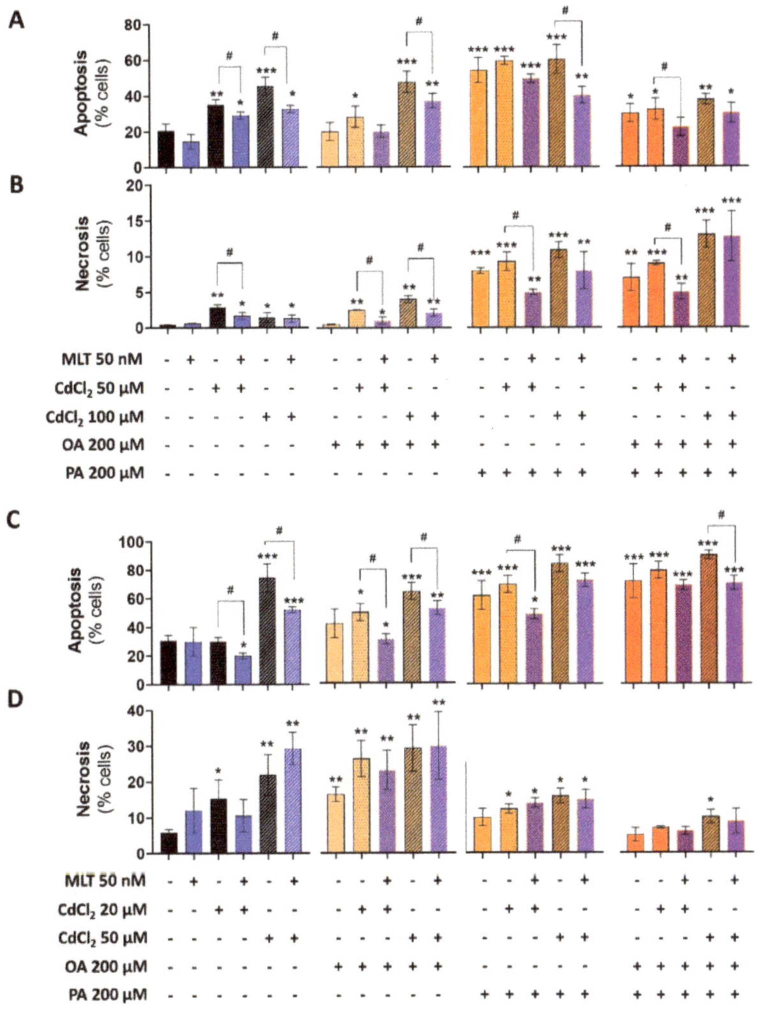

Figure 7. Effect of $CdCl_2$, FFA and MLT on cell death levels of human liver HepaRG cells and CACO-2 intestinal epithelial cells. Apoptosis and necrosis levels were studied in HepaRG (**A**,**B**) and CACO-2 (**C**,**D**) cells exposed to different treatments, as described in EP2 (Figure 1). Apoptosis and necrosis were measured by FACS-scan using Annexin-V and Propidium Iodide probes as described in the methods section. One-way ANOVA test: * $p < 0.05$; ** $p < 0.01$; *** $p < 0.001$ (control vs. all treatments). # $p < 0.05$; ($CdCl_2$ vs. all treatments); # $p < 0.05$ ($CdCl_2$ and/or FFAs vs. $CdCl_2$ and/or FFAs + MLT).

In primary murine hepatocytes, Cd partially synergized with OA + PA to induce cellular oxygen activation (Figure 8C) and H_2O_2 efflux (Figure 8D), and MLT significantly reduced these effects.

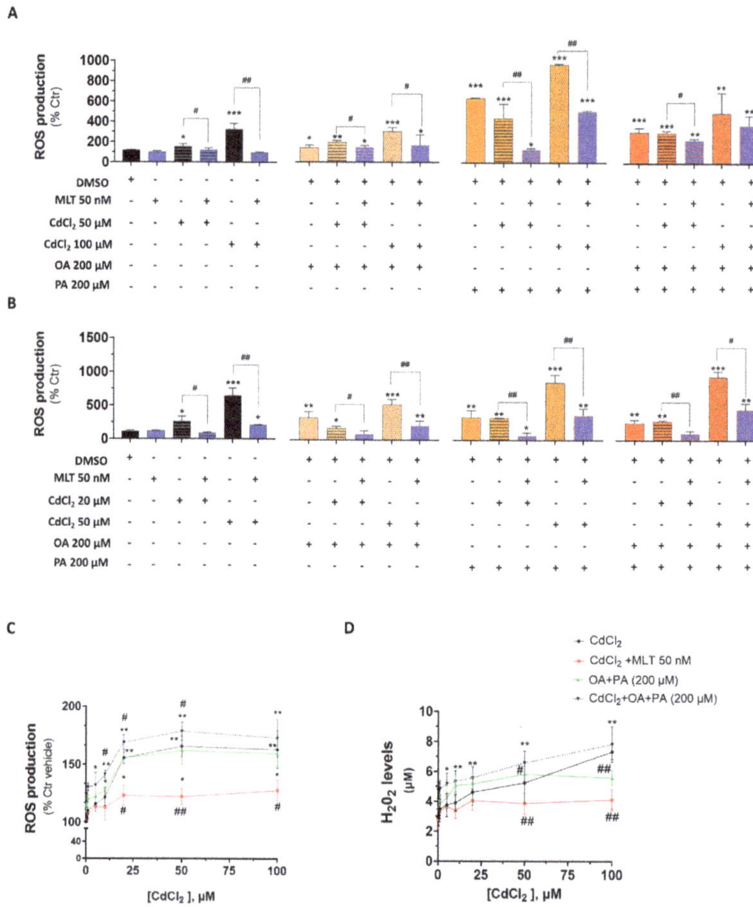

Figure 8. Cellular ROS levels in liver and intestinal cells treated with $CdCl_2$, FFAs and the cytoprotective agent MLT. Cellular ROS were measured in HepaRG (**A**), CACO-2 (**B**) and primary murine hepatocytes (**C**); extracellular levels of H_2O_2 (**D**) were also determined in primary murine hepatocytes only. Cd, FFA and MLT treatments of HepaRG and CACO-2 cells were performed according to EP2 (Figure 1), whereas primary murine hepatocytes were pre-treated for 24 h with increasing concentrations of $CdCl_2$ and then treated for 48 h with FFAs (200 µM final concentration each) in the presence or absence of the cytoprotective agent MLT (50 nm). One-way ANOVA test: (**A**,**B**) * $p < 0.05$; ** $p < 0.01$; *** $p < 0.001$ (control vs. all treatments). # $p < 0.05$ ($CdCl_2$ vs. all treatments); # $p < 0.05$; ## $p < 0.01$ ($CdCl_2$ and/or FFAs vs. $CdCl_2$ and/or FFAs +MLT); (**C**,**D**) # $p < 0.05$; ## $p < 0.01$ ($CdCl_2$ vs. $CdCl_2$ + MLT and OA + PA vs. $CdCl_2$ + OA + PA).

3.5. Effect of $CdCl_2$ and MLT on IL-6 Secretion Levels of HepaRG Cells and Primary Murine Hepatocytes

A dose-dependent induction effect of $CdCl_2$ was observed for IL-6 secretion in both HepaRG cells and primary murine hepatocytes, which was further enhanced by the combination of OA + PA treatment (Figure 9). MLT significantly reduced the effect of Cd and FFA on the levels of this pro-inflammatory cytokine in murine primary hepatocytes.

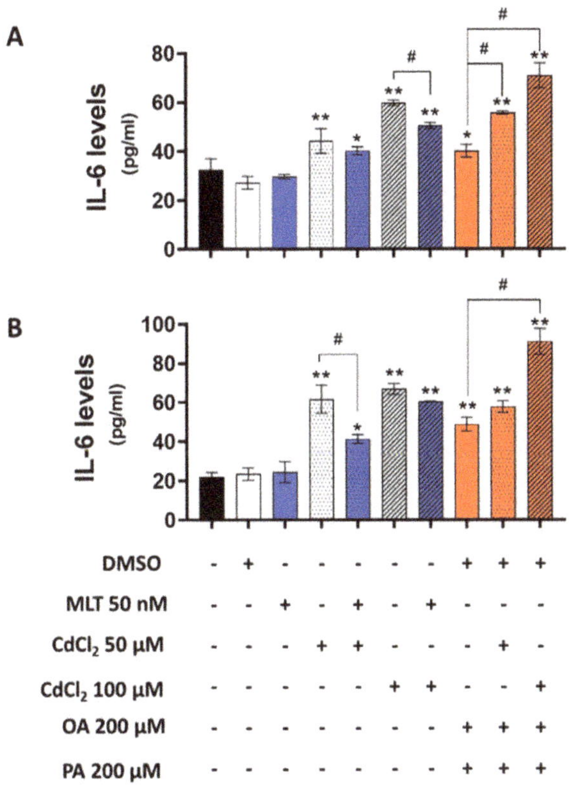

Figure 9. Levels of IL-6 in the culture medium of HepaRG cells (**A**) and primary murine hepatocytes (**B**) treated with $CdCl_2$, FFAs and MLT. The experimental protocol for cell treatments was EP2 of Figure 1. IL-6 (pg/mL) levels were measured by ELISA method in cell culture media. Data were mean ± SD of 3 independent experiments. One way ANOVA test: * $p < 0.05$, ** $p < 0.01$ control (or DMSO) versus all treatments; # $p < 0.05$ $CdCl_2$ versus $CdCl_2$ + MLT or OA + PA versus $CdCl_2$ + OA + PA.

4. Discussion

Cellular toxicity mechanisms of Cd include the induction of mitochondrial damage and increased ROS production, defects of antioxidant defenses, activation of inflammatory genes and cell death programs [4,42]. Another mechanism recently identified for this PTE is the induction of cellular lipotoxicity, a specific process of lipid accumulation deriving from an increased cellular flux of FFA and activation of lipid biosynthesis and redistribution processes. This results in a lipid excess stored in lipid droplets, a functionally specialized vesicular system of lysosomal origin that characterizes the "fatty" phenotype of some tissues, especially the liver [40]. These cellular effects of Cd exposure are associated with the induction of lipid metabolism alterations and oxidative stress, which have been documented in in vitro studies on human and animal liver cells [13–21] as well as in vivo in animal models of chronic exposure to Cd [21,23]. Alterations of the cellular lipidome recently characterized in mouse liver and human hepatocarcinoma cells demonstrate that Cd toxicity induces specific changes in membrane phospholipids, especially in phosphatidylcholine synthesis and remodeling, also increasing the relative abundance of arachidonic acid residues in complex lipids [43], which is a characteristic lipidomic hallmark of inflammation and a potential therapeutic target of human fatty liver disease [44].

These pieces of evidence stimulated us to explore the hypothesis that Cd may synergize with other factors to induce lipotoxicity; these include FFA that have a major pathogenic role in NAFLD (reviewed in [40,44]). In this in vitro study, we utilized OA and PA since their supplementation to human liver cells, performed at different concentrations and times of exposure [38,45], has consistently been demonstrated to induce lipotoxicity and all of the other hepatocellular, molecular and metabolic features of NAFLD. This experimental model has been utilized to explore both lipotoxicity mechanisms and the efficacy of pharmacological therapies and nutraceuticals at the pre-clinical level [46,47]. We also studied Cd and FFA toxicity in CACO-2 intestinal epithelial cells since, to our knowledge, the effects of lipotoxicity on the gut epithelium remain poorly characterized, although potentially relevant for the liver health [40].

Our results confirm that Cd and FFA synergize their lipotoxicity effects in liver cells, including mouse primary hepatocytes and the human pre-hepatocyte cell line HepaRG that was studied either before or after differentiation to mature liver cells. These effects were confirmed for all of the lipotoxicity hallmarks investigated in this study (shown from Figure 4 onward), including lipid droplet formation, ROS production, antioxidant enzyme expression, inflammatory parameters such as 5-LOX expression and IL-6 secretion, a reduction in cell viability, induction of cell death by apoptosis and necrosis, and a loss of cellular elements as assessed through direct microscopy observation.

CACO-2 intestinal epithelial cells also develop cues of lipotoxicity when exposed to Cd or to a combination of Cd and FFA, even if the levels of cellular lipid accumulation were lower compared to those observed in the liver cells (Figure 5).

Furthermore, we utilized MLT to explore whether the synergistic effects of Cd and FFA as lipotoxicity inducers can be prevented in the liver cell. This molecule was selected because it is a cytoprotective agent with proven efficacy in lessening Cd toxicity effects in different experimental models ([31,32] and references therein); moreover, it is effective in preventing liver damage in experimental models of fatty liver disease as well as in clinical trials on NAFLD patients ([34–36] and references therein).

As an important finding in the present study, MLT was found to protect both HepaRG and primary hepatocytes from the pro-oxidant and pro-inflammatory (Figure 7, Figure 9 and Supplementary Figure S6) effect of both individual and combined treatments with Cd and FFA, significantly reducing cytotoxicity in HepaRG and CACO-2 cells (Figure 6 and Supplementary Figure S6) and primary murine hepatocyte (Figure 5E). These results confirm the homeostatic properties of this molecule in Cd toxicity, highlighting its pleiotropic activity as a cytoprotective agent in cellular models of Cd-induced lipotoxicity, which is a main finding of this study. Herein, we demonstrate that MLT controls the effects of Cd exposure via MAPKs modulation (Figure 3). Cd is known to activate ERK and STAT3, and these are key players of the cellular stress responses [10,48–50] that can be modulated by MLT. This hormonal substance is a potent MAPK-ERK1/2 agonist, with a key role in cell cycle regulation and death signaling during the stress response to ROS and inflammatory factors [41]. In this respect, it is worth noting the inhibitory effect of a pharmacological dose of MLT (50 nM) on the induction response to Cd exposure of SAPKs and MAPK-ERK of liver cells (Figure 3 and Supplementary Figure S3); this provides a mechanistic explanation to the antioxidant and anti-inflammatory activity of MLT, as well as to its cytoprotective effect on liver cells exposed to Cd and FFA-induced lipotoxicity.

The use of "supraphysiological" concentrations of Cd and melatonin could be interpreted as a limit of this in vitro study; however, these concentrations are commonly utilized in vitro to obtain significant responses in liver cells and their relevance is discussed considering in vivo data. In accordance with this, MLT was found to improve hepatometabolic indices of NAFLD in obese mice fed a high-fat diet via MAPK-JNK/P38 signaling modulation [36], and reduced levels of pro-inflammatory cytokines (as TNFα, Il-6 and Il-1β) and an inducible form of the NO-generating enzyme iNos in the liver of mice exposed to Cd toxicity [51], which appears to support our mechanistic interpretation.

5. Conclusions

Here, we demonstrate that Cd can synergize with FFA to induce lipotoxicity effects in the liver and intestinal epithelial, which is a potentially relevant pathogenic mechanism of fatty liver disease [45]. Mechanistic aspects of such a toxicological interaction include the stimulation of pro-oxidant and inflammatory pathways that can be mitigated by MLT. The cytoprotective properties of this molecule are worth investigating at the pre-clinical and clinical level for an application protocol of hepatoprotection and therapy of hepatic lipotoxicity in Cd exposure.

Supplementary Materials: The following supporting information can be downloaded at: https://www.mdpi.com/article/10.3390/biom13121758/s1; Supplementary Figures S1–S6. Figure S1. Concentration- and time-dependent effect of $CdCl_2$ and cytoprotective effect of MLT on cell viability of HepaRG cells (A–D), CACO-2 cells (E–H) and primary murine hepatocytes (I–N); Figure S2. Concentration- and time-dependent effect of $CdCl_2$ on cell viability of HepaRG cells differentiated to hepatocyte-like cells, and cytoprotective effect of MLT; Figure S3. Effects of MAPKs ERK1/2 activity MLT-related in HepaRG cells pre-treated with ERK inhibitor and treated with $CdCl_2$ and MLT; Figure S4. Immunoblotting of antioxidant enzymes in HepaRG (A,B) and CACO-2 (C,D) cells treated for 24 h with $CdCl_2$ and/or MLT; Figure S5. Immunoblotting of 5-LOX in HepaRG (A) and CACO-2 (B) cells treated for 4 h with $CdCl_2$ and/or MLT; Figure S6. Effect of $CdCl_2$, FFA and MLT on cell viability of the human liver cell line HepaRG and in CACO-2 intestinal cells.

Author Contributions: Conceptualization, D.B. and F.G.; methodology, D.B., T.B., F.M. and A.M., software, D.B., G.D.S. and F.G.; validation, D.B., F.M. and T.B.; formal analysis, D.B.; investigation, D.B., T.B., F.M., A.M. and G.D.S.; resources, F.G., D.B., T.B. and F.M.; data curation, D.B.; writing—original draft preparation, D.B. and F.G.; writing—review and editing, D.B., F.G. and M.R.; visualization, D.B., T.B., F.M., A.M., G.D.S., M.R. and F.G.; supervision, M.R. and F.G.; project administration, D.B. and F.G.; funding acquisition, F.G. All authors have read and agreed to the published version of the manuscript.

Funding: A.M., D.B. and F.G. have been funded by the "Fondazione Cassa di Risparmio di Perugia", Perugia, Italy (Grant number # 20420-2021.0339).

Institutional Review Board Statement: The animal study protocol was approved by the Ethics Committee of University of Perugia and the ITALIAN MINISTRY OF HEALTH (protocol code 605/2019-PR and 2019) for studies involving animals.

Informed Consent Statement: Not applicable.

Data Availability Statement: The data presented in this study are available on request from the corresponding author.

Acknowledgments: The authors are grateful to Gabriele Galeazzi for his technical support during his thesis training activity.

Conflicts of Interest: The authors declare no conflict of interest.

References

1. Farag, A.M.; Nimick, D.A.; Kimball, B.A.; Church, S.E.; Harper, D.D.; Brumbaugh, W.G. Concentrations of metals in water, sediment, biofilm, benthic macroinvertebrates, and fish in the Boulder River watershed, Montana, and the role of colloids in metal uptake. *Arch. Environ. Contam. Toxicol.* **2007**, 52, 397–409. [CrossRef] [PubMed]
2. Chaumont, A.; De Winter, F.; Dumont, X.; Haufroid, V.; Bernard, A. The threshold level of urinary cadmium associated with increased urinary excretion of retinol-binding protein and β2-microglobulin: A re-assessment in a large cohort of nickel-cadmium battery workers. *Occup. Environ. Med.* **2011**, 68, 257–264. [CrossRef] [PubMed]
3. Moore, W., Jr.; Stara, J.F.; Crocker, W.C.; Malanchuk, M.; Iltis, R. Comparison of 115m cadmium retention in rats following different routes of administration. *Environ. Res.* **1973**, 6, 473–478. [CrossRef]
4. Cannino, G.; Ferruggia, E.; Luparello, C.; Rinaldi, A.M. Cadmium and mitochondria. *Mitochondrion* **2009**, 9, 377–384. [CrossRef]
5. Pizzino, G.; Bitto, A.; Interdonato, M.; Galfo, F.; Irrera, N.; Mecchio, A.; Pallio, G.; Ramistella, V.; De Luca, F.; Minutoli, L.; et al. Oxidative stress and DNA repair and detoxification gene expression in adolescents exposed to heavy metals living in the Milazzo-Valle del Mela area (Sicily, Italy). *Redox Biol.* **2014**, 2, 686–693. [CrossRef] [PubMed]
6. Buha, A.; Matovic, V.; Antonijevic, B.; Bulat, Z.; Curcic, M.; Renieri, E.A.; Tsatsakis, A.M.; Schweitzer, A.; Wallace, D. Overview of Cadmium Thyroid Disrupting Effects and Mechanisms. *Int. J. Mol. Sci.* **2018**, 19, 1501. [CrossRef]

7. Zhou, Z.; Wang, C.; Liu, H.; Huang, Q.; Wang, M.; Lei, Y. Cadmium induced cell apoptosis, DNA damage, decreased DNA repair capacity, and genomic instability during malignant transformation of human bronchial epithelial cells. *Int. J. Med. Sci.* **2013**, *10*, 1485–1496. [CrossRef]
8. Chen, L.; Liu, L.; Luo, Y.; Huang, S. MAPK and mTOR pathways are involved in cadmium-induced neuronal apoptosis. *J. Neurochem.* **2008**, *105*, 251–261. [CrossRef]
9. Chen, L.; Xu, B.; Liu, L.; Luo, Y.; Zhou, H.; Chen, W.; Shen, T.; Han, X.; Kontos, C.D.; Huang, S. Cadmium induction of reactive oxygen species activates the mTOR pathway, leading to neuronal cell death. *Free Radic. Biol. Med.* **2011**, *50*, 624–632. [CrossRef]
10. Moroni-González, D.; Sarmiento-Ortega, V.E.; Diaz, A.; Brambila, E.; Treviño, S. Pancreas-Liver-Adipose Axis: Target of Environmental Cadmium Exposure Linked to Metabolic Diseases. *Toxics* **2023**, *11*, 223. [CrossRef]
11. Nordberg, M.; Nordberg, G.F. Metallothionein and Cadmium Toxicology-Historical Review and Commentary. *Biomolecules* **2022**, *12*, 360. [CrossRef]
12. Prozialeck, W.C.; Edwards, J.R. Mechanisms of cadmium-induced proximal tubule injury: New insights with implications for biomonitoring and therapeutic interventions. *J. Pharmacol. Exp. Ther.* **2012**, *343*, 2–12. [CrossRef]
13. DeBose-Boyd, R.A.; Ye, J. SREBPs in lipid metabolism, insulin signaling, and beyond. *Trends Biochem. Sci.* **2018**, *43*, 358–368. [CrossRef] [PubMed]
14. Go, Y.-M.; Sutliff, R.L.; Chandler, J.D.; Khalidur, R.; Kang, B.-Y.; Anania, F.A.; Orr, M.; Hao, L.; Fowler, B.A.; Jones, D.P. Low-dose cadmium causes metabolic and genetic dysregulation associated with fatty liver disease in mice. *Toxicol. Sci.* **2015**, *147*, 524–534. [CrossRef] [PubMed]
15. He, W.; Guo, W.; Qian, Y.; Zhang, S.; Ren, D.; Liu, S. Synergistic hepatotoxicity by cadmium and chlorpyrifos: Disordered hepatic lipid homeostasis. *Mol. Med. Rep.* **2015**, *12*, 303–308. [CrossRef] [PubMed]
16. Zhang, S.; Jin, Y.; Zeng, Z.; Liu, Z.; Fu, Z. Subchronic exposure of mice to cadmium perturbs their hepatic energy metabolism and gut microbiome. *Chem. Res. Toxicol.* **2015**, *28*, 2000–2009. [CrossRef]
17. Masarone, M.; Rosato, V.; Dallio, M.; Gravina, A.G.; Aglitti, A.; Loguercio, C.; Federico, A.; Persico, M. Role of oxidative stress in pathophysiology of nonalcoholic fatty liver disease. *Oxidative Med. Cell. Longev.* **2018**, *2018*, 9547613. [CrossRef]
18. Chen, Z.; Tian, R.; She, Z.; Cai, J.; Li, H. Role of oxidative stress in the pathogenesis of nonalcoholic fatty liver disease. *Free Radic. Biol. Med.* **2020**, *152*, 116–141. [CrossRef]
19. Ding, R.-B.; Bao, J.; Deng, C.-X. Emerging roles of SIRT1 in fatty liver diseases. *Int. J. Biol. Sci.* **2017**, *13*, 852. [CrossRef]
20. Men, H.; Young, J.L.; Zhou, W.; Zhang, H.; Wang, X.; Xu, J.; Lin, Q.; Tan, Y.; Zheng, Y.; Cai, L. Early-Life Exposure to Low-Dose Cadmium Accelerates Diethylnitrosamine and Diet-Induced Liver Cancer. *Oxidative Med. Cell. Longev.* **2021**, *2021*, 1427787. [CrossRef]
21. Zhu, Y.; Zhao, Y.; Chai, X.X.; Zhou, J.; Shi, M.J.; Tian, Y.; Wang, X.M.; Ying, T.X.; Feng, Q.; Sheng, J.; et al. Chronic exposure to low-dose cadmium facilitated nonalcoholic steatohepatitis in mice by suppressing fatty acid desaturation. *Ecotoxicol. Environ. Saf.* **2022**, *233*, 113306. [CrossRef] [PubMed]
22. Young, J.L.; Yan, X.; Xu, J.; Yin, X.; Zhang, X.; Arteel, G.E.; Barnes, G.N.; States, J.C.; Watson, W.H.; Kong, M.; et al. Cadmium and High-Fat Diet Disrupt Renal, Cardiac and Hepatic Essential Metals. *Sci. Rep.* **2019**, *9*, 14675. [CrossRef] [PubMed]
23. Sarmiento-Ortega, V.E.; Treviño, S.; Flores-Hernández, J.; Aguilar-Alonso, P.; Moroni-González, D.; Aburto-Luna, V.; Diaz, A.; Brambila, E. Changes on serum and hepatic lipidome after a chronic cadmium exposure in Wistar rats. *Arch. Biochem. Biophys.* **2017**, *635*, 52–59. [CrossRef] [PubMed]
24. Hyder, O.; Chung, M.; Cosgrove, D.; Herman, J.M.; Li, Z.; Firoozmand, A.; Gurakar, A.; Koteish, A.; Pawlik, T.M. Cadmium exposure and liver disease among US adults. *J. Gastrointest. Surg.* **2013**, *17*, 1265–1273. [CrossRef] [PubMed]
25. Zhang, J.; Li, F.; Zhang, X.; Xie, T.; Qin, H.; Lv, J.; Gao, Y.; Li, M.; Gao, Y.; Jia, Y. Melatonin Improves Turbot Oocyte Meiotic Maturation and Antioxidant Capacity, Inhibits Apoptosis-Related Genes mRNAs In Vitro. *Antioxidants* **2023**, *12*, 1389. [CrossRef]
26. Yang, Y.; Xu, W.; Du, X.; Ye, Y.; Tian, J.; Li, Y.; Jiang, Q.; Zhao, Y. Effects of dietary melatonin on growth performance, antioxidant capacity, and nonspecific immunity in crayfish, Cherax destructor. *Fish Shellfish. Immunol.* **2023**, *138*, 108846. [CrossRef]
27. Alvarez-Diduk, R.; Galano, A.; Tan, D.X.; Reiter, R.J. N-Acetylserotonin and 6-Hydroxymelatonin against Oxidative Stress: Implications for the Overall Protection Exerted by Melatonin. *J. Phys. Chem. B* **2015**, *119*, 8535–8543. [CrossRef]
28. Galano, A.; Tan, D.X.; Reiter, R.J. Melatonin as a natural ally against oxidative stress: A physicochemical examination. *J. Pineal Res.* **2011**, *51*, 1–16. [CrossRef]
29. Luchetti, F.; Betti, M.; Canonico, B.; Arcangeletti, M.; Ferri, P.; Galli, F.; Papa, S. ERK MAPK activation mediates the antiapoptotic signaling of melatonin in UVB-stressed U937 cells. *Free Radic. Biol. Med.* **2009**, *46*, 339–351. [CrossRef]
30. Moniruzzaman, M.; Ghosal, I.; Das, D.; Chakraborty, S.B. Melatonin ameliorates H_2O_2-induced oxidative stress through modulation of Erk/Akt/NFkB pathway. *Biol. Res.* **2018**, *51*, 17. [CrossRef]
31. Knani, L.; Bartolini, D.; Kechiche, S.; Tortoioli, C.; Murdolo, G.; Moretti, M.; Messaoudi, I.; Reiter, R.J.; Galli, F. Melatonin prevents cadmium-induced bone damage: First evidence on an improved osteogenic/adipogenic differentiation balance of mesenchymal stem cells as underlying mechanism. *J. Pineal Res.* **2019**, *67*, e12597. [CrossRef]
32. Bartolini, D.; Arato, I.; Mancuso, F.; Giustarini, D.; Bellucci, C.; Vacca, C.; Aglietti, M.C.; Stabile, A.M.; Rossi, R.; Cruciani, G.; et al. Melatonin modulates Nrf2 activity to protect porcine pre-pubertal Sertoli cells from the abnormal H_2O_2 generation and reductive stress effects of cadmium. *J. Pineal Res.* **2022**, *73*, e12806. [CrossRef]

33. Migni, A.; Galeazzi, G.; Marcantonini, G.; Zatini, L.; Rende, M.; Galli, F.; Bartolini, D. Cadmium and palmitic acid synergize their pro-oxidant and cytotoxic effects in liver and gut epithelial cells: The cytoprotective role of α-tocopherol (vitamin E) and melatonin. In Proceedings of the 2023 Annual Meeting of the Society for Free Radical Research—Europe (SFRR-E) Hotel Savoyen, Vienna, Austria, 1 May 2023.
34. Bahrami, M.; Cheraghpour, M.; Jafarirad, S.; Alavinejad, P.; Asadi, F.; Hekmatdoost, A.; Mohammadi, M.; Yari, Z. The effect of melatonin on treatment of patients with non-alcoholic fatty liver disease: A randomized double blind clinical trial. *Complement. Ther. Med.* **2020**, *52*, 102452. [CrossRef] [PubMed]
35. Pakravan, H.; Ahmadian, M.; Fani, A.; Aghaee, D.; Brumanad, S.; Pakzad, B. The Effects of Melatonin in Patients with Nonalcoholic Fatty Liver Disease: A Randomized Controlled Trial. *Adv. Biomed. Res.* **2017**, *6*, 40. [CrossRef]
36. Sun, H.; Wang, X.; Chen, J.; Song, K.; Gusdon, A.M.; Li, L.; Bu, L.; Qu, S. Melatonin improves non-alcoholic fatty liver disease via MAPK-JNK/P38 signaling in high-fat-diet-induced obese mice. *Lipids Health Dis.* **2016**, *15*, 202. [CrossRef] [PubMed]
37. Cabral, F.; Miller, C.M.; Kudrna, K.M.; Hass, B.E.; Daubendiek, J.G.; Kellar, B.M.; Harris, E.N. Purification of Hepatocytes and Sinusoidal Endothelial Cells from Mouse Liver Perfusion. *J. Vis. Exp.* **2018**, *132*, e56993. [CrossRef]
38. Bartolini, D.; Torquato, P.; Barola, C.; Russo, A.; Rychlicki, C.; Giusepponi, D.; Bellezza, G.; Sidoni, A.; Galarini, R.; Svegliati-Baroni, G.; et al. Nonalcoholic fatty liver disease impairs the cytochrome P-450-dependent metabolism of alpha-tocopherol (vitamin E). *J. Nutr. Biochem.* **2017**, *47*, 120–131. [CrossRef] [PubMed]
39. Mosmann, T. Rapid colorimetric assay for cellular growth and survival: Application to proliferation and cytotoxicity assays. *J. Immunol. Methods* **1983**, *65*, 55–63. [CrossRef] [PubMed]
40. Svegliati-Baroni, G.; Pierantonelli, I.; Torquato, P.; Marinelli, R.; Ferreri, C.; Chatgilialoglu, C.; Bartolini, D.; Galli, F. Lipidomic biomarkers and mechanisms of lipotoxicity in non-alcoholic fatty liver disease. *Free Radic. Biol. Med.* **2019**, *144*, 293–309. [CrossRef] [PubMed]
41. Luchetti, F.; Canonico, B.; Betti, M.; Arcangeletti, M.; Pilolli, F.; Piroddi, M.; Canesi, L.; Papa, S.; Galli, F. Melatonin signaling and cell protection function. *FASEB J.* **2010**, *24*, 3603–3624. [CrossRef]
42. Cannino, G.; Ferruggia, E.; Luparello, C.; Rinaldi, A.M. Effects of cadmium chloride on some mitochondria-related activity and gene expression of human MDA-MB231 breast tumor cells. *J. Inorg. Biochem.* **2008**, *102*, 1668–1676. [CrossRef]
43. Gu, J.; Kong, A.; Guo, C.; Liu, J.; Li, K.; Ren, Z.; Zhou, Y.; Tang, M.; Shi, H. Cadmium perturbed lipid profile and induced liver dysfunction in mice through phosphatidylcholine remodeling and promoting arachidonic acid synthesis and metabolism. *Ecotoxicol. Environ. Saf.* **2022**, *247*, 114254. [CrossRef]
44. Torquato, P.; Giusepponi, D.; Alisi, A.; Galarini, R.; Bartolini, D.; Piroddi, M.; Goracci, L.; Di Veroli, A.; Cruciani, G.; Crudele, A.; et al. Nutritional and lipidomics biomarkers of docosahexaenoic acid-based multivitamin therapy in pediatric NASH. *Sci. Rep.* **2019**, *9*, 2045. [CrossRef]
45. Eynaudi, A.; Diaz-Castro, F.; Borquez, J.C.; Bravo-Sagua, R.; Parra, V.; Troncoso, R. Differential Effects of Oleic and Palmitic Acids on Lipid Droplet-Mitochondria Interaction in the Hepatic Cell Line HepG2. *Front. Nutr.* **2021**, *8*, 775382. [CrossRef] [PubMed]
46. Bartolini, D.; Zatini, L.; Migni, A.; Frammartino, T.; Guerrini, A.; Garetto, S.; Lucci, J.; Moscardini, I.F.; Marcantonini, G.; Stabile, A.M.; et al. Transcriptomics of natural and synthetic vitamin D in human hepatocyte lipotoxicity. *J. Nutr. Biochem.* **2023**, *117*, 109319. [CrossRef]
47. Bartolini, D.; Marinelli, R.; Stabile, A.M.; Frammartino, T.; Guerrini, A.; Garetto, S.; Lucci, J.; Migni, A.; Zatini, L.; Marcantonini, G.; et al. Wheat germ oil vitamin E cytoprotective effect and its nutrigenomics signature in human hepatocyte lipotoxicity. *Heliyon* **2022**, *8*, e10748. [CrossRef] [PubMed]
48. Souza, V.; del Carmen Escobar, M.; Bucio, L.; Hernández, E.; Gómez-Quiroz, L.E.; Gutiérrez Ruiz, M.C. NADPH oxidase and ERK1/2 are involved in cadmium induced-STAT3 activation in HepG2 cells. *Toxicol. Lett.* **2009**, *187*, 180–186. [CrossRef] [PubMed]
49. Ali, I.; Damdimopoulou, P.; Stenius, U.; Halldin, K. Cadmium at nanomolar concentrations activates Raf-MEK-ERK1/2 MAPKs signaling via EGFR in human cancer cell lines. *Chem. Biol. Interact.* **2015**, *231*, 44–52. [CrossRef]
50. Arroyo, V.; Flores, K.; Ortiz, L.; Gómez-Quiroz, L.; Gutiérrez-Ruiz, M. Liver and Cadmium Toxicity. *J. Drug Metab. Toxicol.* **2012**, *S5*, 001. [CrossRef]
51. Yang, Z.; He, Y.; Wang, H.; Zhang, Q. Protective effect of melatonin against chronic cadmium-induced hepatotoxicity by suppressing oxidative stress, inflammation, and apoptosis in mice. *Ecotoxicol. Environ. Saf.* **2021**, *228*, 112947. [CrossRef]

biomolecules

MDPI

Review

Factors Important in the Use of Fluorescent or Luminescent Probes and Other Chemical Reagents to Measure Oxidative and Radical Stress

Peter Wardman

Formerly of the Gray Cancer Institute, Mount Vernon Hospital/University of Oxford, UK; peterwardman@btinternet.com

Abstract: Numerous chemical probes have been used to measure or image oxidative, nitrosative and related stress induced by free radicals in biology and biochemistry. In many instances, the chemical pathways involved are reasonably well understood. However, the rate constants for key reactions involved are often not yet characterized, and thus it is difficult to ensure the measurements reflect the flux of oxidant/radical species and are not influenced by competing factors. Key questions frequently unanswered are whether the reagents are used under 'saturating' conditions, how specific probes are for particular radicals or oxidants and the extent of the involvement of competing reactions (e.g., with thiols, ascorbate and other antioxidants). The commonest-used probe for 'reactive oxygen species' in biology actually generates superoxide radicals in producing the measured product in aerobic systems. This review emphasizes the need to understand reaction pathways and in particular to quantify the kinetic parameters of key reactions, as well as measure the intracellular levels and localization of probes, if such reagents are to be used with confidence.

Keywords: free radicals; oxidative stress; reactive oxygen species; ROS; superoxide radicals; nitrosative stress; fluorescent probes; luminescent probes; rate constants; pulse radiolysis

Citation: Wardman, P. Factors Important in the Use of Fluorescent or Luminescent Probes and Other Chemical Reagents to Measure Oxidative and Radical Stress. *Biomolecules* **2023**, *13*, 1041. https://doi.org/10.3390/biom13071041

Academic Editor: Jeffrey Stuart

Received: 5 June 2023
Revised: 20 June 2023
Accepted: 24 June 2023
Published: 26 June 2023

1. Introduction

While most papers in this Special Issue focus on modified endogenous biological molecules as markers of oxidative, nitrosative or other changes initiated by free radicals, such damage has also been routinely assessed using exogenous, xenobiotic molecular probes. Thus, as described below, a reduced fluorescein dye has been widely used to detect the production of (usually unspecified) 'reactive oxygen species' in biological material or biomolecular models—astonishingly, in over 6000 studies to date—despite the obvious shortcomings discussed below. Hence, it is important not to take a too narrow or literal viewpoint in considering 'biomarkers' of oxidative stress and radical damage.

There will often be competing pathways in the production of endogenous biomarkers of oxidative or other radical damage, with the yield of biomarker reflecting not only the production of radical reactants but also the concentrations and availability of potentially protective substances such as thiols, ascorbate or phenolic antioxidants (for example). Similarly, the 'signal' reflecting chemical change in exogenous markers or probes, whether measured concentrations or image intensity (e.g., in fluorescence readers or microscopy), may reflect not only the extent of the production of radicals or oxidants but also the involvement of competing reactions. The outcome of such competing processes will be defined by the Law of Mass Action, and so the important characteristics will be the product of concentration *and rate constant* for all the competing reactions. Hence, whether damage is assessed by endogenous biomarkers or by added chemical probes, chemical kinetics is central to an informed assessment of the damage pathways, and the availability of rate constants for specific reactions involving free radicals (in particular) and potential biomolecular targets (or reasonable models for them) is central to an informed discussion about competing pathways in free-radical biology. Of course, the intracellular concentrations of reactants *or* probes are equally important, and cellular heterogeneity is a daunting challenge to the accurate chemical modelling of biological pathways, as discussed elsewhere in the context of radiation damage [1]. As an example, 'molecular crowding' can modulate reaction rates (or effective rate constants), as illustrated recently [2]. As a first step, though, analyzing potential competing reactions by simple homogenous competition kinetics is far better than nothing, even if, ultimately, more sophisticated approaches are desirable, such as applying Monte Carlo numerical simulation methods to a heterogenous model of cellular compartments.

This review aims to illustrate the importance of kinetic factors in using chemical probes for radical-initiated damage, although the concepts apply equally to discussing the production of endogenous biomarkers. It will also help readers to access sources of pertinent kinetic information in both contexts. Kinetic information (rate constants) can be estimated by measuring the final product yields analyzed by competition kinetics in model systems where a reference reaction has been characterized separately, but the direct observation of reactive intermediates—radical or product—is usually the most reliable. This involves the production of sufficient radicals in a time significantly shorter than the reaction timescale to follow the reactions and enable detection, e.g., by kinetic spectrophotometry. The technique of flash photolysis, introduced around 1950, made this possible for photochemically initiated reactions. The analogous technique of pulse radiolysis, introduced a decade later, is much more relevant for free-radical reactions, since ionizing radiation generates free radical pathways which, in the case of water as a solvent, can be easily manipulated to monitor reactions of specific radicals of biological interest. Hence, radiation chemistry has provided, as a 'spinoff', much kinetic information concerning reactions involving free radicals and biomolecules. It is not necessary to understand the complexity of radiation damage to appreciate the application of the specialized techniques to general redox chemistry [3].

It is not intended here to give an extensive overview of the chemistry and/or methodology of markers and/or probes for 'reactive oxygen species' or oxidative damage, which has recently been reviewed with specific recommendations [4]. The present author reviewed the chemistry of fluorescent and luminescent probes for oxidative and nitrosative stress in some detail [5], and numerous more recent reviews in this area updated and expanded the earlier survey [6–49]. Rather, the intention here is to illustrate the importance of understanding the reaction pathways involved in the use of these probes and, especially, the importance of kinetics in controlling these pathways.

We first have to consider the terms used in labelling these probes; following this, a discussion of the chemistry of some of the most commonly used probes serves as a 'template' or 'worked example' against which other probes can be assessed. Finally, the questions which all users of chemical probes should ask themselves before commencing on a study are briefly summarized.

2. 'Oxidative Stress' and 'Reactive Oxygen Species' ('ROS'): Their Definitions (or Lack of)

If we are to discuss sensibly probes for chemical species, then we need to be clear about the molecules involved. While the term 'oxidative stress' is reasonably well understood and quite well defined [50–53], the related but not synonymous term 'reactive oxygen species' and its acronym 'ROS' are mere 'catch-all' phrases lacking usefulness because of their vague, all-encompassing nature. Indeed, experts in free-radical biology are increasingly warning about the indiscriminate use of 'ROS'. Thus, it was recommended: 'The use of ROS or RNS should be ... only when it is clearly stated that the species is unknown or one of several implicated molecules without certainty' [54]; it was commented: 'The ubiquitous use of these terms seems to provide a screen to hide the detailed chemistry of these species' [55]. Another group of experts noted: 'Reactive oxygen species' (ROS) is a generic term that defines a wide variety of oxidant molecules with vastly different properties and biological functions ... The generic term ROS should not be used to describe specific molecular agents' [43]. The first recommendation of an expert review group [4] was '... wherever possible, the actual chemical species involved in a biological process should be stated, and consideration given to whether the observed effect is compatible with its reactivity, lifespan, products generated and fate in vivo'.

The difficulty in beginning to assess the reactions of vague 'ROS' in kinetic terms is apparent if we compare the rate constants for reactions of the common biological antioxidant, the thiol, glutathione (GSH), with specific radical or molecular oxidants, all containing oxygen, at physiological pH. Thus, the rate constants (units of $M^{-1}\ s^{-1}$) for reactions of GSH with different 'ROS' span over *ten orders of magnitude*, with the rate constants for $^{\bullet}OH$ (~1×10^{10} [56]), HOCl (~1×10^{8} [57]), $NO_2^{\bullet}$ (~2×10^{7} [58]) and $CO_3^{\bullet-}$ (~5×10^{6} [59]) being several orders of magnitude higher than those for reactions of $O_2^{\bullet-}$ (~2×10^{2} [60]) and H_2O_2 (~9×10^{-1} [61]).

In the context of chemical probes for oxidizing radicals derived from biomolecules, it should also be noted that radicals centered on *sulfur* (thiyl radicals, $RS^{\bullet}$, and not normally viewed as 'reactive oxygen species') can oxidize targets directly: the mid-point reduction potential of the couple $GS^{\bullet},H^{+}/GSH$ is ~0.90 V at pH ~7 [62], only ~0.04 V lower than that of the tyrosine phenoxyl radical, $TyrO^{\bullet},H^{+}/TyrOH$ [63]. Thiols are often viewed as protective antioxidants, but they do have the *potential* to *elevate* 'ROS'/oxidative stress following the chemical 'repair' of diverse radical sites via the sequence of reactions [64,65]:

$$\text{radical damage} + RSH \rightarrow \text{'repaired' (or modified) damage} + RS^{\bullet}$$

$$RS^{\bullet} + RS^{-} \rightleftharpoons (RSSR)^{\bullet-}$$

$$(RSSR)^{\bullet-} + O_2 \rightarrow RSSR + O_2^{\bullet-}.$$

However, ascorbate ($AscH^{-}$) can disrupt this pathway by intercepting thiyl radicals [66]:

$$RS^{\bullet} + AscH^{-} \rightarrow RSH + Asc^{\bullet-},$$

but it is important to note that many in vitro cell culture models lack ascorbate and so are poor models for tissues. The potential reaction pathways of thiyl radicals in cells are multiple and complex [67–69] and include the catalysis of the *cis/trans* isomerization of unsaturated lipids [70–73]. Since thiyl radicals can form reactive thiylperoxyl ($RSOO^{\bullet}$) and sulphonyl radicals in the presence of oxygen [74,75], thiol radical chemistry should be included in any discussion of 'ROS'.

It has also long been recognized that 'ROS' should not be viewed in isolation, in particular, reactive nitrogen-based oxidants include peroxynitrous acid/peroxynitrite ($ONOOH/ONOO^{-}$, from the reaction between $^{\bullet}NO$ and $O_2^{\bullet-}$), which in turn can source $^{\bullet}OH$ and $^{\bullet}NO_2$ as well as carbonate radicals ($CO_3^{\bullet-}$) [76,77]. $CO_3^{\bullet-}$ and $NO_2^{\bullet}$ are fairly powerful oxidants: the reduction potentials of the couples $CO_3^{\bullet-}/CO_3^{2-}$ and $^{\bullet}NO_2/NO_2^{-}$ are ~1.57 and 1.04 V, respectively [78], higher than the midpoint potentials at pH 7 of the radical/reductant couples of tyrosine, glutathione or ascorbate [3]. Hence, the spectrum of radical oxidants potentially reactive towards chemical probes is quite wide—and of course the non-radical oxidants H_2O_2 and HOCl cannot be ignored, particularly since H_2O_2 can form oxidizing intermediates in peroxidase chemistry or upon reaction with cytochrome *c*, as noted below. Overall, then, a whole battery of oxidants are *potentially* reactive towards *some* chemical probes for 'ROS' or oxidative stress, presenting a major challenge to the informed use of probes.

3. Dichlorodihydrofluorescein ($DCFH_2$): By Far, the Most Widely Used—And Certainly the Most Abused—Probe for 'ROS' or Cellular Oxidative Stress

It is sensible—and instructive—to illustrate the importance of mapping reaction pathways of molecular probes and to characterize them kinetically by discussing the chemistry of the most widely used probe for oxidative events associated with free radicals in biology, not least because the reactivity of this probe has been kinetically characterized quite well.

However, as discussed below, there remain many gaps in our knowledge, and the results of any study using this or similar probes must continue to be viewed with skepticism.

Assays based on the fluorescence of a probe or its product are both sensitive and facilitate imaging. Fluorescein is perhaps the archetypical fluorophore, and its reduced form (dihydrofluorescein) is a non-fluorescent or 'leuco' dye. The variant 2′,7′-dichlorodihydrofluorescein ($DCFH_2$, Figure 1, **1**) is the most commonly used probe for vague 'ROS' or general oxidative damage in biology associated with free radicals, with the fluorescent product dichlorofluorescein (DCF) measured; an appropriate PubMed search revealed almost 7000 'hits' in May 2023 for papers referencing these dyes. Even accounting for the papers which simply mention or review the use of $DCFH_2$ or related probes, it seems likely that well over 6000 studies have utilized such probes. Despite well-known shortcomings, which have been repeatedly stressed by numerous authors, e.g., [5,6,8,9,13,19,23,24,27,30,54,55,79–89], these probes continue to be widely recommended and used in many investigations, e.g., [90–95], including a recent study that attracted quite widespread press coverage [96].

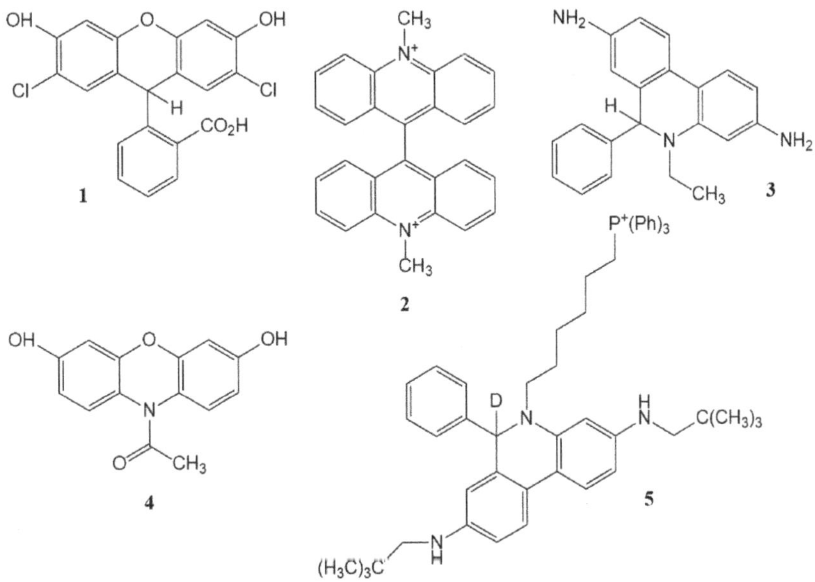

Figure 1. Structures of the probes discussed: **1**, dichlorodihydrofluorescein ($DCFH_2$); **2**, lucigenin; **3**, hydroethidine (HE); **4**, Amplex Red; **5**, 'MitoNeoD'.

There are three basic problems with this and similar 'leuco' dyes such as the chloromethyl analogue and other variants promoted by manufacturers; while $DCFH_2$ is the best characterized, it seems likely that all exhibit many of the problems discussed below. First, oxidation may reflect the level of the catalyst rather than that of the oxidant; second, the rate of oxidation by radicals varies widely, and radicals other than 'ROS' react rapidly with the probe; and third, the radical intermediate produced during oxidation—either directly by radicals or by a catalyst activated by H_2O_2—reacts rapidly with oxygen *to generate superoxide radicals*.

Consider first oxidation by non-radical species, of which H_2O_2 is the most studied. For a start, the direct reaction between H_2O_2 and $DCFH_2$ is very slow, and the reaction is most unlikely to compete with the destruction of H_2O_2 by cellular peroxidases, peroxiredoxins, etc.: oxidation by H_2O_2 requires a *catalyst*, as shown in early studies with $DCFH_2$ or analogues where horseradish peroxidase (HRP) or hematin was used [97], and the signal measured, therefore, may reflect the level of catalyst rather than (or as well as) the level of H_2O_2 generation. Fluorescent probes for catalytic iron(II) were developed [34].

Another potent catalyst is cytochrome *c* [98], very important because it is released from the mitochondria during apoptosis. Further, diverse hydroperoxides can substitute for H_2O_2 in the catalyzed oxidation of $DCFH_2$ [79]. The peroxidase- or cytochrome *c*-catalyzed oxidation is also subject to interference by antioxidants, including uric acid, ascorbate and α-tocopherol [79], as well as NADH and thiols [99]. Overall, then, associating fluorescence with changes in H_2O_2 production rather than with the levels of catalyst or antioxidants is difficult, and confusion with both apoptotic pathways and hydroperoxide formation rather than H_2O_2 is quite possible.

Other non-radical oxidants have been investigated in this context: hypochlorite reacts inefficiently with $DCFH_2$ [100]; peroxynitrite ($ONOO^-$) is much more efficient, but kinetic studies [101] showed this probably reflects radical oxidation involving $NO_2^{\bullet}$, $^{\bullet}OH$ and/or $CO_3^{\bullet-}$ radicals, decomposition products of peroxynitrite under physiological conditions [76,77].

The rates of oxidation of $DCFH_2$ by radical oxidants vary widely: superoxide has very low reactivity, while $^{\bullet}OH$ radicals react at the diffusion-controlled limit (rate constant $k \sim 1.3 \times 10^{10}\ M^{-1}\ s^{-1}$; intermediate reactivity is seen with $NO_2^{\bullet}$ ($k \sim 1.3 \times 10^{7}\ M^{-1}\ s^{-1}$) and $CO_3^{\bullet-}$ ($k \sim 2.6 \times 10^{8}\ M^{-1}\ s^{-1}$, pH 8.2)) [102]. (Unless otherwise indicated, the rate constants here refer to physiological pH; the pH value can be important, since $DCFH_2$ dissociates with pK_a values estimated as ~7.9 and 9.2 [102], and oxidants may show pH-dependent reactivity. $NO_2^{\bullet}$, for example, may oxidize a phenolate species considerably faster than the corresponding phenol [103,104].) Sulfur-centered radicals ($RS^{\bullet}$, thiyl) oxidize $DCFH_2$ with rate constants of ~3.7×10^7 (GSH) or 1.7×10^7 (cysteine) $M^{-1}\ s^{-1}$ [85]; since such radicals are commonly produced in diverse radical 'repair' reactions (e.g., the donation of H to carbon-centered radicals, as noted above), the lack of specificity of $DCFH_2$ towards oxygen-based radicals is obvious. Note, however, that the involvement of thiols in probe chemistry is more complex than the direct oxidation of $DCFH_2$ by thiyl radicals: the oxidized probe DCF has a phenolic moiety and is a substrate for peroxidases, the resulting phenoxyl radical DCF(–$O^{\bullet}$) oxidizing GSH (as well as NADH and ascorbate), generating superoxide radicals via further reactions [81]:

$$\mathrm{DCF + oxidant \rightarrow DCF(\text{-}O^{\bullet})}$$

$$\mathrm{DCF(\text{-}O^{\bullet}) + GSH\ (or\ NADH) \rightarrow DCF + GS^{\bullet}\ (or\ NAD^{\bullet})}$$

$$\mathrm{GS^{\bullet} + GSH \rightleftharpoons (GSSG)^{\bullet-}}$$

$$\mathrm{(GSSG)^{\bullet-} + O_2 \rightarrow GSSG + O_2^{\bullet-}}$$

$$\mathrm{NAD^{\bullet} + O_2 \rightarrow NAD^{+} + O_2^{\bullet-}}.$$

The oxidized probe also reacts rapidly with some radicals, e.g., the rate constant for the reaction of $CO_3^{\bullet-}$ radicals with DCF is almost identical to that for the reaction of $DCFH_2$; however, DCF is much less reactive towards $NO_2^{\bullet}$ than is $DCFH_2$ [105].

By the rule of spin conservation, the oxidation of $DCFH_2$ to the fluorescent product DCF by radicals *must* produce a radical intermediate, $DCFH^{\bullet}$. Further, the catalyzed oxidation of $DCFH_2$ also proceeds via this radical intermediate. In landmark studies [80,81,83], it was shown that this radical also reacts with oxygen to produce superoxide:

$$\mathrm{DCFH^{\bullet} + O_2 \rightarrow DCF + O_2^{\bullet-}\ (+\ H^{+})}$$

leading the authors to describe such probes for oxidative stress as a 'self-fulfilling prophesy' [83]. A direct observation of the reaction of $DCFH^{\bullet}$ with O_2 showed $k \sim 5.3 \times 10^{8}\ M^{-1}\ s^{-1}$ [102]. The rate of the possible competing reaction:

$$2\ DCFH^{\bullet} \rightarrow DCFH_2 + DCF$$

is pH-dependent around physiological pH because of prototropic equilibria, but the effective rate constant is $2k \sim 2.8 \times 10^8\ M^{-1}\ s^{-1}$ at pH 7.4. Since the two rate constants are of the same order, and it is easily calculated that in most biological systems, the concentration of O_2 far exceeds that of $DCFH^{\bullet}$ at a steady state, superoxide formation seems an inevitable consequence of $DCFH_2$ oxidation, whether by (catalyzed) peroxides or directly by radicals [102].

The extent of $DCFH_2$ oxidation by both H_2O_2 (or hydroperoxides) and radicals will reflect competition kinetics (the Law of Mass Action), since there are obviously competing pathways for the species to react, but this presents an immediate difficulty: the reaction pathways reflect the products of rate constant and concentration, but hardly any users of these probes have measured the intracellular concentration of $DCFH_2$. Usually, the diacetate derivative is used, relying on cellular esterases to cleave the diacetate and relying on the resulting negative charge on the dissociated carboxylic acid to trap the probe intracellularly. In the author's Laboratory, L K Folkes measured ~300 μM $DCFH_2$ as the *average* intracellular concentration after incubating hamster fibroblasts in cell suspension (5×10^6/mL) at pH 7.4 with $DCFH_2$ diacetate (10 μM) for 15 min at 37 °C [102].

There will be intracellular concentration gradients typical of a weak acid, as well as lipid/water partitioning of both probe and product [106]; this study reported $DCFH_2$ was 'totally partitioned into the octanol' using octanol/water 1:1 at pH 6, while DCF had a partition coefficient of 2:1 (the octanol:/water partition coefficient of $DCFH_2$ at an unspecified pH was quoted elsewhere as 2.62 [107], but it will be pH-dependent because of prototropic equilibria [102]). There may also be leakage of product into the extracellular medium [108]. Further, the nature of the culture medium must be considered, including the presence of catalytic metals, antioxidants, pH indicators, etc. [109,110].

Considering, as an example, competing reactions between $CO_3^{\bullet-}$ and either $DCFH_2$ or GSH present together, the rate constant for the reaction of $CO_3^{\bullet-}$ and GSH is $\sim 5.3 \times 10^6\ M^{-1}\ s^{-1}$ at pH ~7 [59], i.e., ~50–fold lower than that for the reaction between $CO_3^{\bullet-}$ and $DCFH_2$, but the intracellular concentration of GSH is typically around 10-fold higher than that of $DCFH_2$ *if* the concentration of the latter is similar to that in the measurements with hamster cells described above. (Ascorbate reacts with $CO_3^{\bullet-}$ ~260 × faster than does GSH [111].) Nominally, just considering the one competing reaction with GSH, the probe will 'capture' a good fraction—but not all—$CO_3^{\bullet-}$ radicals, but since other free thiols are present along with other targets for $CO_3^{\bullet-}$ radicals (protein thiols, tyrosine residues, etc., and ascorbate in tissues), it seems likely that the probe is often not being used in 'saturating' conditions, making quantitation difficult. This is more clear-cut if we consider competing reactions of $NO_2^{\bullet}$ radicals. The rate constants for the oxidation of GSH or cysteine at pH 7.4 by $NO_2^{\bullet}$ are ~2 or $5 \times 10^7\ M^{-1}\ s^{-1}$, respectively [58], i.e., both higher than that for the reaction of $NO_2^{\bullet}$ with $DCFH_2$ ($\sim 1.3 \times 10^7\ M^{-1}\ s^{-1}$). Hence, ~0.3 mM $DCFH_2$ will 'pick up' only a very small fraction of any $NO_2^{\bullet}$ generated. The total 'scavenging capacity (Σk[scavenger]) for $^{\bullet}OH$ radicals in mammalian cells has been estimated as $\sim 8 \times 10^8\ s^{-1}$, compared with the corresponding value for $^{\bullet}OH$ reacting with 0.3 mM $DCFH_2$, i.e., $\sim 4 \times 10^6\ s^{-1}$; so, clearly it is most unlikely that $DCFH_2$ (and indeed most probes at realistic intracellular concentrations) can intercept a significant fraction of $^{\bullet}OH$ radicals.

Kinetic competition may also be exhibited between the activated peroxidase intermediate (such as HRP Compound I) reacting with either the probe or endogenous antioxidants; so, for a full understanding of probe chemistry, quite extensive studies must be undertaken. These are likely to involve several sophisticated techniques; in this case, electron paramagnetic resonance spectroscopy to identify reaction intermediates, stopped-flow rapid mixing to measure the rates of non-radical reactions, flash photolysis to monitor reactions of excited states (see below), high-performance liquid chromatography to measure intracellular uptake, preferably in specific organelles, and (especially) pulse radiolysis to generate specific 'ROS' in known amounts and monitor their reactions in real time.

Even leaving aside the obvious contraindication of the generation of one specific 'ROS' (superoxide radical) via the obligate involvement of an oxygen-reactive radical in the formation of the fluorescent product, without measurements of *both* the rate constants of the reactions of putative reactants with a probe *and* the estimates of the intracellular (or better, intra-organelle or localized) concentrations of the probe, it is obvious that users are working in the dark in seeking to quantify reactive oxidants using such probes. Actually, users *had* best work in the dark, because a further complication is that visible light can generate an excited state of DCF, quenched by cellular reductants such as GSH or NADH (and doubtless other reactants such as ascorbate) to generate $DCFH^{\bullet}$ and hence $O_2^{\bullet-}$ [112]. Photooxidation of Amplex® Red has also been reported, once again involving the *production* of superoxide radicals [113].

4. Some Kinetic Studies of Other Chemical Probes for Oxidants and Biological Radicals

$DCFH_2$ is not the only 'ROS' probe to be a 'self-fulfilling prophesy'. As discussed in more detail previously [5], the bis-*N*-methyl acridinium salt, lucigenin (LC^{2+}, Figure 1, **2**), is reduced to a radical $LC^{\bullet+}$, probably by flavoprotein reductases, and exhibits chemiluminescence on reaction of this radical intermediate with superoxide radicals via an unstable intermediate producing an excited state of *N*-methylacridone [114]. However, the obligate radical intermediate $LC^{\bullet+}$ reacts with oxygen *to generate superoxide* [115,116] with a rate constant of ~3×10^6 M^{-1} s^{-1}:

$$LC^{\bullet+} + O_2 \rightleftharpoons LC^{2+} + O_2^{\bullet-}$$

and since the reduction potential of the $LC^{2+}/LC^{\bullet+}$ couple is –0.28 V vs. NHE, the equilibrium constant of the reaction is ~50; the reverse reaction producing $LC^{\bullet+}$ from superoxide must have a rate constant of ~6×10^4 M^{-1} s^{-1} and is unlikely to compete with the reaction of $O_2^{\bullet-}$ with superoxide dismutase in experiments involving loading LC^{2+} into mammalian cells [117]. Again, despite such an obvious contraindication, the probe is still being used: thus, in a recent study '... superoxide production in the brain, heart tissue, and aorta was measured using lucigenin-enhanced chemiluminescence' [118].

Even if repeated warnings about the shortcomings of commonly used probes continue to be ignored, some groups *have* exhibited a clear understanding of the problems; the collaboration between the Institute of Applied Radiation Chemistry at Łodz University of Technology and the Medical College of Wisconsin has been particularly fruitful (e.g., [9,27,30,39]), and considerable progress is being made in the kinetic characterization of some probes. Some examples are outlined very briefly below. Nonetheless, it is not at all encouraging that, while a widely consulted *Handbook of Fluorescent Probes* (see: https://www.thermofisher.com/uk/en/home/global/forms/mp-handbook-download-request-form-2014.html, accessed on 27 May 2023) has a 27-page chapter devoted to 'Probes for Reactive Oxygen Species, Including Nitric Oxide', the terms 'rate' or 'rate constant' are both conspicuously absent.

Progress is also illustrated by a discussion of the extent of intercepting oxidants using the more recently introduced boronate probes for nucleophilic oxidants, which has been the subject of intense activity [36], and the reactivity of probes for hypochlorous acid [44]. Earlier, a much clearer understanding of the reactions involved in the use of hydroethidine as a probe for superoxide radicals emerged after a series of careful studies [119,120], and the insightful review by Zielonka and Kalyanaraman [30] is an essential reading for users of luminescent probes for cellular oxidizing and nitrating species.

In particular, the generation of specific radicals by pulse radiolysis was used to estimate key rate constants involved in the detection of superoxide radicals by dihydroethidium (hydroethidine, HE, Figure 1, **3**), which superficially is similar to that of lucigenin in that the oxidized probe reacts with superoxide to form a detectable product. The oxidation of hydroethidine by $O_2^{\bullet-}$ has a rate constant around 2×10^6 M^{-1} s^{-1}; the rate constant for the reaction of $(HE)^{\bullet+}$ with $O_2^{\bullet-}$ to form the measured product, 2-hydroxyethidium, is orders of magnitude higher; so, the first reaction may be rate-limiting, depending on the concentrations of HE and $(HE)^{•+}$—at least in systems where only $O_2^{•-}$ is the oxidant: obviously, other oxidants can achieve the first step, as directly observed using radiolytically produced radicals [120].

'Amplex® Red', a dihydroxyphenoxazine (Figure 1, **4**), has been quite widely used to assay hydrogen peroxide, peroxidase-catalyzed oxidation yielding the fluorescent dye resorufin. However, other oxidants, including biological oxidants, are also reactive: stopped-flow kinetic spectrophotometry was used to demonstrate that peroxynitrite-derived oxidants, but not peroxynitrite, also oxidized the probe, and pulse radiolysis quantified the reactivity towards carbonate radicals, although at pH 10.3 (where the reactivity might be higher than at pH 7.4 because of prototropic properties that the authors demonstrated [121]). Intermediates formed during the conversion of Amplex® Red into resorufin were characterized. An earlier study [10] discussed possible chain reactions.

The reactivity towards superoxide of a new mitochondria-targeted probe, the phenanthridine derivative 'MitoNeoD' (Figure 1, **5**), was also characterized using pulse radiolysis, with a rate constant for the reaction of the probe with $O_2^{•-}$ of ~1.4×10^7 M^{-1} s^{-1} measured [122]. The details of this study are outside the scope of the present paper but are well worth studying as they demonstrate the depth of understanding now being applied in the development of new probes.

5. Some Compilations of Rate Constants for Reactions of Free Radicals Relevant to Biology

While there are very few rate constants measured for reactions of radicals with the specific chemical probes of the general type discussed here, there are useful, if dated, compilations of expertly evaluated rate constants for reactions of $^•OH$ and $H^•$ [123,124], $HO_2^•/O_2^{•-}$ [125], peroxyl radicals [126], inorganic radicals including $CO_3^{•-}$, $^•NO_2$ and $ClO_2^•$ [103], phenoxyl radicals [127] and aliphatic carbon-centered radicals [128]. It may be possible to estimate the likely rate constants in a few instances by comparison with the simpler chemical models that have been studied, and these compilations are invaluable in considering the chemical kinetics of possible competing reactions with endogenous reactants. However, the publication dates should be noted; further, while the former database of the University of Notre Dame Radiation Chemistry Data Center (which hosted the production of these compilations) has been transferred to the website of the National Institute of Standards and Technology (see: https://kinetics.nist.gov/solution/, accessed on 23 May 2023), it too suffers from not being updated and has strict syntax. It should be noted that PubMed will often miss relevant studies of a purely chemical nature. The majority of the rate constants involving reactions of interest involving free radicals have been obtained using radiation–chemical methods, and the inclusion of the term 'pulse radiolysis' as a search term may help filter literature searches for relevant kinetic data in both PubMed and Chemical Abstracts (SciFinder®). However, in recent years several pulse radiolysis facilities in Europe have closed down, and this is a barrier to the future application of the technique in this area.

6. Key Points to Consider When Using Chemical Probes in Free-Radical Biology

Biochemists familiar with enzyme-catalyzed reactions will take for granted that an assay for an enzyme will involve conditions such that the catalyst is rate-limiting, rather than the concentration of the substrate(s) (or vice versa if a substrate is to be assayed using an enzyme-catalyzed assay): saturating conditions characterized by Michaelis–Menten kinetics. Yet, in attempting to assay 'ROS' and other biological oxidants, the necessary parameters to evaluate whether a probe is used under saturating conditions or to estimate the fraction of the oxidant that is being detected—the products of rate constants and concentrations for reactions of both probe and competing reactants—have often been ignored or are simply not available.

The applications of chemical probes vary widely, and the criteria for using them in cell-free systems may be very different from those for cellular models and especially tissues. However, it may be useful to reiterate briefly the main questions that researchers should ask *before* using a probe, based on earlier discussion [5,13,30]:

- *Where is the probe located, and what is its concentration?* Consider, if appropriate, whether there are likely to be extracellular–intracellular concentration gradients or inter-organelle or other local variations in the concentration of the probe (driven by, e.g., lipid/water partitioning, trans-membrane pH differentials or binding to macromolecules). Measure, or at least estimate, the probe concentration(s) in the region(s) of interest.
- *What are the species likely to react with the probe?* Consider *all* the putative species being generated and their reactivities towards the probe, as estimated by the product of rate constant and concentration, preferably under relevant conditions (pH; solvent polarity; ideally, also temperature, although competing reactions *may* exhibit broadly similar temperature effects).
- *What are the reaction pathways involved in the generation of the final product being measured?* Identify reaction intermediates if possible, noting that spin conservation is likely to dictate the obligate formation of intermediate free radicals from radical oxidants.
- *Is a catalyst involved in the reaction(s)?* Consider whether the concentration of the catalyst may be rate-limiting and whether the presence of the catalyst results from the treatment being assessed, e.g., the release of cytochrome *c* from mitochondria as a result of apoptosis initiated by the oxidative challenge.
- *What are possible competing reactions, firstly involving kinetic competition between the probe and cellular antioxidants for the species of interest (e.g., oxidizing radicals or molecules)?* The Law of Mass Action will dictate the extent of the competition involving reactions of the species being assessed with endogenous reactants, especially with antioxidants: thiols, ascorbate, urate, NADH, α-tocopherol, etc., and other redox-active reagents (including oxygen, which can also modulate thiol radical chemistry).
- *Do the intermediates in probe chemistry react with endogenous molecules in competition with pathways leading to the final, measured product?* Again, the Law of Mass Action is central to an analysis, and quantifying the redox properties of reactive intermediates will help suggest possible reactants. The archetypical example is the reaction with oxygen of the obligate intermediate radical obtained on oxidation of reduced fluorescein dyes—oxidation either by radical species or by intermediates formed in the catalyzed reaction with hydrogen peroxide—to *generate* superoxide radicals.
- *Are there any potential effects of the cell culture medium (if used)?* Consider comparing the results in full medium (proteins, amino acids, redox- or light-sensitive pH indicator, metals, ascorbate, etc.) with those from cells suspended/plated in as pure phosphate-buffered saline as can be obtained.
- *Could there be artefacts arising from the photochemical properties of the probes and products?* Visible light (including ambient room light or instrumental light sources) may initiate photochemical reactions; photochemically induced excited states may be much more reactive towards cellular reductants than the probes themselves. Consider whether the product build-up is sufficient to initiate inner-filter effects by absorbing a significant fraction of the incident light, e.g., in fluorescence plate readers.
- *How far is the measured product likely to diffuse from the site of interest or 'leak' from cells?* This will be time-dependent and, within a specific organelle, can be estimated from the likely diffusion coefficient and the Einstein–Smoluchowski equation, but trans-membrane transport should also be considered.
- *Does the probe itself affect cellular function?* Probe and/or product may bind to macromolecules or change the mitochondrial membrane potential or may initiate apoptosis and lead to potential catalysts released during the experiment (e.g., cytochrome c in apoptosis).

7. Conclusions

To design, synthesize and use with confidence a chemical probe for reactants of interest in free-radical biology is a daunting prospect, requiring skills spanning disciplines, and techniques and instrumentation unlikely to be found in a single laboratory. The chemical obstacles to the use of the commonest probe for 'ROS' have been discussed previously (and repeatedly) but were summarized here with emphasis on kinetic factors, as a pointer to some of the properties which must be considered in the rational use of probes.

Some readers may find the discussion above of the language of free-radical biology and the advice to avoid the term 'ROS' wherever possible to be unrealistic, pedantic and repetitive. Yet, in the view of the present author, the language is a major part of the problem. The term 'reactive oxygen species' may have arisen in the early days of free-radical biology, when the emphasis was on Fenton chemistry and the term simply avoided distinguishing between free hydroxyl radicals as a product of iron/hydrogen peroxide chemistry, or whether an iron-oxo species was the reactant. The index to the second (1989) edition of a standard text discussing free radicals in biology and medicine does not include 'reactive oxygen' or 'ROS' [129]. Now, though, the term and its acronym retrieved over 327,000 'hits' in PubMed at the time of writing, and it would seem impossible to avoid this vague term. One is reminded of the old computer adage, 'GIGO': 'garbage in, garbage out'; in the context of chemical probes for 'ROS', the vagueness of the input term *almost guarantees* a vague output in the use of the probes. As a group of experts recently noted [4], 'Multiple roles of reactive oxygen species (ROS) ... have led researchers unfamiliar with the complexities of ROS and their reactions to employ commercial kits and probes to measure ROS and oxidative damage inappropriately, treating ROS (a generic abbreviation) as if it were a discrete molecular entity'.

In the present article, the author has again stressed the pitfalls in the use of the commonest probe for 'ROS' to illustrate what can go wrong if a careful study of the literature is outweighed by the convenience of purchasing and using a commercial reagent where the manufacturer has glossed over the problems. This was followed by a brief mention of some recent studies which show that leading workers in the field *do* have a clear grasp of the problems to be addressed. Kinetic measurements in which specific chemical entities are generated and their reactions followed directly to quantify reactivity are central to the successful application of chemical probes for oxidants and radicals in biology, but the rate constants are only part of the story; the local concentrations of the probe *and possible endogenous reactants* at specific sites in the cell are equally important. It is hoped the list above of questions that users should first consider in the use of chemical probes may, at the very least, help scientists to be aware of the uncertainties which arise if these factors are ignored.

Funding: This research received no external funding.

Data Availability Statement: No new data were created or analyzed in this study. Data sharing is not applicable to this article.

Conflicts of Interest: The author declares no conflict of interest.

References

1. Wardman, P. Approaches to modeling chemical reaction pathways in radiobiology. *Int. J. Radiat. Biol.* **2022**, *98*, 1399–1413. [CrossRef] [PubMed]
2. Fuentes-Lemus, E.; Reyes, J.S.; Gamon, L.F.; López-Alarcón, C.; Davies, M.J. Effect of macromolecular crowding on protein oxidation: Consequences on the rate, extent and oxidation pathways. *Redox Biol.* **2021**, *48*, 102202. [CrossRef] [PubMed]
3. Wardman, P. Initiating redox reactions by ionizing radiation: A versatile, selective and quantitative tool. *Redox Biochem. Chem.* **2023**, *5-6*, 100004. [CrossRef]
4. Murphy, M.P.; Bayir, H.; Belousov, V.; Chang, C.J.; Davies, K.J.A.; Davies, M.J.; Dick, T.P.; Finkel, T.; Forman, H.J.; Janssen-Heininger, Y.; et al. Guidelines for measuring reactive oxygen species and oxidative damage in cells and in vivo. *Nat. Metab.* **2022**, *4*, 651–662. [CrossRef] [PubMed]

5. Wardman, P. Fluorescent and luminescent probes for measurement of oxidative and nitrosative species in cells and tissues: Progress, pitfalls, and prospects. *Free Radic. Biol. Med.* **2007**, *43*, 995–1022. [CrossRef]
6. Chen, X.; Zhong, Z.; Xu, Z.; Chen, L.; Wang, Y. 2′,7′-Dichlorodihydrofluorescein as a fluorescent probe for reactive oxygen species measurement: Forty years of application and controversy. *Free Radic. Res.* **2010**, *44*, 587–604. [CrossRef]
7. Dickinson, B.C.; Srikun, D.; Chang, C.J. Mitochondrial-targeted fluorescent probes for reactive oxygen species. *Curr. Opin. Chem. Biol.* **2010**, *14*, 50–56. [CrossRef]
8. Kalyanaraman, B. Oxidative chemistry of fluorescent dyes: Implications in the detection of reactive oxygen and nitrogen species. *Biochem. Soc. Trans.* **2011**, *39*, 1221–1225. [CrossRef]
9. Kalyanaraman, B.; Darley-Usmar, V.; Davies, K.J.; Dennery, P.A.; Forman, H.J.; Grisham, M.B.; Mann, G.E.; Moore, K.; Roberts, L.J., 2nd; Ischiropoulos, H. Measuring reactive oxygen and nitrogen species with fluorescent probes: Challenges and limitations. *Free Radic. Biol. Med.* **2012**, *52*, 1–6. [CrossRef]
10. Liochev, S.I. Free radicals: How do we stand them? Anaerobic and aerobic free radical (chain) reactions involved in the use of fluorogenic probes and in biological systems. *Med. Princ. Pract.* **2013**, *23*, 195–203. [CrossRef]
11. Woolley, J.F.; Stanicka, J.; Cotter, T.G. Recent advances in reactive oxygen species measurement in biological systems. *Trends Biochem. Sci.* **2013**, *38*, 556–565. [CrossRef] [PubMed]
12. Zielonka, J.; Joseph, J.; Sikora, A.; Kalyanaraman, B. Real-time monitoring of reactive oxygen and nitrogen species in a multiwell plate using the diagnostic marker products of specific probes. *Methods Enzymol.* **2013**, *526*, 145–157. [CrossRef] [PubMed]
13. Winterbourn, C.C. The challenges of using fluorescent probes to detect and quantify specific reactive oxygen species in living cells. *Biochim. Biophys. Acta* **2014**, *1840*, 730–738. [CrossRef] [PubMed]
14. Dikalov, S.I.; Harrison, D.G. Methods for detection of mitochondrial and cellular reactive oxygen species. *Antioxid. Redox Signal.* **2014**, *20*, 372–382. [CrossRef]
15. Nauseef, W.M. Detection of superoxide anion and hydrogen peroxide production by cellular NADPH oxidases. *Biochim. Biophys. Acta* **2014**, *1840*, 757–767. [CrossRef] [PubMed]
16. Barzegar Amiri Olia, M.; Schiesser, C.H.; Taylor, M.K. New reagents for detecting free radicals and oxidative stress. *Org. Biomol. Chem.* **2014**, *12*, 6757–6766. [CrossRef]
17. Adegoke, O.; Forbes, P.B. Challenges and advances in quantum dot fluorescent probes to detect reactive oxygen and nitrogen species: A review. *Anal. Chim. Acta* **2015**, *862*, 1–13. [CrossRef]
18. Zhang, X.; Gao, F. Imaging mitochondrial reactive oxygen species with fluorescent probes: Current applications and challenges. *Free Radic. Res.* **2015**, *49*, 374–382. [CrossRef]
19. Debowska, K.; Debski, D.; Hardy, M.; Jakubowska, M.; Kalyanaraman, B.; Marcinek, A.; Michalski, R.; Michalowski, B.; Ouari, O.; Sikora, A.; et al. Toward selective detection of reactive oxygen and nitrogen species with the use of fluorogenic probes—Limitations, progress, and perspectives. *Pharmacol. Rep.* **2015**, *67*, 756–764. [CrossRef]
20. Kolanowski, J.L.; Kaur, A.; New, E.J. Selective and reversible approaches toward imaging redox signaling using small-molecule probes. *Antioxid. Redox Signal.* **2016**, *24*, 713–730. [CrossRef]
21. Chen, X.; Wang, F.; Hyun, J.Y.; Wei, T.; Qiang, J.; Ren, X.; Shin, I.; Yoon, J. Recent progress in the development of fluorescent, luminescent and colorimetric probes for detection of reactive oxygen and nitrogen species. *Chem. Soc. Rev.* **2016**, *45*, 2976–3016. [CrossRef]
22. Ribou, A.C. Synthetic sensors for reactive oxygen species detection and quantification: A critical review of current methods. *Antioxid. Redox Signal.* **2016**, *25*, 520–533. [CrossRef] [PubMed]
23. Żamojć, K.; Zdrowowicz, M.; Jacewicz, D.; Wyrzykowski, D.; Chmurzyński, L. Fluorescent probes used for detection of hydrogen peroxide under biological conditions. *Crit. Rev. Anal. Chem.* **2016**, *46*, 171–200. [CrossRef]
24. Kalyanaraman, B.; Hardy, M.; Podsiadly, R.; Cheng, G.; Zielonka, J. Recent developments in detection of superoxide radical anion and hydrogen peroxide: Opportunities, challenges, and implications in redox signaling. *Arch. Biochem. Biophys.* **2017**, *617*, 38–47. [CrossRef]
25. Andina, D.; Leroux, J.C.; Luciani, P. Ratiometric fluorescent probes for the detection of reactive oxygen species. *Chemistry* **2017**, *23*, 13549–13573. [CrossRef]
26. Lü, R. Reaction-based small-molecule fluorescent probes for dynamic detection of ROS and transient redox changes in living cells and small animals. *J. Mol. Cell. Cardiol.* **2017**, *110*, 96–108. [CrossRef]
27. Hardy, M.; Zielonka, J.; Karoui, H.; Sikora, A.; Michalski, R.; Podsiadły, R.; Lopez, M.; Vasquez-Vivar, J.; Kalyanaraman, B.; Ouari, O. Detection and characterization of reactive oxygen and nitrogen species in biological systems by monitoring species-specific products. *Antioxid. Redox Signal.* **2018**, *28*, 1416–1432. [CrossRef]
28. Jiao, X.; Li, Y.; Niu, J.; Xie, X.; Wang, X.; Tang, B. Small-molecule fluorescent probes for imaging and detection of reactive oxygen, nitrogen, and sulfur species in biological systems. *Anal. Chem.* **2018**, *90*, 533–555. [CrossRef]
29. Erard, M.; Dupré-Crochet, S.; Nüße, O. Biosensors for spatiotemporal detection of reactive oxygen species in cells and tissues. *Am. J. Physiol. Regul. Integr. Comp. Physiol.* **2018**, *314*, R667–r683. [CrossRef] [PubMed]
30. Zielonka, J.; Kalyanaraman, B. Small-molecule luminescent probes for the detection of cellular oxidizing and nitrating species. *Free Radic. Biol. Med.* **2018**, *128*, 3–22. [CrossRef] [PubMed]
31. Gao, P.; Pan, W.; Li, N.; Tang, B. Fluorescent probes for organelle-targeted bioactive species imaging. *Chem. Sci.* **2019**, *10*, 6035–6071. [CrossRef] [PubMed]

32. Bai, X.; Ng, K.K.; Hu, J.J.; Ye, S.; Yang, D. Small-molecule-based fluorescent sensors for selective detection of reactive oxygen species in biological systems. *Annu. Rev. Biochem.* **2019**, *88*, 605–633. [CrossRef]
33. Wu, L.; Sedgwick, A.C.; Sun, X.; Bull, S.D.; He, X.P.; James, T.D. Reaction-based fluorescent probes for the detection and imaging of reactive oxygen, nitrogen, and sulfur species. *Acc. Chem. Res.* **2019**, *52*, 2582–2597. [CrossRef] [PubMed]
34. Hirayama, T. Fluorescent probes for the detection of catalytic Fe(II) ion. *Free Radic. Biol. Med.* **2019**, *133*, 38–45. [CrossRef] [PubMed]
35. Xiao, H.; Zhang, W.; Li, P.; Zhang, W.; Wang, X.; Tang, B. Versatile fluorescent probes for imaging the superoxide anion in living cells and in vivo. *Angew. Chem. Int. Ed. Engl.* **2020**, *59*, 4216–4230. [CrossRef]
36. Sikora, A.; Zielonka, J.; Dębowska, K.; Michalski, R.; Smulik-Izydorczyk, R.; Pięta, J.; Podsiadły, R.; Artelska, A.; Pierzchała, K.; Kalyanaraman, B. Boronate-based probes for biological oxidants: A novel class of molecular tools for redox biology. *Front. Chem.* **2020**, *8*, 580899. [CrossRef]
37. Kalyanaraman, B. Pitfalls of reactive oxygen species (ROS) measurements by fluorescent probes and mitochondrial superoxide determination using MitoSox. In *Measuring Oxidants and Oxidative Stress in Biological Systems*; Berliner, L.J., Parinandi, N.L., Eds.; Springer: Cham, Switzerland, 2020; pp. 7–9.
38. Gardiner, B.; Dougherty, J.A.; Ponnalagu, D.; Singh, H.; Angelos, M.; Chen, C.A.; Khan, M. Measurement of oxidative stress markers in vitro using commercially available kits. In *Measuring Oxidants and Oxidative Stress in Biological Systems*; Berliner, L.J., Parinandi, N.L., Eds.; Springer: Cham, Switzerland, 2020; pp. 39–60.
39. Michalski, R.; Thiebaut, D.; Michałowski, B.; Ayhan, M.M.; Hardy, M.; Ouari, O.; Rostkowski, M.; Smulik-Izydorczyk, R.; Artelska, A.; Marcinek, A.; et al. Oxidation of ethidium-based probes by biological radicals: Mechanism, kinetics and implications for the detection of superoxide. *Sci. Rep.* **2020**, *10*, 18626. [CrossRef]
40. Pramanik, S.K.; Das, A. Fluorescent probes for imaging bioactive species in subcellular organelles. *Chem. Commun.* **2021**, *57*, 12058–12073. [CrossRef]
41. Xu, Y.; Yang, W.; Zhang, B. ROS-responsive probes for low-background optical imaging: A review. *Biomed. Mater.* **2021**, *16*, 022002. [CrossRef]
42. Li, J.; LoBue, A.; Heuser, S.K.; Leo, F.; Cortese-Krott, M.M. Using diaminofluoresceins (DAFs) in nitric oxide research. *Nitric Oxide* **2021**, *115*, 44–54. [CrossRef] [PubMed]
43. Sies, H.; Belousov, V.V.; Chandel, N.S.; Davies, M.J.; Jones, D.P.; Mann, G.E.; Murphy, M.P.; Yamamoto, M.; Winterbourn, C. Defining roles of specific reactive oxygen species (ROS) in cell biology and physiology. *Nat. Rev. Mol. Cell Biol.* **2022**, *23*, 499–515. [CrossRef] [PubMed]
44. Pierzchała, K.; Pięta, M.; Rola, M.; Świerczyńska, M.; Artelska, A.; Dębowska, K.; Podsiadły, R.; Pięta, J.; Zielonka, J.; Sikora, A.; et al. Fluorescent probes for monitoring myeloperoxidase-derived hypochlorous acid: A comparative study. *Sci. Rep.* **2022**, *12*, 9314. [CrossRef] [PubMed]
45. Huang, P.; Yue, Y.; Yin, C.; Huo, F. Design of dual-responsive ROS/RSS fluorescent probes and their application in bioimaging. *Chem. Asian J.* **2022**, *17*, e202200907. [CrossRef]
46. Niu, P.; Zhu, J.; Wei, L.; Liu, X. Application of fluorescent probes in reactive oxygen species disease model. *Crit. Rev. Anal. Chem.* **2022**, 1–36. [CrossRef] [PubMed]
47. Sun, J.; Cao, X.; Lu, W.; Wei, Y.; Kong, L.; Chen, W.; Shao, X.; Wang, Y. Recent advances in fluorescent probes of peroxynitrite: Structural, strategies and biological applications. *Theranostics* **2023**, *13*, 1716–1744. [CrossRef]
48. Geng, Y.; Wang, Z.; Zhou, J.; Zhu, M.; Liu, J.; James, T.D. Recent progress in the development of fluorescent probes for imaging pathological oxidative stress. *Chem. Soc. Rev.* **2023**, *52*, 3873–3926. [CrossRef]
49. Yin, J.; Zhan, J.; Hu, Q.; Huang, S.; Lin, W. Fluorescent probes for ferroptosis bioimaging: Advances, challenges, and prospects. *Chem. Soc. Rev.* **2023**, *52*, 2011–2030. [CrossRef]
50. Sies, H. Oxidative stress: A concept in redox biology and medicine. *Redox Biol.* **2015**, *4*, 180–183. [CrossRef]
51. Sies, H.; Berndt, C.; Jones, D.P. Oxidative stress. *Annu. Rev. Biochem.* **2017**, *86*, 715–748. [CrossRef]
52. Gutteridge, J.M.C.; Halliwell, B. Mini-review: Oxidative stress, redox stress or redox success? *Biochem. Biophys. Res. Commun.* **2018**, *502*, 183–186. [CrossRef]
53. Sies, H. Oxidative stress: Concept and some practical aspects. *Antioxidants* **2020**, *9*, 852. [CrossRef]
54. Forman, H.J.; Augusto, O.; Brigelius-Flohe, R.; Dennery, P.A.; Kalyanaraman, B.; Ischiropoulos, H.; Mann, G.E.; Radi, R.; Roberts, L.J.; Vina, J.; et al. Even free radicals should follow some rules: A guide to free radical research terminology and methodology. *Free Radic. Biol. Med.* **2015**, *78*, 233–235. [CrossRef]
55. Buettner, G.R. Moving free radical and redox biology ahead in the next decade(s). *Free Radic. Biol. Med.* **2015**, *78*, 236–238. [CrossRef] [PubMed]
56. Quintiliani, M.; Badiello, R.; Tamba, M.; Esfandi, A.; Gorin, G. Radiolysis of glutathione in oxygen-containing solutions of pH 7. *Int. J. Radiat. Biol.* **1977**, *32*, 195–202. [CrossRef] [PubMed]
57. Storkey, C.; Davies, M.J.; Pattison, D.I. Reevaluation of the rate constants for the reaction of hypochlorous acid (HOCl) with cysteine, methionine, and peptide derivatives using a new competition kinetic approach. *Free Radic. Biol. Med.* **2014**, *73*, 60–66. [CrossRef] [PubMed]
58. Ford, E.; Hughes, M.N.; Wardman, P. Kinetics of the reactions of nitrogen dioxide with glutathione, cysteine, and uric acid at physiological pH. *Free Radic. Biol. Med.* **2002**, *32*, 1314–1323. [CrossRef]

59. Chen, S.N.; Hoffman, M.Z. Rate constants for the reaction of the carbonate radical with compounds of biochemical interest in neutral aqueous solution. *Radiat. Res.* **1973**, *56*, 40–47. [CrossRef]
60. Jones, C.M.; Lawrence, A.; Wardman, P.; Burkitt, M.J. Electron paramagnetic resonance spin-trapping investigation into the kinetics of glutathione oxidation by the superoxide radical: Re-evaluation of the rate constant. *Free Radic. Biol. Med.* **2002**, *32*, 982–990. [CrossRef]
61. Abedinzadeh, Z.; Gardes-Albert, M.; Ferradini, C. Kinetic study of the oxidation mechanism of glutathione by hydrogen peroxide in neutral aqueous medium. *Can. J. Chem.* **1989**, *67*, 1247–1255. [CrossRef]
62. Madej, E.; Wardman, P. The oxidizing power of the glutathione thiyl radical as measured by its electrode potential at physiological pH. *Arch. Biochem. Biophys.* **2007**, *462*, 94–102. [CrossRef]
63. DeFelippis, M.R.; Murthy, C.P.; Faraggi, M.; Klapper, M.H. Pulse radiolytic measurement of redox potentials: The tyrosine and tryptophan radicals. *Biochemistry* **1989**, *28*, 4847–4853. [CrossRef] [PubMed]
64. Adams, G.E.; McNaughton, G.S.; Michael, B.D. Pulse radiolysis of sulphur compounds. Part 2.—Free radical "repair" by hydrogen transfer from sulphydryl compounds. *Trans. Faraday Soc.* **1968**, *64*, 902–910. [CrossRef]
65. Willson, R.L. Pulse radiolysis studies of electron transfer in aqueous disulphide solutions. *Chem. Commun.* **1970**, 1425–1426. [CrossRef]
66. Forni, L.G.; Mönig, J.; Mora-Arellano, V.O.; Willson, R.L. Thiyl free radicals: Direct observations of electron transfer reactions with phenothiazines and ascorbate. *J. Chem. Soc. Perkin Trans.* **1983**, *2*, 961–965. [CrossRef]
67. Wardman, P.; von Sonntag, C. Kinetic factors that control the fate of thiyl radicals in cells. *Methods Enzymol.* **1995**, *251*, 31–45. [CrossRef]
68. Schöneich, C. Thiyl radicals, perthiyl radicals, and oxidative reactions. In *Biothiols in Health and Disease*; Packer, L., Cadenas, E., Eds.; Marcel Dekker: New York, NY, USA, 1995; pp. 21–47.
69. Schöneich, C. Thiyl radicals: Formation, properties, and detection. In *Redox Chemistry and Biology of Thiols*; Alvarez, B., Comini, M.A., Salinas, G., Trujillo, M., Eds.; Academic Press: Cambridge, MA, USA, 2022; pp. 115–132.
70. Chatgilialoglu, C.; Ferreri, C.; Ballestri, M.; Mulazzani, Q.G.; Landi, L. *Cis-trans* isomerization of monounsaturated fatty acid residues in phospholipids by thiyl radicals. *J. Am. Chem. Soc.* **2000**, *122*, 4593–4601. [CrossRef]
71. Chatgilialoglu, C.; Ferreri, C. Trans lipids: The free radical path. *Acc. Chem. Res.* **2005**, *38*, 441–448. [CrossRef]
72. Chatgilialoglu, C.; Ferreri, C.; Lykakis, I.N.; Wardman, P. *Trans*-fatty acids and radical stress: What are the real culprits? *Bioorg. Med. Chem.* **2006**, *14*, 6144–6148. [CrossRef]
73. Mihaljević, B.; Tartaro, I.; Ferreri, C.; Chatgilialoglu, C. Linoleic acid peroxidation vs. isomerization: A biomimetic model of free radical reactivity in the presence of thiols. *Org. Biomol. Chem.* **2011**, *9*, 3541–3548. [CrossRef] [PubMed]
74. Sevilla, M.D.; Becker, D.; Yan, M. The formation and structure of the sulfoxyl radicals $RSO^{\bullet}$, $RSOO^{\bullet}$, $RSO_2^{\bullet}$, and $RSO_2OO^{\bullet}$ from the reaction of cysteine, glutathione and penicillamine thiyl radicals with molecular oxygen. *Int. J. Radiat. Biol.* **1990**, *57*, 65–81. [CrossRef]
75. Schöneich, C.; Dillinger, U.; Bruchhausen, V.; Asmus, K.-D. Oxidation of polyunsaturated fatty acids and lipids through thiyl and sulfonyl radicals: Reaction kinetics, and influence of oxygen and structure of thiyl radicals. *Arch. Biochem. Biophys.* **1992**, *292*, 456–467. [CrossRef] [PubMed]
76. Radi, R. Oxygen radicals, nitric oxide, and peroxynitrite: Redox pathways in molecular medicine. *Proc. Natl. Acad. Sci. USA* **2018**, *115*, 5839–5848. [CrossRef]
77. Augusto, O.; Goldstein, S.; Hurst, J.K.; Lind, J.; Lymar, S.V.; Merenyi, G.; Radi, R. Carbon dioxide-catalyzed peroxynitrite reactivity—The resilience of the radical mechanism after two decades of research. *Free Radic. Biol. Med.* **2019**, *135*, 210–215. [CrossRef]
78. Armstrong, D.A.; Huie, R.E.; Koppenol, W.H.; Lymar, S.V.; Merényi, G.; Neta, P.; Ruscica, B.; Stanburyb, D.M.; Steenken, S.; Wardman, P. Standard electrode potentials involving radicals in aqueous solution: Inorganic radicals (IUPAC technical report). *Pure App. Chem.* **2015**, *87*, 1139–1150. [CrossRef]
79. LeBel, C.P.; Ischiropoulos, H.; Bondy, S.C. Evaluation of the probe 2¢,7¢-dichlorofluorescin as an indicator of reactive oxygen species formation and oxidative stress. *Chem. Res. Toxicol.* **1992**, *5*, 227–231. [CrossRef] [PubMed]
80. Rota, C.; Chignell, C.F.; Mason, R.P. Evidence for free radical formation during the oxidation of 2¢-7¢-dichlorofluorescin to the fluorescent dye 2¢-7¢-dichlorofluorescein by horseradish peroxidase: Possible implications for oxidative stress measurements. *Free Rad. Biol. Med.* **1999**, *27*, 873–881. [CrossRef] [PubMed]
81. Rota, C.; Fann, Y.C.; Mason, R.P. Phenoxyl free radical formation during the oxidation of the fluorescent dye 2¢,7¢-dichlorofluorescein by horseradish peroxidase. *J. Biol. Chem.* **1999**, *274*, 28161–28168. [CrossRef]
82. O'Malley, Y.Q.; Reszka, K.J.; Britigan, B.E. Direct oxidation of 2′,7′-dichlorodihydrofluorescein by pyocyanin and other redox-active compounds independent of reactive oxygen species production. *Free Radic. Biol. Med.* **2004**, *36*, 90–100. [CrossRef]
83. Bonini, M.G.; Rota, C.; Tomasi, A.; Mason, R.P. The oxidation of 2′,7′-dichlorofluorescin to reactive oxygen species: A self-fulfilling prophesy? *Free Radic. Biol. Med.* **2006**, *40*, 968–975. [CrossRef]
84. Wardman, P. Use of the dichlorofluorescein assay to measure "reactive oxygen species". *Radiat. Res.* **2008**, *170*, 406–408. [CrossRef]
85. Wrona, M.; Patel, K.B.; Wardman, P. The roles of thiol-derived radicals in the use of 2¢,7¢-dichlorodihydrofluorescein as a probe for oxidative stress. *Free Radic. Biol. Med.* **2008**, *44*, 56–62, Erratum in *Free Radic. Biol. Med.* **2008**, *45*, 547. [CrossRef] [PubMed]

86. Murphy, M.P.; Holmgren, A.; Larsson, N.G.; Halliwell, B.; Chang, C.J.; Kalyanaraman, B.; Rhee, S.G.; Thornalley, P.J.; Partridge, L.; Gems, D.; et al. Unraveling the biological roles of reactive oxygen species. *Cell Metab.* **2011**, *13*, 361–366. [CrossRef] [PubMed]
87. Yazdani, M. Concerns in the application of fluorescent probes DCDHF-DA, DHR 123 and DHE to measure reactive oxygen species in vitro. *Toxicol. In Vitro* **2015**, *30*, 578–582. [CrossRef] [PubMed]
88. De Haan, L.R.; Reiniers, M.J.; Reeskamp, L.F.; Belkouz, A.; Ao, L.; Cheng, S.; Ding, B.; van Golen, R.F.; Heger, M. Experimental conditions that influence the utility of 2¢,7¢-dichlorodihydrofluorescein diacetate (DCFH$_2$-DA) as a fluorogenic biosensor for mitochondrial redox status. *Antioxidants* **2022**, *11*, 1424. [CrossRef]
89. Kalyanaraman, B. NAC, NAC, Knockin' on heaven's door: Interpreting the mechanism of action of *N*-acetylcysteine in tumor and immune cells. *Redox Biol.* **2022**, *57*, 102497. [CrossRef]
90. Eruslanov, E.; Kusmartsev, S. Identification of ROS using oxidized DCFDA and flow-cytometry. In *Advanced Protocols in Oxidative Stress II*; Armstrong, D., Ed.; Humana Press: Totowa, NJ, USA, 2010; pp. 57–72.
91. Kim, H.; Xue, X. Detection of total reactive oxygen species in adherent cells by 2¢,7¢-dichlorodihydrofluorescein diacetate staining. *J. Vis. Exp.* **2020**, e60682. [CrossRef]
92. Hans, C.; Saini, R.; Sachdeva, M.U.S.; Sharma, P. 2¢,7¢-dichlorofluorescein (DCF) or 2¢,7¢- dichlorodihydrofluorescein diacetate (DCFH$_2$DA) to measure reactive oxygen species in erythrocytes. *Biomed. Pharm.* **2021**, *138*, 111512. [CrossRef]
93. Zhao, Y.; Lin, X.; Zeng, W.; Qin, X.; Miao, B.; Gao, S.; Liu, J.; Li, Z. Berberine inhibits the progression of renal cell carcinoma cells by regulating reactive oxygen species generation and inducing DNA damage. *Mol. Biol. Rep.* **2023**, *50*, 5697–5707. [CrossRef]
94. Binjawhar, D.N.; Alhazmi, A.T.; Bin Jawhar, W.N.; MohammedSaeed, W.; Safi, S.Z. Hyperglycemia-induced oxidative stress and epigenetic regulation of ET-1 gene in endothelial cells. *Front. Genet.* **2023**, *14*, 1167773. [CrossRef]
95. Ezequiel, J.; Nitschke, M.R.; Laviale, M.; Serôdio, J.; Frommlet, J.C. Concurrent bioimaging of microalgal photophysiology and oxidative stress. *Photosynth. Res.* **2023**, *155*, 177–190. [CrossRef]
96. Huang, J.; Yu, P.; Liao, M.; Dong, X.; Xu, J.; Ming, J.; Bin, D.; Wang, Y.; Zhang, F.; Xia, Y. A self-charging salt water battery for antitumor therapy. *Sci. Adv.* **2023**, *9*, eadf3992. [CrossRef]
97. Cathcart, R.; Schwiers, E.; Ames, B.N. Detection of picomole levels of hydroperoxides using a fluorescent dichlorofluorescein assay. *Anal. Biochem.* **1983**, *134*, 111–116. [CrossRef]
98. Burkitt, M.J.; Wardman, P. Cytochrome *c* is a potent catalyst of dichlorofluorescin oxidation: Implications for the role of reactive oxygen species in apoptosis. *Biochem. Biophys. Res. Commun.* **2001**, *282*, 329–333. [CrossRef]
99. Lawrence, A.; Jones, C.M.; Wardman, P.; Burkitt, M.J. Evidence for the role of a peroxidase compound-I type intermediate in the oxidation of glutathione, NADH, ascorbate, and dichlorofluorescin by cytochrome c/H_2O_2. Implications for oxidative stress during apoptosis. *J. Biol. Chem.* **2003**, *278*, 29410–29419. [CrossRef]
100. Crow, J.P. Dichlorodihydrofluorescein and dihydrorhodamine 123 are sensitive indicators of peroxynitrite *in vitro*: Implications for intracellular measurements of reactive nitrogen and oxygen species. *Nitric Oxide* **1997**, *1*, 145–157. [CrossRef]
101. Glebska, J.; Koppenol, W.H. Peroxynitrite-mediated oxidation of dichlorodihydrofluorescein and dihydrorhodamine. *Free Radic. Biol. Med.* **2003**, *35*, 676–682. [CrossRef]
102. Wrona, M.; Wardman, P. Properties of the radical intermediate obtained on oxidation of 2¢,7¢-dichlorodihydrofluorescein, a probe for oxidative stress. *Free Radic.Biol. Med.* **2006**, *41*, 657–667. [CrossRef] [PubMed]
103. Neta, P.; Huie, R.E.; Ross, A.B. Rate constants for reactions of inorganic radicals in aqueous solution. *J. Phys. Chem. Ref. Data* **1988**, *17*, 1027–1284. [CrossRef]
104. Wardman, P. Nitrogen dioxide in biology: Correlating chemical kinetics with biological effects. In *The Chemistry of N-Centered Radicals*; Alfassi, Z.B., Ed.; Wiley: New York, NY, USA, 1998; pp. 155–179.
105. Wrona, M.; Patel, K.B.; Wardman, P. Reactivity of 2¢,7¢-dichlorodihydrofluorescein and dihydrorhodamine 123 and their oxidized forms towards carbonate, nitrogen dioxide, and hydroxyl radicals. *Free Radic. Biol. Med.* **2005**, *38*, 262–270. [CrossRef] [PubMed]
106. Afri, M.; Frimer, A.A.; Cohen, Y. Active oxygen chemistry within the liposomal bilayer. Part IV: Locating 2′,7′-dichlorofluorescein (DCF), 2′,7′-dichlorodihydrofluorescein (DCFH) and 2′,7′-dichlorodihydrofluorescein diacetate (DCFH-DA) in the lipid bilayer. *Chem. Phys. Lipids* **2004**, *131*, 123–133. [CrossRef] [PubMed]
107. Gomes, A.; Fernandes, E.; Lima, J.L. Fluorescence probes used for detection of reactive oxygen species. *J. Biochem. Biophys. Methods.* **2005**, *65*, 45–80. [CrossRef]
108. Royall, J.A.; Ischiropoulos, H. Evaluation of 2′,7′-dichlorofluorescin and dihydrorhodamine 123 as fluorescent probes for intracellular H_2O_2 in cultured endothelial cells. *Arch. Biochem. Biophys.* **1993**, *302*, 348–355. [CrossRef] [PubMed]
109. Buettner, G.R.; Jurkiewicz, B.A. Catalytic metals, ascorbate and free radicals: Combinations to avoid. *Radiat. Res.* **1996**, *145*, 532–541. [CrossRef] [PubMed]
110. Zielonka, J.; Zielonka, M.; Sikora, A.; Adamus, J.; Joseph, J.; Hardy, M.; Ouari, O.; Dranka, B.P.; Kalyanaraman, B. Global profiling of reactive oxygen and nitrogen species in biological systems: High-throughput real-time analyses. *J. Biol. Chem.* **2012**, *287*, 2984–2995. [CrossRef] [PubMed]
111. Huie, R.E.; Shoute, L.C.T.; Neta, P. Temperature dependence of the rate constants for reactions of the carbonate radical with organic and inorganic reductants. *Int. J. Chem. Kinet.* **1991**, *23*, 541–552. [CrossRef]
112. Marchesi, E.; Rota, C.; Fann, Y.C.; Chignell, C.F.; Mason, R.P. Photoreduction of the fluorescent dye 2¢-7¢-dichlorofluorescein: A spin trapping and direct electron spin resonance study with implications for oxidative stress measurements. *Free Radic. Biol. Med.* **1999**, *26*, 148–161. [CrossRef]

113. Summers, F.A.; Zhao, B.; Ganini, D.; Mason, R.P. Photooxidation of Amplex Red to resorufin: Implications of exposing the Amplex Red assay to light. *Methods Enzymol.* **2013**, *526*, 1–17. [CrossRef]
114. Faulkner, K.; Fridovich, I. Luminol and lucigenin as detectors for $O_2^{\bullet-}$. *Free Radic. Biol. Med.* **1993**, *15*, 447–451. [CrossRef]
115. Liochev, S.I.; Fridovich, I. Lucigenin (bis-*N*-methylacridinium) as a mediator of superoxide anion production. *Archi.Biochem. Biophys.* **1996**, *337*, 115–120. [CrossRef]
116. Vásquez-Vivar, J.; Hogg, N.; Pritchard, K.A., Jr.; Martasek, P.; Kalyanaraman, B. Superoxide anion formation from lucigenin: An electron spin resonance spin-trapping study. *FEBS Lett.* **1997**, *403*, 127–130. [CrossRef]
117. Wardman, P.; Burkitt, M.J.; Patel, K.B.; Lawrence, A.; Jones, C.M.; Everett, S.A.; Vojnovic, B. Pitfalls in the use of common luminescent probes for oxidative and nitrosative stress. *J. Fluoresc.* **2002**, *12*, 65–68. [CrossRef]
118. Wu, Q.; Gurpinar, A.; Roberts, M.; Camelliti, P.; Ruggieri, M.R., Sr.; Wu, C. Identification of the NADPH oxidase (Nox) subtype and the source of superoxide production in the micturition centre. *Biology* **2022**, *11*, 183. [CrossRef] [PubMed]
119. Zielonka, J.; Zhao, H.; Xu, Y.; Kalyanaraman, B. Mechanistic similarities between oxidation of hydroethidine by Fremy's salt and superoxide: Stopped-flow optical and EPR studies. *Free Radic. Biol. Med.* **2005**, *39*, 853–863. [CrossRef]
120. Zielonka, J.; Sarna, T.; Roberts, J.E.; Wishart, J.F.; Kalyanaraman, B. Pulse radiolysis and steady-state analyses of the reaction between hydroethidine and superoxide and other oxidants. *Arch. Biochem. Biophys.* **2006**, *456*, 39–47. [CrossRef]
121. Dębski, D.; Smulik, R.; Zielonka, J.; Michałowski, B.; Jakubowska, M.; Dębowska, K.; Adamus, J.; Marcinek, A.; Kalyanaraman, B.; Sikora, A. Mechanism of oxidative conversion of Amplex®Red to resorufin: Pulse radiolysis and enzymatic studies. *Free Radic. Biol. Med.* **2016**, *95*, 323–332. [CrossRef] [PubMed]
122. Shchepinova, M.M.; Cairns, A.G.; Prime, T.A.; Logan, A.; James, A.M.; Hall, A.R.; Vidoni, S.; Arndt, S.; Caldwell, S.T.; Prag, H.A.; et al. MitoNeoD: A mitochondria-targeted superoxide probe. *Cell Chem. Biol.* **2017**, *24*, 1285–1298.e1212. [CrossRef] [PubMed]
123. Buxton, G.V.; Greenstock, C.L.; Helman, W.P.; Ross, A.B. Critical review of rate constants for reactions of hydrated electrons, hydrogen atoms and hydroxyl radicals ($^{\bullet}OH/^{\bullet}O^-$) in aqueous solution. *J. Phys. Chem. Ref. Data* **1988**, *17*, 513–886. [CrossRef]
124. Madden, K.P.; Mezyk, S.P. Critical review of aqueous solution reaction rate constants for hydrogen atoms. *J. Phys. Chem. Ref. Data* **2011**, *40*, 023103. [CrossRef]
125. Bielski, B.H.J.; Cabelli, D.E.; Arudi, R.L. Reactivity of $HO_2^{\bullet}/O_2^{\bullet-}$ radicals in aqueous solution. *J. Phys. Chem. Ref. Data* **1985**, *14*, 1041–1100. [CrossRef]
126. Neta, P.; Huie, R.E.; Ross, A.B. Rate constants for reactions of peroxyl radicals in fluid solutions. *J. Phys. Chem. Ref. Data* **1990**, *19*, 413–513. [CrossRef]
127. Neta, P.; Grodkowski, J. Rate constants for reactions of phenoxyl radicals in solution. *J. Phys. Chem. Ref. Data* **2005**, *34*, 109–199. [CrossRef]
128. Neta, P.; Grodkowski, J.; Ross, A.B. Rate constants for reactions of aliphatic carbon-centered radicals in aqueous solution. *J. Phys. Chem. Ref. Data* **1996**, *25*, 709–1050. [CrossRef]
129. Halliwell, B.; Gutteridge, J.M.C. *Free Radicals in Biology and Medicine*, 2nd ed.; Clarendon Press: Oxford, UK, 1989.

biomolecules

MDPI

Article

Physico-Chemical Changes Induced by Gamma Irradiation on Some Structural Protein Extracts

Maria Stanca [1], Carmen Gaidau [1], Traian Zaharescu [2], George-Alin Balan [3], Iulia Matei [3], Aurica Precupas [3], Anca Ruxandra Leonties [3] and Gabriela Ionita [1,3,*]

[1] Leather Research Department, Research and Development National Institute for Textiles and Leather-Division Leather and Footwear Research Institute, 93, Ion Minulescu Street, 031215 Bucharest, Romania
[2] INCDIE ICPE CA, 313 Splaiul Unirii, 030138 Bucharest, Romania
[3] "Ilie Murgulescu" Institute of Physical Chemistry of the Romanian Academy, 202 Splaiul Independentei, 060021 Bucharest, Romania
* Correspondence: gabi2ionita@yahoo.com or ige@icf.ro

Abstract: In this study, the effect of gamma irradiation (10 kGy) on proteins extracted from animal hide, scales, and wool was evidenced by calorimetric (μDSC) and spectroscopic (IR, circular dichroism, and EPR) methods. Keratin was obtained from sheep wool, collagen and bovine gelatin from bovine hide, and fish gelatin from fish scales. The μDSC experiments evidenced that gamma irradiation influences the thermal stability of these proteins differently. The thermal stability of keratin decreases, while a resistance to thermal denaturation was noticed for collagen and gelatins after gamma irradiation. The analysis of the IR spectra demonstrated that gamma irradiation determines changes in the vibrational modes of the amide groups that are associated with protein denaturation, most meaningfully in the case of keratin. As evidenced by circular dichroism for all proteins considered, exposure to gamma radiation produces changes in the secondary structure that are more significant than those produced by UV irradiation. Riboflavin has different effects on the secondary structure of the investigated proteins, a stabilizing effect for keratin and fish gelatin and a destabilizing effect for bovine gelatin, observed in both irradiated and non-irradiated samples. The EPR spectroscopy evidences the presence, in the gamma-irradiated samples, of free radicals centered on oxygen, and the increase in their EPR signals over time due to the presence of riboflavin.

Keywords: keratin; collagen; bovine gelatin; fish gelatin; riboflavin; μDSC; circular dichroism spectroscopy; EPR spectroscopy; IR spectroscopy

Citation: Stanca, M.; Gaidau, C.; Zaharescu, T.; Balan, G.-A.; Matei, I.; Precupas, A.; Leonties, A.R.; Ionita, G. Physico-Chemical Changes Induced by Gamma Irradiation on Some Structural Protein Extracts. *Biomolecules* **2023**, *13*, 774. https://doi.org/10.3390/biom13050774

Academic Editor: Chryssostomos Chatgilialoglu

Received: 28 March 2023
Revised: 20 April 2023
Accepted: 28 April 2023
Published: 29 April 2023

1. Introduction

Gamma irradiation at low to medium doses of 5–10 kGy is a known microbial control method and safety measure in food and cosmetic industries [1], as well as a method largely used in medical treatments [2], in particular for cancer therapy. There are studies that indicate this procedure increases the production of extracellular matrix (ECM) that contains large amounts of collagen. The presence of collagen confers resistance to radiation, which has an effect on the reduction in cell mortality in some tumors [3]. On the other hand, gamma irradiation causes the fragmentation of the α chains of collagen in tissues such as bones, determining the loss of connectivity in collagen networks and, by consequence, bone fractures [2]. Gamma irradiation also determines the radiolysis of water, which further generates free radicals. These radicals, by interaction with the components of the ECM, including proteins such as collagen, gelatin and keratin, can initiate a series of chemical processes such as deamination, decarboxylation, reduction in the number of disulfide bonds, oxidation of sulfhydryl groups, and hydrolysis of peptide bonds [4].

Nevertheless, the gamma sterilization method has some advantages that lend it to use for large-scale applications, including the low cost, the small increase in temperature of the irradiated materials, and no production of toxic residues [5]. The negative impact on a system or material can be limited by controlling the dose of gamma radiation, or by introducing other compounds that can inhibit the action of free radicals in the system.

In the case of proteins extracted from the ECM, the extraction procedure itself can modify the native structure of these proteins. In addition, the extracts can be modified by grafting other molecules, depending on their envisaged applications. Thus, gamma-irradiated collagen can find application in tissue engineering [6], or can be used as a component of interpenetrated networks along other biocompatible polymers to find application in wound dressing [7]. In fact, combining polysaccharides with proteins from the ECM represents a common procedure to obtain new medical devices [8]. Recently, the characteristics of a collagen/chitosan membrane obtained by the casting method, and the effect of lemon grass oil as an antimicrobial agent, were reported [9]. In similar studies, keratin hydrolysates were physico-chemically characterized and tested for cellular proliferation [10].

In this study, we aimed to find evidence of structural changes induced by gamma irradiation or UVA irradiation in the presence of riboflavin in different samples of known protein constituents of the ECM extracted from bovine hide, fish scales, and sheep wool. The physico-chemical methods involved were differential scanning microcalorimetry (μDSC), IR spectroscopy, circular dichroism (CD), and electron paramagnetic resonance (EPR) spectroscopy. Riboflavin was used in this investigation, considering our previous study regarding its role in generating collagen crosslinking [11]. This process has already found application in the treatment of eye diseases, such as keratoconus and keratitis of different etiologies, as well as vein or dentine diseases [12–14].

2. Materials and Methods

2.1. Materials

Spin traps 5,5-dimethyl-1-pyrroline-N-oxide (DMPO) and N-tert-butyl-α-phenylnitrone (PBN) were purchased from Aldrich (St. Louis, MO, USA). Riboflavin was purchased from Carl Roth (Karlsruhe, Germany).

2.2. Extraction Procedures

The extraction procedures utilized to obtain the proteins used in this study were as follows.

Keratin hydrolysate was obtained by alkaline–enzymatic hydrolysis in two steps. In the first step, the alkaline hydrolysate was obtained at 80 °C by a previously described method [10]. Briefly, sheep wool was washed and degreased using NH_4OH 4% *w/w*, Na_2CO_3 1% *w/w*, and Boron SE 0.6% *w/w* for 2 h at 40 °C, then washed up to neutral pH and cut into small pieces using a grinding machine (La Minerva, Minerva Omega, Bologna, Italy). Wool was mixed with NaOH 2.5% *w/w* solution in a stainless steel vessel equipped with automatic temperature control and mechanically stirred for 4 h at 80 °C. The obtained hydrolysate was used in the second step for the enzymatic hydrolysis using 1% w/w of the enzyme Valkerase (BioResource International, Durham, NC, USA). The keratin hydrolysate was centrifuged for 15 min at 6000 rpm (Eppendorf 5804, Wien, Austria) and then lyophilized by freeze drying using a DELTA 2-24 LSC freeze dryer (Osterode am Harz, Germany).

Bovine collagen was obtained from bovine delimed hide. Bovine hide was cut into small pieces and washed several times with deionized water. The collagen was extracted at neutral pH in a water bath at 90 °C for 6 h. The obtained product was filtered and dried in an oven at 60 °C.

Bovine gelatin was obtained by acid hydrolysis of bovine delimed hide mixed with a certain amount of distilled water, heated at 90 °C for 5 h. The pH was set at 5.5 using a solution of 1 M acetic acid, and was controlled and adjusted every hour. The obtained gelatin was filtered and dried in an oven at 60 °C.

Fish gelatin was obtained from fish scales following a method described by Akagunduz et al. [15]. Briefly, shredded fish scales were washed with NaCl 5% *w/v* and NaOH 4%

w/v to remove non-collagen proteins. After the alkali treatment, scales were treated with n-butanol (10 mL/100 mL) three times for 30 min to remove fat. Scales were washed with distilled water between steps. In order to partially remove the mineral content, scales were treated for 16 h with a solution containing EDTA and, thereafter, with acetic acid 99% for 3 h. The extraction was made in distilled water at 60 °C for 12 h. The obtained gelatin was filtered and dried in an oven at 60 °C.

2.3. Instrumentation and Sample Preparation

Solutions of bovine collagen, bovine gelatin, and fish gelatin were prepared by dissolving the protein to a 0.25 mg/mL concentration in 0.18 M or 15.7 M acetic acid, in the absence or presence of riboflavin 0.01% (0.1 mg/mL). Keratin was dissolved in water to obtain a 0.46 mg/mL solution.

The physico-chemical characterization through IR, circular dichroism, and EPR spectroscopies, as well as μDSC, was performed for samples prior to and after irradiation with UVA light or gamma radiation. A dose of 10 kGy over an irradiation time of 5 h, in air, was used.

2.3.1. Infrared Spectroscopy

The IR spectra of dried samples obtained by the evaporation of protein solutions prior to and after exposure to radiation were recorded on a Nicolet i-S10 FTIR spectrophotometer (Thermo Scientific, Waltham, MA, USA). For secondary structure estimation, the FTIR spectra in the amide I (1600–1700 cm^{-1}) spectral region were deconvoluted to the minimum number of Gaussian components. The negative maxima in the fourth derivative of the FTIR spectra were used as deconvolution input parameters, as indicated in the literature [16,17]. Band positions and widths were constrained within reasonable limits. Correlation coefficients and standard errors were >0.99 and <0.002, respectively. In order to assess the relative contribution of each secondary structure component, the area of the corresponding band was divided into the sum of areas of all the components considered.

2.3.2. Circular Dichroism (CD) Spectroscopy

The CD spectra of protein solutions were recorded on a J-815 spectrometer (JASCO International Co., Ltd., Tokyo, Japan) equipped with a Peltier-type temperature controller. Measurements were conducted at three temperatures in the range of 298–308 K, which is the range of the endothermic signal obtained by μDSC. The CD spectra were recorded using a 0.5 mm path length cuvette, between 190 nm and 260 nm, with standard sensitivity and at a scan rate of 50 nm/min. The other parameters were set as follows: band width 1 nm, response 4 s, and data pitch 1 nm. Each CD spectrum represents the average of three individual scans. The CD spectrum of the baseline (water or acetic acid 0.18 M) was subtracted. The effect of gamma irradiation on the secondary structure was investigated for the protein samples containing riboflavin. After gamma irradiation, no CD signal was detected for riboflavin; therefore, the CD spectrum of riboflavin was only subtracted for the samples recorded prior to irradiation.

2.3.3. Differential Scanning Microcalorimetry (μDSC)

The thermal stability of protein solutions was measured using the μDSC7 evo calorimeter (Setaram, Caluire, France). Two Hastelloy cells, the sample cell containing 100 μL of sample and the reference cell filled with the same volume of solvent, were heated from 278 K to 363 K using a scan rate of 1 K min^{-1}. Data collection and processing were performed with the Calisto v.1077 software supplied by the instrument manufacturer, using a linear baseline. The temperature of denaturation (T_{peak}) and the denaturation enthalpy change (ΔH) of the samples were determined. The influence of gamma irradiation on the samples containing riboflavin was studied.

2.3.4. Electron Paramagnetic Resonance (EPR) Spectroscopy

The EPR spectra of spin adducts were recorded on a JEOL FA 100 spectrometer (Tokyo, Japan) equipped with a cylindrical-type resonator TE011, with a frequency modulation of 100 kHz, microwave power of 0.998 mW, sweep time of 60 s, modulation amplitude of 1 G, time constant of 0.1 s, modulation width of 1 G, and a magnetic field scan range of 100 G. The hyperfine coupling constants of the spin adducts were obtained from simulation of the experimental spectra using the Winsim software [18,19].

2.3.5. UVA and Gamma Irradiation

UVA light exposure of protein samples was performed with a mercury arc lamp (500 W, LOT-Quantum Design, Darmstadt, Germany) at 370 nm for 30 min. The gamma irradiation process was performed with an irradiator M-38 GAMMATOR (Isomedix, Parsippany, NJ, USA) using a ^{137}Cs source, in air, at room temperature, at a dose rate of 2 kGy/h.

3. Results and Discussions

3.1. *IR Spectroscopy*

The investigation of the protein samples by IR spectroscopy was performed in order to identify the characteristic signals of the main functional groups by referring to the literature data [20–23] (see Figure 1, Table 1 and Tables S1–S3 in the Supplementary Materials file), and to estimate the secondary structural content of the protein samples, prior to and after irradiation. We aimed to assess the extent of conformational changes that are induced in the protein structure by gamma irradiation with a dose of 10 kGy. This information can be retrieved by correlating the positions and intensities of characteristic amide vibrational modes to specific secondary structure elements of the polypeptide chain (α-helix, β-sheet, β-turn, random coil).

First, we discuss the results regarding keratin. The amide A band of the three keratin samples under investigation (keratin, keratin/riboflavin, and keratin/riboflavin/gamma-irradiated) has a maximum at 3275 cm^{-1}. It arises from N-H and O-H stretching vibrations of the polypeptide chain, and its broad feature was previously considered an indication for the presence of twisted β-sheet structures formed via hydrogen bonding [24].

The amide I spectral region (1600–1700 cm^{-1}) comprises C=O stretching (~80%), N-H bending, and C-N stretching (20%) vibrations of the polypeptide chain that are correlated to the backbone conformation [21]. This signal is intense and can be used to obtain an estimate of the secondary structural content of the protein. The main amide I peak is observed at 1639 cm^{-1} and 1642 cm^{-1} for the keratin and keratin/riboflavin samples, and at 1644 cm^{-1} for the keratin/riboflavin/gamma-irradiated sample.

The amide II spectral region (1500–1600 cm^{-1}) contains contributions from N-H in-plane bending and C-N stretching vibrations (1540–1560 cm^{-1}), as well as from C-N and C-C stretching (1520–1540 cm^{-1}). It is less sensitive to conformational changes as compared to the amide I vibrational mode, and is mostly indicative of the protonation state of the peptide unit [25].

The amide III band (1220–1330 cm^{-1}) is associated with C-N stretching, N-H bending (~30% each), C-C stretching (~20%), and C-H bending (~10%) vibrations [21]. Despite having low intensity, the amide III mode is sensitive to the nature of the side chains and to hydrogen bonding [25,26], and it can sometimes be used, complementary to the amide I mode, for conformational analysis.

As stated above, the amide I band (1639 cm^{-1}) of keratin slightly shifts to larger wavenumbers in the presence of riboflavin (1642 cm^{-1}), and more so after exposure to gamma radiation (1644 cm^{-1}). When the IR spectra were normalized (using the 1399 cm^{-1} C-H bending band as a reference) in order to better observe any differences in band intensity, position, or shape, we noted a decrease in the intensity of the amide A band after irradiation. This may indicate some degree of degradation of the amide group (primary structure alteration) [9,27]. The most poignant spectral change caused by irradiation is the change in the intensity ratio of the amide I and amide II bands from 0.82 in the keratin/riboflavin sample to 1 in the keratin/riboflavin/irradiated sample. This change evidences the disruption of the intrinsic hydrogen bonding structure of the protein due to conformational changes (secondary structure alteration). An increase in the amide I/amide II intensity ratio was shown to correlate to an increase in the contribution of ordered structures [9]. In our case, the opposite trend is observed; therefore, we expect irradiation to cause an increase in the random coil contribution at the expense of β and α ordered structures. This can be confirmed by a quantitative assessment of the secondary structure.

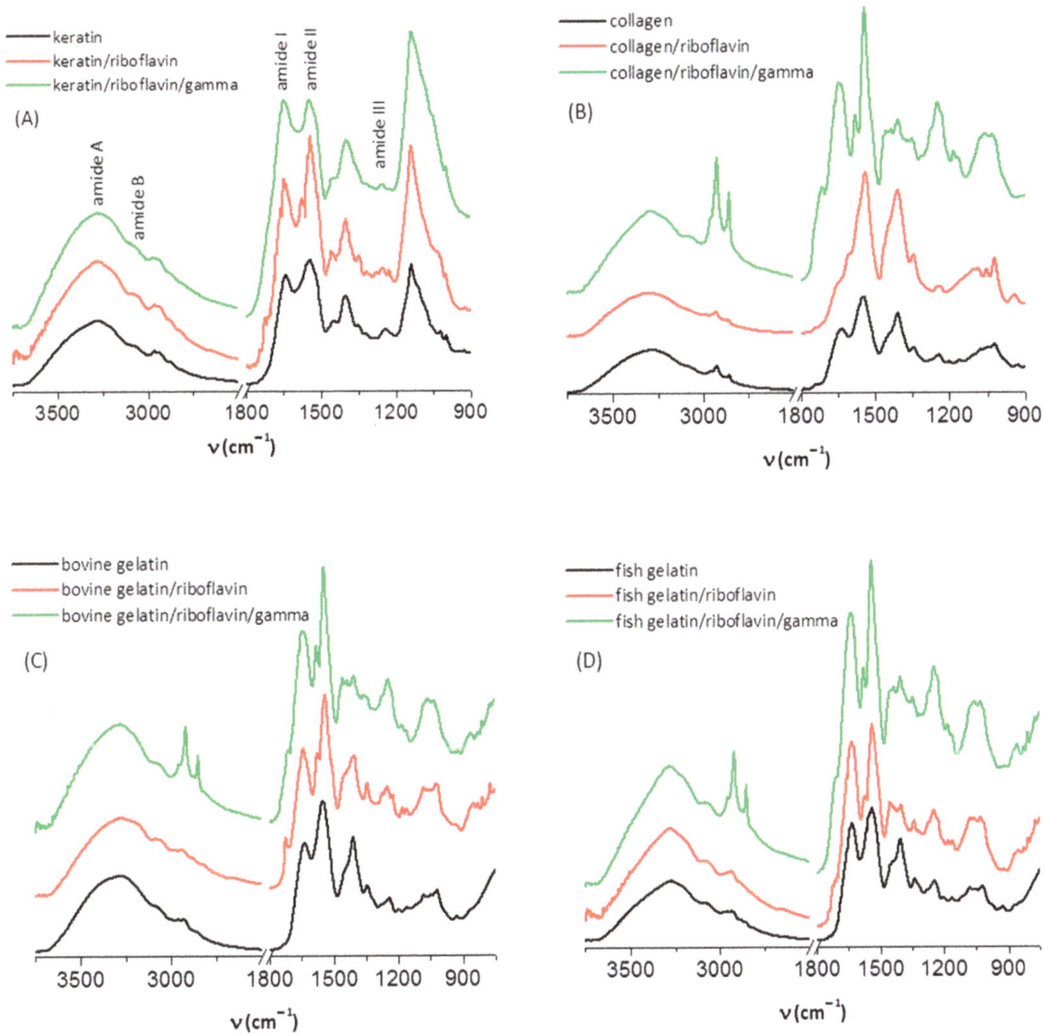

Figure 1. The IR spectra of (**A**) dried keratin, (**B**) collagen, (**C**) bovine gelatin and (**D**) fish gelatin samples.

Table 1. Assignment of the main IR bands of keratin.

Sample	Wavenumber (cm^{-1})	Assignment
keratin	3275, 3060	O-H, N-H stretching (amide A and B)
	2963, 2935	C-H stretching (asymmetric, symmetric)
	1639 (s)	C=O stretching/N-H bending (amide I)
	1544 (s)	C-N stretching/N-H bending (amide II)
	1448 (w), 1399	C-H bending
	1350 (w), 1242 (w)	C-N stretching/N-H bending (amide III)
	1136 (s)	S-O stretching, symmetric (-SO_2-S-, cystine dioxide)
keratin/riboflavin	3275, 3072	N-H stretching (amide A and B)
	2963, 2935	C-H stretching (asymmetric, symmetric)
	1642 (s)	C=O stretching/N-H bending (amide I)
	1540 (s)	C-N stretching/N-H bending (amide II)
	1455 (w), 1398	C-H bending
	1347 (w), 1250	C-N stretching/N-H bending (amide III)
	1135 (s)	S-O stretching, symmetric (-SO_2-S-, cystine dioxide)
keratin/riboflavin/gamma	3275, 3068	N-H stretching (amide A and B)
	2971, 2934	C-H stretching (asymmetric, symmetric)
	1644 (s)	C=O stretching/N-H bending (amide I)
	1542 (s)	C-N stretching/N-H bending (amide II)
	1394	C-H bending
	1251	C-N stretching/N-H bending (amide III)
	1133 (s)	S-O stretching, symmetric (-SO_2-S-, cystine dioxide)

Abbreviations: s—strong, w—weak.

For secondary structure estimation, we focused on the amide I region. In this spectral domain, the secondary structure components of a protein absorb at different positions, as follows: β-sheet (1610–1635 cm^{-1}), random coil (1635–1645 cm^{-1}), α-helix (1645–1665 cm^{-1}), β-turn (1662–1682 cm^{-1}), and β-sheet antiparallel (1682–1689 cm^{-1}) [24,28,29]. It is known that some spectral features appearing as faint in the IR spectrum can become more prominent in the second [30] or fourth [16,17] derivative of the spectrum. For this reason, in order to have a more precise initial estimate of the component bands (in terms of number and position), the negative maxima in the fourth derivative of the IR spectrum of keratin were used. As such, the amide I band was fit with four Gaussian components (initially set at 1607, 1635, 1653, and 1685 cm^{-1}). The bands resolved after deconvolution are presented in Figure 2 and their relative contributions are listed in Table 2, together with their assignment. We note that the presence of riboflavin and the exposure to gamma radiation do not determine significant shifts in the position of these bands, but change their relative contributions substantially. The effect of irradiation is the increase in unordered random coil structures at the expense of β-sheet ordered structures.

In addition to the vibrational modes observed for keratin, the collagen (Figure 1B), bovine gelatin (Figure 1C), and fish gelatin (Figure 1D) samples exhibited bands in the spectral region of 1000–1100 cm^{-1} that have been interpreted considering the glycation of collagen, and thus ascribed to C-O stretching vibrations of carbohydrates [31]. The main spectral changes induced by gamma irradiation of these samples occur in the spectral regions attributed to C-H stretching and C-H bending vibrations. Thus, irradiation produces a strong increase in the intensity of the C-H asymmetric and symmetric stretching bands, accompanied by a strong decrease in the C-H bending mode at ~1405 cm^{-1}. For collagen and bovine gelatins, a shift of the amide II band from ~1547 cm^{-1} in the native samples to ~1540 cm^{-1} in the samples containing riboflavin and in the irradiated samples is also observed. Some authors explain such a shift to lower wavenumbers by the fibrillogenesis of collagen [31].

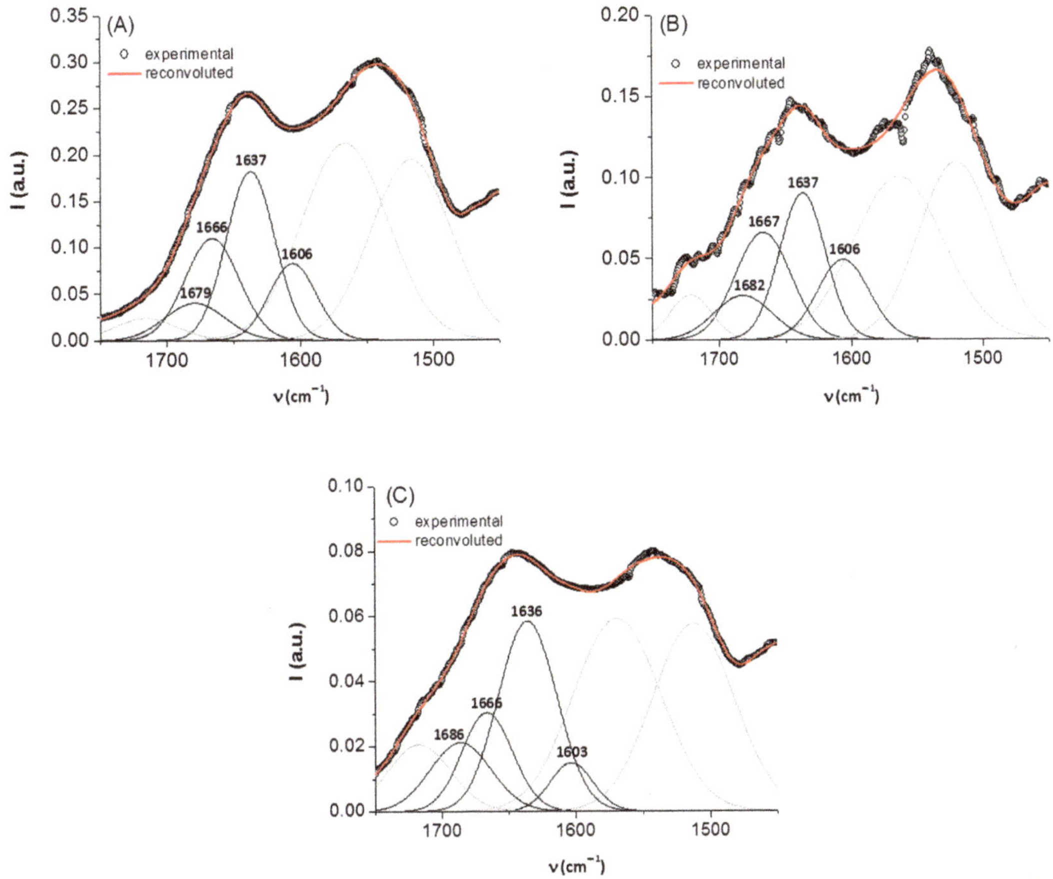

Figure 2. Deconvolution of the amide I region of the IR spectra of keratin samples: (**A**) keratin, (**B**) keratin/riboflavin, (**C**) keratin/riboflavin/gamma-irradiated; for clarity, band components not belonging to the region of interest are depicted in grey.

Table 2. Secondary structural content of keratin samples.

Sample	Band Assignment	ν (cm^{-1})	% Area
keratin	β-sheet	1606	19
	random coil	1637	42
	α-helix	1666	28
	β-turn	1679	11
keratin/riboflavin	β-sheet	1606	22
	random coil	1637	35
	α-helix	1667	30
	β-turn	1682	13
keratin/riboflavin/gamma	β-sheet	1603	9
	random coil	1636	49
	α-helix	1666	23
	β-turn	1686	19

For deconvoluting the spectra of collagen and gelatins, four Gaussian components were initially set at 1615, 1638, 1651, and 1685 cm^{-1}, as predicted by the fourth derivative of the IR spectra. The bands resolved after deconvolution are presented in Figures S1–S3, and their relative contributions are listed in Tables S4–S6. The secondary structural content of bovine gelatin was estimated at 13% β-sheet, 48% random coil, 26% α-helix, and 13% β-turn. No significant changes in these values were observed upon interaction with riboflavin or after gamma irradiation (Table S5). For fish gelatin, the predicted secondary structural content of 16% β-sheet, 47% random coil, 25% α-helix, and 12% β-turn was not altered significantly upon interaction with riboflavin or by gamma irradiation (Table S6). The secondary structural contents of both bovine and fish gelatins evidence a higher degree of denaturation as compared to bovine collagen.

3.2. Circular Dichroism (CD) Spectroscopy

The secondary structures of the protein samples were further analyzed using CD spectroscopy. Below, the changes induced by temperature, the presence of riboflavin, and irradiation with UV or gamma radiation are described.

The CD spectra of aqueous keratin solutions in the absence and presence of riboflavin, prior to and after exposure to UV radiation, at three temperatures, are presented in Figure 3. It is known from the literature [32] that the α-helical CD spectrum of α-keratin shows two negative minima at 222 nm and 208 nm, and a positive maximum at 190 nm. The CD spectra presented in Figure 3 display a single broad negative peak at ~200 nm, indicating a low content of α-helix conformation in the case of our keratin sample. This observation is in agreement with the helical content of only ~28% estimated from IR data.

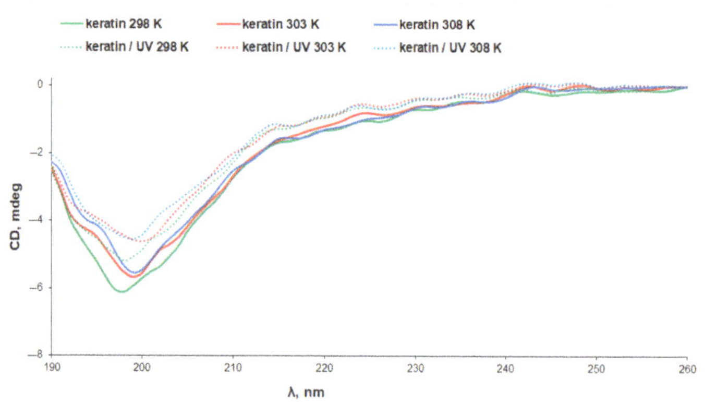

Figure 3. The CD spectra of keratin (line) and UV-irradiated keratin (dotted line) aqueous solutions measured at different temperatures.

Both the increase in temperature and the exposure to UV radiation determine a decrease in the CD signal of keratin, pointing to changes in the protein secondary structure associated with protein denaturation. More intense CD signals were obtained for keratin samples in the presence of riboflavin, indicating a stabilizing effect of riboflavin on the protein structure as a result of interaction (Figure 4A). This observation can be correlated to the slight increase in the content of ordered structures at the expense of the random coils, which decrease from 42% in keratin to 35% in keratin/riboflavin (Table 1).

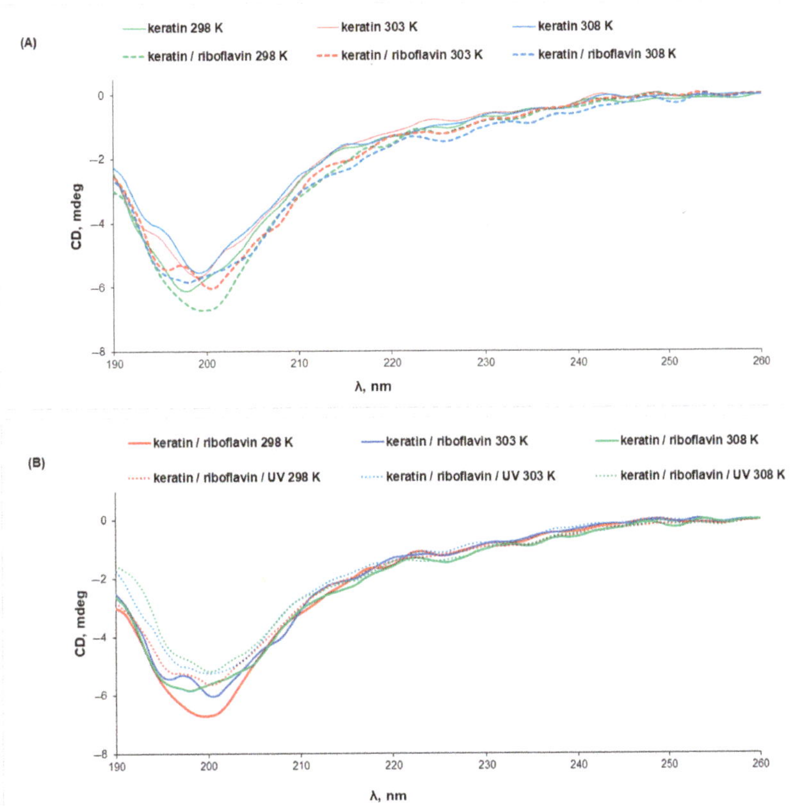

Figure 4. The CD spectra of (**A**) keratin (line) and keratin/riboflavin (dotted line), (**B**) keratin/riboflavin (line) and keratin/riboflavin/UV-irradiated (dotted line) aqueous solutions measured at different temperatures.

UV irradiation determines a 13% decrease in the intensity of the CD signal of the keratin/riboflavin sample, pointing to protein denaturation (Figure 4B). Gamma irradiation of the keratin/riboflavin aqueous solution has an even larger denaturation effect, manifesting in an approximately 50% decrease in the ellipticity value at all temperatures (Figure 5). This correlates to the increase in random coil content at the expense of the β-sheet and α-helix conformations that was observed by IR spectroscopy.

The CD spectrum of native collagen presents a positive maximum peak at 220 nm and a negative minimum peak at around 198 nm, characteristic for the triple-helix conformation [33]. Gelatin is a protein obtained by breaking the triple-helix structure of collagen [34]. It is known that the positive peak at 220–230 nm characteristic to the triple-helix conformation disappears completely as a result of collagen denaturation [35,36]. For bovine and fish gelatin solutions, the CD spectra (Figure 6) present only a minimum peak at ~199 nm, which can be assigned to the unordered structure of the protein, while the maximum at 222 nm is not observed, which points to denatured collagen structures [37], in accordance with the IR data. The temperature increase determines conformational changes associated with the denaturation of both gelatins.

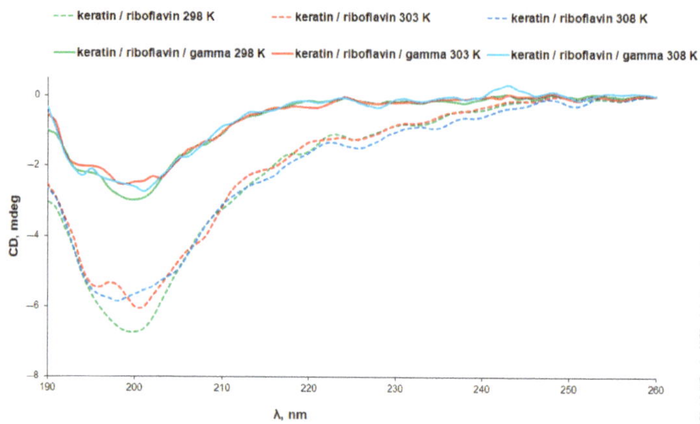

Figure 5. The CD spectra of keratin/riboflavin (line) and keratin/riboflavin/gamma-irradiated (dotted line) aqueous solutions measured at different temperatures.

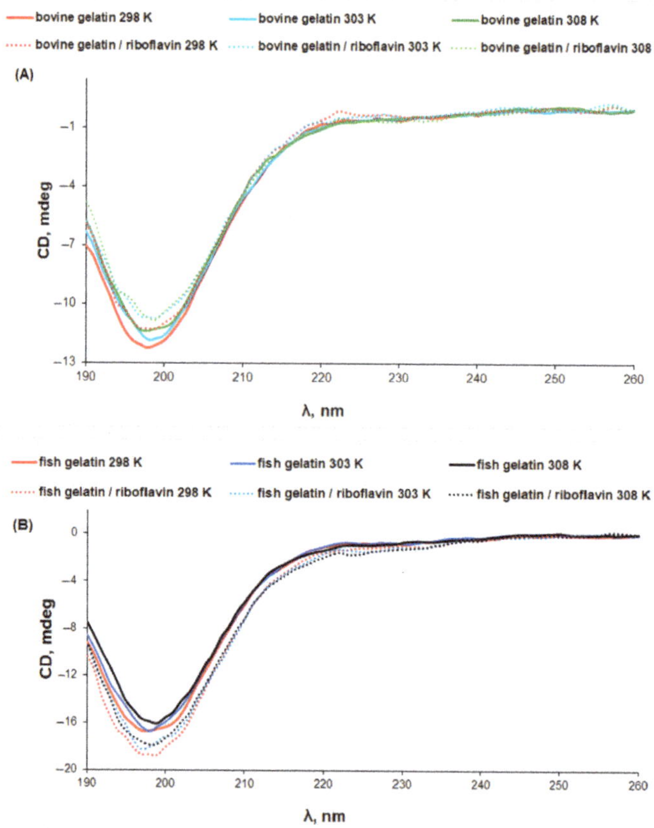

Figure 6. The CD spectra of gelatin (line) and gelatin/riboflavin (dotted line) solutions in acetic acid 0.18 M at different temperatures: (**A**) bovine gelatin and (**B**) fish gelatin.

In the presence of riboflavin, the CD signal presents different trends: the negative ellipticity decreases for bovine gelatin, suggesting denaturation, while the CD signal increases for fish gelatin. Thus, riboflavin presents a two-fold action: (i) it induces a stabilizing effect on the fish gelatin structure and (ii) increases the denaturation of bovine gelatin.

Exposure to gamma radiation generates a further denaturation of gelatin/riboflavin samples, without changing the shape of the CD spectra (Figure 7). The same effect is observed for the collagen/riboflavin solution (Figure S4).

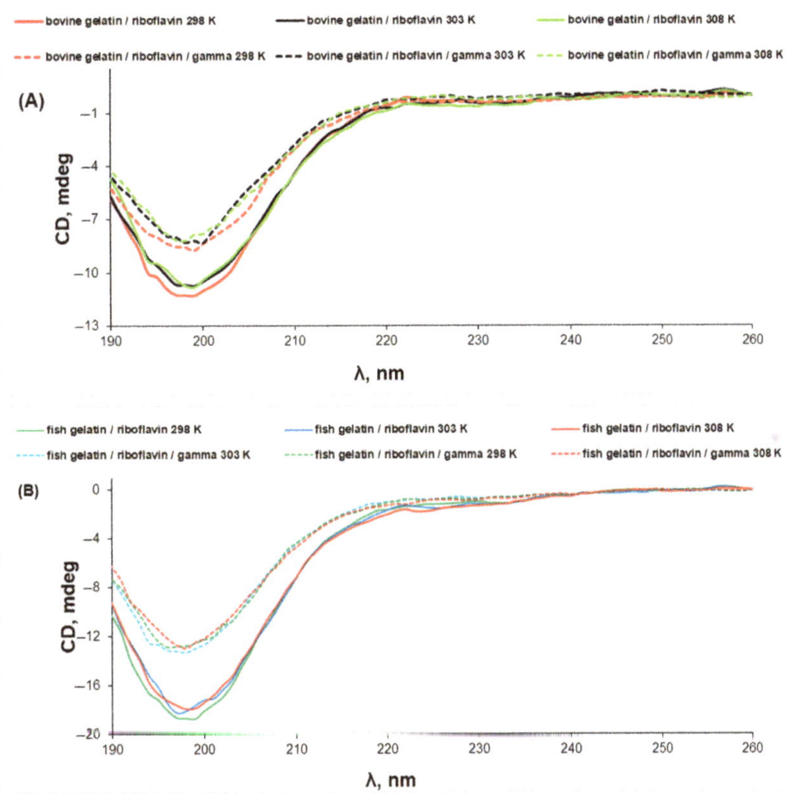

Figure 7. The CD spectra of gelatin/riboflavin (line) and gelatin/riboflavin/gamma-irradiated (dotted line) solutions in acetic acid 0.18 M at different temperatures: (**A**) bovine gelatin and (**B**) fish gelatin.

3.3. *Differential Scanning Microcalorimetry (μDSC)*

The thermal stability of keratin, gelatin, and collagen solutions was investigated using μDSC, and the parameters of denaturation (peak temperature and denaturation enthalpy change) are presented in Table 3. The μDSC signal of all samples presented a broad endothermic peak attributed to the helix to random coil state transition [36,38,39].

Table 3. Peak temperature (T_{peak}) and denaturation enthalpy change (ΔH) obtained from µDSC measurements.

Sample	T_{peak} (K)	ΔH (J/g)
keratin	303.87	2.69
keratin/UV-irradiated	305.50	2.15
keratin/riboflavin	301.68	0.76
keratin/riboflavin/gamma-irradiated	303.44	0.26
collagen/riboflavin	299.09	0.55
collagen/riboflavin/gamma-irradiated	299.59	0.79
bovine gelatin/riboflavin	298.96	0.32
bovine gelatin/riboflavin/gamma-irradiated	301.32	0.66
fish gelatin/riboflavin	299.56	0.35
fish gelatin/riboflavin/gamma-irradiated	299.49	0.43

Note: keratin samples were prepared in water; collagen and gelatin samples were prepared in acetic acid 0.18 M.

The µDSC measurements for the aqueous solution of keratin (Figure 8) indicate a higher thermal stability of keratin after UV irradiation, while the denaturation enthalpy change slightly decreases as a result of the conformational changes observed in the CD spectra. The presence of riboflavin decreases the thermal stability of keratin in water, while gamma irradiation increases the thermal stability of the keratin/riboflavin sample. Although the T_{peak} value increases for the keratin/riboflavin gamma-irradiated sample, the denaturation enthalpy change is significantly reduced. This is a result of the keratin unfolding observed in the CD spectra, which causes the exposure of the hydrophobic domains to the aqueous solution.

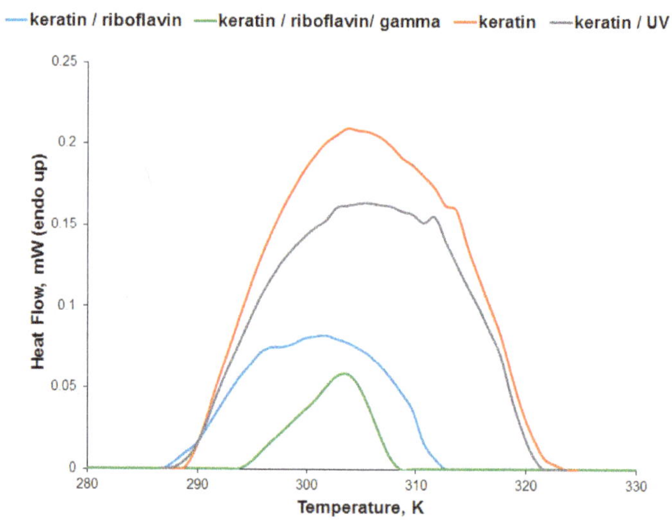

Figure 8. Effect of gamma and UV irradiation on keratin thermal denaturation in water.

A conclusion on the influence of the presence and concentration of acetic acid on the thermal denaturation of keratin can be drawn by comparing the data in Tables 3 and S7. When keratin was dissolved in acetic acid 15.7 M, significant decreases in the T_{peak} and denaturation enthalpy change values were observed, indicating a destabilization effect, induced by the high concentration of co-solvent. Gamma irradiation of the keratin/riboflavin sample prepared in concentrated acetic acid solution did not change the value of T_{peak}, but the denaturation enthalpy change slightly decreased, revealing a somewhat higher exposure of hydrophobic regions of protein to the solvent.

The influence of gamma irradiation on the thermal stability of gelatin/riboflavin samples in acetic acid 0.18 M is presented in Figure 9. The endothermic peak observed for gelatin corresponds to the helix-to-coil transition, as a result of the dissociation of the triple helices upon heating [40]. An increase in the thermal stability of the bovine gelatin/riboflavin sample was observed after irradiation, while no significant change was recorded in the μDSC profile of the fish gelatin/riboflavin sample. These results are in agreement with the IR data that showed no spectral changes for the fish gelatin/riboflavin sample as a result of irradiation.

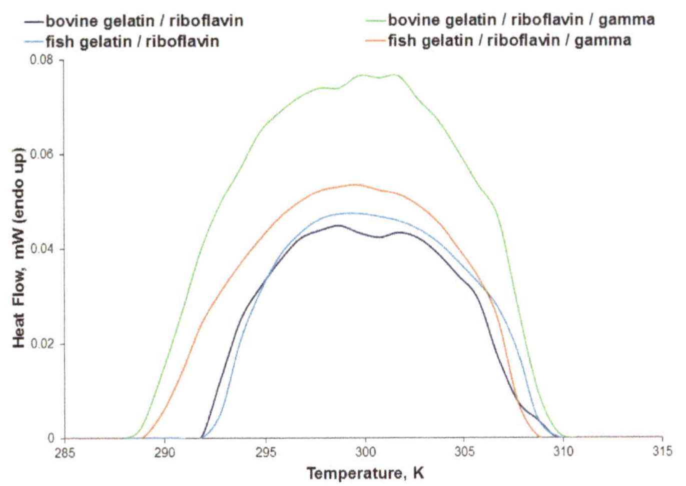

Figure 9. The μDSC signal for the thermal denaturation of bovine and fish gelatin solutions in acetic acid 0.18 M.

Gamma irradiation of bovine gelatin in acetic acid 15.7 M in the presence of riboflavin decreases the thermal stability of the protein (T_{peak} decreases from 302.64 K to 301.45 K, see Table S7), while the same system in acetic acid 0.18 M is more stable to heat after gamma irradiation (T_{peak} increases from 298.96 K to 301.32 K, Table 3). An important change in thermal stability was observed mainly for fish gelatin: when using a higher concentration of acetic acid, the value of T_{peak} increases from 299.49 K to 303.14 K, pointing to a stabilizing effect. An important decrease in the denaturation enthalpy change value was obtained for the fish gelatin/riboflavin sample in acetic acid 15.7 M after gamma irradiation. Gamma irradiation induces a slight increase in the thermal stability of the collagen/riboflavin sample (Figure S5).

3.4. EPR Spectroscopy

Two spin trapping agents, DMPO and PBN, have been used in this study to evidence free radical species that can form in protein/riboflavin systems after gamma irradiation. No EPR signal was recorded for the irradiated solutions in the absence of spin trapping agents. The spin trap was added to the protein solution containing riboflavin immediately after irradiation. The samples were then immersed in liquid nitrogen and the EPR spectra were recorded after 1 h, and again after 2 h. It is worthy of note that the samples were kept on the lab bench between the two measurements, and thus exposed to daylight.

The EPR spectra of the spin adducts of DMPO and PBN are presented in Figures 10 and 11, respectively, and their parameters are provided in Table S8. For all protein samples, it can be observed that the intensity of the EPR signal increases over time, and this is due to the presence of riboflavin that generates $HO^{\bullet}$ radicals. The parameters of the $^{\bullet}$DMPO-OH adducts are similar, with hyperfine coupling constants a_N~14.8 G and a_H~13.9 G.

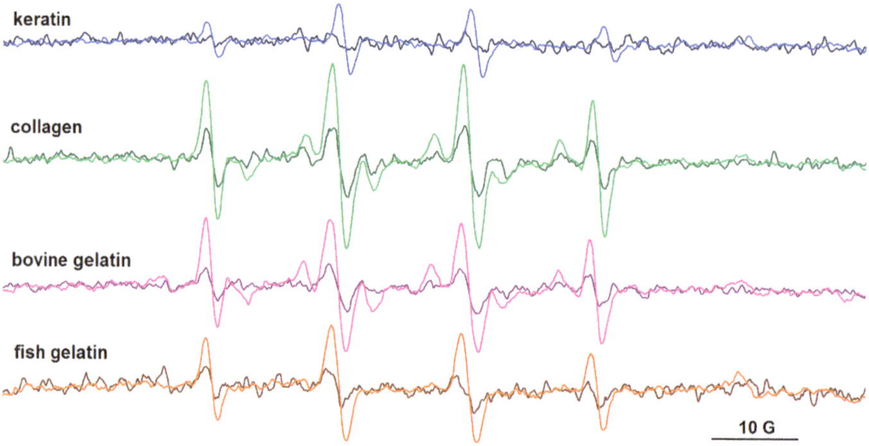

Figure 10. The EPR spectra of DMPO adducts formed after gamma irradiation (10 kGy) in solutions of keratin, collagen, bovine gelatin, and fish gelatin, in the presence of riboflavin. Spectra recorded after 1 h (dark colors) and after 2 h (light colors) from irradiation.

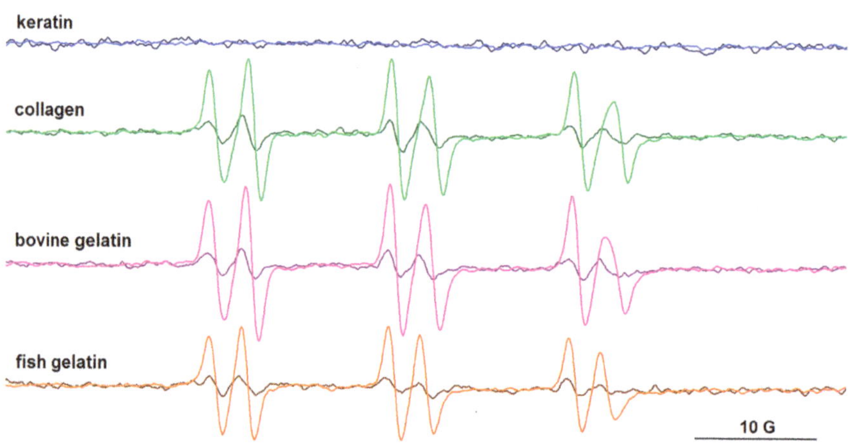

Figure 11. The EPR spectra of PBN adducts formed after gamma irradiation (10 kGy) in solutions of keratin, collagen, bovine gelatin, and fish gelatin, in the presence of riboflavin. Spectra recorded after 1 h (dark colors) and after 2 h (light colors) from irradiation.

The PBN adducts evidence six-line signals, their parameters after one hour from irradiation being a_N ~14.6 G and a_H ~2.6 G. The literature data on carbon-centered radicals formed in aqueous media [41] and in protein samples after gamma irradiation [42] report a_N values > 15.5–16 G and a_H values > 3 aN for PBN adducts. For PBN adducts formed with oxygen-centered radicals, the literature data indicate lower a_N (<15 G) and a_H (<3 G) values [41]. These latter values are closer to those obtained for our systems, which indicates, rather than the presence of carbon-centered radicals, the presence of free radicals centered on oxygen. Nevertheless, the slight increase in time of the a_N and a_H values of the PBN adducts (a_N~14.8 G, a_H~2.9 G after two hours from irradiation) may be an indication of the formation of carbon-centered radicals. They could arise from successive reactions of the HO• radicals with the protein chains or with the acetic acid present in solution. As in the case of the DMPO adducts, the intensity of the signal of PBN adducts increases over time. In the case of the keratin sample, radical formation was not observed using PBN as the spin trap. This can also be correlated with the lower intensity signal of the •DMPO-OH adduct evidenced for the keratin sample.

Exposure of the protein/riboflavin samples to a higher dose of gamma radiation (20 kGy) determined the decomposition of riboflavin in the samples containing collagen or gelatin. The presence of •DMPO-OH spin adducts was only observed for the keratin sample.

4. Conclusions

To summarize, the analysis of changes in protein secondary structures induced by gamma irradiation and the presence of riboflavin revealed the influence of the protein source. The μDSC, IR spectroscopy, and CD spectroscopy evidenced that gamma irradiation and riboflavin influence the secondary structure of the ECM-type protein extracts. Riboflavin facilitates the release of free radicals that further interact with the protein chains. However, the exposure to gamma radiation influences the stability and integrity of the molecules. The proteins with higher degrees of ordered structures are more stable against gamma radiation and temperature. Due to their hydrophilic character, collagen and gelatin, when combined with riboflavin, present more stable secondary structures when exposed to gamma radiation as compared to keratin, which has a hydrophobic character. Further studies may involve similar compounds with riboflavin or investigating the effects of antioxidant compounds on the protein structures exposed to gamma irradiation.

Supplementary Materials: The following supporting information can be downloaded at: https://www.mdpi.com/article/10.3390/biom13050774/s1: Figure S1: Deconvolution of the amide I region of the IR spectra of (A) collagen and (B) collagen/riboflavin/irradiated; for clarity, band components not belonging to the region of interest are depicted in grey; Figure S2: Deconvolution of the amide I region of the IR spectra of (A) bovine gelatin, (B) bovine gelatin/riboflavin and (C) bovine gelatin/riboflavin/gamma-irradiated; for clarity, band components not belonging to the region of interest are depicted in grey; Figure S3: Deconvolution of the amide I region of the IR spectra of (A) fish gelatin, (B) fish gelatin/riboflavin, (C) fish gelatin/riboflavin/gamma-irradiated; for clarity, band components not belonging to the region of interest are depicted in grey; Figure S4: The CD spectra of collagen/riboflavin (line) and collagen/riboflavin/gamma-irradiated (dotted line) solutions in acetic acid 0.18 M at different temperatures; Figure S5: The μDSC signal for the thermal denaturation of collagen/riboflavin solution in acetic acid 0.18 M, prior to and after gamma irradiation; Table S1: Assignment of the main IR bands of collagen; Table S2: Assignment of the main IR bands of bovine gelatin; Table S3: Assignment of the main IR bands of fish gelatin; Table S4: Secondary structural content of collagen samples; Table S5: Secondary structural content of bovine gelatin samples; Table S6: Secondary structural content of fish gelatin samples; Table S7: Peak temperature and denaturation enthalpy change obtained from μDSC measurements for solutions prepared in acetic acid 15.7 M; Table S8: Parameters of the DMPO and PBN spin adducts formed in protein/riboflavin solutions after exposure to gamma radiation.

Author Contributions: Conceptualization, M.S., G.I. and C.G.; Methodology, M.S., G.I. and I.M.; Software, G.-A.B.; Validation, C.G., G.I. and I.M.; Formal analysis, G.-A.B. and A.R.L.; Investigation, T.Z., I.M., A.P. and A.R.L.; Resources, M.S.; Data curation, A.P. and I.M.; Writing—original draft preparation, G.I., I.M. and A.P.; Writing—review and editing, M.S., C.G., G.I. and I.M.; Visualization, C.G.; Supervision, G.I.; Project administration, M.S.; Funding acquisition, M.S., C.G. and G.I. All authors have read and agreed to the published version of the manuscript.

Funding: The present work was supported by the Romanian Ministry of Research, Innovation and Digitalization, CNDI-UEFISCDI, project number PN-III-P1-1.1-PD-2021-0189, GAMMA-COLL, under contract No. 71/2022, and carried out within the research direction "EPR and fluorescence studies on supramolecular interactions in inhomogeneous systems" of the "Ilie Murgulescu" Institute of Physical Chemistry of the Romanian Academy.

Institutional Review Board Statement: Not applicable.

Informed Consent Statement: Not applicable.

Data Availability Statement: Data are contained within the article and Supplementary Materials.

Conflicts of Interest: The authors declare no conflict of interest. The funders had no role in the design of the study; in the collection, analyses, or interpretation of data; in the writing of the manuscript; or in the decision to publish the results.

References

1. Rodrigues, L.M.; Sales, L.A.; Fontes, P.R.; de Almeida Torres Filho, R.; Pereira Dutra Andrade, M.; de Lemos Souza Ramos, A.; Mendes Ramos, E. Combined effects of gamma irradiation and aging on tenderness and quality of beef from Nellore cattle. *Food Chem.* **2020**, *313*, 126137. [CrossRef] [PubMed]
2. Burton, B.; Gaspar, A.; Josey, D.; Tupy, J.; Grynpas, M.D.; Willett, T.L. Bone embrittlement and collagen modifications due to high-dose gamma-irradiation sterilization. *Bone* **2014**, *61*, 71–81. [CrossRef]
3. Azorin, E.; González-Martínez, P.R.; Azorin, J. Collagen I confers gamma radiation resistance. *Appl. Radiat. Isot.* **2012**, *71*, 71–74. [CrossRef] [PubMed]
4. Nguyen, H.; Morgan, D.A.; Forwood, M.R. Sterilization of allograft bone: Effects of gamma irradiation on allograft biology and biomechanics. *Cell Tissue Bank.* **2007**, *8*, 93–105. [CrossRef]
5. Gomes, A.D.; de Oliveira, A.A.R.; Houmard, M.; Nunes, E.H.M. Gamma sterilization of collagen/hydroxyapatite composites: Validation and radiation effects. *Appl. Radiat. Isot.* **2021**, *174*, 109758. [CrossRef] [PubMed]
6. Takitoh, T.; Bessho, M.; Hirose, M.; Ohgushi, H.; Mori, H.; Hara, M. Gamma-cross-linked nonfibrillar collagen gel as a scaffold for osteogenic differentiation of mesenchymal stem cells. *J. Biosci. Bioeng.* **2015**, *119*, 217–225. [CrossRef]
7. Liu, B.; Huang, W.; Yang, G.; An, Y.; Yin, Y.; Wang, N.; Jiang, B. Preparation of gelatin/poly (γ-glutamic acid) hydrogels with stimulated response by hot-pressing preassembly and radiation crosslinking. *Mater. Sci. Eng. C Mater. Biol. Appl.* **2020**, *116*, 111259. [CrossRef]
8. Sun, L.L.; Li, L.H.; Wang, Y.Q.; Li, M.B.; Xu, S.M.; Zhang, C.P. A collagen-based bi-layered composite dressing for accelerated wound healing. *J. Tissue Viability* **2022**, *31*, 180–189. [CrossRef]
9. Rapa, M.; Zaharescu, T.; Stefan, L.M.; Gaidau, C.; Stanculescu, I.; Constantinescu, R.R.; Stanca, M. Bioactivity and thermal stability of collagen–chitosan containing lemongrass essential oil for potential medical applications. *Polymers* **2022**, *14*, 3884. [CrossRef]
10. Olariu, L.; Brindusa, G.D.; Gaidau, C.; Stanca, M.; Tanase, L.M.; Ene, M.D.; Stanculescu, I.R.; Tablet, C. Bioactive low molecular weight keratin hydrolysates for improving skin wound healing. *Polymers* **2022**, *14*, 1125. [CrossRef]
11. Constantin, M.M.; Corbu, C.G.; Mocanu, S.; Popescu, E.I.; Micutz, M.; Staicu, T.; Şomoghi, R.; Trică, B.; Popa, V.T.; Precupas, A.; et al. Model systems for evidencing the mediator role of riboflavin in the UVA cross-linking treatment of keratoconus. *Molecules* **2022**, *27*, 190. [CrossRef] [PubMed]
12. Wollensak, G.; Spoerl, E.; Seiler, T. Riboflavin/ultraviolet-a-induced collagen crosslinking for the treatment of keratoconus. *Am. J. Ophthalmol.* **2003**, *135*, 620–627. [CrossRef] [PubMed]
13. Frullini, A.; Manetti, L.; Di Cicco, E.; Fortuna, D. Photoinduced collagen cross-linking: A new approach to venous insufficiency. *Dermatol. Surg.* **2011**, *37*, 1113–1118. [CrossRef]
14. Hardan, L.; Daood, U.; Bourgi, R.; Cuevas-Suárez, C.E.; Devoto, W.; Zarow, M.; Jakubowicz, N.; Zamarripa-Calderón, J.E.; Radwanski, M.; Orsini, G.; et al. Effect of collagen crosslinkers on dentin bond strength of adhesive systems: A systematic review and meta-analysis. *Cells* **2022**, *11*, 2417. [CrossRef] [PubMed]
15. Akagündüz, Y.; Mosquera, M.; Giménez, B.; Alemán, A.; Montero, P.; Gómez-Guillén, M.C. Sea bream bones and scales as a source of gelatin and ACE inhibitory peptides. *LWT-Food Sci. Technol.* **2014**, *55*, 579–585. [CrossRef]
16. Voicescu, M.; Ionescu, S.; Nistor, C.L. Spectroscopic study of 3-Hydroxyflavone–protein interaction in lipidic bi-layers immobilized on silver nanoparticles. *Spectrochim. Acta A Mol. Biomol. Spectrosc.* **2017**, *170*, 1–8. [CrossRef]
17. Stoian, A.-M.; Matei, I.; Ionescu, S. Global and local conformational changes in albumin–ligand interactions through spectroscopic methods. *Rev. Roum. Chim.* **2017**, *62*, 623–628. Available online: https://revroum.lew.ro/wp-content/uploads/2017/8/Art%2006.pdf (accessed on 27 April 2023).
18. Duling, D.R. *PEST Winsim*; version 0.96; National Institute of Environmental Health Sciences: Triangle Park, NC, USA, 1996.
19. Duling, D.R. Simulation of multiple isotropic spin-trap EPR spectra. *J. Magn. Reson. B* **1994**, *104*, 105–110. [CrossRef]
20. Bart, A. Infrared spectroscopy of proteins. *Biochim. Biophys. Acta* **2007**, *1767*, 1073–1101. [CrossRef]
21. Stani, C.; Vaccari, L.; Mitri, E.; Birarda, G. FTIR investigation of the secondary structure of type I collagen: New insight into the amide III band. *Spectrochim. Acta A Mol. Biomol. Spectrosc.* **2020**, *229*, 118006. [CrossRef]
22. de Campos Vidal, B.; Mello, M.L.S. Collagen type I amide I band infrared spectroscopy. *Micron* **2011**, *42*, 283–289. [CrossRef]
23. Wojciechowska, E.; Włochowicz, A.; Wesełucha-Birczyńska, A. Application of Fourier-transform infrared and Raman spectroscopy to study degradation of the wool fiber keratin. *J. Mol. Struct.* **1999**, *511–512*, 307–318. [CrossRef]
24. Gaidau, C.; Stanca, M.; Niculescu, M.-D.; Alexe, C.-A.; Becheritu, M.; Horoias, R.; Cioineag, C.; Rapa, M.; Stanculescu, I.R. Wool keratin hydrolysates for bioactive additives preparation. *Materials* **2021**, *14*, 4696. [CrossRef] [PubMed]
25. Gaidau, C.; Epure, D.-G.; Enascuta, C.E.; Carsote, C.; Sendrea, C.; Proietti, N.; Chen, W.; Gu, H. Wool keratin total solubilisation for recovery and reintegration—An ecological approach. *J. Clean. Prod.* **2019**, *236*, 117586. [CrossRef]

26. Odlyha, M.; Theodorakopoulos, C.; Campana, R. Studies on woollen threads from historical tapestries. *AUTEX Res. J.* **2007**, *7*, 1.
27. Sionkowska, A.; Skopinska-Wisniewska, J.; Kozlowska, J.; Planecka, A.; Kurzawa, M. Photochemical behaviour of hydrolysed keratin. *Int. J. Cosmet. Sci.* **2011**, *33*, 503–508. [CrossRef]
28. Pelton, J.T.; McLean, L.R. Spectroscopic methods for analysis of protein secondary structure. *Anal. Biochem.* **2000**, *277*, 167–176. [CrossRef]
29. Drobota, M.; Gradinaru, L.M.; Vlad, S.; Bargan, A.; Butnaru, M.; Angheloiu, M.; Aflori, M. Preparation and characterization of electrospun collagen based composites for biomedical applications. *Materials* **2020**, *13*, 3961. [CrossRef]
30. Pielesz, A.; Wlochowicz, A.; Binias, W. The evaluation of structural changes in wool fibre keratin treated with azo dyes by Fourier Transform Infrared Spectroscopy. *Spectrochim. Acta A* **2000**, *56*, 1409–1420. [CrossRef]
31. Muyonga, J.H.; Cole, C.G.B.; Duodu, K.G. Fourier transform infrared (FTIR) spectroscopic study of acid soluble collagen and gelatin from skins and bones of young and adult Nile perch (Lates niloticus). *Food Chem.* **2004**, *86*, 325–332. [CrossRef]
32. Kakkar, P.; Madhan, B.; Shanmugam, G. Extraction and characterization of keratin from bovine hoof: A potential material for biomedical applications. *SpringerPlus* **2014**, *3*, 596. [CrossRef] [PubMed]
33. Gopal, R.; Park, J.S.; Seo, C.H.; Park, Y. Applications of circular dichroism for structural analysis of gelatin and antimicrobial peptides. *Int. J. Mol. Sci.* **2012**, *13*, 3229–3244. [CrossRef] [PubMed]
34. Wu, D.; Xu, G.; Sun, Y.; Zhang, H.; Mao, H.; Feng, Y. Interaction between proteins and cationic gemini surfactant. *Biomacromolecules* **2007**, *8*, 708–712. [CrossRef] [PubMed]
35. Tiffany, M.L.; Krimm, S. Circular dichroism of the "random" polypeptide chain. *Biopolymers* **1969**, *8*, 347–359. [CrossRef]
36. Nikoo, M.; Benjakul, S.; Ocen, D.; Yang, N.; Xu, B.; Zhang, L.; Xu, X. Physical and chemical properties of gelatin from the skin of cultured Amur sturgeon (Acipenser schrenckii). *J. Appl. Ichtyol.* **2013**, *29*, 943–950. [CrossRef]
37. Lois Tifany, M.; Krimm, S. Effect of temperature on the circular dichroism spectra of polypeptides in the extended state. *Biopolymers* **1972**, *11*, 2309–2316. [CrossRef] [PubMed]
38. Komsa-Penkova, R.; Koynova, R.; Kostov, G.; Tenchov, B.G. Thermal stability of calf skin collagen type I in salt solutions. *BBA Protein Struct. Mol. Enzymol.* **1996**, *1297*, 171–181. [CrossRef]
39. Wright, N.T.; Humphrey, J.D. Denaturation of collagen via heating: An irreversible rate process. *Annu. Rev. Biomed. Eng.* **2002**, *4*, 109–128. [CrossRef] [PubMed]
40. Sarbon, N.M.; Badii, F.; Howell, N.K. The effect of chicken skin gelatin and whey protein interactions on rheological and thermal properties. *Food Hydrocoll.* **2015**, *45*, 83–92. [CrossRef]
41. Buettner, G.R. Spin trapping: ESR parameters of spin adducts. *Free Radic. Biol. Med.* **1987**, *3*, 259–303. [CrossRef]
42. Davies, M.J.; Fu, S.; Dean, R.T. Protein hydroperoxides can give rise to reactive free radicals. *Biochem. J.* **1995**, *305*, 643–649. [CrossRef] [PubMed]

biomolecules

MDPI

Review

Primary Processes of Free Radical Formation in Pharmaceutical Formulations of Therapeutic Proteins

Christian Schöneich

Department of Pharmaceutical Chemistry, University of Kansas, 2093 Constant Avenue, Lawrence, KS 66047, USA; schoneic@ku.edu; Tel.: +1-(785)-864-4880

Abstract: Oxidation represents a major pathway for the chemical degradation of pharmaceutical formulations. Few specific details are available on the mechanisms that trigger oxidation reactions in these formulations, specifically with respect to the formation of free radicals. Hence, these mechanisms must be formulated based on information on impurities and stress factors resulting from manufacturing, transportation and storage. In more detail, this article focusses on autoxidation, metal-catalyzed oxidation, photo-degradation and radicals generated from cavitation as a result of mechanical stress. Emphasis is placed on probable rather than theoretically possible pathways.

Keywords: autoxidation; Fenton reaction; free radicals; mechanisms; oxidation; photo-degradation; cavitation

Citation: Schöneich, C. Primary Processes of Free Radical Formation in Pharmaceutical Formulations of Therapeutic Proteins. *Biomolecules* **2023**, *13*, 1142. https://doi.org/10.3390/biom13071142

Academic Editor: Chryssostomos Chatgilialoglu

Received: 29 June 2023
Revised: 12 July 2023
Accepted: 13 July 2023
Published: 17 July 2023

1. Introduction

The advent of biotechnology has enabled the production of recombinant proteins for therapeutic applications. A recent review of the globally highest-selling drugs in 2019 showed that out of ten drug products, seven were proteins [1]. Despite the therapeutic and commercial success of protein therapeutics, the development of stable protein formulations can present challenges [2–7]. Proteins are subject to physical and chemical degradation, potentially compromising the efficacy and safety of drug products. The physical degradation of proteins is often associated with processes such as surface adsorption, aggregation, particle formation and precipitation, while chemical degradation describes the covalent modification of amino acids. Frequently, the physical and chemical degradation of proteins are connected, where, for example, chemical modifications may trigger aggregation or conformational transitions of proteins may facilitate the accrual of chemical modifications.

Oxidation represents a major pathway for the chemical degradation of proteins, which can be carried out by a range of reactive oxygen and nitrogen species, including free radicals [8,9]. The field of redox biology presents many examples of proteins that are subject to oxidative modification in vivo under conditions of oxidative stress. These oxidative modifications may either result in no change in activity, or promote loss or gain of function, depending on the nature of the modifications and the specific proteins. Whether some of these oxidative modifications may be useful as clinical biomarkers will depend on the type, stability and location of the modifications and the pathologies of concern [10–12]. For example, commonly measured protein oxidation products such as protein carbonyls, methionine sulfoxide (MetSO) and some tyrosine-derivatives were poor biomarkers in the biological fluids of rats for either carbon tetrachloride (CCl_4)- or ozone-induced oxidative stress [13,14]. In contrast, some lipid-derived oxidation products, such as malondialdehyde (MDA) or isoprostanes, appeared to be viable biomarkers for CCl_4-induced oxidative stress in rats [13].

Many of the protein oxidation products that have been characterized in vivo can also form as a result of oxidative processes in therapeutic protein formulations in vitro [15]. In addition, these oxidation processes can generate a range of oxidation products from excipients, e.g., from amino acids, especially histidine (His) [16–19], and surfactants [18–24], and likely also carbohydrates. Some of these oxidation products in vitro may correlate with important characteristics of their respective drug products, and are referred to as critical quality attributes (CQAs). For example, oxidation products may have consequences for the shelf-life, bioavailability or immunogenicity of drug products.

The exact nature and sources of oxidants in protein formulations are generally less well defined. For comparison, the biological mechanisms of oxidant production frequently rely on relatively well-characterized enzymes such as xanthine oxidase [25], nitric oxide synthase [26], myeloperoxidase [27] or NADPH oxidase [28]. In contrast, oxidation reactions in therapeutic protein formulations in vitro rely predominantly on adventitious processes promoted by, e.g., mechanical stress, impurities and/or exposure to elevated temperature, light or ionizing radiation. Details on the primary processes that lead to free radical formation and oxidation in pharmaceutical formulations would be highly valuable for the development of mitigation strategies. It is possible to outline the mechanisms of free radical generation in pharmaceutical formulations based on information on impurities and known stress factors that are relevant to pharmaceutical manufacturing, transportation and storage. This is the purpose of this article.

2. Composition of Pharmaceutical Formulations of Therapeutic Proteins

The pharmaceutical formulations of therapeutic proteins display a range of compositions for liquid, frozen and lyophilized forms. Relevant to the potential formation mechanisms of radicals is the fact that these formulations can generally contain several classes of compounds in addition to the protein, such as buffers, surfactants, amino acids, cryoprotectants, chelators and additional tonicifiers [1,29]. Besides the intended functions in the formulations, each of these respective components may play a role in free radical generation through its chemical properties and/or impurities, which may be introduced via chemical synthesis, purification and/or storage.

3. Pathways of Free Radical Formation That Are Relevant to Pharmaceutical Formulations

The following sections will discuss specific pathways of free radical generation with respect to the potential role of impurities and stress factors.

3.1. Autoxidation

Miller et al. define true autoxidation "as the spontaneous oxidation in air of a substance not requiring catalysts" [30]. Hence, autoxidation can be represented by the general reaction (1), where D^- and $O_2^{\bullet-}$ represent an electron donor and superoxide, respectively, and k_1 and k_{-1} are the rate constants for forward and reverse electron transfer.

$$D^- + O_2 \rightleftharpoons D^\bullet + O_2^{\bullet-} \quad (1)$$

A plot of log k_1 vs. log K_1 (where K_1 represents the equilibrium constant for redox equilibrium 1) yields a curve that can be fitted to the Marcus equation, where D^- represents a series of phenolates, indophenolates and other electron donors [31]. This relationship allows us to make an estimate of the sensitivity of amino acid side chains towards autoxidation on the basis of their reduction potentials. Such an estimate suggests that at most cysteine in its deprotonated form, with $E^o{}_2 \approx 0.75$ V for redox equilibrium 2 [32,33], would be susceptible to autoxidation at pH values generally selected for protein formulations.

$$RS^\bullet + e^- \rightleftharpoons RS^- \quad (2)$$

This would limit autoxidation processes to proteins that contain free cysteine residues. Monoclonal antibodies do not contain free cysteine residues, except for small quantities of incompletely folded proteins, implying that autoxidation should be a negligible problem for pharmaceutical formulations containing monoclonal antibodies.

A second target for potential autoxidation would be surfactants [34], especially polysorbate 80, which contains oleic, linoleic and linolenic acid [35,36]. It is possible that autoxidation contributes to the generation of polysorbate radicals [37] and polysorbate oxidation products, including peroxides, in neat polysorbate [38]. However, it is equally likely that oxidation in neat polysorbate is triggered by the homolytic decomposition of reactive fatty acid:oxygen copolymers of the general structure **1** (Figure 1, where residues R_n, n = 1–6, depict moieties of fatty acids that have undergone successive peroxyl radical and oxygen addition to double bonds) [39,40], containing α,β-diperoxide repeats analogous to styrene:oxygen copolymers [41]. Such fatty acid:oxygen copolymers may be generated during polysorbate synthesis and storage.

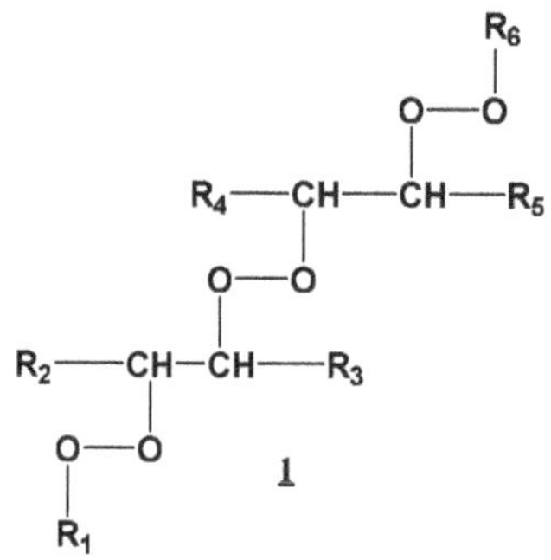

1

Figure 1. General structure of fatty acid:oxygen copolymers (Residues R_n, n = 1–6, depict moieties of fatty acids that have undergone successive peroxyl radical and oxygen addition to double bonds).

Morita and Tokita reported that fatty acid:oxygen copolymers are stronger initiators of lipid peroxidation in model experiments compared to simple hydroperoxides [39]. However, the fact that Bensaid et al. [42] observed that iron levels as low as 20 ppb accelerated polysorbate oxidation in aqueous formulations suggests that, at least in pharmaceutical formulations, metal-catalyzed reactions of hydroperoxides may be kinetically more significant for the formation of free radicals compared to metal-independent decomposition reactions of fatty acid:oxygen copolymers or autoxidation. This is also consistent with data showing the accelerated decomposition of polysorbate when in contact with stainless steel surfaces [19,43]. It is, therefore, unlikely that true autoxidation processes contribute significantly to free radical formation in pharmaceutical formulations.

3.2. *Fenton and Fenton-like Reactions between Metals and Peroxides*

Fenton and Fenton-like reactions represent important pathways for free radical generation. Peroxides can be introduced into formulations through excipients [44], primarily surfactants [38,44,45], and/or as a result of sterilization procedures [46]. Before presenting a detailed discussion of the potential radical-generating reactions of metals and peroxides in pharmaceutical formulations, we need to evaluate which metals and which reactions are most relevant to pharmaceutical formulations. The International Council for Harmonization (ICH) Q3D(R2) guidelines define three classes of elemental impurities based on "their toxicity (PDE) and likelihood of occurrence in the drug product" [47] (PDE = permitted daily exposure). Several elemental impurities in these classes are redox-active and/or catalyze oxidation reactions, such as Co, Ni and V (in class 2A); Ir, Os, Rh and Ru (in class 2B); and Cr, Cu and Mo (in class 3). The ICH Q3D(R2) guidelines list additional elemental impurities "for which PDE values have not been established due to their low inherent toxicity and/or differences in regional regulations" [47]. Of these, Fe, Mn and W are redox-active and/or catalyze oxidation reactions (as, perhaps, does Al [48]; here, Al^{III} does not change its oxidation state but promotes the disproportionation reaction of two complexed H_2O_2 molecules). A representative quantitative analysis of the elemental impurities listed in class 1 and 2A in several formulation components predicts that their levels in therapeutic protein drug products will be significantly below the PDE [49]. These elemental impurities will likely present no toxicological problems; however, even at levels below the PDE, some of these elemental impurities may promote radical formation and/or catalyze oxidation reactions. Therefore, we need to narrow down a selection of metals for further consideration in this article via other means. Class 2B elements "have a reduced probability of occurence in drug product" [47] and will, therefore, not be considered further. Lloyd et al. reported DNA oxidation in the presence of H_2O_2 for Cr(III), Fe(II), V(III) and Cu(II), indicating Fenton or Fenton-like reactivities of these metals [50]. However, no efficient DNA oxidation was observed for Co(II) and Ni(II) in the presence of H_2O_2 [50]. Anipsitakis and Dionysiou surveyed the formation of radicals from the reaction of three oxidants, including hydrogen peroxide (H_2O_2), potassium persulfate ($K_2S_2O_8$) and potassium peroxomonosulfate ($KHSO_5$), with nine metals, including Fe(II), Fe(III), Co(II), Ru(III), Ag(I), Ce(III), V(III), Mn(II) and Ni(II) [51]. Of these, only Fe(II), Fe(III) and Ru(III) generated significant levels of hydroxyl radicals ($HO^{\bullet}$) upon reaction with H_2O_2 [51]. The other metals formed significant yields of inorganic radicals ($SO_4^{\bullet -}$) only upon reaction with $K_2S_2O_8$ and/or $KHSO_5$, oxidants that are likely not present in pharmaceutical formulations. In contrast, Stadtman et al. advocate for the formation of "caged" $HO^{\bullet}$ radicals during the reaction of Mn(II) with H_2O_2, available for the oxidation of substrates [52–54]. The data are consistent with respect to Co(II) and Ni(II), which we do not need to consider further as a source of radicals in pharmaceutical formulations. We will also not further consider any reactions of Mo(IV) as it functions as a co-catalyst in Fe-dependent Fenton reactions [55,56], for example, reducing Fe(III) to Fe(II) [55]. The formation and detection of $HO^{\bullet}$ radicals during reactions of V(III) and Mn(II) with H_2O_2 may depend on the experimental conditions; in a first approximation, the redox reactions of V(III) and Mn(II) with H_2O_2 would be rather comparable to the reactions of the Fe(II)/Fe(III) redox couple with H_2O_2. Hence, a more detailed description of the reaction mechanisms of Fe(II) and Fe(III) would serve as a model for analogous reactions of V(III) and Mn(II), and also of Cr(III) and Cu(II). Therefore, the following will entirely focus on processes of Fe-dependent radical generation through the Fenton reaction that are relevant to pharmaceutical formulations.

Reactions of Ferrous and Ferric Iron

In pharmaceutical formulations of therapeutic proteins, iron impurities can come from multiple sources, including the cell culture medium, manufacturing equipment, containers, proteins and excipients. Iron levels as high as 1–9 μM have been reported for some protein formulations [57,58].

In general, it can be assumed that iron impurities in pharmaceutical formulations will be present as ferric iron, Fe^{III}, coordinated with iron-binding ligands, L [59]. These ligands can originate from the protein as well as excipients such as amino acids and carbohydrates. Based on the formulation composition, it is likely that Fe^{III} may be present in a variety of mixed ligand complexes, i.e., that complexes of Fe^{III} show some heterogeneity. Fe^{III} reacts with H_2O_2 according to equilibrium 3 [60,61].

$$L_xFe^{III} + H_2O_2 \rightleftharpoons L_xFe^{III}(^{-}O_2H) + H^+ \quad (3)$$

Rate constants of $k_3 = 69\ M^{-1}s^{-1}$ and $k_{-3} = 0.11\ s^{-1}$ have been reported for equilibrium 3 in acidic aqueous solution with pH 2.0 (where L = H_2O) [61]. Based on the standard reduction potentials for the couples Fe^{III}/Fe^{II} (0.77 V vs. NHE) [62] and HO_2/HO_2^- (0.79 V), the reduction of Fe^{III} by HO_2^- is feasible [63], so equilibrium 4 is reasonable, where the resulting hydroperoxyl radical is characterized by $pK_a = 4.8$ for equilibrium 5 [64]. The superoxide radical anion ($O_2^{\bullet -}$) can subsequently reduce an additional equivalent of L_xFe^{III} (reaction 6) [62,65].

$$L_xFe^{III}(^{-}O_2H) \rightleftharpoons L_xFe^{II} + HOO^{\bullet} \quad (4)$$

$$HOO^{\bullet} \rightleftharpoons H^{+} + O_2^{\bullet-} \quad (5)$$

$$L_xFe^{III} + O_2^{\bullet-} \rightleftharpoons L_xFe^{II} + O_2 \quad (6)$$

However, it has been pointed out that, specifically, the reduction potential for the couple Fe^{III}/Fe^{II} is very sensitive to pH [59] and the nature and concentration of the ligands [62,63], so the potential reduction of L_xFe^{III} to L_xFe^{II} by HO_2^- must be carefully discussed with respect to these parameters. Specifically, for L_x = EDTA, the reduction potential of Fe^{III}/Fe^{II} decreases to 0.12 V [62], suggesting that the reduction of $(EDTA)Fe^{III}$ by HO_2^- may not be a major pathway of Fe^{II} formation [63] (for L_x = DTPA, the reduction potential decreases even further to 0.03 V [66]). However, this prediction must be compared to experimental results that show that the reaction of H_2O_2 with $(EDTA)Fe^{III}$ yields an oxidant that converts the dipeptide Met-Met (Met = methionine) to products that are also generated via the exposure of Met-Met to a Fenton system, $(EDTA)Fe^{II}/H_2O_2$ [67], suggesting the formation of free or complexed hydroxyl radicals ($HO^{\bullet}$) or higher-valent iron-oxo species such as Fe^{IV}=O [61]. In this respect, the results of Bensaid et al. [42] are important, which show that the levels of iron impurities (20 ppb vs. <2 ppb) in formulations containing a monoclonal antibody, His, sucrose and polysorbate 80 control polysorbate oxidation, which correlates with the oxidation of Met^{255} on the monoclonal antibody. In these formulations, it is likely that $HO^{\bullet}$ and/or L_xFe^{IV}=O are generated via the reaction of L_xFe^{III} with hydrogen peroxide (reactions 3, 4, 7 and 8) and $RO^{\bullet}$ and/or L_xFe^{IV}=O via the reaction of L_xFe^{II} with organic hydroperoxide impurities (reactions 10 and 11). Here, L_xFe^{IV}=O ($E^{o'}_{pH\,7.0} \approx 1.00$ V) is the less powerful and more selective oxidant compared to $HO^{\bullet}$ ($E^{o'}_{pH\,7.0} = 2.18$ V) [68].

$$L_xFe^{II} + H_2O_2 \rightarrow L_xFe^{III} + HO^{\bullet} + HO^- \quad (7)$$

$$L_xFe^{II} + H_2O_2 \rightarrow L_xFe^{IV}{=}O + H_2O \quad (8)$$

$$L_xFe^{III}(^-O_2R) \rightleftharpoons L_xFe^{II} + ROO^{\bullet} \quad (9)$$

$$L_xFe^{II} + HOOR \rightarrow L_xFe^{III} + RO^{\bullet} + HO^- \quad (10)$$

$$L_xFe^{II} + HOOR \rightarrow L_xFe^{IV}{=}O + ROH \quad (11)$$

In this regard, the initial reaction (reaction 3) of L_xFe^{III} with H_2O_2 may become important, as its product, $L_xFe^{III}(^-O_2H)$, reacts significantly more efficiently with L_xFe^{II} ($k_{12} = 7.7 \times 10^5$ $M^{-1}s^{-1}$; L = H_2O, pH 1.0) compared to H_2O_2 ($k \approx 50$ $M^{-1}s^{-1}$) [61].

$$L_xFe^{III}(^-O_2H) + L_xFe^{II} \rightarrow L_xFe^{III} + [L_xFe^{III} + HO^{\bullet}]/L_xFe^{IV}{=}O \quad (12)$$

Ultimately, the resulting oxidizing species, $HO^{\bullet}$ and/or L_xFe^{IV}=O, will have the opportunity to react with formulation constituents such as protein and excipients, generating a plethora of oxidation products and secondary oxidizing species including peroxyl radicals, alkoxyl radicals and peroxides. The initial oxidation reactions of $HO^{\bullet}$ and/or L_xFe^{IV}=O may occur preferentially with the ligands L, coordinating either Fe^{II} or Fe^{III} [61]. Peroxyl and alkoxyl radicals, as well as peroxides, will be generated via the reaction of $HO^{\bullet}$ and/or L_xFe^{IV}=O with the organic constituents of the formulation. An alternative oxidant, the carbonate radical anion ($^{\bullet}CO_3^-$), may be generated if the formulation contains low amounts of bicarbonate (introduced through atmospheric CO_2), which can generate $L_xFe^{II}(CO_3)$ [69]. In such complexes, the initial oxidants, $HO^{\bullet}$ and/or $L_x(CO_3)Fe^{IV}$=O, may oxidize the Fe-bound carbonate to $^{\bullet}CO_3^-$ [69,70], which itself is a powerful yet more selective oxidant.

An important question is that of whether metal chelators can prevent the formation of oxidizing species during the reaction of peroxides with L_xFe^{III} and L_xFe^{II}. Walling et al. [71] demonstrated the oxidation of a variety of organic substrates by $(EDTA)Fe^{III}/H_2O_2$, providing evidence that oxidation reactions prevail in the presence of EDTA. Likewise, Graf et al. showed that EDTA did not prevent the oxidation of dimethylsulfoxide (DMSO) (ultimately to formaldehyde) induced by L_xFe^{III} and hypoxanthine/xanthine oxidase [72]. However, DTPA prevented the oxidation of DMSO, providing evidence that the chelator structure plays an important role in the efficiency of preventing substrate oxidation. These findings can, in part, be rationalized by the complex geometries of $(EDTA)Fe^{II}$ and $(EDTA)Fe^{III}$, where crystal structures demonstrate a distortion from octahedral geometry, resulting in the availability of a seventh binding site for a reaction to take place [73,74]. In aqueous solution, this seventh binding site generally coordinates with water [73,74], also indicated by a dissociable proton of $(EDTA)Fe^{III}(H_2O)$ with $pK_a \approx 7.6$ [75,76]. The bound water can be replaced by H_2O_2 [65,71,77] and, in case of $(EDTA)Fe^{II}$, also by molecular oxygen [78]. In fact, $(EDTA)Fe^{II}$ efficiently reacts with H_2O_2, with $k > 3 \times 10^3\ M^{-1}s^{-1}$ [79,80].

3.3. Photochemical Generation of Radicals

Depending on the manufacturing environment and clinical use, protein formulations can be exposed to UVA and/or visible light [81], and an increasing number of studies show visible- or ambient-light-induced degradation of therapeutic proteins [58,82–92]. In particular, visible light photo-degradation is not easily rationalized with the known absorption characteristics of individual amino acids. This presents a challenge for the mechanistic analysis of processes leading to photo-degradation under visible light exposure, which is addressed in a recent review [93]. It is generally possible that photo-sensitizers are generated from the oxidative degradation of proteins and/or excipients, i.e., protein di-tyrosine from Tyr [94], 6a-hydroxy-2-oxo-octahydropyrollo[2,3-d]imidazole-5-carboxylic acid from His [17], advanced glycation end-products (AGEs) from the breakdown of carbohydrates [95], and cross-links between amino acids and lipid peroxidation products [96]. In addition, certain constituents or impurities present in cell culture media that co-purify with the protein may act as photo-sensitizers, e.g., riboflavin [97–99] or pterin derivatives [100–102]. These photo-sensitizers can generate radicals in pharmaceutical formulations via a type I process, which represents an electron transfer reaction by a photo-sensitizer, subsequent to which a radical intermediate reacts with oxygen [103] (in contrast, a type II process entails the generation of singlet oxygen, $^1O_2$ [103]).

Tryptophan residues can form cation-π complexes [104–106], which absorb visible light [107,108]. In such complexes, the electron density is shared between the Trp π system and the cation, resulting in spectroscopic properties reminiscent of $Trp^\bullet$ radicals [108]. Hence, Trp cation-π complexes may serve as chromophores suitable for the initiation of photo-degradation by visible light, a possibility that should be tested experimentally.

In view of the discussion of Fe^{III}-dependent oxidation reactions in pharmaceutical formulations (see Section 3.2 above), the possibility of photo-Fenton reactions as a source of free radicals is a viable option. Pharmaceutical buffers (e.g., acetate, succinate, citrate) and amino acids contain carboxylate groups, where Fe^{III}-carboxylate complexes are characterized by broad absorption bands in the UVA and visible regions. Under light exposure, these Fe^{III}-carboxylate complexes can undergo ligand-to-metal-charge transfer (LMCT), reducing Fe^{III} to Fe^{II}, and oxidizing the carboxylate ligand, which subsequently decarboxylates reactions (13) and (14) [109–113].

$$RCO_2{}^{-}\text{-}Fe^{III} \rightarrow RCO_2{}^{\bullet}\text{-}Fe^{II} \tag{13}$$

$$RCO_2{}^{\bullet}\text{-}Fe^{II} \rightarrow R^\bullet + CO_2 + Fe^{II} \tag{14}$$

The resulting carbon-centered radical $R^\bullet$ will add oxygen to yield a peroxyl radical, $ROO^\bullet$, unless $R^\bullet$ is $^\bullet CO_2{}^-$ (see below), while Fe^{II} reduces O_2 to $O_2{}^{\bullet -}$ [65] and H_2O_2 [114]. With respect to the necessary concentrations of Fe^{III}, basal levels of Fe^{III} in 10 mM citrate buffer, pH 6.0, were sufficient to promote the photo-oxidation of Met-enkephalin during near-UV photo-irradiation with a light dose of 25.2 Whm^{-2} [115], i.e., ca. 1/8 of the light dose required according the ICH Q1B guidelines for photostability studies [116]. In these experiments, various lots of citrate were tested, and the photo-oxidation yields from Met-enkephalin correlated with the basal Fe^{III} levels [115]. An important detail is the formation of $^{\bullet}CO_2^-$ during the photo-irradiation of citrate-Fe^{III} with either near-UV or visible light, detected via spin-trapping with DMPO [115,117]. The $^{\bullet}CO_2^-$ radical is a powerful reductant ($E^o(CO_2/^{\bullet}CO_2^-) \approx 1.93 \pm 0.22$ V vs. NHE [118]) that reduces Fe^{III} to Fe^{II} [119], O_2 to $O_2^{\bullet -}$ [119,120] and disulfide (RSSR) to a thiyl radical (RS$^{\bullet}$) and thiolate (RS$^-$) [121–123].

Mechanistic studies suggest that the formation of $^{\bullet}CO_2^-$ from citrate involves LMCT from the (deprotonated) citrate hydroxyl group rather than the citrate carboxyl groups, generating an intermediary alkoxyl radical (RO$^{\bullet}$), which undergoes α-β cleavage of the central carboxylate group (Scheme 1; reactions 15 and 16) [117]. In reaction 15, the initial citrate-Fe^{III} complex is drawn with reference to the crystal structure of mononuclear (citrate)$_2Fe^{III}$ [124], which shows that the hydroxyl group is deprotonated.

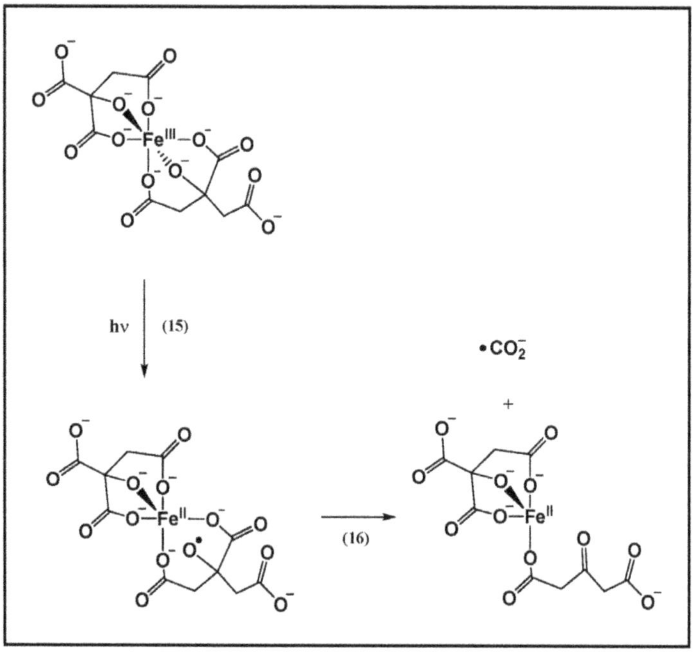

Scheme 1. Formation mechanism of $^{\bullet}CO_2^-$ from citrate-Fe^{III} [117].

A similar mechanism was recently observed for a monoclonal antibody (IgG1) in the presence of Fe^{III} and His buffer. In this case, photo-induced LMCT from a deprotonated Thr residue, Thr^{259}, led to an intermediary Thr side chain alkoxyl radical (Scheme 2, reaction 17), which underwent α-β cleavage, triggering side chain cleavage (Scheme 2, reaction 18) and, ultimately, backbone fragmentation [92].

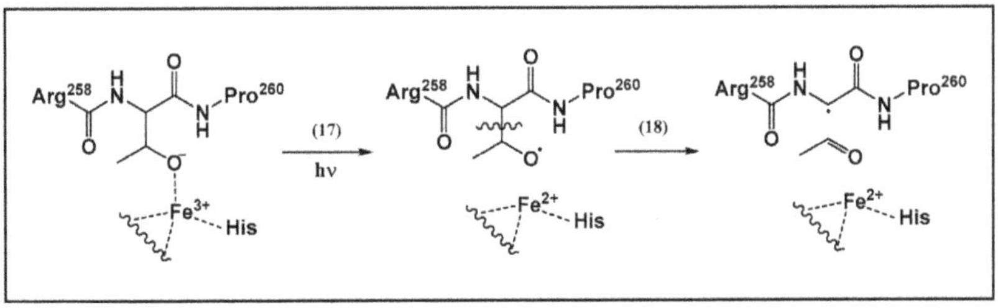

Scheme 2. LMCT mechanism leading to site-specific oxidation of a monoclonal antibody [92]. Photolytic generation of an alkoxyl radical at Thr[259] of IgG1 [92].

3.4. Generation of Radicals via Mechanical Stress

Protein formulations are exposed to various types of mechanical stress during manufacturing and transportation. Under certain circumstances, high shear stresses [125,126], mixing, pumping, filling [127–129] and mechanical shock [130–132] may lead to cavitation [133], a process that can cause the formation of $HO^\bullet$ radicals and even O atoms [134]. Hence, mechanical stresses have the potential to trigger the formation of highly oxidizing radicals, which can subsequently react with formulation constituents.

4. Protein Formulations Containing Additional Excipients

4.1. Formulations Containing Antimicrobial Preservatives

In order to ensure sterility, multidose formulations contain antimicrobial preservatives (APs) such as, e.g., phenol, m-cresol, benzyl alcohol, thimerosal or chlorobutanol [135,136] (for a summary of antimicrobial preservative-containing peptide and protein formulations listed in the *Physicians' Desk Reference*, PDR, see [135]). Some of the common antimicrobial preservatives are susceptible to oxidative degradation, potentially generating radicals in pharmaceutical formulations.

The exposure of benzyl alcohol to air leads to the slow formation of benzaldehyde, Ph-CHO [136]. Benzaldehyde spontaneously oxidizes to benzoic acid [137]. The latter pathway involves the formation of an intermediary benzoylperoxyl radical, Ph-C(O)OO$^\bullet$ (reactions 19 and 20), where In$^\bullet$ represents an initiating radical [137]. However, the presence of benzyl alcohol can suppress benzoic acid formation via the reaction of the benzoylperoxyl radical with benzylalcohol to generate peroxybenzoic acid, Ph-C(O)OOH, and an α-hydroxybenzyl radical, Ph-C$^\bullet$H-OH (reaction 21) [137]. The reaction of the α-hydroxybenzyl radical with molecular oxygen will ultimately generate benzaldehyde and superoxide (reaction 22).

$$\text{Ph-CHO} + \text{In}^\bullet \rightarrow \text{InH} + \text{Ph-}^\bullet\text{C(O)} \quad (19)$$

$$\text{Ph-}^\bullet\text{C(O)} + O_2 \rightarrow \text{Ph-C(O)OO}^\bullet \quad (20)$$

$$\text{Ph-C(O)OO}^\bullet + \text{Ph-CH}_2\text{OH} \rightarrow \text{Ph-C(O)OOH} + \text{Ph-C}^\bullet\text{H-OH} \quad (21)$$

$$\text{Ph-C}^\bullet\text{H-OH} + O_2 \rightarrow \text{Ph-CHO} + H^+/O_2^{\bullet-} \quad (22)$$

Therefore, formulations containing benzyl alcohol bear a potential risk for the formation of oxygen-centered radicals (peroxyl radicals, superoxide) and peroxides (peroxybenzoic acid).

The potential exposure of phenol and m-cresol to hydroxyl radicals (such as those generated by Fenton-type reactions; see Section 3.2 above) will lead to hydroxylation, preferentially in the ortho- or para-position with regard to the existing hydroxy substituent(s) [138–140]. Such hydroxylation reactions generate catechol derivatives, which can further promote Fenton-type reactions through redox cycling [141,142]. During redox cycling, a catechol derivative reduces L_xFe^{III} to L_xFe^{II}, generating a semiquinone radical

(Scheme 3; equilibrium 23), which can further reduce O_2 to $O_2^{\bullet-}$ (equilibrium 24), generating a quinone derivative. The latter can comproportionate with a catechol to regenerate semiquinone derivatives (equilibrium 25) [142]. The dismutation of $O_2^{\bullet-}$ will generate H_2O_2, which will regenerate L_xFe^{III} through a reaction with L_xFe^{II}, generating $HO^{\bullet}$ radicals (see Section 3.2).

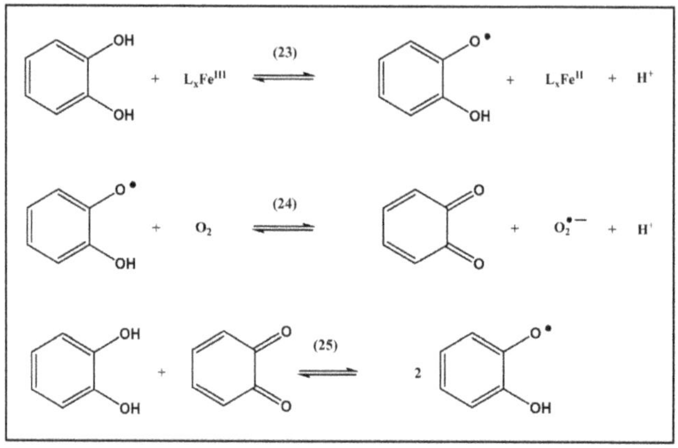

Scheme 3. Redox cycling of catechol.

4.2. *Formulations Containing Zn(II)*

Specifically, insulin formulations, e.g., Humulin R® or Humalog®, contain Zn(II), generally in the form of ZnO (see package inserts for Humulin N® and Humalog®), which releases Zn^{2+} [143]. ZnO confers antimicrobial activity [143], but the released Zn^{2+} ions also support the formation of a native insulin hexamer [144]. Both Humulin N® and Humalog® also contain m-cresol and phenol, which are susceptible to hydroxylation and, subsequently, redox cycling (see Section 4.1). It was demonstrated that Zn^{2+} increased total phenol oxidation (monitored as total organic carbon, TOC) during the Fenton oxidation of phenols, which has been rationalized by a more persistent semiquinone radical as a result of Zn^{2+} complexation, generating more $HO^{\bullet}$ radicals [142]. Hence, the combination of phenols and Zn^{2+} may increase the susceptibility of a formulation to Fenton oxidation.

An alternative mechanism by which ZnO, specifically, may promote oxidation reactions is photo-degradation. ZnO is a semiconductor with a band gap of 3.2–3.7 eV [145], which would require light with wavelengths of λ = 387–335 nm to excite an electron from the valence band to the conduction band. Generally, the conduction band electron can reduce adsorbed O_2 to $O_2^{\bullet-}$, while the remaining positive hole, h^+, in the valence band can oxidize adsorbed H_2O/HO^- to the $HO^{\bullet}$ radical [145,146]. ZnO was tested as a photocatalyst under light exposure with $\lambda > 300$ nm on a coated glass plate [147], showing greater activity than WO_3, an activity comparable to that of brookite (TiO_2), but an activity lower than that of anatase (TiO_2). However, ZnO was more active than anatase in the photocatalytic degradation of humic acid in aqueous solution with pH 7.88 [148].

5. IV Enzyme Formulations for Enzyme Replacement Therapy

A review of IV formulations for enzyme replacement therapy [149,150] reveals that these formulations generally do not contain unusual excipients (for example, see package inserts for Aldurazyme®, Elaprase®, Vimizim®, Naglazyme®, Mepsevii®, VPRIV™ or Nexviazyme™). However, inspection of the active sites of the some of the relevant enzymes shows the presence of Cys residues, e.g., in N-acetylgalactosamine-6-sulfatase [151] and iduronate-2-sulfatase [152]. These Cys residues are post-translationally modified to C_{α}-formylglycine (FGly) and, therefore, are not amenable to Cys oxidation.

6. Conclusions

Based on information on impurities and stress factors that affect pharmaceutical formulations, a series of mechanisms are formulated that could be responsible for free radical formation in pharmaceutical formulations. The focus of this article is on highly probable reactions; additional pathways may be possible in isolated cases when pharmaceutical formulations contain high levels of specific impurities that are not generally present. With respect to the design of stress tests for pharmaceutical formulations, highly probable reactions should be kept in mind. For example, it may be questionable whether the addition of Fe^{II} to a pharmaceutical formulation may generate information about the kinetics of iron-dependent oxidation degradation reactions under storage conditions, as iron impurities will likely be present as Fe^{III}. However, the addition of Fe^{II} may lead to mechanistic information that can be used to predict certain degradation pathways in cases whereby Fe^{III} is converted to Fe^{II}, for example, through reaction with H_2O_2 or hydroperoxides. A limitation of mechanistic investigations of radical-induced oxidation reactions in pharmaceutical formulations will always be that the precise quantity and nature of the radicals specifically generated under storage conditions are usually unknown. Even the monounsaturated oleic acid, the main component of polysorbate 80 fatty acid esters [35,36], can generate a number of different peroxyl radicals [40]. It is unknown to what extent the nature of these different peroxyl radicals would affect the kinetics of chain propagation within polysorbate 80 micelles. This question may be addressed through the quantification of specific reaction products that are representative of individual oxidation pathways, a task that may require the modification or improvement of analytical methodology, potentially supported by artificial intelligence.

Funding: This research received no external funding.

Institutional Review Board Statement: Not applicable.

Informed Consent Statement: Not applicable.

Data Availability Statement: Not applicable.

Conflicts of Interest: The author declares no conflict of interest.

References

1. Strickley, R.G.; Lambert, W.J. A review of Formulations of Commercially Available Antibodies. *J. Pharm. Sci.* **2021**, *110*, 2590–2608. [CrossRef]
2. Manning, M.C.; Chou, D.K.; Murphy, B.M.; Payne, R.W.; Katayama, D.S. Stability of protein pharmaceuticals: An update. *Pharm. Res.* **2010**, *27*, 544–575. [CrossRef] [PubMed]
3. Manning, M.C.; Liu, J.; Li, T.; Holcomb, R.E. Rational Design of Liquid Formulations of Proteins. *Adv. Protein Chem. Struct. Biol.* **2018**, *112*, 1–59. [PubMed]
4. Wang, W.; Ohtake, S. Science and art of protein formulation development. *Int. J. Pharm.* **2019**, *568*, 118505. [CrossRef] [PubMed]
5. Falconer, R.J. Advances in liquid formulations of parenteral therapeutic proteins. *Biotechnol. Adv.* **2019**, *37*, 107412. [CrossRef]
6. Gupta, S.; Jiskoot, W.; Schöneich, C.; Rathore, A.S. Oxidation and Deamidation of Monoclonal Antibody Products: Potential Impact on Stability, Biological Activity, and Efficacy. *J. Pharm. Sci.* **2022**, *111*, 903–918. [CrossRef] [PubMed]
7. Jiskoot, W.; Hawe, A.; Menzen, T.; Volkin, D.B.; Crommelin, D.J.A. Ongoing Challenges to Develop High Concentration Monoclonal Antibody-based Formulations for Subcutaneous Administration: Quo Vadis? *J. Pharm. Sci.* **2022**, *111*, 861–867. [CrossRef] [PubMed]
8. Hawkins, C.L.; Davies, M.J. Detection, identification, and quantification of oxidative protein modifications. *J. Biol. Chem.* **2019**, *294*, 19683–19708. [CrossRef]
9. Fuentes-Lemus, E.; Hagglund, P.; Lopez-Alarcon, C.; Davies, M.J. Oxidative Crosslinking of Peptides and Proteins: Mechanisms of Formation, Detection, Characterization and Quantification. *Molecules* **2021**, *27*, 15. [CrossRef]
10. Ho, E.; Karimi Galougahi, K.; Liu, C.C.; Bhindi, R.; Figtree, G.A. Biological markers of oxidative stress: Applications to cardiovascular research and practice. *Redox Biol.* **2013**, *1*, 483–491. [CrossRef]
11. Tucker, P.S.; Dalbo, V.J.; Han, T.; Kingsley, M.I. Clinical and research markers of oxidative stress in chronic kidney disease. *Biomarkers* **2013**, *18*, 103–115. [CrossRef] [PubMed]
12. Cristani, M.; Speciale, A.; Saija, A.; Gangemi, S.; Minciullo, P.L.; Cimino, F. Circulating Advanced Oxidation Protein Products as Oxidative Stress Biomarkers and Progression Mediators in Pathological Conditions Related to Inflammation and Immune Dysregulation. *Curr. Med. Chem.* **2016**, *23*, 3862–3882. [CrossRef] [PubMed]

13. Kadiiska, M.B.; Gladen, B.C.; Baird, D.D.; Germolec, D.; Graham, L.B.; Parker, C.E.; Nyska, A.; Wachsman, J.T.; Ames, B.N.; Basu, S.; et al. Biomarkers of oxidative stress study II: Are oxidation products of lipids, proteins, and DNA markers of CCl4 poisoning? *Free Radic. Biol. Med.* **2005**, *38*, 698–710. [CrossRef] [PubMed]
14. Kadiiska, M.B.; Basu, S.; Brot, N.; Cooper, C.; Saari Csallany, A.; Davies, M.J.; George, M.M.; Murray, D.M.; Jackson Roberts, L., 2nd; Shigenaga, M.K.; et al. Biomarkers of oxidative stress study V: Ozone exposure of rats and its effect on lipids, proteins, and DNA in plasma and urine. *Free Radic. Biol. Med.* **2013**, *61*, 408–415. [CrossRef] [PubMed]
15. Hipper, E.; Blech, M.; Hinderberger, D.; Garidel, P.; Kaiser, W. Photo-Oxidation of Therapeutic Protein Formulations: From Radical Formation to Analytical Techniques. *Pharmaceutics* **2022**, *14*, 72. [CrossRef]
16. Mason, B.D.; McCracken, M.; Bures, E.J.; Kerwin, B.A. Oxidation of free L-histidine by tert-Butylhydroperoxide. *Pharm. Res.* **2010**, *27*, 447–456. [CrossRef] [PubMed]
17. Stroop, S.D.; Conca, D.M.; Lundgard, R.P.; Renz, M.E.; Peabody, L.M.; Leigh, S.D. Photosensitizers form in histidine buffer and mediate the photodegradation of a monoclonal antibody. *J. Pharm. Sci.* **2011**, *100*, 5142–5155. [CrossRef] [PubMed]
18. Wang, T.; Markham, A.; Thomas, S.J.; Wang, N.; Huang, L.; Clemens, M.; Rajagopalan, N. Solution Stability of Poloxamer 188 Under Stress Conditions. *J. Pharm. Sci.* **2019**, *108*, 1264–1271. [CrossRef]
19. Zheng, X.W.; Sutton, A.T.; Yang, R.S.; Miller, D.V.; Pagels, B.; Rustandi, R.R.; Welch, J.; Payne, A.; Haverick, M. Extensive Characterization of Polysorbate 80 Oxidative Degradation Under Stainless Steel Conditions. *J. Pharm. Sci.* **2023**, *112*, 779–789. [CrossRef] [PubMed]
20. Kishore, R.S.; Kiese, S.; Fischer, S.; Pappenberger, A.; Grauschopf, U.; Mahler, H.C. The degradation of polysorbates 20 and 80 and its potential impact on the stability of biotherapeutics. *Pharm. Res.* **2011**, *28*, 1194–1210. [CrossRef] [PubMed]
21. Kishore, R.S.; Pappenberger, A.; Dauphin, I.B.; Ross, A.; Buergi, B.; Staempfli, A.; Mahler, H.C. Degradation of polysorbates 20 and 80: Studies on thermal autoxidation and hydrolysis. *J. Pharm. Sci.* **2011**, *100*, 721–731. [CrossRef] [PubMed]
22. Borisov, O.V.; Ji, J.A.; Wang, Y.J.; Vega, F.; Ling, V.T. Toward understanding molecular heterogeneity of polysorbates by application of liquid chromatography-mass spectrometry with computer-aided data analysis. *Anal. Chem.* **2011**, *83*, 3934–3942. [CrossRef] [PubMed]
23. Borisov, O.V.; Ji, J.A.; John Wang, Y. Oxidative Degradation of Polysorbate Surfactants Studied by Liquid Chromatography-Mass Spectrometry. *J. Pharm. Sci.* **2015**, *104*, 1005–1018. [CrossRef]
24. Kranz, W.; Wuchner, K.; Corradini, E.; Menzen, T.; Hawe, A. Micelle Driven Oxidation Mechansim and Novel Oxidation Markers for Different Grades of Polysorbate 20 and 80. *J. Pharm. Sci.* **2020**, *109*, 3064–3077. [CrossRef] [PubMed]
25. Hille, R. Xanthine Oxidase-A Personal History. *Molecules* **2023**, *28*, 1921. [CrossRef] [PubMed]
26. Stuehr, D.J.; Haque, M.M. Nitric oxide synthase enzymology in the 20 years after the Nobel Prize. *Br. J. Pharmacol.* **2019**, *176*, 177–188. [CrossRef]
27. Hawkins, C.L.; Davies, M.J. Role of myeloperoxidase and oxidant formation in the extracellular environment in inflammation-induced tissue damage. *Free Radic. Biol. Med.* **2021**, *172*, 633–651. [CrossRef]
28. Schröder, K. NADPH oxidases: Current aspects and tools. *Redox Biol.* **2020**, *34*, 101512. [CrossRef]
29. Mieczkowski, C.A. The Evolution of Commercial Antibody Formulations. *J. Pharm. Sci.* **2023**, *112*, 1801–1810. [CrossRef]
30. Miller, D.M.; Buettner, G.R.; Aust, S.D. Transition metals as catalysts of "autoxidation" reactions. *Free Radic. Biol. Med.* **1990**, *8*, 95–108. [CrossRef]
31. Merenyi, G.; Lind, J.; Jonsson, M. Autoxidation of Closed-Shell Organics—An Outer-Sphere Electron-Transfer. *J. Am. Chem. Soc.* **1993**, *115*, 4945–4946. [CrossRef]
32. Prutz, W.A.; Butler, J.; Land, E.J.; Swallow, A.J. Unpaired electron migration between aromatic and sulfur peptide units. *Free Radic. Res. Commun.* **1986**, *2*, 69–75. [CrossRef] [PubMed]
33. Surdhar, P.S.; Armstrong, D.A. Reduction Potentials and Exchange-Reactions of Thiyl Radicals and Disulfide Anion Radicals. *J. Phys. Chem.* **1987**, *91*, 6532–6537. [CrossRef]
34. Howard, J.A. Homogenous Liquid-Phase Autoxidations. In *Free Radicals*; Kochi, J.K., Ed.; Wiley: New York, NY, USA, 1973; Volume 2, pp. 3–62.
35. Dwivedi, M.; Blech, M.; Presser, I.; Garidel, P. Polysorbate degradation in biotherapeutic formulations: Identification and discussion of current root causes. *Int. J. Pharm.* **2018**, *552*, 422–436. [CrossRef]
36. Konya, Y.; Ochiai, R.; Fujiwara, S.; Tsujino, K.; Okumura, T. Profiling polysorbate 80 components using comprehensive liquid chromatography-tandem mass spectrometry analysis. *Rapid Commun. Mass Spectrom.* **2023**, *37*, e9438. [CrossRef]
37. Mittag, J.J.; Trutschel, M.L.; Kruschwitz, H.; Mader, K.; Buske, J.; Garidel, P. Characterization of radicals in polysorbate 80 using electron paramagnetic resonance (EPR) spectroscopy and spin trapping. *Int. J. Pharm. X* **2022**, *4*, 100123. [CrossRef]
38. Ha, E.; Wang, W.; Wang, Y.J. Peroxide formation in polysorbate 80 and protein stability. *J. Pharm. Sci.* **2002**, *91*, 2252–2264. [CrossRef]
39. Morita, M.; Tokita, M. The real radical generator other than main-product hydroperoxide in lipid autoxidation. *Lipids* **2006**, *41*, 91–95. [CrossRef]
40. Porter, N.A. A Perspective on Free Radical Autoxidation: The Physical Organic Chemistry of Polyunsaturated Fatty Acid and Sterol Peroxidation. *J. Org. Chem.* **2013**, *78*, 3511–3524. [CrossRef]
41. Mayo, F.R.; Miller, A.A. Oxidation of Unsaturated Compounds. 2. Reactions of Styrene Peroxide. *J. Am. Chem. Soc.* **1956**, *78*, 1023–1034. [CrossRef]

42. Bensaid, F.; Dagallier, C.; Authelin, J.R.; Audat, H.; Filipe, V.; Narradon, C.; Guibal, P.; Clavier, S.; Wils, P. Mechanistic understanding of metal-catalyzed oxidation of polysorbate 80 and monoclonal antibody in biotherapeutic formulations. *Int. J. Pharm.* **2022**, *615*, 121496. [CrossRef] [PubMed]
43. Gopalrathnam, G.; Sharma, A.N.; Dodd, S.W.; Huang, L. Impact of Stainless Steel Exposure on the Oxidation of Polysorbate 80 in Histidine Placebo and Active Monoclonal Antibody Formulation. *PDA J. Pharm. Sci. Technol.* **2018**, *72*, 163–175. [CrossRef]
44. Wasylaschuk, W.R.; Harmon, P.A.; Wagner, G.; Harman, A.B.; Templeton, A.C.; Xu, H.; Reed, R.A. Evaluation of hydroperoxides in common pharmaceutical excipients. *J. Pharm. Sci.* **2007**, *96*, 106–116. [CrossRef]
45. Ding, S. Quantitation of hydroperoxides in the aqueous solutions of non-ionic surfactants using polysorbate 80 as the model surfactant. *J. Pharm. Biomed. Anal.* **1993**, *11*, 95–101. [CrossRef] [PubMed]
46. Noh, M.S.; Jung, S.H.; Kwon, O.; Lee, S.I.; Yang, S.J.; Hahm, E.; Jun, B.H. Evaluation of Sterilization Performance for Vaporized-Hydrogen-Peroxide-Based Sterilizer with Diverse Controlled Parameters. *ACS Omega* **2020**, *5*, 29382–29387. [CrossRef]
47. *Q3D(R2) Elemental Impurities. Guidance for Industry*; International Council for Harmonization (ICH): Geneva, Switzerland, 2022.
48. Kuznetsov, M.L.; Kozlov, Y.N.; Mandelli, D.; Pombeiro, A.J.L.; Shul'pin, G.B. Mechanism of Al^{3+}-Catalyzed Oxidations of Hydrocarbons: Dramatic Activation of H_2O_2 toward O-O Homolysis in Complex $[Al(H_2O)_4(OOH)(H_2O_2)]^{2+}$ Explains the Formation of HO center dot Radicals. *Inorg. Chem.* **2011**, *50*, 3996–4005. [CrossRef] [PubMed]
49. Luo, Y.; Sekhar, C.; Lee, H.; Fujimori, K.; Ronk, M.; Semin, D.; Nashed-Samuel, Y. A Risk-Based Approach to Evaluate and Control Elemental Impurities in Therapeutic Proteins. *J. Pharm. Sci.* **2020**, *109*, 3378–3385. [CrossRef]
50. Lloyd, D.R.; Carmichael, P.L.; Phillips, D.H. Comparison of the formation of 8-hydroxy-2′-deoxyguanosine and single- and double-strand breaks in DNA mediated by fenton reactions. *Chem. Res. Toxicol.* **1998**, *11*, 420–427. [CrossRef]
51. Anipsitakis, G.P.; Dionysiou, D.D. Radical generation by the interaction of transition metals with common oxidants. *Environ. Sci. Technol.* **2004**, *38*, 3705–3712. [CrossRef]
52. Stadtman, E.R.; Berlett, B.S.; Chock, P.B. Manganese-dependent disproportionation of hydrogen peroxide in bicarbonate buffer. *Proc. Natl. Acad. Sci. USA* **1990**, *87*, 384–388. [CrossRef]
53. Berlett, B.S.; Chock, P.B.; Yim, M.B.; Stadtman, E.R. Manganese(II) catalyzes the bicarbonate-dependent oxidation of amino acids by hydrogen peroxide and the amino acid-facilitated dismutation of hydrogen peroxide. *Proc. Natl. Acad. Sci. USA* **1990**, *87*, 389–393. [CrossRef] [PubMed]
54. Yim, M.B.; Berlett, B.S.; Chock, P.B.; Stadtman, E.R. Manganese(II)-bicarbonate-mediated catalytic activity for hydrogen peroxide dismutation and amino acid oxidation: Detection of free radical intermediates. *Proc. Natl. Acad. Sci. USA* **1990**, *87*, 394–398. [CrossRef]
55. Liu, J.; Dong, C.C.; Deng, Y.X.; Ji, J.H.; Bao, S.Y.; Chen, C.R.; Shen, B.; Zhang, J.L.; Xing, M.Y. Molybdenum sulfide Co-catalytic Fenton reaction for rapid and efficient inactivation of Escherichia colis. *Water Res.* **2018**, *145*, 312–320. [CrossRef]
56. Yang, J.C.; Yao, H.L.; Guo, Y.D.; Yang, B.W.; Shi, J.L. Enhancing Tumor Catalytic Therapy by Co-Catalysis. *Angew. Chem. Int. Ed.* **2022**, *61*, e202200480.
57. Ouellette, D.; Alessandri, L.; Piparia, R.; Aikhoje, A.; Chin, A.; Radziejewski, C.; Correia, I. Elevated cleavage of human immunoglobulin gamma molecules containing a lambda light chain mediated by iron and histidine. *Anal. Biochem.* **2009**, *389*, 107–117. [CrossRef] [PubMed]
58. Adem, Y.T.; Molina, P.; Liu, H.; Patapoff, T.W.; Sreedhara, A.; Esue, O. Hexyl glucoside and hexyl maltoside inhibit light-induced oxidation of tryptophan. *J. Pharm. Sci.* **2014**, *103*, 409–416. [CrossRef]
59. Koppenol, W.H.; Hider, R.H. Iron and redox cycling. Do's and don'ts. *Free Radic. Biol. Med.* **2019**, *133*, 3–10. [CrossRef]
60. Rachmilovich-Calis, S.; Masarwa, A.; Meyerstein, N.; Meyerstein, D.; van Eldik, R. New Mechanistic Aspects of the Fenton Reaction. *Chem.-Eur. J.* **2009**, *15*, 8303–8309. [CrossRef]
61. Meyerstein, D. What Are the Oxidizing Intermediates in the Fenton and Fenton-like Reactions? A Perspective. *Antioxidants* **2022**, *11*, 1368. [CrossRef]
62. Pierre, J.L.; Fontecave, M.; Crichton, R.R. Chemistry for an essential biological process: The reduction of ferric iron. *Biometals* **2002**, *15*, 341–346. [CrossRef]
63. Chen, H.Y.; Lin, Y.F. DFT study on the catalytic decomposition of hydrogen peroxide by iron complexes of nitrilotriacetate. *J. Comput. Chem.* **2023**. [CrossRef] [PubMed]
64. Bielski, B.H.; Cabelli, D.E. Highlights of current research involving superoxide and perhydroxyl radicals in aqueous solutions. *Int. J. Radiat. Biol.* **1991**, *59*, 291–319. [CrossRef] [PubMed]
65. Bull, C.; McClune, G.J.; Fee, J.A. The Mechanism of Fe-Edta Catalyzed Superoxide Dismutation. *J. Am. Chem. Soc.* **1983**, *105*, 5290–5300. [CrossRef]
66. Vandegaer, J.; Chaberek, S.; Frost, A.E. Iron Chelates of Diethylenetriaminepentaacetic Acid. *J. Inorg. Nucl. Chem.* **1959**, *11*, 210–221. [CrossRef]
67. Hong, J.; Schöneich, C. The metal-catalyzed oxidation of methionine in peptides by Fenton systems involves two consecutive one-electron oxidation processes. *Free Radic. Biol. Med.* **2001**, *31*, 1432–1441. [CrossRef]
68. Koppenol, W.H.; Liebman, J.F. The Oxidizing Nature of the Hydroxyl Radical—A Comparison with the Ferryl Ion (Feo2+). *J. Phys. Chem.* **1984**, *88*, 99–101. [CrossRef]
69. Illes, E.; Mizrahi, A.; Marks, V.; Meyerstein, D. Carbonate-radical-anions, and not hydroxyl radicals, are the products of the Fenton reaction in neutral solutions containing bicarbonate. *Free Radic. Biol. Med.* **2019**, *131*, 1–6. [CrossRef]

70. Patra, S.G.; Mizrahi, A.; Meyerstein, D. The Role of Carbonate in Catalytic Oxidations. *Accounts Chem. Res.* **2020**, *53*, 2189–2200. [CrossRef]
71. Walling, C.; Kurz, M.; Schugar, H.J. Iron(III)-Ethylenediaminetetraacetic Acid-Peroxide System. *Inorg. Chem.* **1970**, *9*, 931. [CrossRef]
72. Graf, E.; Mahoney, J.R.; Bryant, R.G.; Eaton, J.W. Iron-catalyzed hydroxyl radical formation. Stringent requirement for free iron coordination site. *J. Biol. Chem.* **1984**, *259*, 3620–3624. [CrossRef]
73. Lind, M.D.; Hoard, J.L.; Hamor, M.J.; Hamor, T.A. Stereocnemistry of ethylenediaminetetraacetato complexes. II. The structure of crystalline Rb[Fe(OH$_2$)Y].H$_2$O$^{1-3}$. III. The structure of crystalline Li[Fe(OH$_2$Y].2H$_2$O$^{1-3}$. *Inorg. Chem.* **1964**, *3*, 34. [CrossRef]
74. Mizuta, T.; Wang, J.; Miyoshi, K. Molecular-Structures of Fe(II) Complexes with Monoprotonated and Diprotonated Ethylenediamine-N,N,N′,N′-Tetraacetate (Hedta and H2edta), as Determined by X-Ray Crystal Analyses. *Inorg. Chim. Acta* **1995**, *230*, 119–125. [CrossRef]
75. Schwarzenbach, G.; Heller, J. Komplexone. 18. Die Eisen(II) Und Eisen(III)-Komplexe Der Athylendiamin-Tetraessigsaure Und Ihr Redoxgleichgewicht. *Helv. Chim. Acta* **1951**, *34*, 576–591. [CrossRef]
76. Gustafson, R.L.; Martell, A.E. Hydrolytic Tendencies of Ferric Chelates. *J. Phys. Chem.* **1963**, *67*, 576–582. [CrossRef]
77. Brausam, A.; van Eldik, R. Further mechanistic information on the reaction between Fe-III(edta) and hydrogen peroxide: Observation of a second reaction step and importance of pH. *Inorg. Chem.* **2004**, *43*, 5351–5359. [CrossRef]
78. Seibig, S.; vanEldik, R. Kinetics of [Fe-II(edta)] oxidation by molecular oxygen revisited. New evidence for a multistep mechanism. *Inorg. Chem.* **1997**, *36*, 4115–4120. [CrossRef]
79. Rush, J.D.; Koppenol, W.H. Oxidizing intermediates in the reaction of ferrous EDTA with hydrogen peroxide. Reactions with organic molecules and ferrocytochrome c. *J. Biol. Chem.* **1986**, *261*, 6730–6733. [CrossRef]
80. Rush, J.D.; Koppenol, W.H. The reaction between ferrous polyaminocarboxylate complexes and hydrogen peroxide: An investigation of the reaction intermediates by stopped flow spectrophotometry. *J. Inorg. Biochem.* **1987**, *29*, 199–215. [CrossRef]
81. Baertschi, S.W.; Clapham, D.; Foti, C.; Jansen, P.J.; Kristensen, S.; Reed, R.A.; Templeton, A.C.; Tonnesen, H.H. Implications of in-use photostability: Proposed guidance for photostability testing and labeling to support the administration of photosensitive pharmaceutical products, part 1: Drug products administered by injection. *J. Pharm. Sci.* **2013**, *102*, 3888–3899. [CrossRef]
82. Kerwin, B.A.; Remmele, R.L. Protect from light: Photodegradation and protein biologics. *J. Pharm. Sci.* **2007**, *96*, 1468–1479. [CrossRef]
83. Mallaney, M.; Wang, S.H.; Sreedhara, A. Effect of ambient light on monoclonal antibody product quality during small-scale mammalian cell culture process in clear glass bioreactors. *Biotechnol. Prog.* **2014**, *30*, 562–570. [CrossRef]
84. Sreedhara, A.; Yin, J.; Joyce, M.; Lau, K.; Wecksler, A.T.; Deperalta, G.; Yi, L.; John Wang, Y.; Kabakoff, B.; Kishore, R.S. Effect of ambient light on IgG1 monoclonal antibodies during drug product processing and development. *Eur. J. Pharm. Biopharm.* **2016**, *100*, 38–46. [CrossRef]
85. Luis, L.M.; Hu, Y.; Zamiri, C.; Sreedhara, A. Determination of the Acceptable Ambient Light Exposure during Drug Product Manufacturing for Long-Term Stability of Monoclonal Antibodies. *PDA J. Pharm. Sci. Technol.* **2018**, *72*, 393–403. [CrossRef]
86. Liu, M.; Zhang, Z.; Cheetham, J.; Ren, D.; Zhou, Z.S. Discovery and characterization of a photo-oxidative histidine-histidine cross-link in IgG1 antibody utilizing (1)(8)O-labeling and mass spectrometry. *Anal. Chem.* **2014**, *86*, 4940–4948. [CrossRef]
87. Lei, M.; Carcelen, T.; Walters, B.T.; Zamiri, C.; Quan, C.; Hu, Y.; Nishihara, J.; Yip, H.; Woon, N.; Zhang, T.; et al. Structure-Based Correlation of Light-Induced Histidine Reactivity in A Model Protein. *Anal. Chem.* **2017**, *89*, 7225–7231. [CrossRef]
88. Lei, M.; Quan, C.; Wang, Y.J.; Kao, Y.H.; Schöneich, C. Light-Induced Covalent Buffer Adducts to Histidine in a Model Protein. *Pharm. Res.* **2018**, *35*, 67. [CrossRef]
89. Bane, J.; Mozziconacci, O.; Yi, L.; Wang, Y.J.; Sreedhara, A.; Schöneich, C. Photo-oxidation of IgG1 and Model Peptides: Detection and Analysis of Triply Oxidized His and Trp Side Chain Cleavage Products. *Pharm. Res.* **2017**, *34*, 229–242. [CrossRef]
90. Kaiser, W.; Schultz-Fademrecht, T.; Blech, M.; Buske, J.; Garidel, P. Investigating photodegradation of antibodies governed by the light dosage. *Int. J. Pharm.* **2021**, *604*, 120723. [CrossRef]
91. More, H.T.; Bindra, D.S.; Zumba, A.; Zhou, K.; Carvalho, T.; Mantri, R. Effect of light source and UVA quotient on monoclonal antibody stability during ambient light exposure studies. *Eur. J. Pharm. Biopharm.* **2023**, *185*, 177–182. [CrossRef]
92. Zhang, Y.; Schöneich, C. Visible Light Induces Site-Specific Oxidative Heavy Chain Fragmentation of a Monoclonal Antibody (IgG1) Mediated by an Iron(III)-Containing Histidine Buffer. *Mol. Pharm.* **2023**, *20*, 650–662. [CrossRef]
93. Schöneich, C. Photo-Degradation of Therapeutic Proteins: Mechanistic Aspects. *Pharm. Res.* **2020**, *37*, 45. [CrossRef] [PubMed]
94. Reid, L.O.; Vignoni, M.; Martins-Froment, N.; Thomas, A.H.; Dantola, M.L. Photochemistry of tyrosine dimer: When an oxidative lesion of proteins is able to photoinduce further damage. *Photochem. Photobiol. Sci.* **2019**, *18*, 1732–1741. [CrossRef] [PubMed]
95. Masaki, H.; Okano, Y.; Sakurai, H. Generation of active oxygen species from advanced glycation end-products (AGEs) during ultraviolet light A (UVA) irradiation and a possible mechanism for cell damaging. *Biochim. Biophys. Acta* **1999**, *1428*, 45–56. [CrossRef]
96. Lamore, S.D.; Azimian, S.; Horn, D.; Anglin, B.L.; Uchida, K.; Cabello, C.M.; Wondrak, G.T. The malondialdehyde-derived fluorophore DHP-lysine is a potent sensitizer of UVA-induced photooxidative stress in human skin cells. *J. Photochem. Photobiol. B* **2010**, *101*, 251–264. [CrossRef] [PubMed]
97. Karran, P.; Brem, R. Protein oxidation, UVA and human DNA repair. *DNA Repair* **2016**, *44*, 178–185. [CrossRef] [PubMed]

98. Cardoso, D.R.; Franco, D.W.; Olsen, K.; Andersen, M.L.; Skibsted, L.H. Reactivity of bovine whey proteins, peptides, and amino acids toward triplet riboflavin as studied by laser flash photolysis. *J. Agric. Food Chem.* **2004**, *52*, 6602–6606. [CrossRef]
99. Huvaere, K.; Skibsted, L.H. Light-induced oxidation of tryptophan and histidine. Reactivity of aromatic N-heterocycles toward triplet-excited flavins. *J. Am. Chem. Soc.* **2009**, *131*, 8049–8060. [CrossRef] [PubMed]
100. Castano, C.; Dantola, M.L.; Oliveros, E.; Thomas, A.H.; Lorente, C. Oxidation of tyrosine photoinduced by pterin in aqueous solution. *Photochem. Photobiol.* **2013**, *89*, 1448–1455. [CrossRef]
101. Thomas, A.H.; Lorente, C.; Roitman, K.; Morales, M.M.; Dantola, M.L. Photosensitization of bovine serum albumin by pterin: A mechanistic study. *J. Photochem. Photobiol. B* **2013**, *120*, 52–58. [CrossRef]
102. Reid, L.O.; Dantola, M.L.; Petroselli, G.; Erra-Balsells, R.; Miranda, M.A.; Lhiaubet-Vallet, V.; Thomas, A.H. Chemical Modifications of Globular Proteins Phototriggered by an Endogenous Photosensitizer. *Chem. Res. Toxicol.* **2019**, *32*, 2250–2259. [CrossRef]
103. Baptista, M.S.; Cadet, J.; Di Mascio, P.; Ghogare, A.A.; Greer, A.; Hamblin, M.R.; Lorente, C.; Nunez, S.C.; Ribeiro, M.S.; Thomas, A.H.; et al. Type I and Type II Photosensitized Oxidation Reactions: Guidelines and Mechanistic Pathways. *Photochem. Photobiol.* **2017**, *93*, 912–919. [CrossRef]
104. Dougherty, D.A. Cation-pi interactions in chemistry and biology: A new view of benzene, Phe, Tyr, and Trp. *Science* **1996**, *271*, 163–168. [CrossRef] [PubMed]
105. Gallivan, J.P.; Dougherty, D.A. Cation-pi interactions in structural biology. *Proc. Natl. Acad. Sci. USA* **1999**, *96*, 9459–9464. [CrossRef]
106. Dougherty, D.A. The cation-pi interaction. *Acc. Chem. Res.* **2013**, *46*, 885–893. [CrossRef] [PubMed]
107. Roveri, O.A.; Braslavsky, S.E. pi-Cation interactions as the origin of the weak absorption at 532 nm observed in tryptophan-containing polypeptides. *Photochem. Photobiol. Sci.* **2012**, *11*, 962–966. [CrossRef] [PubMed]
108. Juszczak, L.J.; Eisenberg, A.S. The Color of Cation-pi Interactions: Subtleties of Amine-Tryptophan Interaction Energetics Allow for Radical-like Visible Absorbance and Fluorescence. *J. Am. Chem. Soc.* **2017**, *139*, 8302–8311. [CrossRef]
109. Chen, J.; Browne, W.R. Photochemistry of iron complexes. *Coordin. Chem. Rev.* **2018**, *374*, 15–35. [CrossRef]
110. Van der Zee, J.; Krootjes, B.B.; Chignell, C.F.; Dubbelman, T.M.; Van Steveninck, J. Hydroxyl radical generation by a light-dependent Fenton reaction. *Free Radic. Biol. Med.* **1993**, *14*, 105–113.
111. Pozdnyakov, I.P.; Kel, O.V.; Plyusnin, V.F.; Grivin, V.P.; Bazhin, N.M. New insight into photochemistry of ferrioxalate. *J. Phys. Chem. A* **2008**, *112*, 8316–8322. [CrossRef]
112. Nogueira, A.A.; Souza, B.M.; Dezotti, M.W.C.; Boaventura, R.A.R.; Vilar, V.J.P. Ferrioxalate complexes as strategy to drive a photo-FENTON reaction at mild pH conditions: A case study on levofloxacin oxidation. *J. Photoch. Photobio. A* **2017**, *345*, 109–123. [CrossRef]
113. Faust, B.C.; Zepp, R.G. Photochemistry of Aqueous Iron(Iii) Polycarboxylate Complexes—Roles in the Chemistry of Atmospheric and Surface Waters. *Environ. Sci. Technol.* **1993**, *27*, 2517–2522. [CrossRef]
114. Huffman, R.E.; Davidson, N. Kinetics of the Ferrous Iron-Oxygen Reaction in Sulfuric Acid Solution. *J. Am. Chem. Soc.* **1956**, *78*, 4836–4842. [CrossRef]
115. Subelzu, N.; Schöneich, C. Near UV and Visible Light Induce Iron-Dependent Photodegradation Reactions in Pharmaceutical Buffers: Mechanistic and Product Studies. *Mol. Pharm.* **2020**, *17*, 4163–4179. [CrossRef] [PubMed]
116. *Q1B Photostability Testing of New Drug Substances and Products. Guidance for Industry*; International Council for Harmonization (ICH): Geneva, Switzerland, 1996.
117. Zhang, Y.; Richards, D.S.; Grotemeyer, E.N.; Jackson, T.A.; Schöneich, C. Near-UV and Visible Light Degradation of Iron (III)-Containing Citrate Buffer: Formation of Carbon Dioxide Radical Anion via Fragmentation of a Sterically Hindered Alkoxyl Radical. *Mol. Pharm.* **2022**, *19*, 4026–4042. [CrossRef]
118. Koppenol, W.H.; Rush, J.D. Reduction Potential of the Co2/Co2.- Couple—A Comparison with Other C1 Radicals. *J. Phys. Chem.* **1987**, *91*, 4429–4430. [CrossRef]
119. Adams, G.E.; Willson, R.L. Pulse Radiolysis Studies on Oxidation of Organic Radicals in Aqueous Solution. *Trans. Faraday Soc.* **1969**, *65*, 2981. [CrossRef]
120. Fojtik, A.; Czapski, G.; Henglein, A. Pulse Radiolytic Investigation of Carboxyl Radical in Aqueous Solution. *J. Phys. Chem.* **1970**, *74*, 3204. [CrossRef]
121. Willson, R.L. Pulse Radiolysis Studies of Electron Transfer in Aqueous Disulphide Solutions. *J. Chem. Soc. Chem. Comm.* **1970**, *64*, 1425–1426. [CrossRef]
122. Favaudon, V.; Tourbez, H.; Houeelevin, C.; Lhoste, J.M. Co_2.- Radical Induced Cleavage of Disulfide Bonds in Proteins—A Gamma-Ray and Pulse-Radiolysis Mechanistic Investigation. *Biochemistry* **1990**, *29*, 10978–10989. [CrossRef]
123. Joshi, R.; Adhikari, S.; Gopinathan, C.; O'Neill, P. Reduction reactions of bovine serum albumin and lysozyme by CO2- radical in polyvinyl alcohol solution: A pulse radiolysis study. *Radiat. Phys. Chem.* **1998**, *53*, 171–176. [CrossRef]
124. Matzapetakis, M.; Raptopoulou, C.P.; Tsohos, A.; Papaefthymiou, V.; Moon, N.; Salifoglou, A. Synthesis, spectroscopic and structural characterization of the first mononuclear, water soluble iron-citrate complex, $(NH_4)_5Fe(C_6H_4O_7)_2 \cdot 2H_2O$. *J. Am. Chem. Soc.* **1998**, *120*, 13266–13267. [CrossRef]
125. Duerkop, M.; Berger, E.; Durauer, A.; Jungbauer, A. Influence of cavitation and high shear stress on HSA aggregation behavior. *Eng. Life Sci.* **2018**, *18*, 169–178. [CrossRef] [PubMed]

126. Duerkop, M.; Berger, E.; Durauer, A.; Jungbauer, A. Impact of Cavitation, High Shear Stress and Air/Liquid Interfaces on Protein Aggregation. *Biotechnol. J.* **2018**, *13*, 1800062. [CrossRef] [PubMed]
127. Nayak, A.; Colandene, J.; Bradford, V.; Perkins, M. Characterization of subvisible particle formation during the filling pump operation of a monoclonal antibody solution. *J. Pharm. Sci.* **2011**, *100*, 4198–4204. [CrossRef]
128. Gikanga, B.; Eisner, D.R.; Ovadia, R.; Day, E.S.; Stauch, O.B.; Maa, Y.F. Processing Impact on Monoclonal Antibody Drug Products: Protein Subvisible Particulate Formation Induced by Grinding Stress. *PDA J. Pharm. Sci. Technol.* **2017**, *71*, 172–188. [CrossRef]
129. Gikanga, B.; Hui, A.; Maa, Y.F. Mechanistic Investigation on Grinding-Induced Subvisible Particle Formation during Mixing and Filling of Monoclonal Antibody Formulations. *PDA J. Pharm. Sci. Technol.* **2018**, *72*, 117–133. [CrossRef]
130. Randolph, T.W.; Schiltz, E.; Sederstrom, D.; Steinmann, D.; Mozziconacci, O.; Schöneich, C.; Freund, E.; Ricci, M.S.; Carpenter, J.F.; Lengsfeld, C.S. Do not drop: Mechanical shock in vials causes cavitation, protein aggregation, and particle formation. *J. Pharm. Sci.* **2015**, *104*, 602–611. [CrossRef]
131. Torisu, T.; Maruno, T.; Hamaji, Y.; Ohkubo, T.; Uchiyama, S. Synergistic Effect of Cavitation and Agitation on Protein Aggregation. *J. Pharm. Sci.* **2017**, *106*, 521–529. [CrossRef]
132. Wu, H.; Chisholm, C.F.; Puryear, M.; Movafaghi, S.; Smith, S.D.; Pokhilchuk, Y.; Lengsfeld, C.S.; Randolph, T.W. Container Surfaces Control Initiation of Cavitation and Resulting Particle Formation in Protein Formulations After Application of Mechanical Shock. *J. Pharm. Sci.* **2020**, *109*, 1270–1280. [CrossRef]
133. Siavashpouri, M.; Bailey-Hytholt, C.M.; Authelin, J.R.; Patke, S. Quantification and Stability Impact Assessment of Drop Stresses in Biologic Drug Products. *PDA J. Pharm. Sci. Technol.* **2022**, *76*, 461–473. [CrossRef]
134. Yasui, K. Production of O Radicals from Cavitation Bubbles under Ultrasound. *Molecules* **2022**, *27*, 4788. [CrossRef] [PubMed]
135. Meyer, B.K.; Ni, A.; Hu, B.; Shi, L. Antimicrobial preservative use in parenteral products: Past and present. *J. Pharm. Sci.* **2007**, *96*, 3155–3167. [CrossRef]
136. Stroppel, L.; Schultz-Fademrecht, T.; Cebulla, M.; Blech, M.; Marhofer, R.J.; Selzer, P.M.; Garidel, P. Antimicrobial Preservatives for Protein and Peptide Formulations: An Overview. *Pharmaceutics* **2023**, *15*, 563. [CrossRef] [PubMed]
137. Sankar, M.; Nowicka, E.; Carter, E.; Murphy, D.M.; Knight, D.W.; Bethell, D.; Hutchings, G.J. The benzaldehyde oxidation paradox explained by the interception of peroxy radical by benzyl alcohol. *Nat. Commun.* **2014**, *5*, 3332. [CrossRef] [PubMed]
138. Grootveld, M.; Halliwell, B. An aromatic hydroxylation assay for hydroxyl radicals utilizing high-performance liquid chromatography (HPLC). Use to investigate the effect of EDTA on the Fenton reaction. *Free Radic. Res. Commun.* **1986**, *1*, 243–250. [CrossRef]
139. Grootveld, M.; Halliwell, B. Aromatic hydroxylation as a potential measure of hydroxyl-radical formation in vivo. Identification of hydroxylated derivatives of salicylate in human body fluids. *Biochem. J.* **1986**, *237*, 499–504. [CrossRef]
140. Maskos, Z.; Rush, J.D.; Koppenol, W.H. The hydroxylation of the salicylate anion by a Fenton reaction and T-radiolysis: A consideration of the respective mechanisms. *Free Radic. Biol. Med.* **1990**, *8*, 153–162. [CrossRef]
141. Hamilton, G.A.; Friedman, J.P.; Campbell, P.M. Hydroxylation of Anisole by Hydrogen Peroxide in Presence of Catalytic Amounts of Ferric Ion and Catechol. Scope Requirements and Kinetic Studies. *J. Am. Chem. Soc.* **1966**, *88*, 5266. [CrossRef]
142. Friedrich, L.C.; Mendes, M.A.; Silva, V.O.; Zanta, C.L.P.S.; Machulek, A.; Quina, F.H. Mechanistic Implications of Zinc(II) Ions on the Degradation of Phenol by the Fenton Reaction. *J. Brazil Chem. Soc.* **2012**, *23*, 1372–1377. [CrossRef]
143. Pasquet, J.; Chevalier, Y.; Pelletier, J.; Couval, E.; Bouvier, D.; Bolzinger, M.A. The contribution of zinc ions to the antimicrobial activity of zinc oxide. *Colloid. Surf. A* **2014**, *457*, 263–274. [CrossRef]
144. Dodson, G.; Steiner, D. The role of assembly in insulin's biosynthesis. *Curr. Opin. Struct. Biol.* **1998**, *8*, 189–194. [CrossRef] [PubMed]
145. Karthikeyan, C.; Arunachalam, P.; Ramachandran, K.; Al-Mayouf, A.M.; Karuppuchamy, S. Recent advances in semiconductor metal oxides with enhanced methods for solar photocatalytic applications. *J. Alloy. Compd.* **2020**, *828*, 154281. [CrossRef]
146. Qin, H.C.; Li, W.Y.; Xia, Y.J.; He, T. Photocatalytic Activity of Heterostructures Based on ZnO and N-Doped ZnO. *ACS Appl. Mater. Inter.* **2011**, *3*, 3152–3156. [CrossRef] [PubMed]
147. Kubo, W.; Tatsuma, T. Photocatalytic remote oxidation with various photocatalysts and enhancement of its activity. *J. Mater. Chem.* **2005**, *15*, 3104–3108. [CrossRef]
148. Al-Rasheed, R.; Cardin, D.J. Photocatalytic degradation of humic acid in saline waters Part 2. Effects of various photocatalytic materials. *Appl. Catal. A Gen.* **2003**, *246*, 39–48. [CrossRef]
149. Parini, R.; Deodato, F. Intravenous Enzyme Replacement Therapy in Mucopolysaccharidoses: Clinical Effectiveness and Limitations. *Int. J. Mol. Sci.* **2020**, *21*, 2975. [CrossRef]
150. Brennan, G.T.; Saif, M.W. Pancreatic Enzyme Replacement Therapy: A Concise Review. *JOP* **2019**, *20*, 121–125.
151. Rivera-Colon, Y.; Schutsky, E.K.; Kita, A.Z.; Garman, S.C. The structure of human GALNS reveals the molecular basis for mucopolysaccharidosis IV A. *J. Mol. Biol.* **2012**, *423*, 736–751. [CrossRef]
152. Demydchuk, M.; Hill, C.H.; Zhou, A.; Bunkoczi, G.; Stein, P.E.; Marchesan, D.; Deane, J.E.; Read, R.J. Insights into Hunter syndrome from the structure of iduronate-2-sulfatase. *Nat. Commun.* **2017**, *8*, 15786. [CrossRef]

MDPI
St. Alban-Anlage 66
4052 Basel
Switzerland
www.mdpi.com

Biomolecules Editorial Office
E-mail: biomolecules@mdpi.com
www.mdpi.com/journal/biomolecules

www.ingramcontent.com/pod-product-compliance
Lightning Source LLC
Chambersburg PA
CBHW040835250926
51850CB00007BA/1324
* 9 7 8 3 7 2 5 8 0 3 2 8 6 *